Magnetic Excitations and Geometric Confinement

Theory and simulations

Magnetic Excitations and Geometric Confinement

Theory and simulations

Gary Matthew Wysin

Kansas State University, Kansas, USA

IOP Publishing, Bristol, UK

© IOP Publishing Ltd 2015

All rights reserved. No part of this publication may be reproduced, stored in a retrieval system or transmitted in any form or by any means, electronic, mechanical, photocopying, recording or otherwise, without the prior permission of the publisher, or as expressly permitted by law or under terms agreed with the appropriate rights organization. Multiple copying is permitted in accordance with the terms of licences issued by the Copyright Licensing Agency, the Copyright Clearance Centre and other reproduction rights organisations.

Permission to make use of IOP Publishing content other than as set out above may be sought at permissions@iop.org.

Gary Matthew Wysin has asserted his right to be identified as the author of this work in accordance with sections 77 and 78 of the Copyright, Designs and Patents Act 1988.

ISBN 978-0-7503-1074-1 (ebook)
ISBN 978-0-7503-1075-8 (print)
ISBN 978-0-7503-1129-8 (mobi)

DOI 10.1088/978-0-7503-1074-1

Version: 20151201

IOP Expanding Physics
ISSN 2053-2563 (online)
ISSN 2054-7315 (print)

British Library Cataloguing-in-Publication Data: A catalogue record for this book is available from the British Library.

Published by IOP Publishing, wholly owned by The Institute of Physics, London

IOP Publishing, Temple Circus, Temple Way, Bristol, BS1 6HG, UK

US Office: IOP Publishing, Inc., 190 North Independence Mall West, Suite 601, Philadelphia, PA 19106, USA

My father was a chemist who worked for many years at United States Gypsum Corporation in Gypsum, Ohio. He encouraged my interest in taking things apart and putting them back together. This book is dedicated to the memory of Norbert Aloysius Wysin.

Contents

Preface	xv
Acknowledgements	xvii
About the author	xviii
Symbols	xix

1	**Introduction: geometrically confined magnetic systems**	**1-1**
1.1	Magnetic order, dipoles and fields	1-1
1.2	Effects due to geometry	1-3
	1.2.1 Demagnetization energy	1-6
	1.2.2 Demagnetization examples	1-7
1.3	Exchange interactions	1-9
	1.3.1 Exchange of an electron pair	1-10
	1.3.2 Exchange and energy	1-12
	1.3.3 The Heisenberg exchange Hamiltonian	1-14
1.4	Anisotropic exchange couplings	1-15
	1.4.1 Space anisotropy of exchange	1-15
	1.4.2 XY spin symmetry	1-17
	1.4.3 Effects of XY symmetry—vortices	1-17
	1.4.4 More XY symmetry—the plane rotor model	1-20
	1.4.5 Uniaxial to isotropic spin symmetry—skyrmions	1-21
	1.4.6 Quasi-1D magnetism—spin waves and magnetic solitons	1-22
1.5	Local anisotropies	1-23
	1.5.1 Applied magnetic field	1-24
1.6	Theory for linear and nonlinear magnetism	1-25
1.7	Simulations in magnetism	1-26
	Bibliography	1-29

Part I	**Theory and simulation approaches for magnetism**	**2-1**
2	**Magnetism theory: spin models**	**2-3**
2.1	Magnetic dipoles and magnetic ordering	2-3
2.2	Atomic dipoles	2-4
2.3	Local spin interaction models	2-6
	2.3.1 Spin dynamics—atomic scale local spin magnetism	2-7
	2.3.2 Micromagnetics—nanoscale magnetization dynamics	2-7

2.4	Discrete exchange interactions	2-8
2.5	Ferromagnetic spin exchange in a continuum limit	2-8
2.6	Ferromagnetic exchange in micromagnetics	2-10
2.7	FM continuum limit on any lattice	2-13
	2.7.1 Triangular lattice	2-15
	2.7.2 Hexagonal lattice	2-16
2.8	Angular coordinates for classical spin direction	2-18
	2.8.1 Continuum limits in angular notation	2-20
	2.8.2 Models in more dimensions	2-22
2.9	Classical spin mechanics	2-22
	2.9.1 Hamiltonian dynamics	2-23
	2.9.2 Hamiltonian spin dynamics	2-26
	2.9.3 Angular spin coordinates and canonical Poisson brackets	2-27
	2.9.4 Lagrangian spin dynamics	2-28
	2.9.5 Classical dynamics in a system of spins	2-29
	2.9.6 Spin damping	2-30
	2.9.7 A single spin in a fixed field with damping	2-32
2.10	Classical spin torques	2-34
	2.10.1 Continuum limit torques	2-35
	2.10.2 Continuum limit dynamics for angular coordinates	2-39
2.11	Quantum spin mechanics	2-41
	2.11.1 Dynamic equations of an individual quantum spin operator	2-43
	2.11.2 Solution of a quantum spin in a constant field	2-44
	2.11.3 How do discrete quantum spin states exhibit themselves?	2-45
	Bibliography	2-47

3	**Demagnetization effects in thin magnets**	**3-1**
3.1	The magnetostatics problem in a finite magnet	3-1
	3.1.1 Green's functions for magnetostatics	3-2
	3.1.2 Uniform magnetization and effective magnetic surface charges	3-5
3.2	The magnetic field inside a cylindrical magnet	3-6
	3.2.1 Longitudinal magnetization M^z	3-6
	3.2.2 Transverse magnetization M^x	3-9
3.3	Demagnetization fields in thin-film magnets	3-12
	3.3.1 The longitudinal field H_M^z in a thin magnet	3-13
	3.3.2 The transverse field (H_M^x, H_M^y) in a thin magnet	3-15
	3.3.3 Using ρ_M to obtain (H_M^x, H_M^y) in a thin magnet	3-19
	3.3.4 The total field (H_M^x, H_M^y, H_M^z) in a thin magnet	3-19

	3.3.5 Transverse self demagnetization—calculation of $H_\mathrm{M}^x(\mathbf{0})$	3-20
	3.3.6 About the transverse demagnetization for computations	3-23
	3.3.7 Example: applying G^{xx} to a disk with uniform M^x	3-24
3.4	Use of fast Fourier transforms	3-25
	3.4.1 Finite system on a 1D grid	3-27
	3.4.2 Simulation of 1D open boundary conditions	3-29
	3.4.3 Simulation of 2D and 3D open boundaries	3-32
3.5	Demagnetization in a thin permalloy magnet	3-33
	3.5.1 Energy minimization by a spin alignment scheme	3-36
	Bibliography	3-38

4	**Classical Monte Carlo simulation methods**	**4-1**
4.1	Thermal equilibrium and ergodicity	4-1
4.2	Boltzmann distribution for thermal equilibrium	4-2
	4.2.1 Entropy and system energy	4-3
	4.2.2 General derivation of Boltzmann distribution	4-6
	4.2.3 Free energy and statistical averages	4-8
	4.2.4 Averages of magnetic variables	4-10
	4.2.5 Averages from statistical fluctuations	4-12
4.3	Importance sampling and the Metropolis algorithm	4-14
	4.3.1 Averages with importance sampling	4-16
	4.3.2 Why the Metropolis algorithm?	4-17
	4.3.3 Trial states for Monte Carlo simulations of magnets: single-spin updates	4-18
	4.3.4 Over-relaxation updates	4-19
4.4	Monte Carlo simulation of a 2D XY model	4-20
	4.4.1 Setting up the Monte Carlo simulation	4-21
	4.4.2 Monte Carlo averages and errors	4-21
	4.4.3 Another approach for errors of fluctuation quantities	4-23
	4.4.4 Spatial correlation function and susceptibility	4-25
	4.4.5 XY model Monte Carlo results	4-26
4.5	Cluster algorithms for spin updates	4-30
	4.5.1 Ising spins: Swendsen–Wang cluster algorithm	4-32
	4.5.2 Heisenberg spins: the Wolff cluster algorithm	4-36
	4.5.3 Cluster algorithms for Heisenberg spins with XXZ symmetry	4-37
	4.5.4 2D XY model simulation data with cluster updates	4-38
4.6	Microcanonical Monte Carlo	4-39
	4.6.1 Demons	4-40

	4.6.2 Effective temperature and equipartition	4-41
	4.6.3 Equipartition in different systems	4-42
	4.6.4 Effective temperature in Ising models	4-43
	4.6.5 1D spin chain studied with microcanonical Monte Carlo	4-44
	Bibliography	4-45

5 Classical spin dynamics simulations — 5-1

5.1	Landau–Lifshitz–Gilbert spin dynamics	5-1
	5.1.1 Dimensionless units for spin dynamics	5-3
	5.1.2 Dynamic equations for Heisenberg spins—angular variables	5-4
	5.1.3 XY model	5-6
	5.1.4 Planar rotor model	5-7
	5.1.5 Planar rotor in Cartesian components	5-9
5.2	Numerical time evolution for spin dynamics	5-10
	5.2.1 Fourth-order Runge–Kutta time evolution	5-10
	5.2.2 Adams–Bashforth–Moulton fourth-order predictor–corrector method	5-11
5.3	Hybrid Monte Carlo–spin dynamics at $T > 0$	5-12
5.4	Stochastic dynamics—thermal fluctuations in the planar rotor model	5-13
	5.4.1 Langevin equation for a free rotor in a heat bath	5-14
	5.4.2 Planar rotator velocity autocorrelation function	5-15
	5.4.3 Planar rotor diffusion	5-17
5.5	Numerical solutions of Langevin equations	5-19
	5.5.1 Euler method for Langevin equation	5-20
	5.5.2 Developing Langevin second-order methods	5-21
	5.5.3 A Langevin–velocity-Verlet method	5-24
	5.5.4 An out-of-phase second-order Langevin method	5-25
5.6	Langevin spin dynamics	5-29
	5.6.1 Second-order Heun integration scheme	5-31
5.7	Dynamic correlation responses	5-33
	5.7.1 Space Fourier transforms	5-33
	5.7.2 Space correlations	5-37
	5.7.3 Time correlations	5-39
	5.7.4 Time Fourier transforms and correlations	5-40
	5.7.5 Dynamic structure functions	5-41
	5.7.6 Comparing static and dynamic correlations	5-45
	Bibliography	5-47

Part II   Excitations in magnetic systems   6-1

6   Spin waves: extended but low-dimensional systems   6-3
6.1   Spin waves in ferromagnetic models   6-3
    6.1.1   Isotropic Heisenberg models   6-4
    6.1.2   An isotropic ferromagnet with a weak magnetic field   6-8
    6.1.3   Ferromagnetic models with XY-like exchange   6-10
6.2   Spin waves in antiferromagnetic models   6-15
    6.2.1   A 1D anisotropic antiferromagnetic chain   6-17
    6.2.2   2D anisotropic antiferromagnetic model—square lattice   6-21
    6.2.3   2D anisotropic antiferromagnetic model—hexagonal lattice   6-22
6.3   Dynamic correlations of spin wave fluctuations   6-24
    6.3.1   Ferromagnets   6-24
    6.3.2   Antiferromagnets   6-27
6.4   Nonlinear spin waves—ferromagnets   6-29
    Bibliography   6-31

7   Solitons in magnetic chains   7-1
7.1   Nonlinear excitations: solitons in FM magnetic chains   7-1
7.2   Ferromagnetic sine-Gordon kink instability   7-8
    7.2.1   The Liebmann *et al* ansatz   7-10
7.3 π-kinks in ferromagnetic chains   7-15
    7.3.1   Easy-plane ferromagnetic chain   7-15
    7.3.2   Easy-plane ferromagnetic chain with in-plane Ising symmetry   7-17
7.4   Magnetic kinks in antiferromagnetic chains   7-20
    7.4.1   Antiferromagnetic chain ground state   7-21
    7.4.2   Spin waves   7-23
    7.4.3   Continuum limit dynamics in an antiferromagnetic chain   7-24
    7.4.4   XY antiferromagnetic kinks   7-25
    7.4.5   YZ antiferromagnetic kinks   7-27
    7.4.6   Antiferromagnetic YZ kink stability analysis   7-29
    7.4.7   Antiferromagnetic kink ansatz analysis   7-32
    7.4.8   XY kink limit: $\theta_A = \theta_B$ 7-38
    7.4.9   YZ kink limit: $\theta_A + \theta_B = \pi$ 7-39
    7.4.10   Energy–velocity relations from the simplified antiferromagnetic ansatz results   7-41
    Bibliography   7-42

8 Vortices in layered or 2D ferromagnets — 8-1

- 8.1 A 2D ferromagnet with easy-plane exchange anisotropy — 8-1
- 8.2 In-plane and out-of-plane vortices — 8-3
  - 8.2.1 In-plane vortices — 8-5
  - 8.2.2 Out-of-plane vortices — 8-5
  - 8.2.3 Discrete lattice vortex solutions — 8-6
- 8.3 Vortex instability — 8-7
  - 8.3.1 Numerical simulations for stability — 8-8
  - 8.3.2 Discrete energetics of vortex core stability — 8-9
- 8.4 Moving in-plane and out-of-plane vortices — 8-14
- 8.5 The vortex unbinding transition — 8-15
- 8.6 Monte Carlo simulations of the Berezinskii–Kosterlitz–Thouless transition — 8-17
  - 8.6.1 Estimations of the critical temperature—Binder's cumulant — 8-19
  - 8.6.2 Estimations of the critical temperature—scaling of susceptibility — 8-22
  - 8.6.3 A measure of spin twist resistance—the helicity modulus — 8-23
  - 8.6.4 Dependence of critical temperature on anisotropy and vacancies — 8-29
- 8.7 Dynamic correlations in XY models — 8-32
  - 8.7.1 Hybrid Monte Carlo–spin dynamics simulations — 8-33
  - 8.7.2 Low temperature dynamic structure function — 8-34
  - 8.7.3 Higher temperature dynamic structure function and central peak — 8-36
  - 8.7.4 Ideal gas model for vortex thermodynamics — 8-39
  - 8.7.5 Comparison of vortex ideal gas model to simulations — 8-47
  - 8.7.6 Dynamic correlations with vacancies — 8-50
  - Bibliography — 8-54

9 Magnetic vortex core motion and internal dynamics — 9-1

- 9.1 Thiele equations and vortex motion — 9-1
  - 9.1.1 Derivation of Thiele's equation of motion — 9-2
  - 9.1.2 Including vortex mass effects — 9-5
  - 9.1.3 Calculation of vortex gyrovector and other properties — 9-6
- 9.2 Relation of vortex momentum to the Thiele equations — 9-9
  - 9.2.1 Poisson bracket of kinetic momentum components — 9-10
  - 9.2.2 The momentum conjugate to $\mathbf{X}$ — 9-12
  - 9.2.3 Momentum as a generator of translations — 9-12
  - 9.2.4 Time derivative of kinetic momentum $\mathbf{p}$ — 9-13
  - 9.2.5 Comparison of vortex dynamics to electric charge dynamics — 9-14
  - 9.2.6 Lagrangian mechanics for vortex core motion — 9-16

9.3	Vortex forces and motions	9-17
	9.3.1 Force due to applied field	9-18
	9.3.2 Vortex pair forces	9-22
	9.3.3 Vortex image forces	9-23
9.4	Some simple examples of vortex dynamics	9-31
	9.4.1 An individual vortex near a straight boundary	9-31
	9.4.2 A pair of vortices in a large system	9-38
	9.4.3 An individual vortex in a circular system	9-43
9.5	Vortex–spin wave interactions and normal modes	9-45
	9.5.1 The initial discrete lattice vortex solution	9-46
	9.5.2 Perturbations of a static vortex	9-46
	9.5.3 Examples of the spin wave mode spectrum on a vortex	9-51
9.6	Vortex mass obtained from vortex normal modes	9-61
	Bibliography	9-65

10 Vortices in thin ferromagnetic nanodisks — 10-1

10.1	Vortex states in magnetic nanodisks	10-1
10.2	Vortex-in-disk effective potentials	10-4
	10.2.1 Spin-length constraint, as an example	10-5
	10.2.2 Constraint on vortex position	10-7
	10.2.3 Some calculated vortex potentials	10-10
10.3	$T=0$ Gyrotropic vortex motion in thin nanodisks	10-18
	10.3.1 Thiele equation dynamics for a thin-film system	10-19
	10.3.2 Numerical simulation of vortex dynamics at zero temperature	10-20
	10.3.3 Typical gyrotropic motions from simulations	10-23
10.4	Thermalized vortex motion	10-29
	10.4.1 The simulation method	10-29
	10.4.2 Spontaneous gyrotropic vortex motion in stochastic dynamics	10-30
	10.4.3 Hamiltonian dynamics and statistics of vortex core position	10-35
	Bibliography	10-44

11 Spin ices and geometric frustration — 11-1

11.1	Spin ice and frustrated states	11-1
	11.1.1 About frustration	11-3
11.2	Magnetic monopoles and string excitations in spin ice	11-5
11.3	Square lattice spin ice energetics and order parameters	11-8
	11.3.1 Reduction to an Ising limit	11-10

	11.3.2 Measurement of order in square spin ice	11-13
	11.3.3 Monopole charge densities in square spin ice	11-16
11.4	Dynamics in square lattice artificial spin ice	11-18
	11.4.1 Island geometry and coupling parameters	11-20
	11.4.2 Simulations of equilibrium for Wang islands	11-22
	11.4.3 Simulations of square ice with artificial model parameters	11-26
	11.4.4 Magnetic hysteresis in model C with artificial parameters	11-32
11.5	Triangular lattice artificial spin ice	11-36
	11.5.1 Counting and energetics of vertex configurations	11-36
	11.5.2 Comparing frustration in square and triangular spin ice	11-40
	11.5.3 Thermal equilibrium in triangular spin ice	11-41
11.6	Kagomé lattice artificial spin ice	11-42
	11.6.1 Kagomé vertex configurations and energetics	11-44
	11.6.2 Order parameters based on ground state configurations	11-46
	11.6.3 Monopole charge densities in Kagomé spin ice	11-53
11.7	Langevin dynamics for Kagomé ice in thermal equilibrium	11-54
	Bibliography	11-64

Preface

Many intriguing dynamical effects in magnetism take place in systems that are confined in some geometrical sense. The confinement can be as in individual particles of nanometer scale, such as small spheres or disks, or in layered magnets, or in quasi-one-dimensional magnets, to name a few examples. The presence of close boundaries (in contrast to bulk samples) leads to constraints on the magnetic moments within a sample, usually referred to as anisotropies or as demagnetization effects. This, coupled with nonlinearity, tends to allow the presence of various types of magnetic excitations, such as solitons and vortices, whose dynamics can be described with certain charges known as topological charges. In addition, such nonlinear excitations may have motions similar to particles, but possibly with dynamics not described by anything directly like Newtonian mechanics. These nonlinear excitations can also interact with the linear ones, that is, the spin waves. This interaction can be especially important for describing a thermal equilibrium situation in a magnet. Thus, the thermodynamics in various nonlinear magnets of low dimension or constrained geometry is of special interest here.

The particle-like nonlinear excitations are of particular interest, because one hopes to use their topological charges as bits for storing information. These charges might be probed and modified by applied magnetic pulses, and they might be affected to some degree by temperature. One would like to know the degree to which they become unstable due to temperature, or even due to quantum fluctuations. More importantly, researchers want to know how to control and flip the topological charges at will, so that they can be used as data bits.

Confined magnetics also offers great promise for manipulating the material properties of composite media—those made from magnetic particles either arranged randomly or regularly within a matrix. Recently there has arisen the great nano-particle craze, because one hopes to produce new materials with desired properties via control of the geometry and molecular interactions down to nanometer scales. Composites of magnetic or metallic particles in some non-magnetic host material could be used to produce new supermaterials. We can consider some effects, for example, in artificial spin ice arrays, which are known to exhibit geometric frustration due to competing local interactions that cannot be fully satisfied. Nanodots whose lowest energy state is that of a magnetic vortex offer an important scheme for storing data or making magnetic oscillators.

The theory for the kinds of excitations under consideration is based mostly on a set of spin models. For much of the work, classical spin mechanics can be applied, especially if the underlying spin degrees of freedom have $S > \hbar/2$, where $\hbar$ the quantum of action (or spin), Planck's constant divided by 2π. When that is the case, both the theory and simulations can apply classical equations of motion, derived from some classical energy functions (Hamiltonian mechanics). Two of the most important classical spin simulation techniques are spin dynamics, for the time-dependent motion, and classical Monte Carlo, for the thermal equilibrium properties. The low temperature dynamics comes from a torque equation (such as the

Landau–Lifshitz–Gilbert equations), whereas properties in a heat bath can be estimated statistically with Monte Carlo methods. However, at times it is even better to obtain the dynamics in the presence of heat, then there is the very powerful Langevin approach for magnetics, that has been so successful for the analysis of Brownian motion in classical particle systems.

Overall, a set of different magnetic systems displays some great variety of magnetic excitations with unique properties. Here I give an overview of some model systems and the effects that I find to be the most interesting. This book is intended for students and researchers of physics, chemistry and engineering, who are interested in how the theory and simulations are developed. It might be suitable as an auxiliary text in a class on magnetism or solid state physics. Some previous physics knowledge is expected, including understanding of static electric and magnetic fields, classical Hamiltonian mechanics and basic ideas about quantum spins and statistical physics.

Acknowledgements

Many ideas in this work were developed in collaboration with other scientists, including Franz G Mertens (University of Bayreuth, Germany), Alan R Bishop (Los Alamos National Laboratory, NM), Chikao Kawabata (Okayama University, Japan), M Elizabeth Gouvêa and Antônio S T Pires (Universidade Federal de Minas Gerais, Brazil), Dimitre Dimitrov (Tech-X Corp., Boulder, CO), Armin Völkel (Palo Alto Research Center, CA), Boris A Ivanov (Institute of Magnetism, Kiev, Ukraine), Afrânio R Pereira, Winder A Moura-Melo and Lucas A S Mól (Universidade Federal de Viçosa, Brazil), Sidiney A Leonel (Universidade Federal de Juiz de Fora, Brazil) and Wagner Figueiredo (Universidade Federal de Santa Catarina, Brazil). I also have to express my gratitude for the interest and support of my wife, Marcilene Sousa Wysin and daughter, Nayara Sousa Wysin in this project.

About the author

Gary Matthew Wysin

Gary Matthew Wysin has been a professor in the physics department at Kansas State University since 1988. Since before his adolescent years he has been interested in physics, especially anything to do with electricity, magnetism, light and astronomy. He obtained a Bachelor's degree in electrical engineering in 1978 and a Master's degree in physics in 1980, both at the University of Toledo, OH.

In his Master's work he analyzed an electrodynamics problem at the boundary of a metal and a dielectric, where the surface plasmons present there strongly affect the electric fields and produce optical bistability. He earned his doctorate in theoretical physics from Cornell University in 1985 while working on theory, Monte Carlo and spin dynamics simulations for magnetic models with solitons and vortices.

While at Kansas State University, he has lectured on classical electrodynamics, quantum mechanics, computational physics, statistical physics, and introductory physics and astronomy. He has worked on many research projects together with groups at Los Alamos National Laboratory, University of Bayreuth (Germany), Universidade Federal de Minas Gerais (Belo Horizonte, Brazil), Universidade Federal de Viçosa (Viçosa, Brazil) and Universidade Federal de Santa Catarina (Florianópolis, Brazil). He has published more than 75 research articles on various topics of condensed matter physics, including phase transitions, magnetization reversal, spin waves, solitons and vortices in low-dimensional magnetism, and optical effects in dielectrics and metals. Beyond physics, he enjoys hobbies such as making beer and wine, and many outdoor activities, including gardening, hiking, running and fishing.

Symbols

SI units for these quantities are indicated at the right. If listed without units, the quantity is dimensionless. Vector quantities are in **bold**. Tensor or matrix objects are in **bold sans serif** and their components are in sans serif.

α	Magnetic damping coefficient
β	Inverse of absolute temperature times Boltzmann's constant (J^{-1})
γ	Gyromagnetic ratio ($T^{-1}s^{-1}$)
δ	Easy-plane anisotropy strength or $1 - \lambda$
$\delta(\mathbf{r})$	Two-dimensional Dirac delta function (m^{-2})
ϵ_0	Permittivity of free space ($F\,m^{-1}$)
θ	Polar angular coordinate or spin angle
λ	Easy-plane anisotropy constant or $1 - \delta$
λ_{ex}	Ferromagnetic exchange length (m)
λ	Lagrange multiplier for constrained vortex position
Φ_M	Magnetic scalar potential (A)
ϕ	In-plane spin angle
φ	Azimuthal angular coordinate
ρ_e	Electric charge density ($C\,m^{-3}$)
ρ_v	Vorticity charge areal density (m^{-2})
ρ_M	Volume magnetic charge density ($A\,m^{-2}$)
ρ	Magnetic monopole areal density (m^{-2})
σ	Spin density per unit area (spin m^{-2})
σ_M	Magnetic surface charge density ($A\,m^{-1}$)
τ	Dimensionless time variable for simulations
μ_0	Magnetic permeability of free space ($H\,m^{-1}$)
μ_B	Bohr magneton ($J\,T^{-1}$)
$\boldsymbol{\mu}$	Magnetic dipole moment ($A\,m^{-2}$)
Υ	Magnetic helicity modulus (J)
ω	Angular frequency ($rad\,s^{-1}$)
χ_{ab}	Magnetic susceptibility tensor components
A	Ferromagnetic exchange stiffness ($J\,m^{-1}$)
a	Lattice parameter or calculational cell size (m)
$\mathbf{A}$	Vortex core acceleration ($m\,s^{-2}$)
$\mathcal{A}$	Effective vector potential for vortex
$\mathbf{B}$	Magnetic induction vector (T)
C	Heat capacity ($J\,K^{-1}$)
c	Heat capacity per particle ($J\,K^{-1}$)
$C^{xx}(\mathbf{r})$	Static spin–spin correlation function
D	Vortex dissipation tensor
E	Total system energy (J)
$\mathbf{E}$	Electric field vector ($V\,m^{-1}$)
$F(T)$	Helmholtz free energy as a function of temperature (J)
$\mathcal{F}$	Effective field vector acting on spins (s^{-1})
$\mathbf{F}$	Collective force on a spin structure (N)
f	Frequency (Hz)
$\mathbf{G}$	Gyrovector of a spin structure
$G^{ij}(\mathbf{r}, \mathbf{r}')$	Tensor Green's operator for Poisson equation (m^{-3})
$\mathbf{g}(\mathbf{r}, \mathbf{r}')$	Vector Green's function for Poisson equation (m^{-2})

$g_0(\mathbf{r}, \mathbf{r}')$	Scalar Green's function for Poisson equation (m^{-1})
g	Landé g-factor
H	Hamiltonian function (J)
$\mathbf{H}_{\text{ext}}$	Applied magnetic field vector (A m^{-1})
$\tilde{\mathbf{H}}_{\text{ext}}$	Applied magnetic field vector divided by M_s
$\mathbf{H}_M$	Demagnetization field vector (A m^{-1})
$\tilde{\mathbf{H}}_M$	Demagnetization field vector divided by M_s
$\hbar$	Planck's constant over 2π (J s)
JS^2	Atomic exchange energy (J)
J_{cell}	Exchange constant between micromagnetic cells (J)
K	Kinetic term in a Lagrangian (J)
K_A	Local magnetic anisotropy constant (various)
k_B	Boltzmann's constant (J K^{-1})
k_F	Vortex restoring force constant (N m^{-1})
L	Lagrangian function (J)
L	Cylinder length or nanodisk thickness (m)
$\mathbf{L}$	Orbital angular momentum vector (J s)
$\mathbf{M}$	Magnetization vector (A m^{-1})
M_s	Saturation magnetization (A m^{-1})
M	Vortex mass tensor of Thiele equation (s)
m	Vortex mass tensor (kg)
$\mathbf{m}$	Magnetization vector divided by M_s
N_d	Geometric demagnetization factor
$\mathbf{N}$	Torque vector (N m)
$\hat{\mathbf{n}}$	Outward directed normal unit vector
$\mathbf{P}$	Vortex canonical momentum (kg m s^{-1})
p	Polarity charge (integer ± 1)
$\mathbf{p}$	Vortex kinetic momentum (kg m s^{-1})
q	Vorticity charge (integer)
$\mathbf{q}$	Wave vector (m^{-1})
R	Radius of a circular system (m)
$\mathbf{r}$	Position vector (m)
r_v	Vortex core radius (m)
$\mathbf{S}$	Spin angular momentum vector (J s)
$S(E)$	Entropy as a function of system energy (J K^{-1})
$S^{xx}(\mathbf{q}, \omega)$	Dynamic structure function in two dimensions (m^2 s)
$\mathbf{s}$	Dimensionless spin vector
T	Absolute temperature (K)
$\mathcal{T}$	Dimensionless temperature for simulations
t	Time (s)
$U(\mathbf{X})$	Vortex effective potential (J)
V	Volume of a system (m^3)
$\mathbf{V}$	Vortex core velocity (m s^{-1})
v_{cell}	Volume or area of a calculational cell (m^3 or m^2)
$\mathbf{X}$	Vortex core location (m)
Z_a, Z_b	Sublattice spin ice order parameters
Z	Net square lattice spin ice order parameter
$Z(T)$	Canonical partition function at temperature T

Magnetic Excitations and Geometric Confinement
Theory and simulations
Gary Matthew Wysin

Chapter 1

Introduction: geometrically confined magnetic systems

This is a book about magnetism, magnetic order and magnetic excitations under what could be called *geometric confinement*. The basic physical cause of magnetism is the presence of magnetic dipoles, which already produces strong directional properties in magnets. Our world is three-dimensional (3D), but magnets can behave as if they are lower-dimensional. Geometric confinement can mean a system of finite extent, with boundaries and some particular shape. The specific shape and the boundaries can affect what is happening inside the material. Geometric confinement can also refer to a system whose interactions are sufficiently anisotropic that the system behaves as if it has a lower dimensionality. Hence, its geometry is effectively different from 3D. In this chapter we sketch out why geometry matters in magnetism. Geometric factors influence the types of excitations, or deviations from perfect order, usually time-dependent, that can be present in the system. An overview is given here of the types of magnets, excitations and physical effects considered in this book.

1.1 Magnetic order, dipoles and fields

Magnetism naturally has directional properties[1]. Simple experiments with small permanent magnets already reveal the presence of their underlying dipoles. Any such magnet has clearly detected north and south poles. In the theory of classical electricity and magnetism, the magnetic induction vector **B** emerges pointing out of the north pole, passes through space along some field lines and returns back into the south pole. Within the magnet itself, of course, the lines of **B** pass right through and

---

[1] A non-mathematical introduction to magnetism is given in the book by Lee [1]. An introduction with the basic physics and mathematics needed is given in the text by Jiles [2].

make closed loops, which is represented mathematically by Gauss's law for magnetostatics, one of Maxwell's fundamental equations of electromagnetism:

$$\nabla \cdot \mathbf{B} = 0. \tag{1.1}$$

The fundamental laws of electromagnetism, including magnetostatics, can be reviewed in many good texts, such as those by Griffiths [3], Corson and Lorrain [4] and at a more advanced level by Jackson [5]. The zero on the RHS of (1.1) is a way to indicate that the lines of $\mathbf{B}$ are closed loops. The lines do not begin on anything in particular nor end on anything special. Suppose equation (1.1) is integrated over some volume V of interest. If the divergence theorem is used to change the volume integration into a surface integration, one arrives at

$$\int dV\, \nabla \cdot \mathbf{B} = \oint d\mathbf{A} \cdot \mathbf{B} = 0. \tag{1.2}$$

On the RHS, $d\mathbf{A}$ represents an increment of surface area that surrounds the volume region of integration. The fact that the surface integration gives zero shows that any lines of $\mathbf{B}$ which enter the volume must flow back out somewhere else, that is, the total flux of $\mathbf{B}$ through the surface is zero. Equation (1.2) is probably more commonly referred to as Gauss's law for magnetostatics, due to its integral form, but it is equivalent to equation (1.1) in its physical implications.

The zero on the RHS of (1.1) also represents the physical fact that there are no natural scalar charges for magnetic fields. It is popular to state this as 'there are no magnetic monopoles'. Of course, for electric fields, the basic physical entity that produces them is electric charge (the electric version of monopoles), which is the reason that the RHS of Poisson's equation for electrostatics is not zero:

$$\nabla \cdot \mathbf{E} = \rho_e/\epsilon_0. \tag{1.3}$$

The RHS of (1.3) contains the electric charge density ρ_e and the permittivity of free space, ϵ_0. The constant $\epsilon_0 = 8.854 \times 10^{-12}$ C^2 N^{-1} m^{-2} = 8.854 pF m^{-1} defines the efficiency with which electric charge density produces electric field strength. The Poisson equation (1.3) performs the mathematics necessary to describe electric field lines that begin on positive electric charges and end on equal but opposite negative electric charges. Solving (1.3), knowing that $\mathbf{E}$ is a conservative field for electrostatics,

$$\nabla \times \mathbf{E} = 0, \quad \text{or} \quad \oint \mathbf{E} \cdot d\boldsymbol{\ell} = 0, \tag{1.4}$$

the electric field caused by a distribution of charges is then expressed as [5],

$$\mathbf{E}(\mathbf{r}) = \frac{1}{4\pi\epsilon_0} \int d^3r'\, \rho_e(\mathbf{r}') \frac{\mathbf{r} - \mathbf{r}'}{|\mathbf{r} - \mathbf{r}'|^3}. \tag{1.5}$$

The integration is over the fields caused by the increments of charge $d^3r'\, \rho_e(\mathbf{r}')$ at source points $\mathbf{r}'$. Note that a factor of 4π appears in the denominator in the process of finding the solution to (1.3). The last factor in the integrand contains the inverse square law dependence and shows that the field points away from a positive source charge towards the field point $\mathbf{r}$.

No fundamental magnetic monopole charge has yet been discovered, hence, the RHS of the Maxwell equation (1.1) is zero. This means that the basic physical source of magnetic induction field **B** must be magnetic *dipoles*, which is the next higher order multipole that is possible (assuming one does need to go to quadrupoles). An elementary classical magnetic dipole involves an electric current flowing around an area. The direction of the dipole vector is perpendicular to the plane of that area. Since that would give two choices for its direction, the final choice is selected by a well-known right hand rule: with the four fingers of your right hand curled along the current direction, your right thumb points perpendicular to that area in the direction of the magnetic dipole. This simple rule already contains much of the geometry that makes magnetism such an interesting and complex subject.

1.2 Effects due to geometry

Because the magnetic dipoles of a magnet have a directionality, the magnetic fields they produce will have to accommodate to the geometry of the physical shape of the magnet. This could be a large energetic effect. In addition, each dipole of a large magnet produces its own magnetic field, which interacts with other dipoles of the same magnet. This is the dipole–dipole interaction. This can be a very long-range effect across the whole sample of the medium. It leads to the *demagnetization* field in a magnet, that tends to reduce or nearly eliminate the internal magnetic field in a sample. These effects comprise what is called *shape anisotropy* or a kind of *geometric anisotropy*. Much can be learned about how this works by consideration of the macroscopic magnetic field lines around a magnetic sample.

The way in which shape anisotropies come about can be traced to the basic equations of magnetostatics for real magnetic media. In addition to (1.1), there is the relation between magnetic induction **B**, magnetic field **H** and the magnetization **M** (dipole moment per unit volume),

$$\mathbf{B} = \mu_0(\mathbf{H} + \mathbf{M}). \tag{1.6}$$

More details about the definition of **M** will be reviewed in chapter 2. Relation (1.6) relies on the magnetic permeability of free space, $\mu_0 = 4\pi \times 10^{-7}\,\mathrm{T\cdot m\,A^{-1}} = 0.4\pi\,\mu\mathrm{H\,m^{-1}}$. Physically, relation (1.6) displays how lines of magnetic induction are produced either by electric currents (**H** is generated from currents via Ampere's law) or by the presence of magnetic dipoles (the magnetization contains the distribution of dipoles in the sample). The magnetic permeability μ_0 defines the efficiency with which either of these effects produce lines of magnetic induction **B**. When the relation (1.6) is combined with Gauss's law for magnetostatics (1.1), there results

$$\nabla \cdot \mathbf{H} = -\nabla \cdot \mathbf{M}. \tag{1.7}$$

Now in most cases considered in this book, we consider a static or quasi-static situation, without the presence of free electric current density $\mathbf{J}_e$. For a static situation, the magnetic field is governed by Ampere's law in differential form[2],

$$\nabla \times \mathbf{H} = \mathbf{J}_e. \tag{1.8}$$

---

[2] A term involving the time rate of change of the electric displacement, or displacement current, $J_d = \frac{\partial D}{\partial t}$, is not present on the RHS of equation (1.8), because we consider a static situation.

When there are no electric currents present in a region of interest, the magnetic field is curl-free, and thus it can be obtained from an assumed scalar magnetic potential, call it Φ_M. We suppose this potential determines the magnetic field caused by **M** according to a usual relation as would apply to electrostatics, namely,

$$\mathbf{H}_M = -\nabla \Phi_M. \tag{1.9}$$

The subscript on H_M has been introduced to stress that this involves only the part of the magnetic field directly produced by **M**. By combining (1.7) with (1.9), there results a Poisson equation for magnetostatics,

$$-\nabla^2 \Phi_M = \rho_M \tag{1.10}$$

where the *effective magnetic charge density* on the RHS of (1.10) is determined by the state of the magnetization,

$$\rho_M = -\nabla \cdot \mathbf{M}. \tag{1.11}$$

Comparing with (1.7), this also means that H_M is determined from a Gauss's law for magnetostatics,

$$\nabla \cdot \mathbf{H}_M = \rho_M. \tag{1.12}$$

This shows that the distribution of magnetization in a magnet completely determines how the magnetic field lines of $\mathbf{H}_M$ pass through and around the system. The lines of $\mathbf{H}_M$ must begin on positive effective charges and terminate on negative effective charges. This equation has the same form as the Poisson equation (1.3) for electrostatics, hence its solution is in the same form as (1.5), with appropriate substitutions.

Exercise 1.1. Suppose the permanent magnetization of a small magnet of length L points only along the x-direction and is given by

$$\mathbf{M} = M_0 \hat{x} [\tanh(x/w) - \tanh[(x-L)/w]]. \tag{1.13}$$

The parameter w is a small width over which **M** changes from zero to a finite value at the ends of the magnet. (a) Determine and plot the resulting magnetic charge density $\rho_M(x)$ for a choice of L and $w = 0.1L, 0.01L, 0.001L$. (b) For very small w, how much total magnetic charge resides near the end surfaces of the magnet? For example, with ϵ a small parameter (but greater than w), what are the integrals

$$\lim_{\epsilon \to 0} \int_{-\epsilon}^{+\epsilon} dx\, \rho_M(x), \qquad \lim_{\epsilon \to 0} \int_{L-\epsilon}^{L+\epsilon} dx\, \rho_M(x)? \tag{1.14}$$

(c) For the limiting case that $w \to 0$, verify that the charge density can be written using Dirac delta functions (see Weber and Arfken [6], or other texts on mathematical methods) as

$$\rho_M(x) = -M_0 \delta(x) + M_0 \delta(x - L). \tag{1.15}$$

The reason for reviewing these magnetostatics relations, is that one sees that magnetization **M** is responsible for producing an effective magnetic charge density, ρ_M. Although this is not a true monopolar magnetic charge or a magnetic charge associated with elementary particles, it does manifest itself in real magnetic media. The most dramatic manifestation takes place at boundaries. On the inside of the sample, there is some non-zero **M**, but outside the boundary, **M** = 0. The sudden change in **M** necessarily implies the presence of a magnetic surface charge density, σ_M. Gauss's law (the integral form of equation (1.12)) can be applied to a small pillbox of surface area A and vanishingly small height encompassing both sides of the boundary (see figure 1.1). If **H**$_1$ is the field just within the medium, while **H**$_2$ is the field just outside the medium, and $\hat{\mathbf{n}}$ is the outward direction vector normal to the surface, the magnetic boundary condition is found,

$$(\mathbf{H}_2 - \mathbf{H}_1) \cdot \hat{\mathbf{n}} = \sigma_M = \mathbf{M}_1 \cdot \hat{\mathbf{n}}. \tag{1.16}$$

Here **M**$_1$ is the value of magnetization just inside the boundary of the magnet (outside, there is the value **M**$_2$ = 0). This well-known boundary condition shows how the discontinuity in **M** directly leads to a similar discontinuity in the magnetic field intensity. This happens everywhere on the boundary of the magnet, even one with a uniform value of **M** within its interior. With uniform **M**, there would be no magnetic charge density ρ_M in the interior, however, there would still be non-zero surface charge density σ_M. These effective surface charges are responsible for the north or south poles of a permanent magnet. By the derivations above, however, their strengths can be almost anything, depending on the structure of **M(r)**. These surface charges (or poles) can be considered to be the macroscopic sources of the magnetic field lines both inside and outside the magnet. For this reason, the boundaries play a crucial role in the behavior of any magnet. This is primarily due to the fact that magnetic induction and the magnetic field are derived from dipole fields that have inherent directionality. Hence, magnetism has inherently anisotropic properties determined by sample geometry.

Outside the magnet, the **H**$_M$-field generated by magnetic surface charge density tends to point in a direction similar to the way **M** points, on average, in the magnet. Within a magnet, however, the **H**$_M$-field tends to point opposite to the **H**-field of the dipoles themselves. That is, **H**$_M$ points opposite to **M** within a permanent magnet.

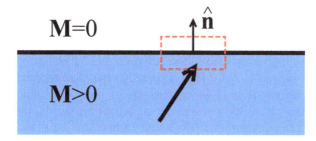

Figure 1.1. Abrupt change in magnetization **M** at a surface with outward unit normal $\hat{\mathbf{n}}$ will lead to an effective magnetic surface charge density $\sigma_M = \mathbf{M} \cdot \hat{\mathbf{n}}$. This can be obtained from Gauss's law (1.12) applied to a small pillbox of surface area A that encompasses the boundary (red dashed line).

Thus, $\mathbf{H}_M$ due to surface charges tends to work against the magnetization and tries to reduce $\mathbf{M}$ in order to reduce the total system energy. This is the *demagnetization effect*, and it is clear that it depends strongly on geometry.

1.2.1 Demagnetization energy

In order to discuss demagnetization effects, it is helpful to note how the energy works out. Any magnetic dipole $\boldsymbol{\mu}$ placed in an externally generated magnetic induction $\mathbf{B}$ has a magnetic potential energy given by

$$E_M = -\mathbf{B} \cdot \boldsymbol{\mu}. \tag{1.17}$$

The negative sign ensures that the energy is minimum when $\boldsymbol{\mu}$ aligns with $\mathbf{B}$. In the case of many dipoles better described by a magnetization, the interaction energy must be an integral over the volume,

$$E_M = -\int dV\, \mathbf{B} \cdot \mathbf{M} \tag{1.18}$$

because $d\boldsymbol{\mu} = dV\, \mathbf{M}$ is an infinitesimal dipole element. Equation (1.18) is correct if the field $\mathbf{B}$ is produced by *other dipoles*. In our case of most interest, however, the $\mathbf{B}$-field is the field actually produced by $\mathbf{M}$. An individual dipole cannot really interact with itself, so when such a self-interaction is removed from (1.18), the result is half as strong[3]. So the demagnetization energy produced by a magnetization is found from

$$E_M = -\frac{1}{2}\mu_0 \int dV\, \mathbf{H}_M \cdot \mathbf{M}. \tag{1.19}$$

Here, it is stressed again that $\mathbf{H}_M$ is the field directly produced by $\mathbf{M}$, rather than the total field, that could include $\mathbf{H}$-fields from other sources external to the magnet. This is only the interaction that takes place between $\mathbf{M}$ and $\mathbf{H}_M$. It is generally a positive energy, because $\mathbf{H}_M$ and $\mathbf{M}$ inside a permanent magnet have the tendency to point in opposite directions. It is sometimes written in various alternative forms, for instance, by supposing a certain demagnetization factor N_d that defines the relation between $\mathbf{M}$ and $\mathbf{H}_M$,

$$\mathbf{H}_M = -N_d \mathbf{M}. \tag{1.20}$$

This should be considered an approximate relation—it may not be precise within the entire magnet, but it gives an idea of the influence that $\mathbf{M}$ and sample geometry have on the internal magnetic field $\mathbf{H}_M$. Typically, the demagnetization factor is a dimensionless number from close to zero up to one. If it makes good sense to describe the whole object by a single demagnetization factor, then the demagnetization energy from equation (1.19) is found to be

---

[3] This is similar to the factor of 1/2 seen in formulas for energy stored in a capacitor, inductor, electric field energy density, magnetic field energy density, and so on.

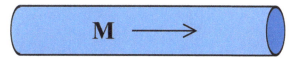

Figure 1.2. A permanently magnetized long thin cylinder will tend to have its magnetization aligned axially. This produces magnetic surface charge densities (or poles), according to (1.16), on the end caps only, which have limited area, reducing its magnetostatic energy.

$$E_M = \frac{1}{2}\mu_0 N_d \int dV\, \mathbf{M} \cdot \mathbf{M}. \tag{1.21}$$

Obviously this form guarantees that the demagnetization energy will be positive. Sometimes this might be called 'magnetostatic energy', however, that term could have various meanings and does not indicate the physical source of the energy.

1.2.2 Demagnetization examples

It is instructive to recall the simplest examples of demagnetization. It is a standard exercise in courses on electricity and magnetism to work out the **B** and **H** fields within a uniformly magnetized circular cylinder of radius R and length L. This solution will be reviewed in more detail later (chapter 3), here we just discuss its gross features. If the cylinder is very long and thin ($L/R \gg 1$), there is very little area at its ends on which any magnetic surface charge could be produced. There is a much more significant area over the long and curved sides, which could support quite a large magnetic surface charge. As a result, if one tries to permanently magnetize the cylinder, which rather closely resembles a long thin needle, then it is extremely energetically favorable for **M** to point along the long axis of the cylinder (in the axial direction), as in figure 1.2. Only small amounts of surface charge will be present on the end caps, leading to a minimum demagnetization energy. The two caps would have surface charges of opposite signs. Taking this example further, if the end caps are made into sharp points, the shape is a needle. The pointed ends do not support the appearance of significant magnetic surface charges, and again **M** will greatly prefer to point parallel to the long direction of the needle. Of course, it will have two opposite directions in which it can point.

If a uniformly magnetized cylinder is very short in length but has a large radius ($L/R \ll 1$), the end caps have substantial area, while the curved edges do not. Now the cylinder is shaped more like a coin or disk. Then, the large and flat end caps could support a very large magnetic surface charge, but only if **M** points axially. That would entail a large demagnetization energy. Hence, it is energetically much more favorable for **M** to point within the plane of the disk (see figure 1.3), only producing small amounts of magnetic surface charges on the curved edge of the disk. Taking this a step more generally, any kind of thin magnetic material will have a strong tendency to remain magnetized with **M** within the plane (i.e. perpendicular to the thin direction).

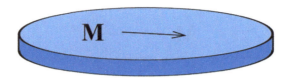

Figure 1.3. A permanently magnetized thin disk will tend to have its magnetization aligned in the plane of the disk. This produces magnetic surface charge densities (or poles), according to (1.16), only on the edges, which have little area, reducing its magnetostatic energy.

Exercise 1.2. Consider a thin circular disk of radius R and thickness $L \ll R$ with uniform magnetization as in figure 1.3, $\mathbf{M} = M\hat{x}$. (a) Determine the magnetic surface charge density σ_M on its curved edge. (b) Surface charges dq_M act as point charges that generate a magnetic field, just as electric charges generate an electric field, e.g.,

$$d\mathbf{H} = \frac{dq_M}{4\pi r^2}\hat{r}. \tag{1.22}$$

Show that the magnetic field at the center of the disk is approximately

$$\mathbf{H}_M(0) = -\frac{L}{4R}\mathbf{M}. \tag{1.23}$$

A thinner disk has a smaller aspect ratio L/R and hence a weaker demagnetization field. That brings more energetic stability to $\mathbf{M}$ in the plane of the disk.

Exercise 1.3. Consider a sphere of radius R with uniform magnetization $\mathbf{M} = M\hat{z}$. Following the example of the previous exercise, show that the magnetic field at its center is

$$\mathbf{H}_M = -\frac{1}{3}\mathbf{M}. \tag{1.24}$$

In fact, the demagnetization field throughout the interior of the sphere is this same value, see chapter 5 of [5]. What then, is the demagnetization factor for a sphere?

We should note that the above examples are idealizations that typically would apply to very small magnetic samples. The phenomenon that the equilibrium direction of $\mathbf{M}$ is greatly determined by the aspect ratio of the cylinder, or other geometric features of any object, is known as *shape anisotropy*. It is due to the overall shape and implied boundaries of the object. This is not what would be considered an intrinsic magnetic anisotropy force or torque, but rather it is derived from how the fields try to achieve minimum demagnetization energy for the entire system.

In magnets of macroscopic scale, demagnetization energy is further minimized when the magnetization field $\mathbf{M}(\mathbf{r})$ breaks into regions called *domains*, with $\mathbf{M}$ pointing in close to one direction in a given domain, but with each domain's magnetic moment aligned differently to its neighbors'. Each domain behaves similarly to a small magnet with its own north and south poles. As long as the neighboring domains are oppositely magnetized, or as long their magnetic moments (as vectors) average out to zero, the demagnetization energy is relatively small. If the magnetizations of all domains can be made more closely parallel, the demagnetization energy will be greatly increased. For that situation to be stable, some other form of energy (or magnetic force or torque) must secure the magnetization in the configuration. That is usually attributed to different types of intrinsic *anisotropy energy* that makes **M** prefer certain particular directions within the material. This kind of local anisotropy refers to torques on local magnetization caused at the atomic level by, for example, electrostatic forces on the atomic dipoles. Intrinsic or local anisotropy energy also plays a major role in controlling how structures in **M** form in a magnetic material. Intrinsic anisotropy can also be present in the way in which magnetization field **M(r)** interacts with itself, that is, in the *exchange interactions*, which are discussed next.

1.3 Exchange interactions

The actual magnetic dipoles of a magnet are ultimately due to atomic magnetic dipoles. In point of fact, the currents needed to produce these dipoles may not be physically detectable, because they are actually intrinsic currents associated with the intrinsic spin angular momentum **S** of electrons. The most correct explanation of magnetism at the fundamental level requires the application of quantum physics. It is known that the quantum states of electrons in atoms determine the magnetic properties and, in particular, the effective magnetic dipoles. Certain quantum rules need to be applied to obtain the possible values of atomic magnetic dipole moments. In quantum theory, there may be many possible states for an atom, including the states of the magnetic dipole moment[4]. In the discussion of elementary atomic physics, there are four important quantum numbers that must be specified to give the state of an electron. They are the principle quantum number n, orbital quantum number l, magnetic quantum number m_l and spin quantum number m_s. Each of these describes a different physical aspect of the atom (or an electron in the atom). The principle quantum number primarily determines the energy, the orbital quantum number determines the magnitude of orbital angular momentum, the magnetic quantum number relates to the direction of orbital angular momentum and the spin quantum number relates to the direction of intrinsic angular momentum of the electron. A specification of these for each electron in an atom can be seen to determine the particular magnetic state.

Not only do the atomic dipoles interact with each other via the magnetic fields they produce (dipolar interactions), but they also interact through a process known as *exchange*. The word refers to the idea that the quantum wave function of a system must reverse sign under the interchange of two electrons. This is the same as saying that the wave function must be *anti-symmetric* with respect to interchange of the two electrons. The reason for this comes from a somewhat bizarre property of electrons. Electrons, being spin-1/2 objects, are known to be fermions (as are other odd-half-spin particles). Fermions have the strange property that two of them cannot occupy

---

[4] Many introductory texts discuss the rules for quantum states of atoms and the ideas of spin exchange, such as Griffiths [7] and Eisberg and Resnick [8].

the same identical quantum state. This is the so-called Pauli exclusion principle (PEP). They could, however, be in nearly the same state, as long as at least one of their quantum numbers is different. For example, two electrons can be in the same orbital (particular values of principle, orbital and magnetic quantum numbers, n, l, m_l, respectively) as long as they have opposite spin quantum numbers, one with $m_s = +1/2$ and the other with $m_s = -1/2$. By insisting on the reversal of the two-electron wave function under interchange of electrons, the PEP is obeyed.

This quantum restriction effectively produces significant forces between electrons, that have no classical counterparts. This does not happen only in magnets, but in many different things. One sees a spectacular display of this type of quantum force in degenerate electron gases (collections of electrons at high enough density that electron wave functions overlap). In a degenerate gas, the overlap of electron wave functions means that the requirements of the PEP must be exhibited. For example, in white dwarf stars the quantum degeneracy force supports the star against gravitational collapse[5]!

For magnetism the quantum effects associated with satisfying the PEP may not seem as spectacular, however, the exchange effect produces a slight difference in energies for an electron pair with aligned spins relative to a pair with misaligned spins. A complete discussion of the exchange effect can be found in many texts on quantum mechanics and on fundamentals of magnetism. Let us give just a brief summary of how the effect comes about.

1.3.1 Exchange of an electron pair

Exchange is a process that requires two atoms and two electrons that might be shared between the atoms. A historical example is the exchange effect in a hydrogen H_2 molecule, which plays a role in explaining the chemical bonding of the atoms. So we consider what happens for two electrons, labeled by some coordinates r_1 and r_2, for two nearby atoms, labeled by a and b. Two coordinates are needed, however, because electrons are indistinguishable, one cannot say which coordinate corresponds to a particular electron. Thus, there is a wave function for the electron pair, $\Psi(r_1, r_2)$, which can be composed from certain single electron states that are solutions to Schrödinger's equation for individual electrons on just one of the atoms under consideration. The square of the total wave function, $\Psi^*\Psi = |\Psi(r_1, r_2)|^2$, gives the probability density for finding the electrons at the chosen positions (in this case, one needs to specify the two positions r_1 and r_2).

The total wave function also must include the effects of electron spin, in addition to the spatial dependence on electron position coordinates. Electrons have quantum spin-1/2, and can occupy only the spin-up ($m_s = +1/2$) or spin-down ($m_s = -1/2$) eigenstates, where up and down refer to the direction of spin angular momentum relative to a selected $\hat{z}$-axis. The condition of anti-symmetry on the total wave function of the electron pair has to include the effects of spin, and this is where there is a significant influence on magnetic energy. It is simplest to suppose that the total wave function is a product of a spatial part and a spin part.

---

[5] The same degenerate fermion gas quantum pressure is present in neutron stars, which are also spin-1/2 fermions, but with a different scale due the difference in neutron mass compared to electron mass.

Consider the spin part, and let $\chi_+(\mathbf{s}_n)$ and $\chi_-(\mathbf{s}_n)$ indicate the unit-normalized up and down states of a quantum spin $\mathbf{s}_n$. Then $\mathbf{s}_1$ and $\mathbf{s}_2$ will be used to refer to the spins of the two different electrons. The notation used here is that an electron spin can be written as

$$\mathbf{S} = \hbar \mathbf{s} \qquad (1.25)$$

where $\mathbf{s}$ is dimensionless (but not a unit vector) and $\hbar$ is the fundamental unit of angular momentum (Planck's constant h divided by 2π). For an electron spin-1/2, the spin angular momentum is described by the spin length parameter $s = 1/2$. Then for our pair of spins, there are four possible ways to choose their states. These can be categorized, however, in terms of the magnitude of the total vector spin, $\mathbf{s} = \mathbf{s}_1 + \mathbf{s}_2$. When one adds two spin-1/2 angular momenta, the rules of quantum mechanics show that the total spin s will be either spin-0 (i.e. $s = 0$) or spin-1 ($s = 1$).[6]

There are three different ways to obtain $s = 1$, which are distinguished by different eigenvalues of the z-component, $s^z = -1, 0, +1$. The states taken together form a *spin-triplet*. The normalized state can be written as follows,

$$\chi_{\text{triplet}}^{+1} = \chi_+(\mathbf{s}_1)\chi_+(\mathbf{s}_2) \qquad (1.26a)$$

$$\chi_{\text{triplet}}^{0} = \frac{1}{\sqrt{2}}\big[\chi_+(\mathbf{s}_1)\chi_-(\mathbf{s}_2) + \chi_-(\mathbf{s}_1)\chi_+(\mathbf{s}_2)\big] \qquad (1.26b)$$

$$\chi_{\text{triplet}}^{-1} = \chi_-(\mathbf{s}_1)\chi_-(\mathbf{s}_2). \qquad (1.26c)$$

The superscripts indicate the three different values of s^z. Note how all of these are *symmetric* with respect to interchange of the electrons. It is typical to say that these states have *parallel spins*, although a study of quantum mechanics leads one not to take this description too literally, because of quantum fluctuations!

There is only one way to combine two spin-1/2s to obtain spin length $s = 0$, and of course it then has the total z-component, $s^z = 0$. This is the *spin singlet*, and it can be written in normalized form as

$$\chi_{\text{singlet}}^{0} = \frac{1}{\sqrt{2}}\big[\chi_+(\mathbf{s}_1)\chi_-(\mathbf{s}_2) - \chi_-(\mathbf{s}_1)\chi_+(\mathbf{s}_2)\big]. \qquad (1.27)$$

This is *anti-symmetric* with respect to interchange of the electrons. It is typical to say that the electron spins are *anti-parallel*, because their spin angular momenta have canceled out. In principle, the electron pair may be either in this anti-symmetric spin state, or in one of the symmetric states of the triplet. One needs to consider the spatial wave function together with the spin state, and see how one influences the other.

We want to concentrate on electrons in some particular electronic state (the space parts), only distinguished by whether that state is centered on atom a or atom b. Suppose there are unit normalized single electron states called $\psi_a(\mathbf{r})$ and $\psi_b(\mathbf{r})$, where the electron probability is concentrated either on atom a or on atom b. It is assumed

---

[6] Of course, the physical magnitude of the total spin angular momentum $\mathbf{S} = \hbar\mathbf{s}$ is given by the quantum mechanical formula, $|\mathbf{S}| = \sqrt{s(s+1)}\,\hbar$, while the eigenvalues of the z-component are $m_s\hbar$, with m_s ranging from $-s$ to $+s$ for the general case.

that both electrons cannot be on the same atom. Thus one needs products of $\psi_a(\mathbf{r})$ and $\psi_b(\mathbf{r})$. But there are two allowed possibilities, different only by the electron coordinates within them,

$$\psi \sim \psi_a(\mathbf{r}_1)\psi_b(\mathbf{r}_2), \qquad \psi \sim \psi_a(\mathbf{r}_2)\psi_b(\mathbf{r}_1). \tag{1.28}$$

Since these are both possible, they must be combined into a linear combination. A little consideration shows that may be combined either *symmetrically* or *anti-symmetrically*, just as the spin parts. Including the normalization constant, the symmetric combination is

$$\psi_{\text{sym}}(\mathbf{r}_1, \mathbf{r}_2) = \frac{1}{\sqrt{2}}[\psi_a(\mathbf{r}_1)\psi_b(\mathbf{r}_2) + \psi_a(\mathbf{r}_2)\psi_b(\mathbf{r}_1)]. \tag{1.29}$$

When $\mathbf{r}_1$ and $\mathbf{r}_2$ are swapped (interchanging the electrons), the wave function in (1.29) does not change. This reflects the electron indistinguishability.

Alternatively, the anti-symmetric spatial state is different only by a sign change on the second term,

$$\psi_{\text{anti}}(\mathbf{r}_1, \mathbf{r}_2) = \frac{1}{\sqrt{2}}[\psi_a(\mathbf{r}_1)\psi_b(\mathbf{r}_2) - \psi_a(\mathbf{r}_2)\psi_b(\mathbf{r}_1)]. \tag{1.30}$$

This wave function encounters a change in sign when the electrons are swapped. But that is not an observable change, because only the square of a wave function is physically detectable. Indeed, for either the symmetric or anti-symmetric wave function, the squared magnitude remains the same and the probability distribution remains the same under electron interchange, indicative of the fact that the interchange of the electrons is not physically observable (they are indistinguishable, after all). Mathematically, we are imagining this process. However, it is not really possible to select the electrons and interchange them!

Now the spin part and space parts must be combined to obtain a net wave function that is anti-symmetric with respect to electron interchange. The interplay between symmetry and spin leads to the quantum exchange interaction energy.

1.3.2 Exchange and energy

There is some Hamiltonian that defines the electron states. In the Heitler–London approximation, summarized in chapter 11 of [2], it is supposed that there is a Hamiltonian for each electron ($H_1(\mathbf{r}_1)$ that acts on $\mathbf{r}_1$ and $H_2(\mathbf{r}_2)$ that acts on $\mathbf{r}_2$) and, in addition, a Hamiltonian $H_{\text{int}}(\mathbf{r}_1, \mathbf{r}_2)$ exhibiting the interaction between the electrons that acts on both coordinates. None of these Hamiltonians has an explicit dependence on spin! Then the total Hamiltonian is taken as a sum,

$$H = H_1(\mathbf{r}_1) + H_2(\mathbf{r}_2) + H_{\text{int}}(\mathbf{r}_1, \mathbf{r}_2). \tag{1.31}$$

The Hamiltonians H_1 and H_2 are single electron operators whose eigenstates are the single electron states such as ψ_a and ψ_b, which are particular states mostly localized on one of the two atoms. That is, these wave functions solve the eigenvalue problems,

$$H_1\psi_a(\mathbf{r}_1) = E_a\psi_a(\mathbf{r}_1), \qquad H_1\psi_b(\mathbf{r}_1) = E_b\psi_b(\mathbf{r}_1), \tag{1.32a}$$

$$H_2\psi_a(\mathbf{r}_2) = E_a\psi_a(\mathbf{r}_2), \qquad H_2\psi_b(\mathbf{r}_2) = E_b\psi_b(\mathbf{r}_2). \tag{1.32b}$$

Operators H_1 and H_2 have the same spectra, with energy eigenvalues E_a and E_b for the two states localized on the two different atoms.

The spatial wave function could be either the symmetric state ψ_{sym} (1.29) or the anti-symmetric state ψ_{anti} (1.30). One way to have a net state that is anti-symmetric under interchange of electrons is to combine the symmetric space part ψ_{sym} with the anti-symmetric spin part χ^0_{singlet} (1.27). This net state has spin $s = 0$ (anti-parallel spins). Let us write this as

$$\Psi_0 = \psi_{\text{sym}}(\mathbf{r}_1, \mathbf{r}_2)\chi^0_{\text{singlet}}. \tag{1.33}$$

The subscript on Ψ shows the total spin $s = 0$. Another way to obtain a totally anti-symmetric state is to combine the anti-symmetric space part ψ_{anti} (1.30) with the symmetric spin part χ_{triplet} (1.26a), forming a state with spin $s = 1$ (parallel spins). We write it as

$$\Psi_1 = \psi_{\text{anti}}(\mathbf{r}_1, \mathbf{r}_2)\chi^{s^z}_{\text{triplet}} \tag{1.34}$$

Obviously the state is threefold degenerate with the three choices of s^z. By evaluating the system energy for the two allowed configurations Ψ_0 and Ψ_1, we obtain the difference in system energy when $\mathbf{s}_1$ and $\mathbf{s}_2$ are anti-parallel, compared to parallel. This is the essence of the exchange interaction and the basis for the Heisenberg exchange Hamiltonian applied so often in magnetism.

The symmetric and anti-symmetric spatial wave functions can be written as one formula,

$$\psi_\alpha(\mathbf{r}_1, \mathbf{r}_2) = \frac{1}{\sqrt{2}}\left[\psi_a(\mathbf{r}_1)\psi_b(\mathbf{r}_2) + \alpha\psi_a(\mathbf{r}_2)\psi_b(\mathbf{r}_1)\right]. \tag{1.35}$$

The parameter α takes the value $\alpha = +1$ for the symmetric wave function and $\alpha = -1$ for the anti-symmetric wave function. This helps us to find the *expectation value* of the energy of both states in (1.33) and (1.34) by the usual procedure of quantum mechanics, doing the following integral:

$$E = \langle H \rangle = \int d\mathbf{r}_1 \int d\mathbf{r}_2 \; \Psi^*_\alpha(\mathbf{r}_1, \mathbf{r}_2) H \Psi_\alpha(\mathbf{r}_1, \mathbf{r}_2). \tag{1.36}$$

As spin does not appear *explicitly* in the Hamiltonian (1.31), the unit-normalized spin parts have disappeared from the energy evaluation! The integral becomes the following, assuming the states ψ_a and ψ_b are unit normalized,

$$\begin{aligned}E = &\int d\mathbf{r}_1 \; \psi^*_a(\mathbf{r}_1)H_1\psi_a(\mathbf{r}_1) + \int d\mathbf{r}_2 \; \psi^*_b(\mathbf{r}_2)H_2\psi_b(\mathbf{r}_2) \\ &+ \int d\mathbf{r}_1 \int d\mathbf{r}_2 \; \psi^*_a(\mathbf{r}_1)\psi^*_b(\mathbf{r}_2)H_{\text{int}}(\mathbf{r}_1, \mathbf{r}_2)\psi_a(\mathbf{r}_1)\psi_b(\mathbf{r}_2) \\ &+ \alpha \int d\mathbf{r}_1 \int d\mathbf{r}_2 \; \psi^*_a(\mathbf{r}_1)\psi^*_b(\mathbf{r}_2)H_{\text{int}}(\mathbf{r}_1, \mathbf{r}_2)\psi_a(\mathbf{r}_2)\psi_b(\mathbf{r}_1).\end{aligned} \tag{1.37}$$

The first two terms in (1.37) are the energies of the single electron states, as they become E_a and E_b, respectively. The third term is known as the *direct term*, because the arguments within ψ_a and ψ_b do not change before and after H_{int} acts. The last term, however, is different and it is known as the *exchange term*, because the arguments in ψ_a and ψ_b are interchanged after H_{int} acts, corresponding to the interchange of the identity of the electrons in some quantum process. To obtain this result, one needed to use that fact that all the parts of the Hamiltonian are Hermitian. If the direct term integral is denoted as Q and the exchange term integral is denoted as J, then this result is summarized as[7]

$$E = E_a + E_b + Q\mathbf{S}^2 + \frac{1}{2}\alpha J \mathbf{S}^2. \qquad (1.38)$$

The integral J will be positive in many situations, because H_{int} is primarily the repulsive electrostatic interaction between electrons. Keep in mind that $\alpha = +1$ holds for the spatially symmetric state (with anti-symmetric spin singlet), Ψ_0, and $\alpha = -1$ holds for the spatially anti-symmetric state (with symmetric spin triplet), Ψ_1. These two states differ in energy by the exchange energy,

$$\Delta E_{\text{ex}} = J\mathbf{S}^2 \qquad (1.39)$$

with the singlet being higher (if J is positive). We can imagine that the singlet has anti-parallel spins and the triplet has parallel spins. Then, this has shown that the *energy of the system depends on the relative orientation of the spins*. That is, the exchange energy effect. The extra factor of $\mathbf{S}^2$ has been included on J for consistency with other equations involving exchange terms.

1.3.3 The Heisenberg exchange Hamiltonian

The above discussion has been used to motivate the reasons for exchange energy between electron pairs. It represents only a brief introduction to the cause of this exchange energy. For most of the practical applications discussed in this text, we consider a Hamiltonian that summarizes this effect and concentrate only on the spin effects, or the spin couplings. That is, the Heisenberg Hamiltonian for exchange energy and, in its isotropic form, it is written here for general spin length as

$$H_{\text{ex}} = -\sum_{i,j} J_{i,j}\, \mathbf{S}_i \cdot \mathbf{S}_j. \qquad (1.40)$$

The equation represents only the exchange energy, taken as the sum over all spin pairs (i, j). The energy constants $J_{i,j}$ are the individual exchange energies between pairs. The minus sign gives two spins with a positive exchange constant the tendency to become aligned. This is consistent with our derivation in the previous paragraph, where the triplet, with a pair of aligned spins, has the lower exchange energy. A positive exchange constant for this Hamiltonian is called *ferromagnetic* (FM) exchange, because of its tendency to align spin pairs. On the other hand, when $J_{i,j}$ is negative here, the tendency will be for the spins to be anti-aligned, and then the interaction is called *antiferromagnetic* (AFM). This is an example of a local spin interaction model, where we can identify the particular spins (perhaps by location on a lattice). This will be discussed further in the next chapter. The combination $J_{ij}\mathbf{S}^2$ is

---

[7] Factors of squared spin length, $\mathbf{S}^2 = s(s + 1)\hbar^2$, have been included on Q and J to be consistent in the units for other equations of classical spin models that follow.

an energy unit, which gives the order of magnitude of exchange energies in the model. Spin vectors are considered here to have dimensions of $\hbar$, even when we consider classical spin vectors.

1.4 Anisotropic exchange couplings

The Heisenberg Hamiltonian (1.40) is considered an *isotropic* interaction, because all the spin components couple equally. As long as a spin pair tends to align (or anti-align), it does not matter in which direction the combined pair tends to point, the energy will be the same. However, electron spins in a material may be subjected to other interactions, perhaps of electrostatic nature due to neighboring atoms, that break this isotropy. Thus, a more general exchange interaction Hamiltonian, based on the Heisenberg Hamiltonian, includes different couplings for the different Cartesian components (indicated by superscripts):

$$H_{\text{ex}} = -\sum_{i,j} \left(J_{i,j}^x S_i^x S_j^x + J_{i,j}^y S_i^y S_j^y + J_{i,j}^z S_i^z S_j^z \right). \tag{1.41}$$

Here we still have a dependence on the particular spin pairs. There could be exchange couplings even for distantly separated pairs of spins, in general.

To make for simpler analysis and discussion, it is common to limit the interactions to those between nearest neighbors on some lattice. Furthermore, we expect the exchange interactions to depend only on pair separations. For atoms on an isotropic lattice, all near neighbor pairs have identical separations and hence identical exchange couplings. In these near neighbor models, we can use a simplified Hamiltonian,

$$H_{\text{ex}} = -\sum_{(i,j)} \left(J^x S_i^x S_j^x + J^y S_i^y S_j^y + J^z S_i^z S_j^z \right) \tag{1.42}$$

where the notation (i,j) indicates that the only terms in the sum are those for the nearest neighbor pairs of sites. This Hamiltonian only contains anisotropy due to the fact that different Cartesian components of spin have different couplings. Next we consider another kind of exchange anisotropy, one due to the particular spatial arrangement that magnetic sites might have in a lattice.

1.4.1 Space anisotropy of exchange

So far, the interactions considered are assumed to take place among spins with neighbors in any space directions, equally. However, in many materials, the arrangement of magnetically active atoms could be in layers, or in chains. The particular geometric arrangement of the magnetic ions will play a strong role in the magnetic behavior.

Consider the case of layers. If the magnetic ions are arranged in well-separated layers, then the typical spacing between magnetic ions within a layer is smaller than the spacing from layer to layer. As a result, the exchange interactions within a layer (intra-layer) are stronger than those between neighboring layers (inter-layer). This is a very strong geometric effect. Effectively, each layer might act mostly independently of the neighboring layers. Then an individual layer is now a *quasi-two-dimensional* system. The magnetic properties of the crystal will be that of a nearly two-dimensional (2D) material, which can be considerably different than one of 3D.

For the case of magnetic ions in chains, one can suppose that the spacing of magnetic ions in a particular chain is small compared to the spacing between neighboring chains. I am assuming that the chains are all parallel, for simplicity. Then the intra-chain exchange interactions will be much stronger than the inter-chain exchange interactions. Each chain can then act nearly independently of its neighboring chain. Then the individual chains behave as *quasi-one-dimensional* systems. The magnetic properties of the crystal will be those of a nearly one-dimensional (1D) material. This could be quite different behavior from that found in 3D.

To make this clearer mathematically, consider an alternative way to write an exchange Hamiltonian, including both anisotropy in spin components and *space anisotropy*. Rather than simply using i, j to indicate the different spins, it is necessary to use a notation that indicates something about spatial positions. Consider an example where the magnetic ions of the full 3D material occupy sites of a 3D lattice, with lattice displacements to the neighboring ions of a chosen site being $\mathbf{a}_n$, for $n = 1, 2, 3, \ldots z$, where z is the number of neighbors (coordination number of the lattice). To have a model that will reduce to 2D, assume that some of the displacements define an xy-plane, and the others point perpendicular to that plane (or they have components along $\hat{z}$). There can be different exchange constants for the different directions, call them $\mathbf{J}_n$. Each is a vector, whose components give the strengths of couplings between different Cartesian spin components. Then a Hamiltonian including *exchange with space anisotropy*, for N spins, can be written as

$$H_{\text{ex}} = -\frac{1}{2} \sum_{i=1}^{N} \sum_{n=1}^{z} \left(J_n^x S_{\mathbf{r}_i}^x S_{\mathbf{r}_i+\mathbf{a}_n}^x + J_n^y S_{\mathbf{r}_i}^y S_{\mathbf{r}_i+\mathbf{a}_n}^y + J_n^z S_{\mathbf{r}_i}^z S_{\mathbf{r}_i+\mathbf{a}_n}^z \right). \quad (1.43)$$

The factor of $\frac{1}{2}$ is necessary, because every interaction pair is counted twice in the summations. Each $\mathbf{r}_i$ is the location of the ith lattice site.

For a model for a quasi-2D magnet, it must be the case that some of the couplings are small enough to be irrelevant. The same Hamiltonian can apply to a quasi-2D magnet! The only thing that would be necessary, is that only the neighbor displacements $\mathbf{a}_n$ and couplings $\mathbf{J}_n$ that lie within the required layer or xy-plane need be retained. The couplings out of that plane will be dropped.

For a model of a quasi-1D magnet, again Hamiltonian (1.43) can still be applied. Only the couplings along the chain direction would be retained. However, it is convenient to take the chain direction as the x-axis. Assuming that a 1D chain of magnetic ions has a uniform spacing, and all neighboring spin pairs have the same coupling, the Hamiltonian can be written in a simpler form,

$$H_{\text{ex}} = -\sum_{i=1}^{N-1} \left(J^x S_i^x S_{i+1}^x + J^y S_i^y S_{i+1}^y + J^z S_i^z S_{i+1}^z \right). \quad (1.44)$$

This is written for a finite chain of N sites, with open ends. One only needs to specify the three components of the exchange energy couplings, J^x, J^y, J^z, assumed to be the same for all near neighbor pairs along a chain. Also note that we would only analyze a single chain, and take the view that couplings from chain to chain is a second-order (weaker) effect.

1.4.2 XY spin symmetry

A great deal of attention is placed on spin models with a planar spin symmetry. Different materials exist where the exchange couplings are of one strength for the spin components within some plane, usually taken as the xy-plane of the coordinate system. Staying within the limits of a near neighbor model, an XY-model Hamiltonian is

$$H_{\text{ex}} = -\sum_{(i,j)} \left[J^x \left(S_i^x S_j^x + S_i^y S_j^y \right) + J^z S_i^z S_j^z \right]. \tag{1.45}$$

It is characterized by two exchange constants, one for *in-plane* interactions (J^x) and one for *out-of-plane* interactions (J^z). This model also goes by other names such as the XXZ model, as a way of indicating the sizes of the J constants. However, in order to make it actually exhibit a planar spin symmetry, it is necessary to have

$$J^x > J^z \quad \text{(XY symmetry)}. \tag{1.46}$$

This will give the spins a tendency to have larger xy-components than z-components, meaning that they are predominantly moving within the xy-plane. Knowing the symmetry of the model is XY-like important, because it not only determines the kinds of ground states in the system, but also, the types of excitations out of those ground states.

1.4.3 Effects of XY symmetry—vortices

As the spins will tend to lower their energy by staying in the xy-plane, this has a strong effect on the type of ground state for the system. While that will depend on the overall specifics of the model, the shape of the system, and so on, we can give some general ideas. Due to the symmetry in the xy-plane, it could be that a ground state will have all the spins aligned, but pointing in any randomly chosen direction within the xy-plane. If all spins are nearly aligned, the state would be referred to as a *quasi-single-domain* (QSD). It is unlikely that they are perfectly aligned, due to demagnetization effects. Also, if demagnetization is strong enough, the system will break up into a set of domains, each with its magnet moment in a different direction, such that their average direction is null. The QDS state will probably be present for smaller systems. In a QSD state, there will be magnetic poles at some of the surfaces, but they should be of weak enough strength such that their field energies do not cause the system to re-organize into different domains.

Another possibility, of much more interest, is that the planar system interacts with surface demagnetization effects, such that along the boundaries of the system the spins prefer to align parallel to a boundary. This is an example of *strong geometric confinement*. In a disk-shaped magnetic system then, the competition between exchange and demagnetization will tend to force the magnetization to follow a boundary. It could be forced to follow the curved boundary of a relatively thin disk, forming a closed loop, while **M** still remains mostly within the *xy*-plane. Further within the disk, away from the boundary, a closed flow of **M** also forms, within **M** nearly planar. This is the state of a *vortex*, an example of the in-plane profile is shown in figure 1.4. The magnetization in a vortex state exhibits a circulation around a center called the *vortex core*. In the core of the vortex, the magnetization may tilt up out of the *xy*-plane, because if it did not, there would be a singular point at the core with an undefined value of **M**. See the out-of-plane view in figure 1.5. A vortex state could be the ground state of some disk-shaped magnets, depending on their particular size, and thickness compared to radius. The size of the core region will be

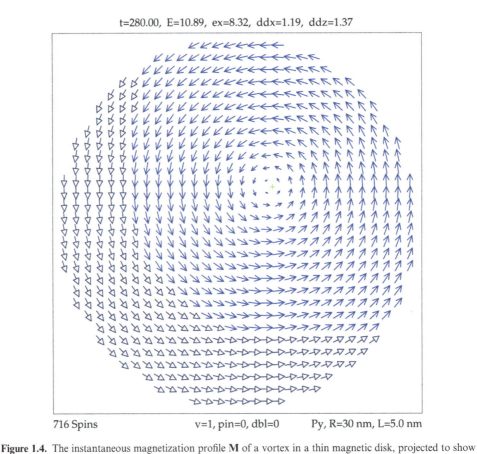

Figure 1.4. The instantaneous magnetization profile **M** of a vortex in a thin magnetic disk, projected to show its *xy*-components. Note that they follow the circular edge to minimize demagnetization energy. Arrows drawn as →▷ (—▷) indicate positive (negative) out-of-plane spin components. The core is visible as the region where the spins point out of the plane; the vorticity core is indicated with a plus sign. This is a configuration of a vortex in clockwise circular motion, from a simulation of the dynamics. Its out-of-plane profile appears in figure 1.5.

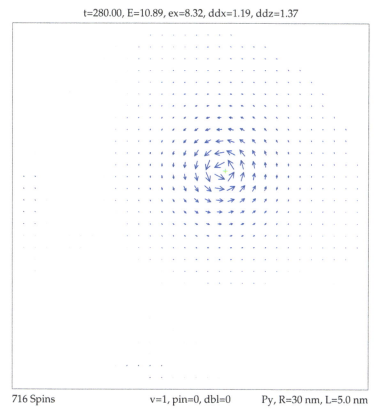

Figure 1.5. The magnetization profile **M** of the vortex in a thin magnetic disk of figure 1.4, in a different view. The lengths of the arrows are proportional to the component of magnetization perpendicular to the disk (M_z). Arrows drawn as → (—▷) indicate positive (negative) out-of-plane spin components. The vorticity core is marked with a plus sign. The vortex is moving in a clockwise circular motion.

determined by the XY-like exchange energy and the strength of demagnetization effects.

Alternatively, vortices may be excitations away from a ground state in other models with XY symmetry [9, 10]. They may be generated with increasing temperature in the XY model and are responsible for an unusual type of phase transition [9, 11, 12]. Not only are there vortices, but *anti-vortices* that act as the antiparticle of vortices. The main distinction of an anti-vortex compared to a vortex is the sense of how the magnetization circulates around the core. In the most studied XY model, whenever a vortex is created from thermal fluctuations, an equivalent anti-vortex is also created. Vortices and anti-vortices have a kind of geometric or *topological charge* and the sign of that charge on vortices is positive while on anti-vortices it is negative. Thus, we will see that XY models which generate vortices thermally always conserve the total topological charge.

Vortices and anti-vortices in XY models are examples of *topological excitations* partly because they possess this topological charge. The other reason has to do with their stability. A topological excitation cannot be easily destroyed (by itself). For this example with vortices/anti-vortices, the only way that a vortex can be destroyed within the system is by annihilation with an anti-vortex. That destruction removes both the vortex and the anti-vortex! Said another way, it is not possible to gradually deform an isolated vortex and arrive at a state of uniform spin alignment. The reason this is impossible, is that these two configurations have different values of the conserved topological charge (non-zero for the vortex, zero for the uniform alignment state). The order in XY-like systems is strongly controlled by vortex–anti-vortex pairs and is associated with the so-called Berezinskii–Kosterlitz–Thouless transition of vortex unbinding [9, 11, 12], to be discussed in further detail in chapter 8.

Vortices are expected to act as digital objects, useful for storing a data bit, because the topological charge in a vortex is a conserved quantity. As a result they have attracted a great deal of attention for technological reasons. A single vortex in a magnetic disk with a size on the order of hundreds of nanometers actually will have the topological charge (two opposite signs), a circulation direction (also two signs) and also a polarization charge (also two signs!). The circulation (or curling) indicates the direction in which the magnetization field flows around the disk (clockwise or counterclockwise, when viewed from a chosen side). The polarization charge is the orientation of the magnetization at the vortex core (either along the $+\hat{z}$- or $-\hat{z}$-directions). There are energy barriers to change the sign of any of these, hence they can be expected to be rather stable against thermal fluctuations.

1.4.4 More XY symmetry—the plane rotor model

So far it has been assumed that the magnetic spins have three Cartesian components. The presence of three components is natural and necessary for real magnetic dipoles. However, it is important to mention another model with XY symmetry, known as the plane rotor (PR) model [13], which is sometimes confused with the XY model. In the PR model, the spins are assumed to have only two Cartesian components. They live within a chosen xy-plane. Instead of specifying the components (S_i^x, S_i^y) for each spin, now called a *rotor*, one could instead just give its angle ϕ_i relative to a reference x-axis in the chosen plane. The rotors only move within that plane, forcing it to have an XY symmetry. They also are assumed to interact with their neighbors via exchange interactions.

The exchange energy in the PR model resembles that for the XY model, but with $J^z = 0$. Including only nearest neighbor interactions, one has

$$H_{\text{ex}} = -J^x \sum_{(i,j)} \left(S_i^x S_j^x + S_i^y S_j^y \right) \tag{1.47}$$

where there is one exchange constant, J^x, for all interacting nearest neighbor pairs. If written in terms of the rotor angles ϕ_i, this is even simpler,

$$H_{\text{ex}} = -J^x S^2 \sum_{(i,j)} \cos\left(\phi_i - \phi_j\right). \tag{1.48}$$

This form makes it fairly obvious that a FM exchange constant $J^x > 0$ will tend to align the neighboring pairs (alignment will reduce their energy). This represents only the exchange energy, which acts as a potential energy in the system. However, it is not sufficient for describing dynamics in the model.

In order for there to be real physical dynamics in the PR model, it is necessary to also include a kinetic energy term for the rotors. They can be imagined as having some rotational inertia I (the same for all sites), and then the kinetic energy in the system must be

$$H_{\text{KE}} = \frac{1}{2} I \sum_i \dot{\phi}_i^2. \tag{1.49}$$

The dot over ϕ_i indicates the time derivative, as $\dot{\phi}_i = \omega_i$ is the angular velocity of the ith rotor, which has kinetic energy $\frac{1}{2}I\omega_i^2$. By combining both the exchange Hamiltonians together with this kinetic energy, real equations of motion can be determined and the dynamics can be studied. The model is simple but interesting because it exhibits vortex type excitations.

1.4.5 Uniaxial to isotropic spin symmetry—skyrmions

The same Hamiltonian (1.45) above could also exhibit a *uniaxial* symmetry, for

$$J^x < J^z \quad \text{(uniaxial symmetry)}. \tag{1.50}$$

This will cause the z-components of spins to tend to dominate over the xy-components, meaning that the spins tend to point out of the xy-plane and, instead, moving more so near the z-direction. The uniaxial anisotropy could compete against any demagnetization effects that depend on the overall shape of the magnetic system. Here we suppose that the uniaxial effects win out. However, there could be considerable complications, if the axis of symmetry for the uniaxial interaction does not align with any symmetry of the geometrical shape of the magnetic material.

With uniaxial spin anisotropy, the overall ground state could be that where all spins would point along either of the $\pm \hat{z}$-axis directions. This has a two-fold degeneracy. The presence of this discrete degeneracy allows the system to have topological excitations called *instantons* or *skyrmions* [14], that can connect between the two ground states. In a skyrmion, the magnetization could point along $+\hat{z}$ at a center point (radius $r = 0$), and then slowly tilt towards $-\hat{z}$ as the magnetization is measured at points far away ($r \to \infty$). Because of the rotational symmetry in the xy-plane, there will also be xy spin components that have a vortex-like circulation around the center point of the skyrmion. However, far from the central point of the skyrmion there are no xy spin components; the vortex-like aspect is present only over a limited range of radius. That range will be determined by the strength of the uniaxial anisotropy.

Interestingly, skyrmions would be present in the isotropic spin model [15, 16] with $J^x = J^z$. This would be the isotropic Heisenberg model. It has neither XY symmetry nor uniaxial symmetry. By having no particular anisotropy, the skyrmions that would be present have no particular length scale. This means that the distance over which **M** changes from $+\hat{z}$ to $-\hat{z}$ can be any distance. The model is free of a natural length scale.

1.4.6 Quasi-1D magnetism—spin waves and magnetic solitons

In the Hamiltonian (1.44) for a chain of magnetic ions with nearest neighbor interactions, it is interesting to consider the case of XY symmetry, taking $J^y = J^x$, with $J^z < J^x$. This will give the model XY spin symmetry, that is, full rotational symmetry around the $\hat{z}$-axis. That in itself could be of interest. However, if a small but uniform magnetic field is applied to the system along, say, the x spin direction, the XY symmetry is slightly broken. Consider the corresponding total Hamiltonian,

$$H = -\sum_{i=1}^{N-1}\left[J^x\left(S_i^x S_{i+1}^x + S_i^y S_{i+1}^y\right) + J^z S_i^z S_{i+1}^z\right] - g\mu_B BS\sum_{i=1}^{N} s_i^x. \quad (1.51)$$

The first part is the exchange Hamiltonian with XY symmetry. The second part is an interaction with a magnetic induction field of strength B along the x-direction. The factors g and μ_B are the Landé g-factor and the Bohr magneton, respectively, while S is a spin magnitude. More will be mentioned about these numerical factors in chapter 2. They determine the efficiency with which an applied external field orients the spins. Studied as a *classical* spin model, the spin length is taken as $S = 1$ and the spin vectors $\mathbf{s}_i$ are assumed to be unit vectors.

The applied magnetic induction B will select the $\hat{x}$-axis as the direction of lowest energy for the spins. The ground state is then the state where all spins point in the $\hat{x}$-direction. However, the presence of fluctuations due to temperature, for instance, can cause deviations away from that ground state. The deviations from the ground state can be of two general types: 1) linear excitations (small amplitude) and 2) nonlinear excitations (large amplitude).

For the linear excitations, we mean small wave-like deviations, usually referred to as spin waves (if classical) or magnons (when quantized). They can be arrived at theoretically by performing an analysis of the equations of motion, *linearized* around the ground state. Spin waves are considered small amplitude so that this linearization procedure is a good approximation. By performing this linearized analysis, one can find the *dispersion relations* for the spin waves, which tells us the frequencies of the oscillations for chosen wave lengths or wave vectors.

For the nonlinear excitations, it is impossible to obtain them from the small-amplitude analysis[8]. Nonlinear excitations require careful solution of the dynamic equations of motion, without making a linearization approximation. Solitons are large-amplitude excitations that can appear in many condensed matter systems [20]. For the model (1.51), careful analysis leads to an equation of motion known as the sine-Gordon (sG) equation, a play on words because the model is similar to the Klein–Gordon equation of particle theory[9]. The solution of the sG equation for this

---

[8] Some 1D magnetic models with solitons are those studied by Tjon and Wright [17] and by Mikeska [18], see also [19].
[9] See advanced texts on relativistic quantum mechanics, such as Dirac [21].

problem leads to a kind of excitation known as a *magnetic soliton*. In this magnetic soliton, the magnetization makes a rotation of 360° (or 2π radians) around the $\hat{z}$-axis as one moves down the chain. Figure 1.6 shows how the spin components vary with position in a moving sG magnetic soliton. The spin orientation at a site in the chain is determined from an angle ϕ measured from the x-axis, and the component s^z of the spin out of the xy-plane. The rotation occurs over a well-defined length scale that is determined by the competition between exchange and the applied magnetic induction. When B is larger, the length scale is smaller, because the magnetization points against the direction of $\mathbf{B}$ in the central region of the soliton. The tilting of the spins away from the xy-plane (s^z) increases with the speed of the soliton. This type of soliton might also be called a *domain wall*, although the two domains that it connects have the same direction of magnetization (parallel to $\mathbf{B}$)!

The sG equation that predicts the properties of these magnetic solitons is Lorentz invariant. That means that the magnetic solitons can also be moving objects, and yet they undergo a Lorentz contraction when moving! The soliton also appears to behave something like a particle. The mathematics predicts interesting particle-like properties for these nonlinear excitations. Although we are referring to a 1D example here, similar magnetic soliton excitations exist in models in 2D.

1.5 Local anisotropies

So far we have focused on the anisotropy and geometric effects in the exchange interactions. The atomic neighborhood of a magnetic ion may also lead to additional anisotropic energies, typically those caused by anisotropic electric fields of the neighboring atoms. Thus, a magnetic spin is likely to be influenced by a *local anisotropy* energy. If the type of anisotropy is *uniaxial*, it will tend to cause a spin to

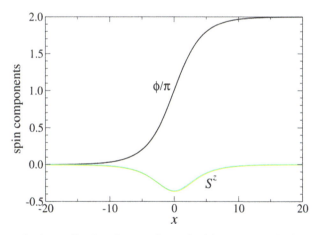

Figure 1.6. The magnetization profile of a soliton moving to the right on a magnetic chain with XY symmetry. The in-plane projection of the spins makes a full 2π rotation over some distance, coming back to its original alignment with a field applied along the x-axis. ϕ is the angle of the spins measured from the x-axis, within the xy-plane. The spin component out of the xy-plane is determined by the speed of the soliton.

point along a particular axis of symmetry. A Hamiltonian term for this interaction acting on an individual spin is

$$H_{\text{ani}} = -K_A (\mathbf{S}_i \cdot \hat{\mathbf{u}})^2 \tag{1.52}$$

The constant $K_A > 0$ is the energy constant for this, and $\hat{\mathbf{u}}$ is a unit vector that determines the axis of symmetry. The energy involves the squared component of the spin along $\hat{\mathbf{u}}$. This may better be called biaxial anisotropy, because actually it gives two equivalent directions of equal low energies ($\mathbf{S}_i$ along $\pm\hat{\mathbf{u}}$). One could also refer to the direction $\hat{\mathbf{u}}$ as an *easy axis*, meaning it is easier for the spin to point there, although in reality both $\hat{\mathbf{u}}$ and $-\hat{\mathbf{u}}$ are easy directions.

If the sign of K_A is negative, this Hamiltonian (1.52) then describes a different kind of anisotropy. The directions $\pm\hat{\mathbf{u}}$ now are relatively higher energy directions for the spin to point, so they are now referred to as *hard directions*. The anisotropy causes the spin to tend to remain close to the plane perpendicular to $\hat{\mathbf{u}}$. That plane is then called an *easy plane*, and the anisotropy would be called *easy-plane anisotropy*. This is an interaction that also leads to physics similar to that for an XY model.

It might be convenient in considering the statics and dynamics of spins with this interaction, to express the energy in terms of an angle θ_i between the spin $\mathbf{S}_i$ and $\hat{\mathbf{u}}$. The dot product leads to a factor of $\cos\theta$. Then an alternative way to express (1.52) is

$$H_{\text{ani}} = -K_A S^2 \cos^2 \theta_i. \tag{1.53}$$

1.5.1 Applied magnetic field

Magnets respond to an externally applied field, which breaks any symmetry that might be present. This gives an anisotropic effect that depends on the type of magnetic field distribution. In a spin model the Hamiltonian due to the applied magnetic induction field $\mathbf{B}(\mathbf{r})$ is a generalization of that used in the 1D mode, equation (1.51),

$$H_{\text{ext}} = -g\mu_B \sum_i \mathbf{B}(\mathbf{r}_i) \cdot \mathbf{S}_i. \tag{1.54}$$

This is written for unit spin operators $\mathbf{S}_i$ that are located at some lattice sites $\mathbf{r}_i$.

For a uniform applied field, the magnetic torques associated with this energy will tend to align all spins along the field. Even when the field is not uniform, the corresponding alignment with the spatially varying field can be an essential effect on a magnet. An example would be the magnetic induction caused by a current-carrying wire, referred to as an Oersted field. The magnetic field lines around a wire current are circles. If a vortex in a magnetic sample were subjected to this kind of circular field, it experiences a set of torques on the underlying spins that tends to move the vortex to become centered on the wire (if that is possible from the geometry). A current-carrying wire contacting a small magnetic sample can then be expected to work to control a vortex position. This is just one example where a field (and a current!) can be an important control parameter for a nonlinear excitation such as a vortex.

1.6 Theory for linear and nonlinear magnetism

The theory for magnetism has two primary goals. The first is to describe the magnetostatic configurations and the second is to describe the dynamic excitations. In the chapters that follow different theoretical approaches will be described for studying both of these.

For statics, one is most interested in the *stable* and *metastable* magnetic configurations of a system. Also, statics is usually concerned with the system's configurations at zero temperature. Any physical system will have a ground state or state of lowest energy for zero temperature. However, in a magnetic system, it is possible that the ground state is not unique, due to whatever symmetries are present, that make a certain set of magnetic configurations physically equivalent. The different configurations would be related to each other, typically by rotations around axes of symmetry that make no change in the total energy. There could be a continuous set of ground states, in the case of a continuous symmetry (such as an XY-like system), or a discrete set of equivalent ground states if there is a discrete symmetry (models on certain lattices). Thus it is the job of a theorist to seek ways to find and describe the ground states expected for a given magnetic model.

We have also mentioned metastable states. These are states of low energy relative to other very similar states that are somehow nearby in the phase space of the system. They would be considered local minima of the energy, not global minima as expected of a ground state. There will be an *energy barrier* that must be surpassed to move out of a metastable state and into another one or perhaps into the ground state. The depth of such energy barriers could be quite large when compared with thermal energy $k_B T$, where k_B is Boltzmann's constant and T is the absolute temperature, making metastable states almost as important as the ground states. This is because a system left in one of the metastable states will be able to remain in that state indefinitely unless it is forced out in some way, say by an external field, a temperature pulse, or some other stimulus. Any of the states of a magnetic system that are local minima of the energy (as viewed in a complex phase space) will be important in the general magnetostatics of the material.

Quite generally, there is no approach that will magically give us the ground states or the metastable states of a given model. In principal, one needs to find a way to obtain the minimum energy solutions (or local minima such as vortices) of the Hamiltonian. In certain cases, because of the symmetry, this may be obvious. In other cases, one needs to resort to numerical simulations that carry out some kind of energy minimization.

For the studies presented in this text, the most important type of metastable states are the topological excitations such as vortices, and others such as solitons and domain walls. These are not ground state configurations and they cannot be obtained by a small perturbation away from a ground state. It is impossible to take a ground state of an XY magnet, for example, and make small variations away from that ground state to arrive at a vortex state. On the other hand, one can start from a ground state and make small (linearized) variations away from it that are the spin waves in the system. Spin waves are just the small-amplitude time-dependent oscillations around any state, but usually considered around a ground state. Because they can be obtained mathematically by carrying out a procedure known as *linearization* of the Hamiltonian, they are called *linear excitations*. The procedure of linearization means taking the equations of motion from a Hamiltonian and keeping only the terms to linear order in the relevant dynamic variables (such as the Cartesian spin components). Linear excitations are the first example also of the *dynamic* states in a system.

In contrast, topological excitations such as vortices and solitons are considered to be *nonlinear excitations*, because they would not be found in the linearized theory. One needs the full equations of motion from the system Hamiltonian to arrive at the topological excitation solutions. There is some simplification, however, if we are initially just interested in static solutions or in solutions that are stationary in time. This would allow the removal of some time-dependent terms from equations of motion, making it easier to find solutions. Once a stationary solution is determined analytically, it may be possible to modify that result slightly and obtain the full time-dependent result. If not, it could be necessary to resort to numerical simulations. Generally, finding a complete solution for the dynamics of a magnetic system, including the nonlinear effects, is not possible from analytics, and simulations play a major role in the further analysis of the physics.

Understanding the dynamics of a magnetic system means to understand its time-dependent behavior. This could be a task such as obtaining the time-dependent motion of a vortex in an otherwise empty system. But more generally, dynamics refers to the overall response of a system to all perturbations, including the temperature. The ground states we have mentioned can be accessed, strictly speaking, only at zero temperature. There will always be spin waves on top of any configuration that are excited thermally. In addition, the topological excitations themselves are usually generated due to thermal fluctuations. Topological excitations also interact with the linearized excitations, and both are modified. Mostly we will be concerned with this situation of *thermal equilibrium* dynamics, although one could also consider a system out of equilibrium (the response due to a magnetic field impulse, for example). Temperature plays a very strong role in the dynamics and much of the theoretical work in magnetism is concerned with how to include it. Again, analytic theory can only go so far (it works best at a linearized level) and it is extremely helpful to develop numerical simulations to elucidate the role of temperature in magnetism.

1.7 Simulations in magnetism

A numerical simulation is an attempt to take into account all the linear and nonlinear effects in a model, and predict what states are expected to occur for the chosen model. Some simulations could be geared more towards finding only stable or metastable states, i.e. statics. Others will address the dynamics. The next several chapters in part I of this book describe how some important simulation techniques work, and we briefly mention these here.

For static configurations, there are different simulation techniques that can be called *energy minimization* schemes. The procedure is fairly simple. Start with some initial state of the system (a configuration of the spins). It could be a randomly selected state. Then make small modifications of that state which lower its energy and check the progress by calculating the total Hamiltonian. If this can be repeated sufficiently, the expectation is that a good approximation to a ground state or an important metastable state can be found. The main difficulty is that without a specially selected initial configuration, there is some uncertainty about the final configuration. Further, always going downhill in energy may not always be the best plan, which could land the system in an unimportant local energy minimum. It may actually pay in the long run to use an algorithm that is allowed to sometimes go uphill in energy, to obtain to the minimum you are looking for.

A very well-known scheme that sometimes allows the system to go uphill in energy, is the *Monte Carlo* (MC) method. It also goes by other names such as *importance sampling*. Importance sampling is a way to obtain a set of configurations of the system (your sample), that mimics how often those same configurations would show up in a real physical system (for the assumed Hamiltonian). Technically, importance sampling is a way to calculate averages of physical quantities such as magnetization, total energy, and so on. But it works by generating a sequence of states of your system, known as a *Markov chain*, that is somewhat random and yet looks as the real system would if you observed it every now and then. By taking averages of physical quantities, based on the set of states from this Markov chain, you obtain averages very much as you would obtain in real experiments, including noisy results (statistical errors). Randomness is built into the simulation, so that it can sample the phase space as well as possible. This includes the ability to go both up and down in energy, just depending on which region of phase space is more likely. An analytically calculated average usually requires an integration (or very large sum) over all of phase space. In MC importance sampling, the averages are calculated using only a tiny fraction of all of phase space, and that *sample* must be selected in such a way that the most relevant states show up more than unimportant states.

MC sampling is such a large developed field now, and comes in a great variety of different but related approaches. We will describe a few in chapter 4. The main thing they have in common, however, is that they are controlled by the assumed temperature T. When the temperature is higher, the system is allowed to wander more within the phase space and it is given a greater tendency to go uphill in energy (as well as downhill). The tendency is controlled by a Boltzmann factor, $\exp\{-\Delta E/k_B T\}$, where ΔE is the energy difference between two states. For a large energy change, this factor becomes smaller and smaller for low temperatures and the uphill wandering is reduced, while downhill wandering is still permitted. This gives the system the tendency to seek a ground state when the MC simulation is performed for very low temperature. This gives another way to energy minimization, as well as much more.

MC is usually considered to be a simulation for thermal equilibrium. It is not considered a simulation for the dynamics, *per se*, because it does not give the time-dependent excitations in the system. The system is only occasionally sampled and time is not a parameter in most versions of MC simulation. However, certain versions of MC are meant to represent the time development of the system, but the 'time' in MC simulations is in MC-steps, and it takes some unusual mapping to translate that into real time.

For dynamics, including the ability to see how the system in equilibrium behaves in time, there is a general set of simulation approaches known as *spin dynamics* (SD). For atomistic calculations the analogous well-known term is *molecular dynamics*, used to find dynamic properties of gases, liquids, *etc*. In SD one computes the time development that is expected from the magnetic Hamiltonian, starting from some initial configuration. There are different ways to do this, in particular depending on how the temperature is included (or not). Many magnetic simulations rely on solving the equations of motion without temperature, known as the Landau–Lifshitz–Gilbert (LLG) equations. The basic simulation is to integrate numerically the LLG equations forward in time. Ignoring the effects of temperature may be reasonable, depending on the different energy scales in the problem. If temperature needs to be included, a simple approach is a hybrid approach. In a hybrid scheme, MC is used to produce initial states for equilibrium at the desired temperature. Then, those equilibrium states can be evolved forward in time by applying the LLG equations (or their equivalent), integrating them forward in time. The drawback of this hybrid approach is that it requires two separate steps. But it produces the dynamics expected in equilibrium, when the whole procedure is averaged over many states from the MC part of the scheme.

Another way to include temperature in the dynamics is to apply a Langevin equation for the SD. The Langevin equation is known as a way to describe Brownian motion (pollen grains being pushed around randomly in a solution). In a Langevin equation the dynamics includes a random force due to temperature (the thermal kicks from atoms) and a damping force (the average effect of the fluid to slow things down), as well as any forces from the Hamiltonian of the system. For a spin system, a similar treatment is possible, making an approach known as *Langevin SD*. Any spin will experience random torques due to temperature and a damping torque. These are added to the usual torques described with the LLG equations (or the spin Hamiltonian). The result is a set of equations that can be integrated forward in time, describing the dynamics of the spin system, including all the thermal effects. Averaging over long time sequences is expected to be equivalent to averaging over phase space.

Once the time evolution is obtained, different time-dependent averages can be calculated, such as *spin correlation functions*. The spin correlations give a sense of the space- and time-varying order in the excitations.

Other types of dynamics results can be found, including the stability limit for some types of vortices, the interactions between topological excitations and spin waves, which is related to the spectrum of spin waves when a topological excitation is present. There are many different applications of different SD simulations; a few will be presented in part II of this book. This will help to sketch out a general impression of how theory and simulation can be used to understand geometric confinement effects in magnetic materials.

Bibliography

[1] Lee E W 1970 *Magnetism: An Introductory Survey* (New York: Dover)
[2] Jiles D 1991 *Introduction to Magnetism and Magnetic Materials* (London: Chapman and Hall)
[3] Griffiths D J 1999 *Introduction to Electrodynamics* 3rd edn (Englewood Cliffs, NJ: Prentice Hall)
[4] Lorrain P and Corson D 1970 *Electromagnetic Fields and Waves* 2nd edn (San Fransisco: CA Freeman)
[5] Jackson J D 1999 *Classical Electrodynamics* 3rd edn (New York: Wiley)
[6] Weber H J and Arfken G B 2004 *Essential Mathematical Methods for Physicists* (New York: Academic)
[7] Griffiths D J 2004 *Introduction to Quantum Mechanics* 2nd edn (Englewood Cliffs, NJ: Prentice-Hall)
[8] Eisberg R M and Resnick R 1985 *Quantum Physics of Atoms, Molecules, Solids, Nuclei, and Particles* 2nd edn (New York: Wiley)
[9] Kosterlitz J M and Thouless D J 1973 Ordering, metastability and phase transitions in two-dimensional systems *J. Phys. C: Solid State. Phys.* **6** 1181
[10] Hikami S and Tsuneto T 1980 Phase transition of quasi-two-dimensional planar system *Prog. Theor. Phys.* **63** 387
[11] Berezinskii V L 1971 Destruction of long-range order in one-dimensional and two-dimensional systems having a continuous symmetry group I. Classical systems *Sov. Phys. JETP* **32** 493
[12] Berezinskii V L 1972 Destruction of long-range order in one-dimensional and two-dimensional systems having a continuous symmetry group II. Quantum systems *Sov. Phys. JETP* **34** 610
[13] Tobochnik J and Chester G V 1979 Monte Carlo study of the planar spin model *Phys. Rev. B* **20** 3761
[14] Skyrme T H 1961 A non-linear field theory *Proc. R. Soc. Lond.* **260** 127
[15] Belavin A A and Polyakov A M 1975 Metastable states of two-dimensional isotropic ferromagnets *JETP Lett.* **22** 245 www.jetpletters.ac.ru/ps/1529/article_23383.shtml
[16] Trimper S 1979 Two-dimensional Heisenberg model and pseudoparticle solutions *Phys. Lett. A* **70** 114
[17] Tjon J and Wright J 1977 Solitons in the continuous Heisenberg spin chain *Phys. Rev. B* **15** 3470
[18] Mikeska H J 1981 Solitons in one-dimensional magnets (invited) *J. Appl. Phys.* **52** 1950
[19] Bernasconi J and Schneider T (ed) 1981 *Physics in One Dimension* (Berlin: Springer)
[20] Bishop A R and Schneider T (ed) 1978 *Solitons in Condensed Matter Physics* (Berlin: Springer)
[21] Dirac P A M 1982 *The Principles of Quantum Mechanics* 4th edn (New York: Oxford University Press)

Part I

Theory and simulation approaches for magnetism

IOP Publishing

Magnetic Excitations and Geometric Confinement
Theory and simulations
Gary Matthew Wysin

Chapter 2

Magnetism theory: spin models

Magnets are atomically ordered due to the quantum interactions between magnetic dipoles of neighboring atoms or ions. The quantities that actually interact are usually the electronic spins. Although the basic *exchange interactions* result from the weird behavior of quantum mechanics, in many instances classical spin models are more than adequate. In this chapter some basic models are given for different types of magnets and magnetic ordering. The models include ferromagnets, antiferromagnets and ferrimagnets, both with isotropic and anisotropic couplings. Quantum and classical spin dynamics (SD) approaches are compared.

2.1 Magnetic dipoles and magnetic ordering

We know that objects such as permanent magnets can attract or repel each other, depending on the proximity of a north or south pole of one magnet to a pole of the other. These poles are due to the dipoles present in the medium, which are the fundamental physical objects that produce forces in magnets. Also, the direction of the force (attractive or repulsive) is determined by whether the poles are opposites (such as north and south) or the same (such as both north poles). The forces are described by the well-known rules that *opposite poles attract* and *like poles repel*. Of course, a permanent magnet will also attract unmagnetized but magnetically permeable materials, such as soft iron, because the field of the permanent magnet induces magnetic order (and poles) in the soft iron. The soft iron can become magnetically polarized, but only as long as the permanent magnet is nearby. When the permanent magnet is removed, the soft iron loses its magnetic order and induced magnetic poles. One piece of soft iron does not usually attract or repel another piece of soft iron. For instance, the nails in a box of nails do not stick together, even though we think of iron as magnetic. They do not attract or repel each other because they are magnetically disordered.

North and south magnetic poles in a magnetic material reflect the internal magnetic ordering. Magnetic order is usually described by talking about the arrangement of magnetic dipoles in the material. Recall that the magnetic dipole of a current-carrying coil, for example, is the product of the current times the area, multiplied by the number of turns (units of $A \cdot m^2$). At an atomic level, charges in atoms also possess currents, hence, there are atomically defined magnetic dipoles. A basic macroscopic quantity to describe the overall magnetic order is the *magnetization* $\mathbf{M}(\mathbf{r})$, which is the magnetic dipole moment per unit volume (units of $A \cdot m^{-1}$) at position $\mathbf{r}$. At the atomic scale, this may be due to a very large number of atomic magnetic dipoles, denoted as $\boldsymbol{\mu}_n$, where n labels the different dipoles at different positions. To obtain a macroscopic definition for $\mathbf{M}$ at some position $\mathbf{r} = (x, y, z)$, we can consider a macroscopically small volume ΔV centered at $\mathbf{r}$, that happens to contain $N(\Delta V)$ dipoles. Then the volume-averaged dipole moment $\mathbf{M}(\mathbf{r})$ per volume is the magnetization [1, 2]:

$$\mathbf{M}(\mathbf{r}) = \frac{1}{\Delta V} \sum_{n=1}^{N(\Delta V)} \boldsymbol{\mu}_n. \qquad (2.1)$$

The sum is only over the dipoles in the considered volume element. Provided the volume is somehow small compared to a volume over which a measurement might average, but contains sufficient atoms, then one can obtain a smoothly defined macroscopic magnetization. The magnetization should contain the most important information about the magnetic order in the system. It has the physical sense of the averaged dipole moment of the magnetic dipoles in each small cell.

2.2 Atomic dipoles

Considering further an atomic picture, it is possible to think that this macroscopic magnetization is not the most fundamental object to be analyzed. Rather, the atomic dipoles actually result from quantum mechanical effects in atoms. For example, an electronic total angular momentum $\mathbf{J}$ corresponds to a certain rate at which the electronic charge flows around an atom. Here $\mathbf{J}$ is considered to be some number times $\hbar$, the basic quantum of angular momentum (Planck's constant h divided by 2π). Vector angular momentum $\mathbf{J}$ should not be confused with an exchange integral J (a scalar). $\mathbf{J}$ is found from certain rules for the vector combination of orbital and spin angular momenta. This leads directly to an atomic magnetic dipole moment. Because electrons carry a negative charge $-e$, the direction of the magnetic dipole moment is opposite to the direction of $\mathbf{J}$.

The total angular momentum has contributions from orbital motion and from intrinsic spin. Suppose there is an orbital angular momentum $\mathbf{L}$ only. Using only *classical mechanics* definitions, it is a simple exercise to find the magnetic dipole moment of an electron (mass m_e, charge $-e$) in a circular orbit of radius r and frequency f, with this amount of angular momentum. The angular momentum magnitude is $L = m_e v r = m_e (2\pi f r) r$, where velocity $v = \omega r$ was found from the angular frequency $\omega = 2\pi f$, while the magnetic moment is $\mu_L = iA = (-ef)(\pi r^2)$. Then the classical relation between orbital angular momentum and its magnetic dipole moment is [3]

$$\mu_L = -\frac{e}{2m_e}\mathbf{L}. \tag{2.2}$$

In quantum theory, however, any component of the orbital angular momentum is quantized in units of $\hbar$. For instance, the component along a z-axis is $L_z = m_l \hbar$, where azimuthal quantum number m_l takes on $2l + 1$ integer values, $m_l = -l$, $-l + 1, \ldots, l - 1, l$, and the quantum number l is an integer representing the size of the orbital angular momentum. Then the orbital magnetic dipole moment is quantized along any axis in a well-known amount,

$$\mu_{L,z} = -m_l \frac{e\hbar}{2m_e} = -m_l \mu_B. \tag{2.3}$$

Because the quantum index m_l only takes integer values, then this has led to the *Bohr magneton* μ_B as the basic quantum of orbital magnetic dipole moment,

$$\mu_B = \frac{e\hbar}{2m_e} = 9.27 \times 10^{-24} \text{A} \cdot \text{m}^2. \tag{2.4}$$

The intrinsic spin angular momentum $\mathbf{S}$ of electrons also contributes to the magnetic dipole moment. An individual electron has spin $s = 1/2$, which means that the component of spin angular momentum along some chosen z-axis is $S_z = m_s \hbar$, with $m_s = -1/2, +1/2$. Now, due to a relativistic effect, the spin-only contribution to the magnetic dipole moment turns out to be [4, 5]

$$\mu_{S,z} = -2m_s \frac{e\hbar}{2m_e} = -2m_s \mu_B. \tag{2.5}$$

There is an extra factor of 2 in the formula, however, because m_s only takes on values that differ by integers, again the Bohr magneton plays the role of the basic unit of quantized magnetic dipoles due to spin. As a vector relation, the magnetic dipole due to spin angular momentum also contains an extra factor of 2,

$$\mu_S = -2\frac{e}{2m_e}\mathbf{S}. \tag{2.6}$$

This is usually described by saying that the spin angular momentum has a *g*-factor of $g = 2$ (whereas the *g*-factor for orbital angular momentum is $g = 1$). Indeed, if both orbital and spin angular momenta are present simultaneously, then the relation is written in a general way,

$$\mu = -g\frac{e}{2m_e}\mathbf{J}. \tag{2.7}$$

The actual *g*-factor is a result of the vector combination of $\mathbf{L}$ and $\mathbf{S}$.

Oftentimes the relation between angular momentum and magnetic dipole moment is expressed instead using the *gyromagnetic ratio* γ of the electron. γ is the ratio of magnetic dipole moment to angular momentum. Then we replace the last relation by an equivalent one,

$$\boldsymbol{\mu} = \gamma \mathbf{J}. \tag{2.8}$$

In the case of purely orbital angular momentum ($\mathbf{J} = \mathbf{L}$), one has $\gamma = -e/2m_e$. On the other hand, for purely spin angular momentum $\mathbf{J} = \mathbf{S}$, one has $\gamma = -e/m_e = -1.76 \times 10^{11}$ (T s)$^{-1}$.

2.3 Local spin interaction models

The atomic dipoles under consideration could be those of nearly free atoms (as in a dilute gas) or those of more strongly interacting atoms, especially such as those in solids. For free atoms, one need not consider interactions among atoms, and magnetic ordering is not a relevant question for one free atom. For solids, however, the atomic interactions essentially determine the resulting macroscopic magnetic ordering effects. The particular electronic structure determines how those atoms interact, and whether it be weakly or strongly. If the electrons that contribute the main part of the magnetic dipole moments are inner shell electrons, they can tend to be *localized* on individual atoms in a crystal lattice. This is true for lanthanides where orbital contributions are quenched, and the 4f electrons mainly contribute spin angular momentum. In that case, a reasonable model for inter-atomic interactions considers that there are atomic magnetic dipoles of fixed magnitude, localized at sites of a lattice, that interact with each other. This is called the local moment model, and is the primary approach used in this book. The individual dipoles are assumed to be distinguishable and fixed in position. This is a simplification that may not be absolutely true in many materials, yet it provides a good starting point to study magnetic physics.

In ferromagnetic (FM) transition metals such as iron, cobalt and nickel, however, the localized moment model may be called into question, primarily because the magnetically important electrons are in the outer unfilled 3d orbital, and they occupy it as a band. Each electron is only weakly attracted to any particular atomic nucleus. Thus, these electrons cannot be considered as belonging to any particular atom, but rather, spread out in the lattice. Furthermore, the effective magnitude of the net atomic dipole moment becomes a fractional value of the Bohr magneton. This is known as the itinerant electron model [5]. It requires calculations of the properties of electron bands in order to find the magnetic ordering effects. Although it may be more realistic than the localized moment model, the greater difficulty of calculation results in some loss of basic intuition about the physical properties when the results are interpreted. As a result, it is preferable to make some analysis even for these materials, based on a localized moment model. While a certain element of physical realism is lost, considerable understanding can still be obtained.

In this text we consider the magnetic dynamics at two different levels of space detail. The first is literally at the atomic level, appropriate for small numbers of atoms. It will be referred to in this text as SD. The second is at a coarse-grained level more appropriate for larger samples that would contain a number of atoms too large to analyze down to the atomic scale. It is called *micromagnetics*. These two approaches have a lot of similarities, but it is good to keep the distinctions between them in mind.

2.3.1 Spin dynamics—atomic scale local spin magnetism

For analysis of a system at the atomic scale, it is typical to develop the Hamiltonian, equations of motion, and so on, for a set of spins $\mathbf{S}_n$, each of which corresponds to an individual atomic dipole. In simulations, computers may be able to keep track of the dynamics of the order of a hundred quantum spins, at most, or of the order of hundreds to perhaps a million classical spins. Even so, the physical size of a system with a million spins may still be much smaller than what can be fabricated for real magnetic applications. The goal of numerical simulations is to find the time evolution of $\mathbf{S}_n(t)$ for all the spins in the system. In general, the 'spins' are any quantum angular momenta that are present which determine the magnetic properties of the overall sample.

2.3.2 Micromagnetics—nanoscale magnetization dynamics

Another scale of magnetic simulation has been developed, referred to as *micromagnetics* [6–8]. Micromagnetics is a way to perform simulations of magnetization in a sample usually hundreds of nanometers up to a few microns in size. The number of spins in such a sample would be so large as to make simulation of the dynamics of all of them impossible. A coarse grain partitioning of the system into working cells is a way to reduce the number of degrees of freedom that are simulated, without losing the essential physics. Micromagnetic numerical simulations are developed to find the time development of the magnetization, $\mathbf{M}(\mathbf{r}_n, t)$, where at different points $\mathbf{r}_n$ in space (on a grid) there are cells in which $\mathbf{M}$ has been defined. Each cell's $\mathbf{M}$ represents the averaged out motion of a large group of underlying atomic spins. In micromagnetics, the individual motion of these underlying spins is blurred out and only the behavior over longer length scales is found.

When discussing micromagnetics we will be concerned with finding $\mathbf{M}(\mathbf{r}_n, t)$, in contrast to SD, where we will find $\mathbf{S}_n(t)$ for discrete spins. Both types of simulations take place on a lattice of sites. Both can be considered also in a continuum limit, which is helpful for comparing with analytic calculations. There is not a particularly strong distinction between these two magnetic simulation approaches. The primary difference is in the length scales being considered and associated with each site of the grid or lattice. The other distinction is that SD could, in principal, be for quantum or for classical spins. Many numerical approaches are being developed for quantum spin thermodynamics (various methods of quantum Monte Carlo (MC)) and quantum SD (also based on quantum MC). On the other hand, the averaging procedure to arrive at the micromagnetics approach makes it a classical simulation from the start. In this text we will mainly consider simulations of classical spin models, especially when interested in the time evolution.

2.4 Discrete exchange interactions

For discrete atomic spins (or effective angular momenta higher than spin-1/2), using the localized moment approach, the basic interaction between two dipoles is that known as quantum exchange. The physical source of exchange was discussed in chapter 1. Exchange is usually considered as the interaction between quantum spins $\mathbf{S}_n$ and $\mathbf{S}_m$, with a strength given from some exchange integral J_{nm} (not to be confused with the total angular momentum $\mathbf{J}$). For a pair of interacting spins, the basic Heisenberg exchange interaction energy (Hamiltonian operator) is expressed,

$$H_{nm} = -J_{nm} \mathbf{S}_n \cdot \mathbf{S}_m. \tag{2.9}$$

The exchange integral represents how the system energy is affected by swapping two electrons, which are fermions. It is an overlap integral of wave functions for possible states of the two electrons. The result comes from the requirement that the total system wave function change sign under interchange of the two electrons, as needed for fermions. When the exchange integral is positive ($J_{nm} > 0$), the interaction is known as FM and tends to align the two spins. Aligned spins would have lower energy than anti-aligned spins. A strong FM interaction in an extended system clearly leads to long-range magnetic ordering, where large regions of the system have their dipoles tending towards alignment. On the other hand, when the exchange integral is negative ($J_{nm} < 0$), it is called an antiferromagnetic (AFM) interaction, and will tend to cause the two spins to point in opposite directions. While this may seem to cause disorder in a system, that may not be the real case, depending on the type of crystal lattice where the spins are located. Indeed, some of the most interesting magnetic ordering can take place when the interactions are AFM. Generally, these exchange integrals will depend on the distance between the two spins.

2.5 Ferromagnetic spin exchange in a continuum limit

In chapter 1 equation (1.42) was developed as the Hamiltonian for a system with anisotropic nearest neighbor exchange. The anisotropy comes about due to the different exchange couplings J^x, J^y, J^z. The spins themselves are assumed to be defined on a lattice, which is not specified for this Hamiltonian. Here we consider more about this model and how to compare it with continuum theory for its Hamiltonian.

Suppose that the spins occupy a uniform linear chain (a 1D system), such as that shown in figure 2.1. The exchange Hamiltonian then resembles equation (1.51) for a 1D XY chain, but with more general exchange couplings,

$$H = -\sum_{n=1}^{N-1} \left[J^x S_n^x S_{n+1}^x + J^y S_n^y S_{n+1}^y + J^z S_n^z S_{n+1}^z \right]. \tag{2.10}$$

Figure 2.1. Grid cells of width a used for a uniform 1D magnetic chain.

The nearest neighbor sites have some uniform separation distance, a (the lattice constant). The positions of the sites are then points along an x-axis, given by $x_n = (n - 1/2)a$, for $n = 1, 2, 3 \ldots N$. This places each spin at the center of a grid cell of width a, and the total width of all cells combined is the length of the system, $L = Na$. We can change to an approximate continuum description and notation for the spins, by the replacements $\mathbf{S}_n \to \mathbf{S}(x)$ and $S_{n+1} \to \mathbf{S}(x + a)$. The Hamiltonian involves pair interactions. When they are assumed to be FM, neighboring spins will tend to be nearly aligned. Then it helps to perform Taylor expansions of the 'neighbors',

$$\mathbf{S}(x \pm a) \approx \mathbf{S}(x) \pm a\frac{\partial}{\partial x}\mathbf{S}(x) + \frac{1}{2}a^2\frac{\partial^2}{\partial x^2}\mathbf{S}(x) + \ldots. \tag{2.11}$$

Assuming that the spins vary rather slowly with position (they also vary with time), we cut off this expansion with the second-order term. It is simplest to suppose that the spins are classical, although that may not be essential. Then one pair interaction in the chain, corresponding to the J^x coupling, is

$$H^x_{n,n+1} = -J^x S^x(x) S^x(x + a) = -J^x S^x(x)\left[S^x(x) + a\frac{\partial}{\partial x}S^x(x) + \frac{1}{2}a^2\frac{\partial^2}{\partial x^2}S^x(x)\right]. \tag{2.12}$$

The spin at x interacts with ones at $x + a$ and $x - a$, which have opposite signs on the term linear in the gradient. When combined, the terms linear in the gradient cancel. Thus the pair interaction averaged over right and left neighbors is

$$H^x_n = \frac{1}{2}\left(H^x_{n,n+1} + H^x_{n,n-1}\right) = -\frac{J^x}{2}S^x(x)[S^x(x - a) + S^x(x + a)]$$

$$= -J^x\left[S^x(x)^2 + \frac{1}{2}a^2 S^x(x)\frac{\partial^2}{\partial x^2}S^x(x)\right]. \tag{2.13}$$

The factor of 1/2 on the LHS can also be interpreted to undo the double counting of exchange pair interactions. This needs to be summed over all sites.

The summation in equation (2.10) (or any similar expression) can be changed to an integration by the rule of integral calculus,

$$\sum_{n=1}^{N} H^x_n \longrightarrow \int_{x=0}^{L} \frac{dx}{a} H^x_n. \tag{2.14}$$

This can be used together with (2.13) to obtain the part of the continuum Hamiltonian due to the J^x coupling,

$$H^x = \sum_n H^x_n \longrightarrow -J^x \int \frac{dx}{a}\left[S^x(x)^2 + \frac{1}{2}a^2 S^x(x)\frac{\partial^2}{\partial x^2}S^x(x)\right]$$

$$H^x = J^x \int \frac{dx}{a}\left[-S^x(x)^2 + \frac{1}{2}a^2\left(\frac{\partial S^x}{\partial x}\right)^2\right]. \tag{2.15}$$

The last line was obtained using integration by parts. A similar procedure applies to the other parts of H involving the couplings J^y and J^z. Then the total continuum Hamiltonian becomes

$$H = \int \frac{dx}{a} \left\{ J^x \left[-S^x(x)^2 + \frac{1}{2}a^2 \left(\frac{\partial S^x}{\partial x}\right)^2 \right] \right.$$
$$\left. + J^y \left[-S^y(x)^2 + \frac{1}{2}a^2 \left(\frac{\partial S^y}{\partial x}\right)^2 \right] + J^z \left[-S^z(x)^2 + \frac{1}{2}a^2 \left(\frac{\partial S^z}{\partial x}\right)^2 \right] \right\}. \quad (2.16)$$

The result includes anisotropic effects involving the squared spin components and involving their squared gradients.

If the couplings are the same, $J^x = J^y = J^z \equiv J$, the model represents *isotropic Heisenberg exchange*, and it simplifies because the squared spin length is a constant:

$$\mathbf{S}^2 = S^{x2} + S^{y2} + S^{z2} = \begin{cases} S(S+1) & \text{for quantum spins} \\ S^2 & \text{for classical spins.} \end{cases} \quad (2.17)$$

The classical case is obtained from the quantum length in the limit of very large spin S. Then for isotropic exchange the 1D chain has the Hamiltonian

$$H_{\text{isotropic}} = J \int \frac{dx}{a} \left\{ -\mathbf{S}^2 + \frac{1}{2}a^2 \left[\left(\frac{\partial S^x}{\partial x}\right)^2 + \left(\frac{\partial S^y}{\partial x}\right)^2 + \left(\frac{\partial S^z}{\partial x}\right)^2 \right] \right\}. \quad (2.18)$$

The first term of the integrand is a constant that depends on the length of the chain; it can usually be dropped as an unimportant shift of the energy scale. The other parts involve the squared gradient of the spins, which can be expressed alternatively as

$$H_{\text{isotropic}} = J \int \frac{dx}{a} \frac{a^2}{2} \left(\frac{\partial \mathbf{S}}{\partial x}\right)^2. \quad (2.19)$$

It is easy to see that this makes physical sense for the exchange effect. If the spin field $\mathbf{S}(x)$ is a constant, all spins are aligned, and the expression gives $H_{\text{isotropic}} = 0$. This will be the minimum system energy, hence it is the ground state. Any kind of spatial variation in $\mathbf{S}(x)$ corresponds to misaligned spins and leads to a positive exchange energy.

2.6 Ferromagnetic exchange in micromagnetics

The above discussion of exchange Hamiltonians applies to atomic modeling of the spins. For some problems it makes more sense to perform a macroscopic or continuum modeling, where the physical object of more relevance is the continuum magnetization $\mathbf{M}(\mathbf{r})$. This is especially important for analysis of FM materials in an approach known as micromagnetics, where the system is divided into a set of cells (figure 2.2), each thought to contain thousands of magnetic atoms [9]. The continuum magnetization $\mathbf{M}(\mathbf{r})$ can be considered to be a classical object, obtained

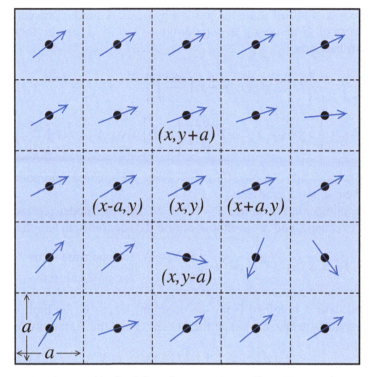

Figure 2.2. Square grid cells of dimensions $a \times a$ for a uniform 2D magnetic system.

from averaging over many quantum spins, so we avoid the complications of quantum mechanics. In this kind of continuum modeling, the magnetization can be assumed to be locally saturated in an individual cell. That is, the FM couplings within a cell are assumed to be so strong that all dipoles are ferromagnetically aligned within a small cell, such as those in figure 2.2 for a 2D system. Suppose the cells for 3D are cubes of side a, volume $v_{\text{cell}} = a^3$, obtained by an extension of figure 2.2 along the perpendicular (z) direction. The magnitude of the magnetization is some value M_s in every cell (the saturation magnetization, in units of $\text{A} \cdot \text{m}^{-1}$), while its direction is variable and depends slowly on position $\mathbf{r}$. Thus one writes

$$\mathbf{M}(\mathbf{r}) = M_s \, \hat{\mathbf{m}}(\mathbf{r}), \tag{2.20}$$

where $\hat{\mathbf{m}}(\mathbf{r})$ is a direction vector of unit magnitude. That enforces the assumption that $\mathbf{M}$ retains a fixed magnitude.

Now the exchange interactions act between two neighboring cells. Consider the interaction of a cell at position $\mathbf{r}$ with a neighboring cell at position $\mathbf{r} + a\hat{\mathbf{x}}$ (one cell away along the x-axis). Each cell has a total dipole moment of magnitude $\mu_{\text{cell}} = v_{\text{cell}} M_s$. Let their actual dipoles be $\boldsymbol{\mu}(\mathbf{r}) = \mu_{\text{cell}} \hat{\mathbf{m}}(\mathbf{r})$ and $\boldsymbol{\mu}(\mathbf{r} + a\hat{\mathbf{x}}) = \mu_{\text{cell}} \hat{\mathbf{m}}(\mathbf{r} + a\hat{\mathbf{x}})$. Suppose these cells interact with a certain isotropic exchange coupling, call it J_{cell}, such that the interaction energy between only these two cells is given by

$$H_{\text{pair}} = -J_{\text{cell}} \, \hat{\mathbf{m}}(\mathbf{r}) \cdot \hat{\mathbf{m}}(\mathbf{r} + a\hat{\mathbf{x}}). \tag{2.21}$$

Any two neighboring cells will have this interaction. Note that J_{cell} is taken to have energy units[1]. To obtain a continuum limit, the usual and simplest procedure is to expand the second dipole in a Taylor series, just as we did for spin models,

$$\hat{\mathbf{m}}(\mathbf{r} + a\hat{\mathbf{x}}) \approx \hat{\mathbf{m}}(\mathbf{r}) + (a\hat{\mathbf{x}} \cdot \nabla)\hat{\mathbf{m}}(\mathbf{r}) + \frac{1}{2}(a\hat{\mathbf{x}} \cdot \nabla)^2 \hat{\mathbf{m}}(\mathbf{r}) + \ldots. \quad (2.22)$$

A similar expansion holds for the opposite neighboring dipole at $\mathbf{r} - a\hat{\mathbf{x}}$, with the opposite sign on the second term on the RHS. When the energies of these two pairs are combined, with a factor of 1/2 to undo double counting, the linear terms cancel, and the result is familiar,

$$H_{\text{pair}} \propto \frac{1}{2} \hat{\mathbf{m}}(\mathbf{r}) \cdot [\hat{\mathbf{m}}(\mathbf{r} - a\hat{\mathbf{x}}) + \hat{\mathbf{m}}(\mathbf{r} + a\hat{\mathbf{x}})] = 1 + \frac{1}{2} \hat{\mathbf{m}}(\mathbf{r}) \cdot \left(a\frac{\partial}{\partial x}\right)^2 \hat{\mathbf{m}}(\mathbf{r}). \quad (2.23)$$

The 1 on the RHS results because $\hat{\mathbf{m}}(\mathbf{r})$ is a unit vector and gives an irrelevant constant. Now the cell at $\mathbf{r}$ also interacts with neighboring cells at $\mathbf{r} \pm a\hat{\mathbf{y}}$ and $\mathbf{r} \pm a\hat{\mathbf{z}}$ (for a 3D set of cells on a cubic lattice). Furthermore, we need to add the contributions of all cell pairs. If the cells are small enough, the sum can be replaced by a 3D integral over space. For 3D we perform the following replacement for cubic cells:

$$\sum_{\text{pairs}} H_{\text{pair}} \longrightarrow \int \frac{dx}{a} \int \frac{dy}{a} \int \frac{dz}{a} H_{\text{pair}} = \int \frac{dV}{a^3} H_{\text{pair}}. \quad (2.24)$$

A short exercise will bring this to a standard form.

Exercise 2.1. Assuming a cubic arrangement of micromagnetics cells, generalize H_{pair} in (2.23) to apply to all nearest neighbor pairs, and convert to an integration by (2.24). Show that a sum over neighboring pairs leads to the result for the isotropic exchange Hamiltonian H_{ex} in micromagnetics:

$$\sum_{\text{pairs}} H_{\text{pair}} \longrightarrow H_{\text{ex}} = \frac{1}{2} J_{\text{cell}}\, a^{-1} \int dV \left[\left(\frac{\partial \hat{\mathbf{m}}}{\partial x}\right)^2 + \left(\frac{\partial \hat{\mathbf{m}}}{\partial y}\right)^2 + \left(\frac{\partial \hat{\mathbf{m}}}{\partial z}\right)^2\right]. \quad (2.25)$$

An integration by parts will be helpful. An arbitrary constant has been dropped and terms that would come from the boundary of the system have been dropped, assuming a volume of integration large enough to include all of the magnet, out to a region where there is no more magnetization.

Note that result (2.25) is the 3D generalization of (2.19) for a 1D spin chain, renaming $\mathbf{S}$ as $\hat{\mathbf{m}}$ and adjusting the exchange constant. Thus there is not a great distinction between SD and micromagnetics.

---

[1] For micromagnetics it is convenient to let the exchange constant J_{cell} have dimensions of energy, because it is combined with interactions of dimensionless direction vectors $\hat{\mathbf{m}}(\mathbf{r})$ for local magnetization. In expressions for exchange energy involving spins, the quantity JS^2 has been taken to have energy dimensions, which is more convenient for discussion of Hamiltonian and quantum dynamics.

It is customary to rewrite the exchange energy (2.25) with some different shorthand notations,

$$H_{ex} = A \int dV \, \nabla \hat{m} \cdot \nabla \hat{m} = A \int dV \, |\nabla \hat{m}|^2 = A \int dV \sum_{i=1}^{3} \sum_{j=1}^{3} \partial_i \hat{m}_j \, \partial_i \hat{m}_j. \quad (2.26)$$

The last form uses the notation for partial derivatives, $\partial_i = \frac{\partial}{\partial x_i}$ for $x_i = x, y, z$. The constant A is known as the *exchange stiffness*. This result reflects how exchange forces tend to align all the dipoles in a region, while working against randomizing forces such as the temperature. For 3D, exchange stiffness A can be related to the assumed cell to cell interaction as [9]

$$A = \frac{1}{2} J_{cell} \, a^{-1}. \quad (2.27)$$

The units of A are J m^{-1}. In practice, A is a constant determined by the magnetic medium. For instance, a common FM material of much practical application is permalloy-79 (an alloy of nickel and iron, with 79% nickel, 21% iron, denoted as Py for short), which has $A \approx 13$ pJ m^{-1}. It is also interesting to note that the saturation magnetization for Py is around $M_s = 860$ kA m^{-1}. This last relation can be useful in the design of micromagnetic simulations, as it determines the cell interaction strength needed for a material of known parameter A and cubic cell edge a.

There is also great interest in micromagnetics in 2D, or especially, in thin-film magnetism with quasi-2D behavior. In this case a 2D grid of cells is used, whose dimensions could be $a \times a \times L$, where L is their dimension perpendicular to the 2D film. There is integration only over x and y, and the exchange Hamiltonian transforms into

$$H_{ex} = \frac{1}{2} J_{cell} \int dx \, dy \left[\left(\frac{\partial \hat{m}}{\partial x} \right)^2 + \left(\frac{\partial \hat{m}}{\partial y} \right)^2 \right] = A \int L \, dx \, dy \sum_{i=1}^{2} \sum_{j=1}^{3} \partial_i \hat{m}_j \, \partial_i \hat{m}_j. \quad (2.28)$$

The magnetization is assumed to be independent of the z-coordinate, perpendicular to the film. Then for 2D magnetics the relation between cell interaction and the continuum exchange stiffness is

$$A = \frac{1}{2} J_{cell} \, L^{-1}. \quad (2.29)$$

This will be applied later (chapters 3 and 10) to problems involving the competition between exchange and demagnetization in 2D.

2.7 FM continuum limit on any lattice

The derivations above have been performed on a uniform linear chain and on a cubic lattice of cells. In a magnetic crystal the lattice structure could be different, such as hexagonal, triangular and others of more complexity. We can then ask whether the continuum limit for a spin system on a lattice other than cubic might have a different result.

At this point we confine the derivations to FM interactions and assume a 2D lattice of spins. A motivation for this is that later in in this text we will consider 2D systems. With FM interactions, neighboring spins will be nearly aligned. We obtain some general results and then check the details for different lattices.

The spins $\mathbf{S}_n$ occupy the sites of some lattice, interacting via pair exchange according to Hamiltonian (1.42). Isotropic couplings $J^x = J^y = J^z \equiv J$ are assumed. For any lattice, it is important to be able to specify the locations of the neighbors of any of its sites. Then the basic mathematics already developed in the earlier examples can be applied with very few changes. If $\mathbf{r}$ is the location of the spin $\mathbf{S}_n$, then there are z neighboring sites, at displacements written as $\mathbf{a}_j$, $j = 1, 2, 3 \ldots z$, where z is called the *coordination number* of the lattice. One has $z = 4$ for a square lattice, $z = 3$ for a hexagonal or honeycomb lattice, and $z = 6$ for a triangular lattice. The positions of neighboring sites are written $\mathbf{r} + \mathbf{a}_j$.

Following the same procedure used in sections 2.5 and 2.6 for a continuum limit, we need the averaged pair exchange interaction. A neighbor spin is obtained from a Taylor expansion to second-order in gradients, such as

$$\mathbf{S}(\mathbf{r} + \mathbf{a}_j) \approx \mathbf{S}(\mathbf{r}) + (\mathbf{a}_j \cdot \nabla)\mathbf{S}(\mathbf{r}) + \frac{1}{2}(\mathbf{a}_j \cdot \nabla)^2 \mathbf{S}(\mathbf{r}) + \ldots. \tag{2.30}$$

The average isotropic exchange interaction with all nearest neighbors is then

$$H_{\text{pair}} \propto \frac{1}{2}\sum_{j=1}^{z} \mathbf{S}(\mathbf{r}) \cdot \mathbf{S}(\mathbf{r} + \mathbf{a}_j) = \frac{z}{2}\mathbf{S}^2 + \frac{1}{2}\sum_{j=1}^{z} \mathbf{S}(\mathbf{r}) \cdot \frac{1}{2}(\mathbf{a}_j \cdot \nabla)^2 \mathbf{S}(\mathbf{r}). \tag{2.31}$$

Note that the term linear in gradients is gone, because the neighboring displacements will sum to zero.

To transform the exchange sum into a 2D integral, a little consideration shows that in place of (2.14) for a 1D chain or (2.24) for 3D cubic cells, the most general replacement is

$$\sum_{\text{pairs}} H_{\text{pair}} \longrightarrow \int \frac{dV}{v_{\text{cell}}} H_{\text{pair}} \tag{2.32}$$

where dV is an integration volume (or area) element and v_{cell} is the volume (or area) of the cell associated with one spin site. The value of v_{cell} depends on the lattice. For the simple case of cubic cells, we used $v_{\text{cell}} = a^3$. After an integration by parts but keeping the constant depending on $\mathbf{S}^2$ for comparison with the results on a 1D spin chain, the general continuum expression for an isotropic FM exchange model is found to be:

$$H = \frac{1}{2}J \sum_{j=1}^{z} \int \frac{dV}{v_{\text{cell}}} \left[-\mathbf{S}^2 + \frac{1}{2}\left|(\mathbf{a}_j \cdot \nabla)\mathbf{S}(\mathbf{r})\right|^2 \right]. \tag{2.33}$$

Exercise 2.2. Verify that (2.33) reproduces the result (2.25) in section 2.6 (with $\mathbf{S}$ in place of $\hat{\mathbf{m}}$ and J in place of J_{cell}) for a cubic lattice of cells, with $v_{\text{cell}} = a^3$ and $z = 6$, and displacements $\mathbf{a}_j = \pm a\hat{\mathbf{x}}, \pm a\hat{\mathbf{y}}, \pm a\hat{\mathbf{z}}$.

Exercise 2.3. Investigate whether it is possible to generalize (2.33) to the case of spin-anisotropic exchange couplings, $J^x \neq J^y \neq J^z$, on an arbitrary lattice of cells. The calculation may produce a result with terms like that in (2.16) for a 1D chain.

2.7.1 Triangular lattice

In a triangular lattice, each site has six nearest neighbors at a distance equal to the lattice constant a, as shown in figure 2.3. Starting from a central site n at position $\mathbf{r}$, the displacements to its neighbors $\mathbf{a}_j$, $j = 1, 2, 3\ldots 6$, are along two basis vectors $\mathbf{a}$ and $\mathbf{b}$, which make a 60° degree angle with each other, and a third derived basis vector $\mathbf{c} = \mathbf{b} - \mathbf{a}$. They can be taken as follows:

$$\mathbf{a} = a\hat{\mathbf{x}}, \qquad \mathbf{b} = a\left(\frac{1}{2}\hat{\mathbf{x}} + \frac{\sqrt{3}}{2}\hat{\mathbf{y}}\right), \qquad \mathbf{c} = a\left(-\frac{1}{2}\hat{\mathbf{x}} + \frac{\sqrt{3}}{2}\hat{\mathbf{y}}\right). \qquad (2.34)$$

Then the six neighbors of the site at $\mathbf{r}$ are located at positions $\mathbf{r} \pm \mathbf{a}$, $\mathbf{r} \pm \mathbf{b}$, and $\mathbf{r} \pm \mathbf{c}$, or at $\mathbf{r} + \mathbf{a}_j$, $j = 1, 2, 3\ldots 6$.

The integration area element can be $dV = dx\, dy$. The system can be broken up into hexagonal Wigner–Seitz unit cells centered on each site, with some area v_{cell} per site. These are separated by the dashed lines in figure 2.3. Each of the triangles of

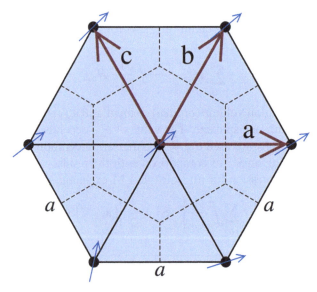

Figure 2.3. Geometry for a triangular lattice of spins, with lattice constant a and displacements to nearest neighbors $\mathbf{a}$, $\mathbf{b}$, $\mathbf{c}$. The dashed lines show the hexagonal Wigner–Seitz unit cell for a central spin site and for the neighboring sites. The edge length of that hexagon is $\frac{a}{\sqrt{3}}$ and its area is the factor $v_{\text{cell}} = \frac{\sqrt{3}}{2}a^2$.

edge a in a triangular lattice has an area $A_{\text{tri}} = \frac{\sqrt{3}}{4}a^2$. Any given spin site is touching six different triangles, and each triangle has three vertices. Thus A_{tri} is the area associated with half a site. Then the area per spin site (the dashed hexagons in figure 2.3) is seen to be $v_{\text{cell}} = 2 \times A_{\text{tri}} = \frac{\sqrt{3}}{2}a^2$. This means that summations on the triangular lattice are transformed to integrations by

$$\sum_{\text{pairs}} H_{\text{pair}} \longrightarrow \frac{2}{\sqrt{3}} \int \frac{\mathrm{d}x\,\mathrm{d}y}{a^2}\, H_{\text{pair}}. \tag{2.35}$$

To apply the result (2.33) to obtain H for the triangular lattice, we need the following algebra for the neighbor displacements:

$$\mathbf{a} \cdot \nabla = a\frac{\partial}{\partial x}, \quad \mathbf{b} \cdot \nabla = a\left(\frac{1}{2}\frac{\partial}{\partial x} + \frac{\sqrt{3}}{2}\frac{\partial}{\partial y}\right), \quad \mathbf{c} \cdot \nabla = a\left(-\frac{1}{2}\frac{\partial}{\partial x} + \frac{\sqrt{3}}{2}\frac{\partial}{\partial y}\right). \tag{2.36}$$

Then performing the sum in (2.33), the continuum isotropic exchange Hamiltonian on a triangular lattice becomes

$$H = J \int \frac{2\,\mathrm{d}x\,\mathrm{d}y}{\sqrt{3}\,a^2} \left\{-3S^2 + \frac{3}{4}a^2\left[\left(\frac{\partial \mathbf{S}}{\partial x}\right)^2 + \left(\frac{\partial \mathbf{S}}{\partial y}\right)^2\right]\right\} \tag{2.37}$$

This has an isotropic structure (equal factors on both gradient components), similar to our previous results, except for some of the constants. The factor $v_{\text{cell}} = \frac{\sqrt{3}}{2}a^2$ has been grouped with the integration element, so that the combination measures the number of spins inside $\mathrm{d}x\,\mathrm{d}y$ on the original discrete model. Then every original spin gives a ground state constant energy of $-3JS^2$, corresponding to the six nearest neighbor bonds (divided by two so they are not counted twice). One can also see that the factor of v_{cell} is the *Jacobian* of the transformation between coordinates (x, y) and $(\mathbf{a}, \mathbf{b})$, by inspecting equation (2.36). Factor v_{cell} was introduced here, however, based on geometric reasoning for correct counting of spins. The factor of $\frac{3}{4}a^2$ on the squared gradient is different from a factor of $\frac{1}{2}$ for a cubic grid, and is characteristic of the different density of spins on the triangular lattice. Finally, a curiosity of 2D exchange is that the factor of a^2 in the cell area cancels with another factor of a^2 on the squared gradient.

Exercise 2.4. One can perform similar analysis for other lattices and other models. For example, find the continuum Hamiltonian analogous to (2.37) but for anisotropic couplings ($J^x \neq J^y \neq J^z$). Compare with the 1D chain result in (2.16) and also with the general result for isotropic coupling (2.33).

2.7.2 Hexagonal lattice

Each site of a hexagonal (or honeycomb) lattice has three nearest neighbors ($z = 3$), however, one can see in figure 2.4 that the lattice is composed from two *sublattices*

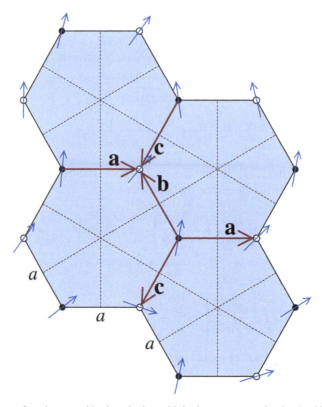

Figure 2.4. Geometry for a hexagonal lattice of spins, with lattice constant a. On the A-sublattice (solid dots), the displacements to nearest neighbors are **a**, **b** and **c**. On the B-sublattice (open dots), the near neighbor displacements are $-$**a**, $-$**b**, and $-$**c**. The dashed lines show the triangular Wigner–Seitz unit cells around sites. The edge length of those triangles is $\sqrt{3}\,a$ and their area is the factor $v_{cell} = \frac{3\sqrt{3}}{4}a^2$.

with different displacements to the neighbors of a site[2]. Half of the sites are on the A-sublattice, and the displacements to the neighbors are

$$\mathbf{a} = a\hat{\mathbf{x}}, \quad \mathbf{b} = a\left(-\frac{1}{2}\hat{\mathbf{x}} + \frac{\sqrt{3}}{2}\hat{\mathbf{y}}\right), \quad \mathbf{c} = -(\mathbf{a} + \mathbf{b}) = a\left(-\frac{1}{2}\hat{\mathbf{x}} - \frac{\sqrt{3}}{2}\hat{\mathbf{y}}\right). \quad (2.38)$$

These satisfy $\mathbf{a} + \mathbf{b} + \mathbf{c} = 0$ identically. The other half of the sites on the hexagonal lattice are on the B-sublattice, whose displacements to the neighbors are $-$**a**, $-$**b** and $-$**c**. Any site on the A-sublattice has three neighbors on the B-sublattice, and vice versa. Although the A- and B-sites are not equivalent, it makes no difference whether equation (2.33) is applied to the A-sites or the B-sites, due to this symmetry between them.

---

[2] A hexagonal lattice could also be generated from a triangular lattice with a two-atom basis, but that approach will not be used here.

Each site on a hexagonal lattice is contained in an equilateral triangular Wigner–Seitz unit cell (the hexagonal and triangular lattices are dual lattices). The center to center distance between the hexagons of the lattice is $\sqrt{3}\,a$, which is also the side of one of the triangles. Then the area of one triangular cell is found to be $v_{\text{cell}} = \frac{3\sqrt{3}}{4}a^2$. On an A-site, we have the following algebra:

$$\mathbf{a}\cdot\nabla = a\frac{\partial}{\partial x}, \quad \mathbf{b}\cdot\nabla = a\left(-\frac{1}{2}\frac{\partial}{\partial x} + \frac{\sqrt{3}}{2}\frac{\partial}{\partial y}\right), \quad \mathbf{c}\cdot\nabla = a\left(-\frac{1}{2}\frac{\partial}{\partial x} - \frac{\sqrt{3}}{2}\frac{\partial}{\partial y}\right). \tag{2.39}$$

The corresponding factors on a B-site are oppositely signed, but that is irrelevant in the Hamiltonian (2.33). Squaring gradient factors and combining, the continuum Hamiltonian for isotropic FM exchange is found to be

$$H = J\int\frac{4\,dx\,dy}{3\sqrt{3}\,a^2}\left\{-\frac{3}{2}\mathbf{S}^2 + \frac{3}{8}a^2\left[\left(\frac{\partial\mathbf{S}}{\partial x}\right)^2 + \left(\frac{\partial\mathbf{S}}{\partial y}\right)^2\right]\right\}. \tag{2.40}$$

As expected, the result has isotropic coupling in the gradients, but with particular factors weighting the integration element and the squared gradients that are characteristic of the density of sites on the hexagonal lattice.

2.8 Angular coordinates for classical spin direction

In FM models, nearest neighbor spins have the tendency to remain nearly aligned, due to the exchange interactions. Above we noted the distinction between a discrete spin system and a continuum limit, using the Cartesian spin components in both cases. Indeed, Cartesian coordinates may be the best for development of numerical simulations, however, better insight into the physics might come from angular coordinates for some analysis.

For systems with an XY symmetry or a uniaxial symmetry ($J^x = J^y \neq J^z$), the motion of spins within the xy-plane is either preferred or discouraged, respectively. There is some conceptual advantage to thinking of the spin motion as a part within the xy-plane (due primarily to S^x, S^y) and another part out of the xy-plane (due primarily to S^z). This can be achieved using spherical coordinates for the spins. Even with isotropic exchange, there can be a certain advantage for using spherical coordinates instead of Cartesian coordinates, isolating the in-plane motions from the out-of-plane motions.

In polar spherical coordinates, an individual classical spin $\mathbf{S}_n$ is described by an in-plane angle ϕ_n and a polar angle θ_n, in the usual definition,

$$\mathbf{S}_n = \left(S_n^x, S_n^y, S_n^z\right) = S\left(\sin\theta_n\cos\phi_n,\ \sin\theta_n\sin\phi_n,\ \cos\theta_n\right). \tag{2.41}$$

The polar angle θ_n is measured from the z-axis to the spin. The polar angle is indicated as θ_z in figure 2.5. These are the spherical coordinates used in most common physics applications. For a system with XY symmetry, however, the spins have the tendency to remain near the non-zero value, $\theta_n \approx \frac{\pi}{2}$. This non-zero value

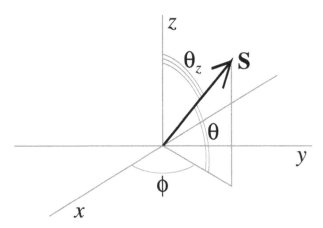

Figure 2.5. Angular coordinates ϕ, θ for a spin **S** in the planar system and ϕ, θ_z in the polar system.

would be present for ground states, and needs to be subtracted out of some quantities when measuring deviations from a ground state.

If one wants to focus on the strength of deviations out of the xy-plane, it is better to use a slight modification of spherical coordinates, making them *planar* spherical coordinates. In this alternative definition, which is used in many places throughout this book, we take

$$\mathbf{S}_n = (S_n^x, S_n^y, S_n^z) = S(\cos\theta_n \cos\phi_n, \cos\theta_n \sin\phi_n, \sin\theta_n). \qquad (2.42)$$

Then the angle θ_n measures the angle of the spin tilting out of the xy-plane, rather than from the polar z-axis, as indicated in figure 2.5. With XY symmetry, the tendency will be for $\theta_n \approx 0$. This notation has the further advantage that out-of-plane motions towards positive/negative S_n^z also have opposite signs of θ_n. θ_n will be called the *out-of-plane angle*. Obviously the angle ϕ_n is the usual azimuthal angle of the spin relative to the x-axis for both spherical coordinates. We will usually need to indicate which type of spherical coordinate is being used in any problem.

Consider again the Hamiltonian (2.10) for a 1D chain of spins with anisotropic nearest neighbor exchange. In terms of planar spherical coordinates (2.42), the pair interaction is

$$H_{n,n+1} = -S^2 \Big\{ \cos\theta_n \cos\theta_{n+1} \big[J^x \cos\phi_n \cos\phi_{n+1} + J^y \sin\phi_n \sin\phi_{n+1} \big] + J^z \sin\theta_n \sin\theta_{n+1} \Big\}. \qquad (2.43)$$

Note that in the case of XY symmetry with $J^x = J^y$, this simplifies to

$$H_{n,n+1} = -S^2 \Big\{ J^x \cos\theta_n \cos\theta_{n+1} \cos(\phi_n - \phi_{n+1}) + J^z \sin\theta_n \sin\theta_{n+1} \Big\}. \qquad (2.44)$$

The term proportional to J^x indicates how the in-plane angles will tend to have small differences between neighboring sites.

2.8.1 Continuum limits in angular notation

We can obtain the continuum limit of the anisotropic Hamiltonian (2.43) by expanding the angles in the usual way, for instance, with $\theta_n \to \theta(x)$, $\theta_{n+1} \to \theta(x+a)$, and

$$\theta(x+a) = \theta(x) + a\frac{\partial}{\partial x}\theta(x) + \frac{1}{2}a^2\frac{\partial^2}{\partial x^2}\theta(x) + \ldots \quad (2.45)$$

with a similar notation and expansion for $\phi_n \to \phi(x)$. Then the pair Hamiltonian can be found with the help of some expanding of the trigonometric functions,

$$\cos\theta_{n+1} \to \cos\theta(x+a) = \cos\left(\theta + a\frac{\partial\theta}{\partial x} + \frac{a^2}{2}\frac{\partial^2\theta}{\partial x^2}\right)$$
$$= \cos\theta\cos\left(a\frac{\partial\theta}{\partial x} + \frac{a^2}{2}\frac{\partial^2\theta}{\partial x^2}\right) - \sin\theta\sin\left(a\frac{\partial\theta}{\partial x} + \frac{a^2}{2}\frac{\partial^2\theta}{\partial x^2}\right). \quad (2.46)$$

Now making small angle approximations for the arguments of sine and cosine,

$$\cos\theta(x+a) \approx \cos\theta\left[1 - \frac{1}{2}a^2\left(\frac{\partial\theta}{\partial x}\right)^2\right] - \sin\theta\left[a\frac{\partial\theta}{\partial x} + \frac{a^2}{2}\frac{\partial^2\theta}{\partial x^2}\right]. \quad (2.47)$$

Also for the J^z term one obtains by similar algebra

$$\sin\theta(x+a) \approx \sin\theta\left[1 - \frac{1}{2}a^2\left(\frac{\partial\theta}{\partial x}\right)^2\right] + \cos\theta\left[a\frac{\partial\theta}{\partial x} + \frac{a^2}{2}\frac{\partial^2\theta}{\partial x^2}\right]. \quad (2.48)$$

Then the exchange coupling of z-components involves the term

$$\sin\theta(x)\sin\theta(x+a) = \sin^2\theta\left[1 - \frac{1}{2}a^2\left(\frac{\partial\theta}{\partial x}\right)^2\right] + \frac{1}{2}\sin(2\theta)\left[a\frac{\partial\theta}{\partial x} + \frac{a^2}{2}\frac{\partial^2\theta}{\partial x^2}\right]. \quad (2.49)$$

This will be combined with a term for the neighbor $\mathbf{S}_{n-1}$, which will cancel out the linear term in the gradient. Then it must be integrated over the whole chain, which is left as an exercise.

Exercise 2.5. With the help of an integration by parts, show that the part of the Hamiltonian involving only the z spin components is

$$H^z = J^z S^2 \int \frac{dx}{a}\left[-\sin^2\theta + \frac{a^2}{2}\cos^2\theta\left(\frac{\partial\theta}{\partial x}\right)^2\right]. \quad (2.50)$$

One can check that this is consistent with an equivalent term in (2.16) for the 1D chain.

Now for the couplings of the x- and y-components, we need

$$\cos\theta(x)\cos\theta(x+a) = \cos^2\theta\left[1 - \frac{1}{2}a^2\left(\frac{\partial\theta}{\partial x}\right)^2\right] - \frac{1}{2}\sin(2\theta)\left[a\frac{\partial\theta}{\partial x} + \frac{a^2}{2}\frac{\partial^2\theta}{\partial x^2}\right]. \tag{2.51}$$

Then for the J^x coupling some algebra leads to the surviving terms (terms linear in gradients cancel out)

$$\cos\theta(x)\cos\theta(x+a)\cos\phi(x)\cos\phi(x+a)$$

$$= \cos^2\theta\cos^2\phi\left\{1 - \frac{1}{2}a^2\left[\left(\frac{\partial\theta}{\partial x}\right)^2 + \left(\frac{\partial\phi}{\partial x}\right)^2\right]\right\} - \frac{1}{2}\sin(2\theta)\cos^2\phi\left[\frac{a^2}{2}\frac{\partial^2\theta}{\partial x^2}\right]$$

$$- \frac{1}{2}\sin(2\phi)\cos^2\theta\left[\frac{a^2}{2}\frac{\partial^2\phi}{\partial x^2}\right] + \frac{1}{2}\sin(2\theta)\sin(2\phi)\left[\frac{a^2}{2}\frac{\partial\theta}{\partial x}\frac{\partial\phi}{\partial x}\right]. \tag{2.52}$$

Using an integration by parts leads to the Hamiltonian for the J^x coupling,

$$H^x = J^x S^2 \int \frac{dx}{a}\left\{-\cos^2\theta\cos^2\phi + \frac{a^2}{2}\left[\sin^2\theta\cos^2\phi\left(\frac{\partial\theta}{\partial x}\right)^2\right.\right.$$

$$\left.\left. + \cos^2\theta\sin^2\phi\left(\frac{\partial\phi}{\partial x}\right)^2 + \frac{1}{2}\sin(2\theta)\sin(2\phi)\frac{\partial\theta}{\partial x}\frac{\partial\phi}{\partial x}\right]\right\}. \tag{2.53}$$

A similar approach also gives the Hamiltonian for the J^y coupling,

$$H^y = J^y S^2 \int \frac{dx}{a}\left\{-\cos^2\theta\sin^2\phi + \frac{a^2}{2}\left[\sin^2\theta\sin^2\phi\left(\frac{\partial\theta}{\partial x}\right)^2\right.\right.$$

$$\left.\left. + \cos^2\theta\cos^2\phi\left(\frac{\partial\phi}{\partial x}\right)^2 - \frac{1}{2}\sin(2\theta)\sin(2\phi)\frac{\partial\theta}{\partial x}\frac{\partial\phi}{\partial x}\right]\right\}. \tag{2.54}$$

While these are somewhat complicated expressions, they can be verified by inserting the angular coordinate definition (2.42) into the 1D chain result, (2.16), with much less effort!

The case of XY symmetry is especially interesting. For $J^x = J^y \neq J^z$, these results combine into the total Hamiltonian $H_{XY} = H^x + H^y + H^z$ given by

$$H_{XY} = S^2 \int \frac{dx}{a}\left\{J^x\left(-\cos^2\theta + \frac{a^2}{2}\left[\sin^2\theta\left(\frac{\partial\theta}{\partial x}\right)^2 + \cos^2\theta\left(\frac{\partial\phi}{\partial x}\right)^2\right]\right)\right.$$

$$\left. + J^z\left(-\sin^2\theta + \frac{a^2}{2}\cos^2\theta\left(\frac{\partial\theta}{\partial x}\right)^2\right)\right\}. \tag{2.55}$$

It is interesting that there are terms not depending on gradients, yet they carry anisotropy even for a uniform state:

$$S^2\int\frac{dx}{a}(-J^x\cos^2\theta - J^z\sin^2\theta) = S^2\int\frac{dx}{a}[-J^x + (J^x - J^z)\sin^2\theta]. \tag{2.56}$$

With $J^x > J^z$ required for FM XY symmetry, the term $\sin^2\theta$ helps to confine spins to the xy-plane. The remaining terms in (2.55) have anisotropic effects (the dependence on θ) and exchange effects (dependence on squared gradients), that are responsible for a lot of interesting physical phenomena, such as vortices.

One also sees that the full isotropic limit will be arrived at with $J^x = J^z$. Then for a 1D ferromagnet with isotropic exchange in the continuum limit, the Hamiltonian can be represented as

$$H_{\text{isotropic}} = JS^2 \int \frac{dx}{a}\left(-1 + \frac{a^2}{2}\left[\left(\frac{\partial\theta}{\partial x}\right)^2 + \cos^2\theta\left(\frac{\partial\phi}{\partial x}\right)^2\right]\right). \tag{2.57}$$

Exercise 2.6. Verify that (2.57) also results by substituting the angular coordinates (2.42) into the previously found continuum spin Hamiltonian, (2.19).

2.8.2 Models in more dimensions

The last derivations were found for a 1D chain, with only dependence on location on the chain, x. Clearly if the model is one for 3D, then the partial derivatives with respect to x will be replaced by the 3D gradient, $\nabla = (\frac{\partial}{\partial x}, \frac{\partial}{\partial y}, \frac{\partial}{\partial z})$. This is a simple replacement to go to related models in higher dimensions. However, it should be kept in mind that if the underlying discrete lattice is different than cubic, different numerical factors can appear, related to the density of spins sites as seen in section 2.7.

2.9 Classical spin mechanics

Dynamics is an essential component of magnetism theory. With much of the theory being applied to classical spin models, we can summarize here the basic mechanical rules for classical spins, that are based on either Hamiltonian or Lagrangian mechanics [10].

The ratio of a magnetic moment μ to its angular momentum is the gyromagnetic ratio γ. For electrons γ is negative, due to the negative electron charge, and has a value of about $\gamma \approx -1.76 \times 10^{11}$ $(\text{Ts})^{-1}$ when considering spin angular momentum. The particular value is not as important as keeping in mind that μ is obtained from **S** by including a factor of γ. Also, the distinction between orbital and spin angular momentum is irrelevant for the dynamics, and any angular momentum will be called a 'spin' for simplicity. Then generally one assumes a relation for an atomic dipole,

$$\boldsymbol{\mu} = \gamma \mathbf{S}. \tag{2.58}$$

Note that the negative value of γ for electrons means that $\boldsymbol{\mu}$ and **S** have time derivatives of opposite sign when dynamics is considered.

Consider first just one spin **S** and its dynamics in some local magnetic induction **B**. The local magnetic induction determines an energy or Hamiltonian H

for this one spin, which is just a scalar product of its magnetic moment with the field,

$$H = -\mu \cdot \mathbf{B} = -\gamma \mathbf{S} \cdot \mathbf{B}. \tag{2.59}$$

The elementary dynamics is governed by Newton's second law for rotation, which says that the time rate of change of angular momentum is equal to the torque. The torque on a magnetic dipole tends to push it perpendicular to $\mathbf{B}$ and to $\mathbf{S}$, see figure 2.6. Indeed, the infinitesimal change in $\mathbf{S}$ over some time interval Δt is $\Delta \mathbf{S} = \mathbf{N}\Delta t$. The torque $\mathbf{N}$ is the cross product of μ with the magnetic induction. The equation of motion for the dynamics of $\mathbf{S}$ can be written as

$$\frac{d\mathbf{S}}{dt} = \mathbf{N} = \mu \times \mathbf{B} = \gamma \mathbf{S} \times \mathbf{B} = \mathbf{S} \times \gamma \mathbf{B}. \tag{2.60}$$

This also implies a similar equation for the magnetic moment,

$$\frac{d\mu}{dt} = \mu \times \gamma \mathbf{B}. \tag{2.61}$$

This shows that both $\mathbf{S}(t)$ and $\mu(t)$ follow an identical mathematical equation controlled by the applied field scaled by the gyromagnetic ratio $\gamma \mathbf{B}$.

2.9.1 Hamiltonian dynamics

One can verify the SD equation (2.60) by using classical Hamiltonian dynamics [10], where the equation of motion for any dynamical variable $A(t)$ is given by

$$\dot{A} = \{A, H\} + \frac{\partial A}{\partial t} \tag{2.62}$$

where the dot represents the time derivative $\frac{d}{dt}$, the partial time derivative on the RHS is determined by the *explicit time-dependence* of A and the braces $\{,\}$

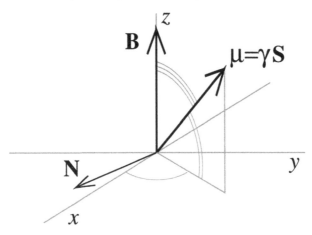

Figure 2.6. With a spin's magnetic moment μ being acted on by a magnetic induction $\mathbf{B}$ along the z-axis, the resulting torque $\mathbf{N} = \mu \times \mathbf{B}$ lies in the xy-plane perpendicular to both μ and $\mathbf{B}$.

indicate a Poisson bracket (PB). The PB between two quantities A and B is defined as

$$\{A, B\} \equiv \sum_i \left(\frac{\partial A}{\partial q_i} \frac{\partial B}{\partial p_i} - \frac{\partial A}{\partial p_i} \frac{\partial B}{\partial q_i} \right) \tag{2.63}$$

where there is some set of canonical variables $\{q_i\}$, or coordinates and their conjugate momenta $\{p_i\}$. These canonical variables and momenta have the simple PBs,

$$\{q_i, p_j\} = \delta_{i,j}, \tag{2.64}$$

i.e. unit values only when the conjugate momentum is paired with its coordinate. Then for these canonical coordinates and momenta, the Hamilton equations of motion are very simple:

$$\dot{q}_i = \{q_i, H\} = \frac{\partial H}{\partial p_i}, \tag{2.65a}$$

$$\dot{p}_i = \{p_i, H\} = -\frac{\partial H}{\partial q_i}. \tag{2.65b}$$

The quantities A and B in (2.63) have some functional dependence on the canonical coordinates and momenta. Thus it is important to realize another property of PBs by rewriting the definition, with the help of inserting some '1's (e.g. $\frac{\partial q_i}{\partial q_i}$) and '0's (e.g. $\frac{\partial q_i}{\partial p_i}$) in each part:

$$\{A, B\} = \sum_i \left[\left(\frac{\partial A}{\partial q_i} \frac{\partial p_i}{\partial p_i} - \frac{\partial A}{\partial p_i} \frac{\partial p_i}{\partial q_i} \right) \frac{\partial B}{\partial p_i} + \left(\frac{\partial A}{\partial q_i} \frac{\partial q_i}{\partial p_i} - \frac{\partial A}{\partial p_i} \frac{\partial q_i}{\partial q_i} \right) \frac{\partial B}{\partial q_i} \right] \tag{2.66a}$$

$$\{A, B\} = \sum_i \left[\{A, p_i\} \frac{\partial B}{\partial p_i} + \{A, q_i\} \frac{\partial B}{\partial q_i} \right]. \tag{2.66b}$$

The RHS simply has a sum over variations of B with respect to all coordinates and momenta of the system. In the special case that B is the Hamiltonian, this gives a time derivative relation for only the implicit time-dependence (due to $q_i(t)$ and $p_i(t)$),

$$\dot{A} = \{A, H\} = \sum_i \left[\{A, p_i\} \dot{q}_i - \{A, q_i\} \dot{p}_i \right]. \tag{2.67}$$

Seeing that $\{A, p_i\} = \frac{\partial A}{\partial q_i}$ and $\{A, q_i\} = -\frac{\partial A}{\partial p_i}$, this verifies the fundamental equation of motion (2.62):

$$\{A, H\} = \sum_i \left[\frac{\partial A}{\partial q_i} \dot{q}_i + \frac{\partial A}{\partial p_i} \dot{p}_i \right] = \frac{d}{dt} A(\{q_i\}, \{p_i\}). \tag{2.68}$$

It is an interesting exercise to use Hamiltonian dynamics to find the torque on a classical angular momentum written as $\mathbf{S} = \mathbf{r} \times \mathbf{p}$, by determining $\dot{\mathbf{S}}$ from equation (2.62). Start by using the following notation for the cross product, using Einstein's summation convention for repeated indices,

$$\mathbf{S} = \mathbf{r} \times \mathbf{p} = \varepsilon_{ijk}\, \hat{\mathbf{e}}_i x_j p_k, \tag{2.69}$$

where i, j, k refer to Cartesian components (123 for xyz) and ε_{ijk} is the Levi–Civita symbol, which has the non-zero values $\varepsilon_{123} = \varepsilon_{231} = \varepsilon_{312} = 1$ (the even permutations), and $\varepsilon_{321} = \varepsilon_{213} = \varepsilon_{132} = -1$ (the odd permutations). If two indices are the same, $\varepsilon_{ijk} = 0$. The set $\hat{\mathbf{e}}_i$, $i = 1, 2, 3$ is the same as the unit vectors $\hat{\mathbf{x}}$, $\hat{\mathbf{y}}$, $\hat{\mathbf{z}}$. Then the time derivative is obtained by

$$\dot{\mathbf{S}} = \varepsilon_{ijk}\, \hat{\mathbf{e}}_i \frac{d}{dt}(x_j p_k) = \varepsilon_{ijk}\, \hat{\mathbf{e}}_i \left(\dot{x}_j p_k + x_j \dot{p}_k \right). \tag{2.70}$$

The Hamiltonian is

$$H = -\boldsymbol{\mu} \cdot \mathbf{B} = -\gamma \mathbf{S} \cdot \mathbf{B} = -\gamma B_l S_l = -\gamma \varepsilon_{lmn} B_l x_m p_n \tag{2.71}$$

again with implied sums over l, m, n from 1 to 3. This will help to give the required time derivatives,

$$\dot{x}_j = +\frac{\partial H}{\partial p_j} = -\gamma \varepsilon_{lmj} B_l x_m \tag{2.72a}$$

$$\dot{p}_k = -\frac{\partial H}{\partial x_k} = \gamma \varepsilon_{lkn} B_l p_n. \tag{2.72b}$$

Then these can be substituted into (2.70) to give the derivative of one component,

$$\dot{S}^i = \varepsilon_{ijk}\left(\dot{x}_j p_k + x_j \dot{p}_k \right) = \gamma \varepsilon_{ijk}\left(-\varepsilon_{lmj} B_l x_m p_k + x_j \varepsilon_{lkn} B_l p_n \right). \tag{2.73}$$

The indices on the Levi–Civita symbol can be shifted in even permutations with no change in the value. It is helpful to perform the shift until the index in common in a product of these symbols falls into the first position. Once that happens, it is a straightforward exercise to verify the well-known identity,

$$\varepsilon_{ijk}\varepsilon_{ilm} = \delta_{jl}\delta_{km} - \delta_{jm}\delta_{kl}. \tag{2.74}$$

Then applying this, and renaming the dummy indices, we have the following algebra for one spin component,

$$\begin{aligned}\dot{S}^i &= \gamma\left(-\varepsilon_{jki}\varepsilon_{jlm}\,B_l x_m p_k + \varepsilon_{kij}\varepsilon_{knl}\,B_l x_j p_n\right)\\ &= \gamma\left(-B_k x_i p_k + B_i x_k p_k + B_l x_l p_i - B_i x_j p_j\right)\\ &= \gamma\left(-B_k x_i p_k + B_k x_k p_i\right)\\ &= \gamma\left[(\mathbf{r}\cdot\mathbf{B})p_i - (\mathbf{p}\cdot\mathbf{B})x_i\right].\end{aligned} \qquad (2.75)$$

Compare to the expression for $\mathbf{S}\times\mathbf{B}$,

$$\mathbf{S}\times\mathbf{B} = (\mathbf{r}\times\mathbf{p})\times\mathbf{B} = (\mathbf{r}\cdot\mathbf{B})\mathbf{p} - (\mathbf{p}\cdot\mathbf{B})\mathbf{r}. \qquad (2.76)$$

Then it is clear that the torque is given by

$$\mathbf{N} = \dot{\mathbf{S}} = \gamma\mathbf{S}\times\mathbf{B} = \boldsymbol{\mu}\times\mathbf{B} \qquad (2.77)$$

and this is completely equivalent to (2.60) obtained from elementary methods.

2.9.2 Hamiltonian spin dynamics

Luckily there is a more direct way to arrive at $\dot{\mathbf{S}}$, via the fundamental PBs

$$\{S^x, S^y\} = S^z \qquad (2.78)$$

and two other relations obtained from permuting the indices as $xyz \to yzx \to zxy$. These could be proven starting from angular momentum definitions as in (2.69). For SD, the spin components become the canonical coordinates and momenta, however, they are constrained to give some required spin length $\mathbf{S}^2$, so all three are not independent. The relations among them can be combined into a single equation,

$$\{S^i, S^j\} = \varepsilon_{ijk} S^k. \qquad (2.79)$$

Based on this, the spin equation of motion is obtained promptly. As for any dynamic quantity, the SD equation of motion is

$$\dot{\mathbf{S}} = \{\mathbf{S}, H\}. \qquad (2.80)$$

Consider applying (2.80) to the S^x-component, using a relation equivalent to (2.66a), but where spin degrees of freedom S^x, S^y, S^z replace the set of (q_i, p_i),

$$\dot{S}^x = \{S^x, H\} = \{S^x, S^x\}\frac{\partial H}{\partial S^x} + \{S^x, S^y\}\frac{\partial H}{\partial S^y} + \{S^x, S^z\}\frac{\partial H}{\partial S^z}. \qquad (2.81)$$

Of course, the first PB is $\{S^x, S^x\} = 0$, and the dynamics of $\dot{S}^x$ is determined only from $\{S^x, S^y\} = S^z$ and $\{S^x, S^z\} = -\{S^z, S^x\} = -S^y$. This gives

$$\dot{S}^x = S^z\frac{\partial H}{\partial S^y} - S^y\frac{\partial H}{\partial S^z}. \qquad (2.82)$$

The same algebra can be followed for $\dot{S}^y$, and $\dot{S}^z$, or by shifting the indices. The results for all components can be written as follows

$$\dot{S}^x = -S^y \frac{\partial H}{\partial S^z} + S^z \frac{\partial H}{\partial S^y} = S^y \mathcal{F}^z - S^z \mathcal{F}^y \qquad (2.83a)$$

$$\dot{S}^y = -S^z \frac{\partial H}{\partial S^x} + S^x \frac{\partial H}{\partial S^z} = S^z \mathcal{F}^x - S^x \mathcal{F}^z \qquad (2.83b)$$

$$\dot{S}^z = -S^x \frac{\partial H}{\partial S^y} + S^y \frac{\partial H}{\partial S^x} = S^x \mathcal{F}^y - S^y \mathcal{F}^x. \qquad (2.83c)$$

This is clearly in the form of a cross product with an *effective field* $\mathcal{F}$, defined from the variation of the Hamiltonian with respect to its spin coordinates,

$$\mathcal{F} \equiv -\frac{\partial H}{\partial \mathbf{S}}, \quad \text{or} \quad \mathcal{F} = (\mathcal{F}^x, \mathcal{F}^y, \mathcal{F}^z) = \left(-\frac{\partial H}{\partial S^x}, -\frac{\partial H}{\partial S^y}, -\frac{\partial H}{\partial S^z}\right). \qquad (2.84)$$

Then the dynamics is described from

$$\dot{\mathbf{S}} = \mathbf{S} \times \mathcal{F}. \qquad (2.85)$$

This is a very general result and applies to the SD of any Hamiltonian system.

For the special case of the spin in a magnetic induction **B**, the Hamiltonian (2.59) gives a very simple result for the effective field,

$$\mathcal{F} = -\frac{\partial}{\partial \mathbf{S}}(-\gamma \mathbf{S} \cdot \mathbf{B}) = \gamma \mathbf{B}. \qquad (2.86)$$

Then we again recover the dynamic equation (2.60) that was obtained by elementary methods. Note that the gyromagnetic ratio only enters when considering a magnetic induction **B** measured in tesla.

2.9.3 Angular spin coordinates and canonical Poisson brackets

Suppose a spin **S** is represented in polar spherical coordinates (ϕ, θ) as defined in (2.41). Let us consider the dynamic equations for their time derivatives. Keep in mind that $S^z = S \cos \theta$ points along the axis around which ϕ is measured. Consider the following PB:

$$\{\phi, S^z\}, \quad \text{where } \phi = \tan^{-1}\left(\frac{S^y}{S^x}\right). \qquad (2.87)$$

The chain rule can be used to expand this operation, and combined with the fundamental PBs (2.79) of the spin components. One has

$$\{\phi, S^z\} = \frac{\partial \phi}{\partial S^x}\{S^x, S^z\} + \frac{\partial \phi}{\partial S^y}\{S^y, S^z\} = \frac{-S^y(-S^y)}{S^{x2} + S^{y2}} + \frac{S^x(S^x)}{S^{x2} + S^{y2}} = 1. \qquad (2.88)$$

This shows that the S^z spin component is the momentum that is canonically conjugate to the in-plane angle ϕ. This result is analogous to the fundamental classical PB in (2.64). This is an important relationship, because it offers another approach for finding dynamic equations of motion.

If we work in the mixed coordinates (ϕ, S^z) for a spin, the equations of motion take a particularly simple form, using the chain rule,

$$\dot{\phi} = \{\phi, H\} = \{\phi, S^z\}\frac{\partial H}{\partial S^z} = \frac{\partial H}{\partial S^z} \tag{2.89a}$$

$$\dot{S^z} = \{S^z, H\} = \{S^z, \phi\}\frac{\partial H}{\partial \phi} = -\frac{\partial H}{\partial \phi}. \tag{2.89b}$$

Of course, that is the standard form one expects for the dynamics of a coordinate and its conjugate momentum. If the z-component is expressed as $S^z = S \cos\theta$ where θ is the polar angle to the z-axis, then one has

$$\frac{\partial H}{\partial \theta} = \frac{\partial H}{\partial S^z}\frac{\partial S^z}{\partial \theta} = \frac{\partial H}{\partial S^z}(-S \sin\theta) \tag{2.90}$$

and then the dynamics in *polar spherical coordinates* follows

$$S\dot{\phi} \sin\theta = -\frac{\partial H}{\partial \theta} \tag{2.91a}$$

$$S\dot{\theta} \sin\theta = \frac{\partial H}{\partial \phi}. \tag{2.91b}$$

If *planar spherical coordinates* are used with $S^z = S \sin\theta$, the dynamics follows

$$S\dot{\phi} \cos\theta = \frac{\partial H}{\partial \theta} \tag{2.92a}$$

$$S\dot{\theta} \cos\theta = -\frac{\partial H}{\partial \phi}. \tag{2.92b}$$

In a certain sense this is preferable because the signs are the same as in the canonical time derivatives (2.89a), because S^z increases monotonically with θ for planar spherical coordinates.

2.9.4 Lagrangian spin dynamics

For a single spin we know that ϕ and S^z are canonically conjugate variables, with the fundamental PB, $\{\phi, S^z\} = 1$. We can write that S^z is the momentum conjugate to coordinate ϕ:

$$p_\phi = S^z, \tag{2.93}$$

and also one has from Lagrangian mechanics,

$$p_\phi \equiv \frac{\partial L}{\partial \dot\phi} \tag{2.94}$$

where L is the Lagrangian, whose time integral determines the classical action,

$$A = \int dt\, L\left[\phi(t),\, \dot\phi(t),\, S^z(t)\right]. \tag{2.95}$$

In classical mechanics, the action integral is required to be an extremum for the true motion. Then the Lagrangian is obtained from the Hamiltonian $H[\phi, S^z]$ for this spin by

$$L\left[\phi, \dot\phi, S^z\right] = \dot\phi p_\phi - H = \dot\phi S^z - H. \tag{2.96}$$

For SD the use of Hamiltonian equations of motion is uncomplicated, with H depending on the in-plane angle ϕ and the spin component perpendicular to that plane. The Lagrangian, conversely, does not involve $\dot S^z$ explicitly, and because S^z is already the momentum conjugate to ϕ, there is no momentum conjugate to S^z. Even so, there are occasions when Lagrangian spin mechanics is easier to work with than Hamiltonian mechanics. Besides that, the canonical momentum is seen to be well defined within the Lagrangian formalism. Another advantage of Lagrangian mechanics becomes obvious when a continuum system is considered, later in this chapter.

We can check the equations of motion for ϕ and for S^z derived from L. The Euler–Lagrange equations [10] for L that make the action an extremum are

$$\frac{\partial L}{\partial S^z} - \frac{d}{dt}\frac{\partial L}{\partial \dot S^z} = 0 \implies \dot\phi = \frac{\partial H}{\partial S^z} \tag{2.97a}$$

$$\frac{\partial L}{\partial \phi} - \frac{d}{dt}\frac{\partial L}{\partial \dot\phi} = 0 \implies \dot S^z = -\frac{\partial H}{\partial \phi}. \tag{2.97b}$$

These are identical to those in (2.89a) derived from PBs with H.

2.9.5 Classical dynamics in a system of spins

In a problem with many spins, each one has an equation of motion determined by the variation of H with respect to that spin $\mathbf{S}_i$, as derived from Hamiltonian dynamics for the spin components,

$$\dot{\mathbf{S}}_i = \mathbf{S}_i \times \mathcal{F}_i, \quad \mathcal{F}_i = -\frac{\partial H}{\partial \mathbf{S}_i}. \tag{2.98}$$

Physically, some may prefer to think that each spin receives torques from magnetic fields due to its surroundings (i.e. its neighbors, anisotropy forces, an external field, and so on). Then one could define the effective magnetic induction acting on some site by

$$\mathbf{B}_i \equiv \frac{1}{\gamma}\mathcal{F}_i = -\frac{1}{\gamma}\frac{\partial H}{\partial \mathbf{S}_i}. \tag{2.99}$$

Then the equivalent equations of motion are

$$\dot{\mathbf{S}}_i = \mathbf{S}_i \times \gamma \mathbf{B}_i. \qquad (2.100)$$

This is the fundamental form of the undamped Landau–Lifshitz (LL) equation of motion [11]. To completely specify the problem, the components of $\mathbf{B}_i$ or $\mathcal{F}_i$ must be determined from H, depending on the type of model being considered. It can be noted also that γ can be removed from the problem by a rescaling of the time variable.

The problem could also be analyzed in terms of a Lagrangian L for the system, using the canonically conjugate pairs (ϕ_n, S_n^z) for each site as the coordinates. Based on (2.88), those pairs have the fundamental PB relations,

$$\{\phi_i, S_j^z\} = \delta_{ij}. \qquad (2.101)$$

It is convenient to think of L as composed from a *kinetic term* K and the Hamiltonian,

$$L = K - H = \sum_n \dot{\phi}_n S_n^z - H. \qquad (2.102)$$

The equations of motion for each site are those derived from the Euler–Lagrange variation, and it is obvious that they are similar to (2.97), but with a site index:

$$\frac{\partial L}{\partial S_n^z} - \frac{d}{dt}\frac{\partial L}{\partial \dot{S}_n^z} = 0 \implies \dot{\phi}_n = +\frac{\partial H}{\partial S_n^z} \qquad (2.103a)$$

$$\frac{\partial L}{\partial \phi_n} - \frac{d}{dt}\frac{\partial L}{\partial \dot{\phi}_n} = 0 \implies \dot{S}_n^z = -\frac{\partial H}{\partial \phi_n}. \qquad (2.103b)$$

Of course, the interactions among different spins, according to the structure of H, will lead to the interesting dynamics caused by a set of competing interactions.

2.9.6 Spin damping

The torque implied in (2.85) is perpendicular to $\mathbf{S}$ and to $\mathcal{F}$, and causes a precessional motion of the spin around $\mathcal{F}$, see figure 2.6. The spin equations can be damped, making them energy non-conserving, by adding a term whose torque has a component along the effective field $\mathcal{F}$. There are two ways to do this that are essentially equivalent, up to a rescaling of the time variable. One can see in figure 2.7 that $(\mathbf{S} \times \mathcal{F}) \times \mathbf{S}$ has a positive component along $\mathcal{F}$. In LL damping [11], a torque along $\mathcal{F}$ is produced by a term $+(\alpha/S)(\mathbf{S} \times \mathcal{F}) \times \mathbf{S}$, where α is a dimensionless damping constant. From figure 2.7 or by some vector algebra, this damping torque has the greatest component along $\mathcal{F}$ when $\mathbf{S}$ is perpendicular to $\mathcal{F}$, and the least component along $\mathcal{F}$ when $\mathbf{S}$ is parallel to $\mathcal{F}$. Then this type of term will have a large effect only when the spin is far from parallel to $\mathcal{F}$. Once the spin has relaxed to be nearly parallel to $\mathcal{F}$, there is no more damping that can take place anyway.

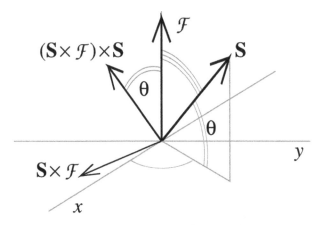

Figure 2.7. A spin S acted on by effective field $\mathcal{F} = \gamma \mathbf{B}$ along the z-axis. The torque without damping lies in the direction of $\mathbf{S} \times \mathcal{F}$. If LL damping is present, there is an additional damping torque equal to α/S multiplied by $(\mathbf{S} \times \mathcal{F}) \times \mathbf{S}$, which has a component along $\mathcal{F}$. The damping torque makes an angle θ with $\mathcal{F}$ which is the same as the out-of-plane angle of the spin.

In another form of damping known as Gilbert damping [12, 13], a term $+(\alpha'/S)(\dot{\mathbf{S}} \times \mathbf{S})$ is used, where α' is another dimensionless damping constant.

With LL damping, the dynamics of a single spin in the LL equation [11, 14] is assumed to be

$$\dot{\mathbf{S}} = \mathbf{S} \times \mathcal{F} + \frac{\alpha}{S}(\mathbf{S} \times \mathcal{F}) \times \mathbf{S} \tag{2.104}$$

and with Gilbert damping [12, 13] the dynamics is taken as

$$\dot{\mathbf{S}} = \mathbf{S} \times \mathcal{F} + \frac{\alpha'}{S}(\dot{\mathbf{S}} \times \mathbf{S}). \tag{2.105}$$

To show these are related, substitute this last equation into itself for the damping term,

$$\dot{\mathbf{S}} = \mathbf{S} \times \mathcal{F} + \frac{\alpha'}{S}\left[\mathbf{S} \times \mathcal{F} + \frac{\alpha'}{S}(\dot{\mathbf{S}} \times \mathbf{S})\right] \times \mathbf{S}. \tag{2.106}$$

Expand out the last double cross product:

$$(\dot{\mathbf{S}} \times \mathbf{S}) \times \mathbf{S} = (\dot{\mathbf{S}} \cdot \mathbf{S})\mathbf{S} - S^2\dot{\mathbf{S}} = -S^2\dot{\mathbf{S}}. \tag{2.107}$$

The term $(\dot{\mathbf{S}} \cdot \mathbf{S}) = 0$, as can be verified from the equation of motion, which guarantees a conserved spin length. Now the dynamics with Gilbert damping follows

$$\dot{\mathbf{S}} = \mathbf{S} \times \mathcal{F} + \frac{\alpha'}{S}\left[(\mathbf{S} \times \mathcal{F}) \times \mathbf{S} + \frac{\alpha'}{S}(-S^2\dot{\mathbf{S}})\right] \tag{2.108a}$$

$$(1 + \alpha'^2)\dot{\mathbf{S}} = \mathbf{S} \times \mathcal{F} + \frac{\alpha'}{S}(\mathbf{S} \times \mathcal{F}) \times \mathbf{S}. \tag{2.108b}$$

Then one can see that this is almost the same as LL damping, equation (2.104), except for a rescaling of the time due to the factor on the LHS. At a small damping parameter, this will not make any significant difference. At a large parameter, Gilbert damping is 'slower', because it results in a smaller net time derivative of the spin (for the same damping constant, $\alpha' = \alpha$). It also slows down the original torque effect (first term on the RHS). Mostly, the LL form of damping will be applied in this text. If Gilbert damping is applied the dynamic equation is called the Landau–Lifshitz–Gilbert equation.

2.9.7 A single spin in a fixed field with damping

A problem that can be solved exactly is an individual spin that is precessing in a fixed magnetic field, with damping acting. We assume the spin has some initial direction $\mathbf{S}(0)$ and the constant field is taken to be along the z-axis, with $\boldsymbol{\mathcal{F}} = \mathcal{F}\hat{\mathbf{z}} = \gamma B^z\hat{\mathbf{z}}$. The dynamic equation with LL damping is

$$\dot{\mathbf{S}} = \mathbf{S} \times \boldsymbol{\mathcal{F}} + \frac{\alpha}{S}[S^2\boldsymbol{\mathcal{F}} - (\mathbf{S} \cdot \boldsymbol{\mathcal{F}})\mathbf{S}]. \tag{2.109}$$

The extra effect of the damping terms is to push the spin towards the direction of $\boldsymbol{\mathcal{F}}$. In terms of components the dynamics is determined by

$$\dot{S}^x = S^y\mathcal{F} - \frac{\alpha}{S}(S^z\mathcal{F})S^x, \tag{2.110a}$$

$$\dot{S}^y = -S^x\mathcal{F} - \frac{\alpha}{S}(S^z\mathcal{F})S^y, \tag{2.110b}$$

$$\dot{S}^z = \alpha S\mathcal{F} - \frac{\alpha}{S}(S^z\mathcal{F})S^z. \tag{2.110c}$$

It is helpful that the z-component equation is separated from the other two. Then it can be solved first although it is nonlinear. It separates and is solved with the help of the transformation, $S^z = S \tanh\theta$:

$$\int_{S^z(0)}^{S^z(t)} \frac{\mathrm{d}S^z/S}{1 - (S^z/S)^2} = \int_0^t \alpha\mathcal{F}\,\mathrm{d}t, \tag{2.111a}$$

$$\tanh^{-1}\frac{S^z(t)}{S} - \tanh^{-1}\frac{S^z(0)}{S} = \alpha\mathcal{F}t. \tag{2.111b}$$

The solution is then

$$S^z(t) = S\tanh\left[\tanh^{-1}\left(\frac{S^z(0)}{S}\right) + \alpha\mathcal{F}t\right]. \tag{2.112}$$

The spin goes smoothly towards the applied field direction. In the limit of infinite time the hyperbolic tangent goes to 1 and thus $S^z \to S$, as it should. The time scale over which the greatest changes occur is $(\alpha\mathcal{F})^{-1}$, which means the relaxation is all the faster for large damping and large field, as can be expected on physical intuition.

Now to obtain the in-plane motion, look again at the differential equations (2.110a). They have an obvious symmetry that allows one to cancel out the damping terms for the xy motion. Perform the following combination:

$$S^y \dot{S}^x - S^x \dot{S}^y = \mathcal{F}(S^{x2} + S^{y2}), \quad \text{or} \quad \frac{S^y \dot{S}^x - S^x \dot{S}^y}{S^{x2} + S^{y2}} = \mathcal{F}. \tag{2.113}$$

This combination of derivatives is related to the derivative of the in-plane angle ϕ, assuming that the spin can be described by planar spherical coordinates, equation (2.42). Check the algebra for a time derivative,

$$\phi = \tan^{-1}\left(\frac{S^y}{S^x}\right), \quad \dot{\phi} = \frac{1}{1 + \left(\frac{S^y}{S^x}\right)^2} \frac{S^x \dot{S}^y - S^y \dot{S}^x}{S^{x2}} = \frac{S^x \dot{S}^y - S^y \dot{S}^x}{S^{x2} + S^{y2}}. \tag{2.114}$$

So the motion in-plane is very simple and, surprisingly, *unaffected* by the damping!

$$\dot{\phi} = \frac{S^x \dot{S}^y - S^y \dot{S}^x}{S^{x2} + S^{y2}} = -\mathcal{F}, \quad \Longrightarrow \quad \phi(t) = -\mathcal{F}t + \phi_0. \tag{2.115}$$

The total motion is summarized by giving $S^z(t)$ and $\phi(t)$. There is a uniform precession around the z-axis at frequency $\omega = -\mathcal{F} = -\gamma B$, while the S^z-component slowly goes towards its maximum value of S. Note that the precession of **S** goes around $\mathcal{F}$ in the negative sense, however, for a negative gyromagnetic ratio, **S** precesses around **B** in the positive sense (given by a right hand rule with the right thumb along **B** and the right hand fingers pointing in the precessional direction). The S^z equation may also be expressed in terms of the out-of-plane angle $\theta(t)$ for the spin, measured from to the plane perpendicular to the field $\mathcal{F}$.

Exercise 2.7. Show that another way to express $S^z(t)$ for a spin with damping in a constant magnetic field is

$$S^z(t) = S \sin\theta(t) = S \frac{\sinh \alpha\mathcal{F}t + \sin\theta_0 \cosh \alpha\mathcal{F}t}{\cosh \alpha\mathcal{F}t + \sin\theta_0 \sinh \alpha\mathcal{F}t}, \tag{2.116}$$

where θ_0 is the out-of-plane angle at $t = 0$.

An example of the relaxation of a spin towards $\mathcal{F}$ is shown in figure 2.8. For LL damping with $\alpha = 0.1$, figure 2.8 shows how the S^z-component rises from its initial value of $S^z = 0.2S$ and heads towards $S^z \to S$, while the in-plane S^x-component exhibits the spiral motion of **S** around the field.

The above discussion was applied to an individual spin. In a system with many spins, damping is still considered a local effect, that can be considered on each spin individually as was done in this section.

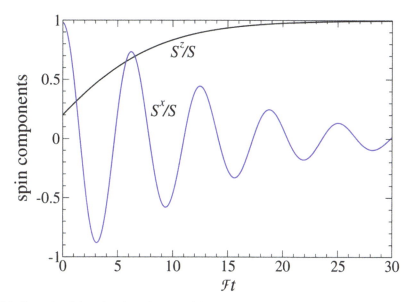

Figure 2.8. Example of the relaxation of a spin with LL damping of strength $\alpha = 0.1$, in an effective field $\mathcal{F} = \mathcal{F}\hat{z}$. The spin was initiated pointing at $S^z(0) = 0.2$ and in-plane angle $\phi(0) = 0$. There is a spiral precession of the spin around $\mathcal{F}$ as it becomes ever more aligned with $\mathcal{F}$ over time.

2.10 Classical spin torques

The previous paragraphs give an account of the damping torques, which are non-conservative torques within the system, usually reducing the total magnetic energy. Here we give some examples of the conservative torques that can be associated with the Hamiltonian of the system. The conservative torques are those derived from equation (2.98), via a cross product of $\mathbf{S}_i$ with its effective field $\mathcal{F}_i$.

The torque due to exchange energy is the first example. Consider the total exchange torque on site n of a 1D magnetic chain with the discrete Hamiltonian (2.10). Site n interacts with sites $n - 1$ and $n + 1$, i.e. there are two exchange bonds that determine the torque. Applying (2.98), we obtain the x-component of the effective field acting on $\mathbf{S}_n$ to be

$$\mathcal{F}_n^x = -\frac{\partial H}{\partial S_n^x} = -\frac{\partial}{\partial S_n^x}[-J^x S_n^x (S_{n-1}^x + S_{n+1}^x)] = J^x (S_{n-1}^x + S_{n+1}^x). \quad (2.117)$$

The effective field is simply the neighboring spins scaled by the exchange constant. The vector relation is the generalization,

$$\mathcal{F}_n = \mathbf{J} \cdot (\mathbf{S}_{n-1} + \mathbf{S}_{n+1}), \quad (2.118)$$

Here a matrix notation for the exchange constants is implied, meaning that $\mathbf{J}$ is a tensor of exchange constants, in a 3×3 array:

$$\mathbf{J} = \begin{pmatrix} J^x & 0 & 0 \\ 0 & J^y & 0 \\ 0 & 0 & J^z \end{pmatrix}. \tag{2.119}$$

Written this way, one could even consider off-diagonal exchange effects. This notation assumes that $\mathcal{F}_n$ and $\mathbf{S}_{n\pm 1}$ are represented as column vectors. Then the continuum limit dynamic equation is the discrete LL equation

$$\dot{\mathbf{S}}_n = \mathbf{S}_n \times \mathcal{F}_n = \mathbf{S}_n \times [\mathbf{J} \cdot (\mathbf{S}_{n-1} + \mathbf{S}_{n+1})]. \tag{2.120}$$

In the simplest case of isotropic exchange, $J^x = J^y = J^z = J$, the effective field reduces to a scaled sum of the neighboring spins:

$$\mathcal{F}_n = J(\mathbf{S}_{n-1} + \mathbf{S}_{n+1}) \tag{2.121}$$

In higher-dimensional models, the sum would be over all possible neighbors of a site n.

A second example of discrete torques is that due to a local anisotropy with energy constant K_A, such as that introduced in equation (1.52), with a local anisotropy axis $\hat{\mathbf{u}}_n$ for spin $\mathbf{S}_n$. Using the definition for $\mathcal{F}_n$, equation (2.84), we have for only the local anisotropy field,

$$\mathcal{F}_n = -\frac{\partial}{\partial \mathbf{S}_n}[-K_A(\mathbf{S}_n \cdot \hat{\mathbf{u}}_n)^2] = 2K_A(\mathbf{S}_n \cdot \hat{\mathbf{u}}_n)\hat{\mathbf{u}}_n. \tag{2.122}$$

If this were the only torque acting on the spin, the equation of motion would be

$$\dot{\mathbf{S}}_n = \mathbf{S}_n \times \mathcal{F}_n = 2K_A(\mathbf{S}_n \cdot \hat{\mathbf{u}}_n)(\mathbf{S}_n \times \hat{\mathbf{u}}_n). \tag{2.123}$$

This causes generally a precession around the $\pm\hat{\mathbf{u}}_n$-axis (note that the RHS is an even function of $\hat{\mathbf{u}}$), depending on the initial state. With damping included and a positive value of K_A, the torque would cause the spin to relax towards either $\hat{\mathbf{u}}_n$ or $-\hat{\mathbf{u}}_n$ (uniaxial anisotropy). In the case where $K_A < 0$, this torque will have an opposite effect and the spin (with damping) would precess until moving into a plane perpendicular to $\hat{\mathbf{u}}_n$.

2.10.1 Continuum limit torques

In the case of a continuum limit Hamiltonian, the effective field $\mathcal{F}$ needs to be found with slightly more care, although the results can be checked against the continuum limit of the discrete torque result (2.118). A 1D system is considered here; it is not difficult to extend the results to a higher-dimensional lattice. For a continuum system, $\mathbf{S}_n$ will be replaced with $\mathbf{S}(x)$. The total spin in the system is obtained correctly as long as summation is changed to integration with

$$\sum_n \mathbf{S}_n \to \int \frac{\mathrm{d}x}{a} \mathbf{S}(x). \tag{2.124}$$

The sum over n will then map to the same total as the integral, by including the weighting factor $\frac{dx}{a}$. For an increment of length $dx = a$, there will be one of the original spins on the 1D chain[3]. In a similar way, the Hamiltonian for the continuum model is an integral over an effective energy density $\mathcal{H}$,

$$H = \int \frac{dx}{a} \mathcal{H}(\mathbf{S}, \partial_x \mathbf{S}). \tag{2.125}$$

By keeping the factor of a in the measure, the dimensions on H and $\mathcal{H}$ are both energy, and this simplifies the formulas. The Hamiltonian density $\mathcal{H}$ generally depends on the spin field and its space gradients. Here we have $\mathcal{H}$ expressed in terms of Cartesian components, but it could just as well be written using the canonical pairs $\phi(x)$ and $S^z(x)$, and their gradients. If that is the case, one can transform to the Lagrangian of the continuum system, in the same way as for discrete spins, equation (2.102), including a time derivative $\partial_t \phi$ in a kinetic integral term. The Lagrangian will be an integral also over a continuous density function $\mathcal{L}$, according to

$$L = \int \frac{dx}{a} \mathcal{L} \tag{2.126a}$$

$$\mathcal{L} = \mathcal{L}\left(\phi, \phi_x, \dot{\phi}, S^z, S^z_x\right) = S^z \dot{\phi} - \mathcal{H}. \tag{2.126b}$$

The subscripts are used here to indicate partial derivatives, i.e. $\phi_x = \partial_x \phi$, and so on; the dot in $\dot{\phi}$ is the partial time derivative. The action that is to be made an extremum for the true motion, is an integration of $\mathcal{L}$ over *time* and *space*,

$$A = \int dt \, L = \int dt \int \frac{dx}{a} \mathcal{L}. \tag{2.127}$$

This is stationary with respect to arbitrary variations in $\phi(x, t)$ and $S^z(x, t)$, provided they satisfy Euler–Lagrange equations that treat time and space equivalently. The Euler–Lagrange equations, found via calculus of variations, are

$$\frac{\partial \mathcal{L}}{\partial S^z} - \frac{\partial}{\partial x} \frac{\partial \mathcal{L}}{\partial S^z_x} - \frac{\partial}{\partial t} \frac{\partial \mathcal{L}}{\partial \dot{S}^z} = 0 \implies \dot{\phi} = +\left(\frac{\partial \mathcal{H}}{\partial S^z} - \frac{\partial}{\partial x} \frac{\partial \mathcal{H}}{\partial S^z_x}\right) \tag{2.128a}$$

$$\frac{\partial \mathcal{L}}{\partial \phi} - \frac{\partial}{\partial x} \frac{\partial \mathcal{L}}{\partial \phi_x} - \frac{\partial}{\partial t} \frac{\partial \mathcal{L}}{\partial \dot{\phi}} = 0 \implies \dot{S}^z = -\left(\frac{\partial \mathcal{H}}{\partial \phi} - \frac{\partial}{\partial x} \frac{\partial \mathcal{H}}{\partial \phi_x}\right). \tag{2.128b}$$

The definition (2.126b) of the Lagrangian density $\mathcal{L}$ in terms of $\mathcal{H}$ was used to arrive at this result. Note that it is very similar in structure to the discrete equations (2.103).

---

[3] For a higher-dimensional system, the factor dx/a should be changed to dV/v_{cell}, where v_{cell} is the volume occupied by one spin on the original lattice.

The operations on $\mathcal{H}$ are used to define the *functional derivatives* of H, which describe how H changes when one of the functions that defines it changes:

$$\frac{\delta H}{\delta S^z} \equiv \frac{\partial \mathcal{H}}{\partial S^z} - \frac{\partial}{\partial x}\frac{\partial \mathcal{H}}{\partial S_x^z} \qquad (2.129a)$$

$$\frac{\delta H}{\delta \phi} \equiv \frac{\partial \mathcal{H}}{\partial \phi} - \frac{\partial}{\partial x}\frac{\partial \mathcal{H}}{\partial \phi_x}. \qquad (2.129b)$$

In this way, the continuum dynamic equations look even more similar to (2.103a),

$$\dot{\phi} = +\frac{\delta H}{\delta S^z} \qquad (2.130a)$$

$$\dot{S}^z = -\frac{\delta H}{\delta \phi}. \qquad (2.130b)$$

As long as (2.129a) are used to define these variations in the Hamiltonian, the dynamics for a continuum system looks just like that for the discrete lattice, taking into account that exchange interactions appear involving space derivatives in the continuum Hamiltonian.

The time derivatives of the canonical fields $\phi(x)$ and momenta $S^z(x)$ can also be found by generalization of the PB definition (2.63) to a continuum form, replacing the sum by an integration over canonical degrees of freedom, and using functional derivatives,

$$\{A, B\} \equiv \int \frac{dx}{a}\left(\frac{\delta A}{\delta \phi}\frac{\delta B}{\delta S^z} - \frac{\delta A}{\delta S^z}\frac{\delta B}{\delta \phi}\right). \qquad (2.131)$$

The fundamental PB that accompanies this definition is a generalization of (2.64),

$$\{\phi(x), S^z(x')\} = a\,\delta(x - x'). \qquad (2.132)$$

These can easily be extended to higher space dimensions. Then the Hamilton equations of motion (2.130a) directly result from the time derivative of any dynamical variable,

$$\dot{\phi}(x) = \{\phi(x), H\} \qquad (2.133a)$$

$$\dot{S}^z(x) = \{S^z(x), H\}. \qquad (2.133b)$$

The equation of motion in terms of Cartesian components will also be the continuum generalization of the PB in (2.98), which can be demonstrated by transforming the above results from (ϕ, S^z) into all Cartesian coordinates,

$$\dot{\mathbf{S}}(x) = \{\mathbf{S}(x), H\} = \mathbf{S}(x) \times \mathcal{F}(x). \qquad (2.134)$$

This will be correct provided that the effective field is defined not by a usual gradient of H with respect to $\mathbf{S}$, as in (2.98), but instead by the functional derivative,

$$\mathcal{F}(x) = -\frac{\delta H}{\delta \mathbf{S}} = -\left[\frac{\partial \mathcal{H}}{\partial \mathbf{S}} - \frac{\partial}{\partial x}\left(\frac{\partial \mathcal{H}}{\partial(\partial_x \mathbf{S})}\right)\right]. \tag{2.135}$$

The field $\mathcal{F}(x)$ is determined mainly by the contributions to H that come from a region of width dx around x. The functional derivative includes the derivative of $\mathcal{H}$ with respect to the field $\mathbf{S}$ and its space derivatives. For more space dimensions, one should add additional terms on the RHS depending on the gradients with respect to the additional coordinates. Then finally, once the effective torque producing field is known, the continuum spin equation of motion is

$$\dot{\mathbf{S}}(x) = \mathbf{S}(x) \times \mathcal{F}(x) = \mathbf{S}(x) \times \frac{-\delta H}{\delta \mathbf{S}}. \tag{2.136}$$

This does not contain any damping. Damping can be added, if needed, by a local term in the same way as for discrete systems.

As an example, these results can be applied to the continuum limit of a 1D chain, Hamiltonian (2.16). The parts of $\mathcal{H}$ that depend on S^x and its gradient are

$$\mathcal{H}\left(\mathbf{S}, \frac{\partial \mathbf{S}}{\partial x}\right) = J^x\left[-S^{x2} + \frac{1}{2}a^2\left(\frac{\partial S^x}{\partial x}\right)^2\right] + \ldots \tag{2.137}$$

and there are terms of the same structure for the dependence on S^y and S^z. One has

$$\frac{\partial \mathcal{H}}{\partial S^x} = -2J^x S^x \tag{2.138a}$$

$$\frac{\partial}{\partial x}\left(\frac{\partial \mathcal{H}}{\partial(\partial_x S^x)}\right) = a^2 J^x \frac{\partial^2 S^x}{\partial x^2}. \tag{2.138b}$$

Using (2.135), these combine into the net effective continuum field (extending to all components),

$$\mathcal{F}(x) = \mathbf{J} \cdot \left(2\mathbf{S} + a^2 \frac{\partial^2 \mathbf{S}}{\partial x^2}\right). \tag{2.139}$$

This can be checked with a Taylor expansion of the discrete result (2.118) for $\mathcal{F}_n$. Expanding as in equation (2.11) gives

$$\mathcal{F}_n \rightarrow \mathcal{F}(x) = \mathbf{J} \cdot [\mathbf{S}(x+a) + \mathbf{S}(x-a)] \approx \mathbf{J} \cdot \left(2\mathbf{S}(x) + a^2 \frac{\partial^2 \mathbf{S}}{\partial x^2}\right) \tag{2.140}$$

identical to the result from using the functional derivative.

With these results for the effective field, the continuum equation of motion is then

$$\dot{\mathbf{S}}(x) = \mathbf{S} \times \mathcal{F}(x) = \mathbf{S}(x) \times \left[\mathbf{J} \cdot \left(2\mathbf{S} + a^2 \frac{\partial^2 \mathbf{S}}{\partial x^2} \right) \right]. \qquad (2.141)$$

Although the exchange constants form a diagonal matrix, with $\mathbf{J}_{ij} = J_i \delta_{ij}$, the first term on the RHS is not zero unless the exchange is isotropic. For isotropic exchange, this gives the continuum LL equation for a 1D chain,

$$\dot{\mathbf{S}} = Ja^2 \, \mathbf{S} \times \frac{\partial^2 \mathbf{S}}{\partial x^2}. \qquad (2.142)$$

Note that the original exchange constant J refers to an atomic pair interaction. In this last equation it has been rescaled by the factor a^2, which was the cell size associated with one site (or, the pair separation). The derivations here could be repeated in more space dimensions, starting from interacting spins on different lattices. It is clear that if one started with a cubic lattice of sites, the result for 3D isotropic exchange will be

$$\dot{\mathbf{S}} = Ja^2 \, \mathbf{S} \times \nabla^2 \mathbf{S}, \qquad (2.143)$$

where the Laplacian ∇^2 appears on the RHS, defined in the usual way

$$\nabla^2 = \frac{\partial^2}{\partial x^2} + \frac{\partial^2}{\partial y^2} + \frac{\partial^2}{\partial z^2}. \qquad (2.144)$$

In section 2.5 the continuum isotropic exchange Hamiltonian was derived for different lattices. In all of the 2D lattices considered (cubic, triangular, hexagonal), the energy density depends on the squared divergence of the spin field, e.g.

$$\mathcal{H} \propto Ja^2 \left[\left(\frac{\partial \mathbf{S}}{\partial x} \right)^2 + \left(\frac{\partial \mathbf{S}}{\partial y} \right)^2 \right] = Ja^2 (\nabla \cdot \mathbf{S})^2. \qquad (2.145)$$

The different lattices have different numerical factors multiplying J. But whether a 2D or 3D lattice is considered, this result holds for $\mathcal{H}$ and leads to an equation of motion of the form in (2.143).

2.10.2 Continuum limit dynamics for angular coordinates

For a Hamiltonian expressed with angular spin variables (ϕ, θ) one can use the results of the previous section and apply the functional derivative to obtain the equations of motion. The result for $\dot{\phi}$ is related to the component of torque around (or along) the z-axis. The result for $\dot{\theta}$ is not as simple to interpret, and depends on whether one is using polar or planar spherical coordinates. However, it gives the rate at which the spin is moving away from or towards the xy-plane.

For concreteness, assume θ is defined with planar spherical coordinates, with $S^z = S \sin \theta$. Then because S^z is the momentum conjugate to ϕ, the generalization of (2.92) to a continuum Hamiltonian, requires using the functional derivative as in (2.130) instead of the partial derivative. One needs to apply

$$S\dot\phi \cos\theta = +\frac{\delta H}{\delta\theta} \qquad (2.146a)$$

$$S\dot\theta \cos\theta = -\frac{\delta H}{\delta\phi}. \qquad (2.146b)$$

These are the canonical equations of motion.

An important application of this is using it for the isotropic continuum limit 1D ferromagnet, with Hamiltonian given in (2.57). The Hamiltonian density function (units of energy) is

$$\mathcal{H} = JS^2\left[-1 + \frac{a^2}{2}\left(\theta_x^2 + \phi_x^2 \cos^2\theta\right)\right] \qquad (2.147)$$

where subscripts indicate partial derivatives. Note that the -1 is irrelevant and it is usually removed, setting the ground state energy to zero. Then the functional derivatives needed are

$$\frac{\delta H}{\delta\theta} = \frac{\partial\mathcal{H}}{\partial\theta} - \frac{\partial}{\partial x}\left(\frac{\partial\mathcal{H}}{\partial\theta_x}\right) = JS^2 a^2\left(-\phi_x^2 \cos\theta\sin\theta - \theta_{xx}\right) \qquad (2.148a)$$

$$\frac{\delta H}{\delta\phi} = \frac{\partial\mathcal{H}}{\partial\phi} - \frac{\partial}{\partial x}\left(\frac{\partial\mathcal{H}}{\partial\phi_x}\right) = JS^2 a^2\left(-\phi_{xx}\cos^2\theta + 2\phi_x\theta_x\cos\theta\sin\theta\right). \qquad (2.148b)$$

Using (2.146) then gives the equations of motion,

$$\dot\phi \cos\theta = JSa^2\left(-\theta_{xx} - \phi_x^2 \cos\theta\sin\theta\right) \qquad (2.149a)$$

$$\dot\theta \cos\theta = JSa^2\left(\phi_{xx}\cos^2\theta - 2\phi_x\theta_x\cos\theta\sin\theta\right). \qquad (2.149b)$$

This is a very efficient way to obtain the dynamic equations, provided one already has the continuum Hamiltonian. The extra factor of $\cos\theta$ in (2.149b) has been kept because it makes the RHS a perfect differential, which is often useful for the development of its solutions. It is easy to add other torque terms obtained from additional interactions in the Hamiltonian. Similar dynamic equations, including a local anisotropy and an in-plane applied field, are derived in a different way in chapter 7 for the discussion of magnetic solitons.

Exercise 2.8. For the XY symmetry continuum chain whose Hamiltonian is given in (2.55): (a) Show that the Hamiltonian can be written

$$H_{XY} = J^x S^2 \int \frac{dx}{a}\left\{-1 + (1-\lambda)\sin^2\theta \right.$$

$$\left. + \frac{a^2}{2}\left[\theta_x^2 \sin^2\theta + \phi_x^2 \cos^2\theta + \lambda\theta_x^2\cos^2\theta\right]\right\}, \qquad (2.150)$$

where $\lambda \equiv J^z/J^x < 1$ is an anisotropy constant. (b) Use functional derivatives to determine the equations of motion for $\dot\phi$ and $\dot\theta$.

2.11 Quantum spin mechanics

Many of the studies performed in this text will be concerned with classical spin mechanics. The reason for this is that classical simulations of time-dependence are much easier than a treatment of quantum time evolution, and make reasonable sense at large enough values of quantum spin length S. Even so, it is important to summarize the differences that can be expected for a quantum description, when needed.

The mapping from classical to quantum mechanics (QM) can be obtained from the basic dynamic equation (2.62), for the time derivative of any dynamic quantity. In this section, QM operators will be indicated with a caret, such as $\hat{A}$, to be distinguished from their eigenvalues or classical quantities. For quantum mechanics, a simple procedure is to let the PBs go over into commutators of the operators, divided by $i\hbar$, where $\hbar$ is Planck's constant divided by 2π. As a rule, the replacement is

$$\{A, B\} \implies \frac{1}{i\hbar}\left[\hat{A}, \hat{B}\right] = \frac{1}{i\hbar}\left(\hat{A}\hat{B} - \hat{B}\hat{A}\right). \tag{2.151}$$

The factor of $\hbar$ (the quantum of action) is necessary to preserve the units, as commutation does not change units whereas a PB adds units of inverse action. A pair of conjugate quantum operators $\hat{q}_i, \hat{p}_i$ do not commute, just as the PB of the corresponding classical quantities is non-zero. From a classical PB, $\{q_i, p_i\} = 1$, the corresponding quantum relation is

$$\left[\hat{q}_i, \hat{p}_i\right] = \hat{q}_i\hat{p}_i - \hat{p}_i\hat{q}_i = i\hbar. \tag{2.152}$$

Then by a similar replacement, the quantum equation of motion for an operator is obtained from (2.62), using the Hamiltonian $\hat{H}$,

$$\dot{\hat{A}} = \frac{1}{i\hbar}\left[\hat{A}, \hat{H}\right] + \frac{\partial \hat{A}}{\partial t}. \tag{2.153}$$

Note that this is the dynamic equation of motion in the Heisenberg picture of QM, where the time-dependence is carried in the operators, in contrast to the Schrödinger picture where time-dependence is in the wave functions or state vectors. One can briefly recall the QM derivation of (2.153), based on the time derivative of an expectation value of operator $\hat{A}$ for some state $|\psi\rangle$. An expectation value is defined by a scalar or inner product, using a normalized state vector with $\langle \psi | \psi \rangle = 1$,

$$\left\langle \hat{A} \right\rangle \equiv \langle \psi | \hat{A} | \psi \rangle. \tag{2.154}$$

The time-dependence of the state ket-vector in the Schrödinger picture follows the basic Schrödinger equation (taken as one of the postulates of QM),

$$\frac{\partial}{\partial t}|\psi\rangle = \frac{1}{i\hbar}\hat{H}|\psi\rangle. \tag{2.155}$$

Then for the bra-vector there is the conjugate relation,

$$\frac{\partial}{\partial t}\langle\psi| = \frac{-1}{i\hbar}\langle\psi|\hat{H}^\dagger, \tag{2.156}$$

however, the Hermitian conjugate of the Hamiltonian is the same as the Hamiltonian, so $\hat{H}^\dagger = \hat{H}$. This allows one to find the time derivative of $\langle \hat{A} \rangle$, according to the chain rule,

$$\begin{aligned}
\frac{d}{dt}\langle \hat{A} \rangle &= \left(\frac{\partial}{\partial t}\langle \psi|\right)\hat{A}|\psi\rangle + \langle\psi|\frac{\partial \hat{A}}{\partial t}|\psi\rangle + \langle\psi|\hat{A}\left(\frac{\partial}{\partial t}|\psi\rangle\right) \\
&= \frac{-1}{i\hbar}\langle\psi|\hat{H}\hat{A}|\psi\rangle + \langle\psi|\frac{\partial \hat{A}}{\partial t}|\psi\rangle + \frac{1}{i\hbar}\langle\psi|\hat{A}\hat{H}|\psi\rangle \\
&= \frac{1}{i\hbar}\langle\psi|\left(\hat{A}\hat{H} - \hat{H}\hat{A}\right)|\psi\rangle + \langle\psi|\frac{\partial \hat{A}}{\partial t}|\psi\rangle.
\end{aligned} \quad (2.157)$$

Since the operator $\hat{A}$ was arbitrary, one can see that any expectation values (averaged quantum measurements) will be correctly described in time if $\hat{A}$ follows the equation of motion (2.153), as stated above. In many cases, there will not be an explicit dependence on time and the term $\frac{\partial \hat{A}}{\partial t} = 0$. That will usually be the case when trying to find the time-dependence of spins.

In the absence of explicit time-dependence, it is a simple exercise to show that the formal solution for the time-dependent operator in the Heisenberg picture is

$$\hat{A}(t) = e^{i\hat{H}t/\hbar}\hat{A}(0)e^{-i\hat{H}t/\hbar}. \quad (2.158)$$

$\hat{A}(0)$ is the operator at the initial time $t = 0$. This expression is a direct result of the formal solution of the Schrödinger equation,

$$|\psi(t)\rangle = e^{-i\hat{H}t/\hbar}|\psi(0)\rangle. \quad (2.159)$$

Depending on the types of interactions present in $\hat{H}$, the solution for $\hat{A}(t)$ could either be found from the differential equation (2.153) or from this formal solution (2.158).

For magnetism, we are interested in time-dependent quantum spins $\hat{\mathbf{S}}(t)$. The previously found PB of spin components, equation (2.78), is mapped into the following quantum commutator (which can be verified by several different ways for any quantum angular momentum operators),

$$\left[\hat{S}^x, \hat{S}^y\right] = i\hbar\hat{S}^z. \quad (2.160)$$

This is written for all possible components using the Levi–Civita symbol,

$$\left[\hat{S}^i, \hat{S}^j\right] = i\hbar\varepsilon_{ijk}\hat{S}^k. \quad (2.161)$$

This can be applied to obtain the equation of motion for spins, setting $\hat{A} = \hat{\mathbf{S}}$ in equation (2.153), where the spin has no explicit time-dependence,

$$\dot{\hat{\mathbf{S}}} = \frac{1}{i\hbar}\left[\hat{\mathbf{S}}, \hat{H}\right]. \quad (2.162)$$

This is an equation for each Cartesian spin component. The particular behavior will depend on the Hamiltonian.

2.11.1 Dynamic equations of an individual quantum spin operator

Consider the simple example of spin in a fixed magnetic induction **B**,

$$\hat{H} = -\hat{\mathbf{S}} \cdot \gamma \mathbf{B}. \tag{2.163}$$

The dynamics of $\hat{S}^x$ is then

$$\dot{\hat{S}}^x = \frac{1}{i\hbar}\left[\hat{S}^x, \hat{H}\right] = \frac{-\gamma}{i\hbar}\left[\hat{S}^x, \hat{S}^x B^x + \hat{S}^y B^y + \hat{S}^z B^z\right]$$
$$= \frac{-\gamma}{i\hbar}\left(\left[\hat{S}^x, \hat{S}^y\right]B^y + \left[\hat{S}^x, \hat{S}^z\right]B^z\right). \tag{2.164}$$

When the commutators are evaluated using (2.161), the factors of $i\hbar$ cancel out, leaving

$$\dot{\hat{S}}^x = \gamma\left(\hat{S}^y B^z - \hat{S}^z B^y\right). \tag{2.165}$$

Then it is clear that this is the same cross product found for classical SD,

$$\dot{\hat{\mathbf{S}}} = \gamma \hat{\mathbf{S}} \times \mathbf{B} = \hat{\mathbf{S}} \times \gamma \mathbf{B}. \tag{2.166}$$

Again, the motion is controlled by an effective field, $\mathcal{F} = \gamma \mathbf{B}$. It is somewhat surprising that the quantum dynamics for this simple case is governed by the same equation as classical dynamics. However, the interpretation of the spin needs to be different in quantum theory (it is an operator here). Further, with a different interaction, the quantum equation of motion need not have this simple form.

To see that the form of the QM dynamical equation can be different, consider instead a Hamiltonian involving only uniaxial anisotropy,

$$\hat{H} = -K_A\left(\hat{\mathbf{S}} \cdot \hat{\mathbf{u}}\right)^2 = -K_A\left(\hat{S}^z\right)^2. \tag{2.167}$$

For simplicity, it is assumed that the unit anisotropy axis is $\hat{\mathbf{u}} = \hat{\mathbf{z}}$ (not a QM operator). The $\hat{S}^z$-component commutes with this Hamiltonian and is a conserved operator. For transverse components $\hat{S}^x$ and $\hat{S}^y$, the algebra leading to their equations of motion from (2.162) is

$$\dot{\hat{S}}^x = \frac{-K_A}{i\hbar}\left[\hat{S}^x, \left(\hat{S}^z\right)^2\right] = -\frac{K_A}{i\hbar}\left(\hat{S}^z\left[\hat{S}^x, \hat{S}^z\right] + \left[\hat{S}^x, \hat{S}^z\right]\hat{S}^z\right) = K_A\left(\hat{S}^z\hat{S}^y + \hat{S}^y\hat{S}^z\right) \tag{2.168a}$$

$$\dot{\hat{S}}^y = \frac{-K_A}{i\hbar}\left[\hat{S}^y, \left(\hat{S}^z\right)^2\right] = -\frac{K_A}{i\hbar}\left(\hat{S}^z\left[\hat{S}^y, \hat{S}^z\right] + \left[\hat{S}^y, \hat{S}^z\right]\hat{S}^z\right) = -K_A\left(\hat{S}^z\hat{S}^x + \hat{S}^x\hat{S}^z\right). \tag{2.168b}$$

These are in the form of a symmetrized cross product. One can see that the equations of motion are equivalent to

$$\dot{\hat{\mathbf{S}}} = K_A\left[\left(\hat{\mathbf{S}} \cdot \hat{\mathbf{u}}\right)\left(\hat{\mathbf{S}} \times \hat{\mathbf{u}}\right) + \left(\hat{\mathbf{S}} \times \hat{\mathbf{u}}\right)\left(\hat{\mathbf{S}} \cdot \hat{\mathbf{u}}\right)\right]. \tag{2.169}$$

This is the QM generalization of the classical results in (2.123), obtained by making the operations on the RHS symmetric.

2.11.2 Solution of a quantum spin in a constant field

Suppose a spin finds itself undergoing the dynamics of a constant applied field, Hamiltonian (2.163). For simplicity, we take the magnetic field along $\hat{z}$, that is, $\mathbf{B} = B\hat{z}$. The equations are then

$$\dot{\hat{S}}^x = \gamma B \hat{S}^y \tag{2.170a}$$

$$\dot{\hat{S}}^y = -\gamma B \hat{S}^x. \tag{2.170b}$$

$\hat{S}^z$ commutes with $\hat{H}$ and hence it is conserved. The solution for these operators must be based on their initial states, some operators $\hat{S}^x(0)$ and $\hat{S}^y(0)$, from which the time-evolving operators $\hat{S}^x(t)$ and $\hat{S}^y(t)$ are obtained. As this is a linear equation, the solution can be found straightforwardly. One approach is to introduce a two-component state vector for the instantaneous transverse spin components,

$$\psi(t) = \begin{pmatrix} \hat{S}^x(t) \\ \hat{S}^y(t) \end{pmatrix}. \tag{2.171}$$

The equation of motion for ψ is

$$\dot{\psi} = \gamma B \begin{pmatrix} 0 & 1 \\ -1 & 0 \end{pmatrix} \psi = \gamma B \mathbf{M} \psi. \tag{2.172}$$

One might note that the matrix $\mathbf{M}$ on the RHS happens to be $\mathbf{M} = i\sigma_y$, where σ_y is one of the Pauli spin matrices. For our purpose here, however, we only need to note that it has a simple property:

$$\mathbf{M} = \begin{pmatrix} 0 & 1 \\ -1 & 0 \end{pmatrix}, \quad \mathbf{M}^2 = \begin{pmatrix} -1 & 0 \\ 0 & -1 \end{pmatrix} = -\mathbf{I}, \quad \mathbf{M}^3 = -\mathbf{M}, \quad \mathbf{M}^4 = \mathbf{I}, \tag{2.173}$$

where $\mathbf{I}$ is the 2×2 unit matrix. The equation of motion formally has an exponential solution,

$$\psi(t) = \exp\{\gamma B \mathbf{M} t\} \psi(0). \tag{2.174}$$

Expanding the exponential in a power series,

$$\exp\{\gamma B \mathbf{M} t\} = \left[1 - \frac{(\gamma Bt)^2}{2!} + \frac{(\gamma Bt)^4}{4!} - \cdots\right]\mathbf{I} + \left[\gamma Bt - \frac{(\gamma Bt)^3}{3!} + \frac{(\gamma Bt)^5}{5!} - \cdots\right]\mathbf{M} \tag{2.175}$$

the result for $\psi(t)$ is

$$\psi(t) = [\cos(\gamma Bt)\mathbf{I} + \sin(\gamma Bt)\mathbf{M}]\psi(0). \tag{2.176}$$

Expressed in terms of the original spin components, this is just a continuous rotation of the operators:

$$\hat{S}^x(t) = \cos(\gamma B t)\hat{S}^x(0) + \sin(\gamma B t)\hat{S}^y(0), \tag{2.177a}$$

$$\hat{S}^y(t) = -\sin(\gamma B t)\hat{S}^x(0) + \cos(\gamma B t)\hat{S}^y(0). \tag{2.177b}$$

This rotation with time takes place at the Larmor precession frequency, $\omega = \gamma B$. With time, the x and y spin operators continuously transform from one to the other and back again. Yet, the discreteness of the spin states still does not appear explicitly. There is nothing in these derivations that depends on the quantum spin length s.

2.11.3 How do discrete quantum spin states exhibit themselves?

The effects of quantum spin manifest themselves differently according to the value of dimensionless quantum spin length $s = 1/2, 1, 3/2$, *etc*. The classical limit corresponds to a very large value of s. Small s will mean that the separation of quantized states is obvious. In the simple examples above, the spin length does not appear explicitly. Thus it is instructive to consider how the discreteness of the spin states enters in the dynamics.

In a quantum measurement, say, of a spin component, we know that the rules of QM state that the measurements can only give one of the eigenvalues for the corresponding operator. The discussion above has considered only the expectation values, which predict averages over many physical measurements, that do not need to equal allowed eigenvalues. In a particular measurement, a spin component will instead have a discrete value equal to one of the eigenvalues for a spin component. This is where the effects of discrete QM states become apparent.

QM theory is a probabilistic theory, and the square of a wave function such as $\psi(x) \equiv \langle x|\psi\rangle$ gives a result proportional to the probability of finding the system in the position eigenstate $|x\rangle$. Instead of position eigenstates, we are interested in eigenstates of one of the spin components. For instance, the eigenvalues of $\hat{S}^z$ are $m\hbar$, with quantum number m ranging from $-s$ to $+s$ in increments of 1. This basic rule gives the $2s + 1$ independent states, whether one discusses eigenstates of $\hat{S}^z$, $\hat{S}^x$, or $\hat{S}^y$, for any spin length s. Because there are more possible states for large s and fewer for small s, the distribution of probabilities is where quantum discreteness most strongly manifests itself.

Let the eigenstates of $\hat{S}^z$ be denoted as $|m\rangle$, corresponding to the eigenvalue $m\hbar$. The eigenstates of $\hat{S}^x$ can be denoted $|s^x\rangle$ and those of $\hat{S}^y$ are denoted $|s^y\rangle$, with eigenvalues $s^x\hbar$ and $s^y\hbar$, respectively. Again, the quantum numbers s^x and s^y have the possibility to range from $-s$ to $+s$ in unit increments. Then for some particular instantaneous quantum state $|\psi(t)\rangle$, one can define, for example, the amplitude to be found in the eigenstate $|m\rangle$ for $\hat{S}^z$,

$$C_m \equiv \langle m|\psi(t)\rangle. \tag{2.178}$$

Then the probability to be found in this eigenstate of $\hat{S}^z$ is its square,

$$P_m = |C_m|^2 = |\langle m|\psi(t)\rangle|^2 = \langle \psi(t)|m\rangle\langle m|\psi(t)\rangle. \tag{2.179}$$

By similar reasoning, the probability to be found in some particular eigenstate $|s^x\rangle$ of $\hat{S}^x$ would be

$$P_{s^x} = |\langle s^x|\psi(t)\rangle|^2 = \langle \psi(t)|s^x\rangle\langle s^x|\psi(t)\rangle. \tag{2.180}$$

There would be a similar expression for states of $\hat{S}^y$. Due to the time-dependence of the state vector, this last probability can be written as

$$P_{s^x} = \langle \psi(0)|e^{i\hat{H}t/\hbar}|s^x\rangle\langle s^x|e^{-i\hat{H}t/\hbar}|\psi(0)\rangle. \tag{2.181}$$

One sees that this is an expectation value of a time-dependent projection operator, with the typical time evolution due to $\hat{H}$ for any operator in the Heisenberg picture,

$$\hat{Q}_{s^x}(t) \equiv e^{i\hat{H}t/\hbar} |s^x\rangle\langle s^x| e^{-i\hat{H}t/\hbar}. \tag{2.182}$$

The probability for being in state $|s^x\rangle$ is then

$$P_{s^x}(t) = \langle \psi(0)| \hat{Q}_{s^x}(t) |\psi(0)\rangle. \tag{2.183}$$

Due to the definition for $\hat{Q}_{s^x}$, one can see that it must follow the standard equation of motion for an operator,

$$\dot{\hat{Q}}_{s^x} = \frac{1}{i\hbar}\left[\hat{Q}_{s^x}, \hat{H}\right], \tag{2.184}$$

and of course similar equations for projection operators to give probabilities $P_m(t)$ and $P_{s^y}(t)$. For some simpler models it should be possible to solve the dynamics from this last equation, obtaining the time-dependent projection operator, leading to a discrete set of time-dependent state probabilities. The probability distribution is a discrete set, while the time-dependence can be expected to be described by smoothly changing functions of time. The time-dependence will essentially reflect the precessional motion of spins in, for example, the simple model of a quantum spin in a constant magnetic field, as considered earlier.

For different spin lengths s, there will be different sets of these time-dependent probabilities. For the simplest case of spin-1/2, there are only two states of any spin component, and just two probabilities to keep track of for that component. For spin-1, of course, there will be three probabilities, and so on for higher spin-s. Furthermore, at some point one will want to keep track of transitions between the spin states. For spin-1/2 with only two states, the transitions are very limited in type and quantum discreteness plays an important role. As the value of spin length becomes great, the presence of many possible transitions leads one to a dynamics more similar to that for classical spins (Bohr's correspondence principle) as one approaches the classical limit. Here we have just given a very brief introduction of the most basic aspects for quantum SD.

For spin models with many degrees of freedom (i.e. many spins), the types of quantum transitions to be considered will involve collective dynamics of many spins, such as in quantized spin waves, or *magnons*, or other more complex dynamical excitations. Those topics will be discussed in more detail in a later chapter, when spin wave excitations will be described.

Bibliography

[1] Lorrain P and Corson D 1970 *Electromagnetic Fields and Waves* 2nd edn (San Fransisco, CA: Freeman)
[2] Jackson J D 1999 *Classical Electrodynamics* 3rd edn (New York: Wiley)
[3] Griffiths D J 2004 *Introduction to Quantum Mechanics* 2nd edn (Englewood Cliffs, NJ: Prentice-Hall)
[4] Eisberg R M and Resnick R 1985 *Quantum Physics of Atoms, Molecules, Solids, Nuclei, and Particles* 2nd edn (New York: Wiley)
[5] Jiles D 1991 *Introduction to Magnetism and Magnetic Materials* (London: Chapman and Hall)
[6] García-Cervera C J 1999 Magnetic domains and magnetic domain walls *PhD thesis* New York University
[7] Suessa D, Fidlera J and Schrefl T 2006 *Handbook magn. mater.* **16** 41
[8] García-Cervera C J, Gimbutas Z and Weinan E 2003 Accurate numerical methods for micromagnetics simulations with general geometries *J. Comput. Phys.* **184** 37
[9] Wysin G M 2010 Vortex-in-nanodot potentials in thin circular magnetic dots *J. Phys.: Condens. Matter* **22** 376002
[10] Goldstein H, Poole C P and Safko J L 2001 *Classical Mechanics* 3rd edn (Reading, MA: Addison-Wesley)
[11] Landau L D and Lifshitz E 1935 On the theory of the dispersion of magnetic permeability in ferromagnetic bodies *Phys. Z. Sov.* **8** 153 http://ujp.bitp.kiev.ua/files/journals/53/si/53SI06p.pdf
[12] Gilbert T L 1955 A Lagrangian formulation of the gyromagnetic equation of the magnetic field *Phys. Rev.* **100** 1235
[13] Gilbert T L 2004 A phenomenological theory of damping in ferromagnetic materials *IEEE Trans. Magn.* **40** 3443
[14] Kosevich A M, Ivanov B A and Kovalev A S 1990 Magnetic solitons *Phys. Rep.* **194** 117–238

IOP Publishing

Magnetic Excitations and Geometric Confinement
Theory and simulations
Gary Matthew Wysin

Chapter 3

Demagnetization effects in thin magnets

A magnet could have isotropic exchange couplings and no crystal field anisotropies, but still exhibit a geometric anisotropy. This is due to the internal magnetic field $\mathbf{H}_M$ inside a macroscopic magnet that is generated by the magnetization $\mathbf{M}$ (the dipole moment per unit volume) of the magnet itself. This is a basic magnetostatics problem. Some ideas about incorporating these long-range dipolar effects in 2D magnets are presented here. The discussion is based on a continuum description of the magnet and the field, although it can be connected to an alternative analysis that considers the superposition of the many fields from a multitude of individual magnetic dipoles. The techniques here are especially useful in the realistic simulation of magnetization dynamics in nanomagnets, for example.

3.1 The magnetostatics problem in a finite magnet

In solving magnetostatics, and even electrodynamics, there are no magnetic monopoles. The implications of this were pointed out in chapter 1, and now more details of how that leads to internal demagnetization fields are presented. We consider the demagnetization of a magnetic sample with some internal magnetization profile $\mathbf{M}$. We begin with a general discussion that does not assume a thin magnet. Later the results will be specialized to the case of thin magnets, obtaining procedures that are advantageous for simulations of thin systems.

It has been mentioned in chapter 1 that the internal magnetization generates its own magnetic field, $\mathbf{H}_M$, which itself contains an energy and tends to work to reduce the original magnetization. This energy and the associated field have different properties depending on the geometry of the sample. Here we consider the mathematics of finding the demagnetization field $\mathbf{H}_M$ in general and for specific geometries that might be applied in simulations. The magnetostatic theory used here is described at an elementary level in chapter 9 of the book by Lorrain and Corson [1] and at a more advanced level in chapter 5 of [2].

The magnetic induction **B** obeys a Gauss law where there is no fundamental source charge:

$$\nabla \cdot \mathbf{B} = 0, \quad \mathbf{B} = \mu_0(\mathbf{H}_M + \mathbf{M}). \tag{3.1}$$

Here **M** is the dipole moment per unit volume (magnetization) and $\mathbf{H}_M$ is the demagnetization field that is directly caused by **M**. A situation is considered where no field enters the sample from an outside source. The demagnetization field determines a part of the magnetic energy in the system, sometimes called *magnetostatic energy*, according to a volume integral,

$$E_M = -\frac{1}{2}\mu_0 \int dV\ \mathbf{M} \cdot \mathbf{H}_M. \tag{3.2}$$

Normally, the interaction action energy density of **M** in a field is just $-\mathbf{M} \cdot \mu_0 \mathbf{H}$. However, demagnetization energy is equivalent to the summation of dipole–dipole interactions in the magnet, see (3.19) later in this chapter. There is an extra factor of 1/2 on the demagnetization energy density, because of the fact that $\mathbf{H}_M$ is produced by and proportional to **M**, and the 1/2 removes the double counting of the dipole–dipole pair interactions that are counted twice in the above integral.

The starting point for determining $\mathbf{H}_M$ is the relation for divergence-free magnetic induction **B**, which can be rearranged as

$$\nabla \cdot \mathbf{H}_M = -\nabla \cdot \mathbf{M}. \tag{3.3}$$

Then magnetic field $\mathbf{H}_M$ will be generated by an effective magnetic charge density, defined as

$$\rho_M = -\nabla \cdot \mathbf{M}. \tag{3.4}$$

In a situation where there are no free electric currents (current density, $\mathbf{J}_e = 0$), the magnetic field obeys $\nabla \times \mathbf{H}_M = 0$, implying that it can be found from a magnetic potential Φ_M,

$$\mathbf{H}_M = -\nabla \Phi_M \tag{3.5}$$

which leads to the magnetostatic Poisson equation presented earlier in chapter 1,

$$\nabla^2 \Phi_M = -\rho_M. \tag{3.6}$$

The magnetic charge density ρ_M (not a true monopolar density!) acts as the source for lines of $\mathbf{H}_M$.

3.1.1 Green's functions for magnetostatics

If there happened to be a single effective magnetic charge q_M at the origin, its charge density would be $\rho_M = q_M \delta(\mathbf{r})$, where $\delta(\mathbf{r})$ is a 3D Dirac delta function. Then from electrostatics it is known that the solution of (3.6) for the potential in 3D would be the well-known potential of a point charge, just as in electrostatics,

$$\Phi_M(\mathbf{r}) = \frac{q_M}{4\pi |\mathbf{r}|}. \tag{3.7}$$

The Green's function being the solution for a charge of unit magnitude at some source point $\mathbf{r}'$, is obtained directly from this. For 3D this scalar Green's function, denoted as $g_0(\mathbf{r}, \mathbf{r}')$, solves the Poisson equation with a delta function source,

$$\nabla^2 g_0(\mathbf{r}, \mathbf{r}') = -\delta(\mathbf{r} - \mathbf{r}'). \tag{3.8}$$

The solution is

$$g_0(\mathbf{r}, \mathbf{r}') = \frac{1}{4\pi |\mathbf{r} - \mathbf{r}'|} \tag{3.9}$$

which actually depends only on the distance between source point $\mathbf{r}'$ and field point $\mathbf{r}$. Then the solution of the Poisson equation for the potential in an unbounded system with an arbitrary magnetic charge density (i.e. summing over the effects of many point charges) is:

$$\Phi_M(\mathbf{r}) = \int d^3 r' \, g_0(\mathbf{r}, \mathbf{r}') \rho_M(\mathbf{r}') = \int d^3 r' \frac{\rho_M(\mathbf{r}')}{4\pi |\mathbf{r} - \mathbf{r}'|} = \int d^3 r' \frac{-\nabla' \cdot \mathbf{M}(\mathbf{r}')}{4\pi |\mathbf{r} - \mathbf{r}'|}. \tag{3.10}$$

This gives the solution directly from ρ_M, or equivalently, from the divergence of $\mathbf{M}$.

The final step is to take the gradient of Φ_M to obtain the demagnetization field. That gives a result which can be written using another Green's function, which is a vector operator $\mathbf{g}(\mathbf{r}, \mathbf{r}')$:

$$\mathbf{H}_M(\mathbf{r}) = -\nabla \Phi_M(\mathbf{r}) = \int d^3 r' \, \mathbf{g}(\mathbf{r}, \mathbf{r}') \rho_M(\mathbf{r}'). \tag{3.11}$$

This new Green's operator is just the negative gradient of g_0,

$$\mathbf{g}(\mathbf{r}, \mathbf{r}') = -\nabla g_0(\mathbf{r}, \mathbf{r}') = \frac{\mathbf{r} - \mathbf{r}'}{4\pi |\mathbf{r} - \mathbf{r}'|^3}. \tag{3.12}$$

Note that ∇ involves only the derivative with respect to $\mathbf{r}$, and not $\mathbf{r}'$. The Green function $\mathbf{g}(\mathbf{r}, \mathbf{r}')$ is a vector directed along $\mathbf{r} - \mathbf{r}'$. It acts on a source charge element $dq_M = \rho_M(\mathbf{r}') d^3 r'$ and represents the increment of magnetic field generated by that source element. Equations (3.11) and (3.12) are the same as the basic rule for generation of electric field in electrostatics. Equation (3.11) offers one good approach for determining the demagnetization field in numerical simulations, once the effective magnetic charge density has been determined.

In some situations it may be convenient to obtain Φ_M directly from $\mathbf{M}$, avoiding the extra step of first finding ρ_M. One can perform an integration by parts in (3.10) using vector calculus manipulations, letting the gradient act instead on the Green's function g_0:

$$\Phi_M(\mathbf{r}) = \int d^3 r' \, \nabla'\left(\frac{1}{4\pi |\mathbf{r} - \mathbf{r}'|}\right) \cdot \mathbf{M}(\mathbf{r}'). \tag{3.13}$$

There also would have been a surface term, but by taking that surface outside of the magnet, its contribution is zero. Letting the gradient act, this is

$$\Phi_M(\mathbf{r}) = \int d^3r' \left(\frac{\mathbf{r} - \mathbf{r}'}{4\pi |\mathbf{r} - \mathbf{r}'|^3} \right) \cdot \mathbf{M}(\mathbf{r}'). \tag{3.14}$$

Again, the same Green's function $\mathbf{g} = (g^x, g^y, g^z)$ defined in (3.12) is encountered here. $\mathbf{g}$ points radially outward from the source point and acts directly on the components of $\mathbf{M}$, but it only produces Φ_M,

$$\Phi_M(\mathbf{r}) = \int d^3r' \, \mathbf{g}(\mathbf{r} - \mathbf{r}') \cdot \mathbf{M}(\mathbf{r}'). \tag{3.15}$$

One sees that $\mathbf{g}$ only has a radial component along a unit vector $(\mathbf{r} - \mathbf{r}')/|\mathbf{r} - \mathbf{r}'|$, with magnitude

$$g_r(\mathbf{r}, \mathbf{r}') = -\frac{\partial}{\partial |\mathbf{r} - \mathbf{r}'|} g_0(\mathbf{r} - \mathbf{r}') = \frac{1}{4\pi |\mathbf{r} - \mathbf{r}'|^2}. \tag{3.16}$$

Using $\mathbf{g}$ to calculate the field without going through the intermediate step of obtaining the charge density also has a conceptual advantage. It does not require the calculation of ρ_M, which could be considered something impossible to measure directly.

In this last form (3.15), the Green's operator $\mathbf{g}$ acts directly on the magnetization. It is interesting to realize that equation (3.14) is simply a representation of the effective magnetostatic potential of a collection of dipoles. In (3.14), each infinitesimal dipole is $d\mathbf{p}' = d^3r' \, \mathbf{M}(\mathbf{r}')$, and it generates a response according to the displacement from its position, $\mathbf{r} - \mathbf{r}'$. We could finally take the gradient with respect to $\mathbf{r}$ to obtain $\mathbf{H}_M$, taking care to avoid the possibility of a singular integrand. The potential of a point dipole $\mathbf{p}$ at the origin (put $\mathbf{r}' = 0$) is

$$\Phi(\mathbf{r}) = \frac{\mathbf{r} \cdot \mathbf{p}}{4\pi r^3} = \mathbf{g}(\mathbf{r}, 0) \cdot \mathbf{p}. \tag{3.17}$$

The gradient leads to the well-known expression for the field caused by that point dipole:

$$\mathbf{H} = -\nabla \Phi = \frac{1}{4\pi r^3} [3\hat{\mathbf{r}}(\hat{\mathbf{r}} \cdot \mathbf{p}) - \mathbf{p}]. \tag{3.18}$$

Performing these same operations for expression (3.14), one obtains this alternative way to express the total demagnetization field,

$$\mathbf{H}_M(\mathbf{r}) = \int d^3r' \left\{ \frac{3[(\mathbf{r} - \mathbf{r}')(\mathbf{r} - \mathbf{r}') \cdot] - |\mathbf{r} - \mathbf{r}'|^2}{4\pi |\mathbf{r} - \mathbf{r}'|^5} \right\} \mathbf{M}(\mathbf{r}'). \tag{3.19}$$

Note that the first contribution points along $\mathbf{r} - \mathbf{r}'$ while the second is in the same direction as $\mathbf{M}(\mathbf{r}')$. These results show that the demagnetization field $\mathbf{H}_M$ is equivalent to the total magnetic field of the magnetic dipoles in the sample. Thus, taking into account magnetic demagnetization is the same as taking into account the long-range dipolar interactions. Note that this last expression may not be the most efficient to use in for a uniformly magnetized sample. In that case, the expression (3.10) is a better approach.

Expression (3.19) can be seen to be equivalent to the following operation,

$$\mathbf{H}_M(\mathbf{r}) = -\nabla \int d^3 r' \, \mathbf{g}(\mathbf{r}, \mathbf{r}') \cdot \mathbf{M}(\mathbf{r}') = \int d^3 r' \sum_{j=1}^{3} [-\nabla g^j(\mathbf{r}, \mathbf{r}')] M^j(\mathbf{r}'). \quad (3.20)$$

This involves yet another Green's operator G, which acting on **M** leads directly to $\mathbf{H}_M$. It is a 3 × 3 tensor operator, whose elements are given by mixed gradients of the original g_0 Green's operator.

$$\mathsf{G}^{ij}(\mathbf{r}, \mathbf{r}') = -\partial_i g^j(\mathbf{r}, \mathbf{r}') = \partial_i \partial_j g_0(\mathbf{r}, \mathbf{r}'). \quad (3.21)$$

Then once this has been determined, the components of the demagnetization field can also be expressed as

$$H_M^i(\mathbf{r}) = \int d^3 r' \sum_{j=1}^{3} \mathsf{G}^{ij}(\mathbf{r}, \mathbf{r}') M^j(\mathbf{r}'). \quad (3.22)$$

Later this approach will be applied in the determination of $\mathbf{H}_M$ in thin-film magnets. Note that it can be very powerful because it generates the demagnetization field directly from the magnetization, without the need for finding ρ_M first. Furthermore, the tensor Green's operator is only calculated once at the beginning of a simulation, and repeatedly re-used.

3.1.2 Uniform magnetization and effective magnetic surface charges

In a magnet that is uniformly magnetized with **M** = constant, the internal charge density ρ_M is zero *within* the magnet. So how can there be any $\mathbf{H}_M$ if equation (3.10) is evaluated? The answer is that at the surface of the magnet, there is a discontinuous change in the magnetization, which suddenly goes from some non-zero value inside the magnet to zero on the outside. This change corresponds to an effective volume charge density that can be represented as a delta function on the surface. Stated otherwise, the Gauss law used on (3.4) (the divergence theorem applied to a pillbox at the surface) will tell us that there is a local surface charge density, σ_M given by

$$\sigma_M = \mathbf{M} \cdot \hat{\mathbf{n}}, \quad (3.23)$$

where $\hat{\mathbf{n}}$ is the outward normal vector from the surface of the magnet. This surface charge density is greatly responsible for generating the demagnetization field even more so that the volume charge density, because the spatial variations in **M** within the volume are usually much less drastic than the sudden change at the surface.

In a case where there is surface charge density, the element of effective charge is

$$dq_M = \sigma_M dA = \mathbf{M} \cdot \hat{\mathbf{n}} \, dA, \quad (3.24)$$

where dA is a surface area element. The contribution to the potential only from surface charge can be written as

$$\Phi_M^S(\mathbf{r}) = \int_A \mathrm{d}A'\, g_0(\mathbf{r}-\mathbf{r}')\sigma_M(\mathbf{r}') = \int_A \mathrm{d}A'\, \frac{\mathbf{M}(\mathbf{r}')\cdot\hat{\mathbf{n}}'}{4\pi\,|\mathbf{r}-\mathbf{r}'|}. \quad (3.25)$$

Indeed, in any real problem, if **M** is present, then surface charge density is present, and this result should be combined with the fields from volume charge density, equation (3.10), to obtain the total demagnetization field. The general solution for the potential can always be written as

$$\Phi_M(\mathbf{r}) = \int \mathrm{d}^3 r'\, \frac{\rho_M(\mathbf{r}')}{4\pi\,|\mathbf{r}-\mathbf{r}'|} \quad (3.26)$$

as long as $\mathrm{d}^3 r'\rho_M(\mathbf{r}')$ includes both volume and surface charges in the sample.

Next the demagnetization effect is analyzed in detail for some simple geometries. A cylinder of circular cross section is considered due to its symmetry and it being a common shape for magnets. Then a cylinder of square cross section is analyzed, because that shape can be used as an element in numerical simulations. Finally the demagnetization effects in a thin system will be analyzed, due to the application in quasi-2D magnetic models.

3.2 The magnetic field inside a cylindrical magnet

First, consider a magnet of length L with its long axis the z-axis. The ends of the cylinder lie at $z = \pm\delta$, so that the length is $L = 2\delta$, and $z = 0$ is at the middle of the cylinder. The cross section is initially taken to be a circle of radius R. There is no special assumption about the size of the radius R compared to the cylinder length L. In principle, the cylinder could have any arbitrary magnetization distribution, however, it is more practical to consider a uniform magnetization either along its axis (longitudinal magnetization) or perpendicular to the axis (transverse magnetization).

3.2.1 Longitudinal magnetization M^z

Initially, suppose the cylinder is magnetized in the z-direction, that is, along its axis of symmetry. Then $\mathbf{M} = M^z\hat{\mathbf{z}}$, for any point within the magnet. There will be no volume magnetic charge density, however, there will be surface charge densities at the ends, $\sigma = \pm M^z$ at $z = \pm\delta$, respectively. The top end ($z = +\delta$) has positive magnetic charge (it is a source of $\mathbf{H}_M$), the bottom end ($z = -\delta$) has negative magnetic charge (it is a sink of $\mathbf{H}_M$).

To find the potential at an observer point $\mathbf{r} = (x, y, z)$ inside the magnet, consider the positive source charges at $\mathbf{r}' = (x', y', \delta)$ and the negative source charges at $\mathbf{r}' = (x', y', -\delta)$. From the Green function integral (3.26) over charge density, one has now only surface integrals on the ends, as in (3.25),

$$\Phi_M(x, y, z) = \frac{M^z}{4\pi}\int \mathrm{d}x'\mathrm{d}y' \left\{ \frac{1}{\sqrt{r_{xy}^2 + (z-\delta)^2}} - \frac{1}{\sqrt{r_{xy}^2 + (z+\delta)^2}} \right\}. \quad (3.27)$$

To save space, the notation

$$r_{xy}^2 \equiv (x - x')^2 + (y - y')^2 \tag{3.28}$$

is used for the squared displacement within the xy-plane. The integral is over the cross section of the cylinder's end caps. Using cylindrical coordinates $\mathbf{r} = (\rho, \theta, z)$ and $\mathbf{r}' = (\rho', \theta', z')$, this squared displacement is

$$r_{xy}^2 = \rho^2 + \rho'^2 - 2\rho\rho' \cos(\theta - \theta'). \tag{3.29}$$

Then one can evaluate the above integrals in terms of Bessel functions, if desired.

The field down the axis of the cylinder can be found from simpler expressions, putting $x = y = 0$, and thus $\rho = 0$. This avoids tedious mathematics while providing the basic physical solution. By symmetry there will only be a longitudinal field component, H^z, which is a function of z. In this case there is dependence only on $\rho' = \sqrt{x'^2 + y'^2}$ and only a radial integration is needed, because the angular integration is trivial,

$$\Phi_M(z) = \frac{M^z}{4\pi} \int_0^{2\pi} d\theta' \int_0^R \rho' d\rho' \left\{ \frac{1}{\sqrt{\rho'^2 + (z - \delta)^2}} - \frac{1}{\sqrt{\rho'^2 + (z + \delta)^2}} \right\}. \tag{3.30}$$

The integration is quite simple, due to the circular cross section. There results

$$\Phi_M(z) = \frac{M^z}{2} \left[\sqrt{\rho'^2 + (z - \delta)^2} - \sqrt{\rho'^2 + (z + \delta)^2} \right]_0^R$$

$$= \frac{M^z}{2} \left[\sqrt{R^2 + (z - \delta)^2} - \sqrt{R^2 + (z + \delta)^2} - |z - \delta| + |z + \delta| \right]. \tag{3.31}$$

The resulting field $H_M^z(z)$ should be an even function of z, meaning that $\Phi_M(z)$ is expected to be an odd function. Indeed, for any z inside the cylinder, $-\delta < z < \delta$, this result becomes

$$\Phi_M(z) = \frac{M^z}{2} \left[\sqrt{R^2 + (z - \delta)^2} - \sqrt{R^2 + (z + \delta)^2} + 2z \right]. \tag{3.32}$$

Then the field on the axis of the cylinder is found quickly,

$$H_M^z(z) = -\frac{\partial \Phi_M}{\partial z} = -M^z \left[1 + \frac{1}{2} \frac{z - \delta}{\sqrt{R^2 + (z - \delta)^2}} - \frac{1}{2} \frac{z + \delta}{\sqrt{R^2 + (z + \delta)^2}} \right]. \tag{3.33}$$

Note the somewhat surprising result. At $z = 0$, the last terms equal each other and combine. With the cylinder length $L = 2\delta$, this gives

$$H_M^z(0) = -M^z \left(1 - \frac{\delta}{\sqrt{R^2 + \delta^2}} \right) = -M^z \left(1 - \frac{L/2}{\sqrt{R^2 + L^2/4}} \right). \tag{3.34}$$

The field points *opposite* to **M**, which is why it is *demagnetization*. Further, its strength depends on the aspect ratio of the cylinder, L/R. Note the limiting behaviors:

$$H_M^z(0) \approx \begin{cases} -\dfrac{2R^2}{L^2} M^z & \text{for } L \gg R \\ -\left(1 - \dfrac{L}{2R}\right) M^z & \text{for } L \ll R. \end{cases} \quad (3.35)$$

When the cylinder is long and thin ($L \gg R$), the longitudinal field at its center becomes very small. The center point is far from the end caps that possess non-zero ρ_M. On the other hand, if the cylinder is short and wide ($L \ll R$), the longitudinal field at its center is maximized, nearly equal to the strength of its magnetization. This corresponds to the strongest demagnetization that can take place.

Putting $z = \pm\delta = \pm L/2$, the field at either end cap is

$$H_M^z\left(\pm\frac{L}{2}\right) = -M^z\left(1 - \frac{L/2}{\sqrt{R^2 + L^2}}\right) \approx \begin{cases} -\left(\dfrac{1}{2} + \dfrac{R^2}{4L^2}\right) M^z & \text{for } L \gg R \\ -\left(1 - \dfrac{L}{2R}\right) M^z & \text{for } L \ll R. \end{cases} \quad (3.36)$$

For the long thin cylinder ($L \gg R$), the demagnetization field at the ends becomes half of that in the center. For the short wide cylinder ($L \ll R$), The value at the ends is nearly the same as at the center; the demagnetization field is nearly independent of z.

Note further the reason for the name, demagnetization. In some situations, **M** could be generated by the action of an externally applied field, according to $\mathbf{M} \sim \chi \mathbf{H}_{\text{ext}}$, where $\chi > 0$ is a paramagnetic susceptibility. Then the total magnetic field in the sample would be the combination of applied field and this demagnetization field. They oppose each other, hence, the demagnetization field tends to reduce the internal effect of the applied field. It tends to prevent the applied field from entering the sample.

Average of H_M^z

It is useful to know the average magnetic field in the sample. This can be achieved easily for the field on the axis of the circular cylinder. The average over z is:

$$\overline{H_M^z} = \frac{1}{L} \int_{-\delta}^{\delta} dz \, H_M^z(z) = -M^z - \frac{M^z}{2L} \int_{-\delta}^{\delta} dz \left[\frac{z-\delta}{\sqrt{R^2 + (z-\delta)^2}} - \frac{z+\delta}{\sqrt{R^2 + (z+\delta)^2}}\right]$$

$$= -M^z - \frac{M^z}{2L}\left[\sqrt{R^2 + (z-\delta)^2} - \sqrt{R^2 + (z+\delta)^2}\right]_{-\delta}^{\delta}$$

$$= -\frac{M^z}{L}\left(L + R - \sqrt{R^2 + L^2}\right) \approx \begin{cases} -M^z \dfrac{R}{L} & \text{for } L \gg R \\ -M^z\left(1 - \dfrac{L}{2R}\right) & \text{for } L \ll R. \end{cases} \quad (3.37)$$

The average has a behavior similar to that for the value at $z = 0$. This is summarized by saying that there is a longitudinal demagnetization factor, N_z, defined from

$$N_z = \frac{1}{L}\left(L + R - \sqrt{L^2 + R^2}\right), \quad \overline{H_M^z} = -N_z M^z. \tag{3.38}$$

The behavior of N_z with aspect ratio R/L is shown in figure 3.1. When N_z takes a larger value, there is a strong longitudinal demagnetization field $\overline{H_M^z}$ within the magnet, and this corresponds to a large magnetostatic energy E_M as defined in (3.2). On the other hand, when N_z is small, the longitudinal demagnetization field is small, and there is much less magnetostatic energy. One sees the limiting behaviors,

$$N_z \approx \begin{cases} \dfrac{R}{L} & \text{for } L \gg R \\ \left(1 - \dfrac{L}{2R}\right) & \text{for } L \ll R. \end{cases} \tag{3.39}$$

This shows that a cylinder magnetized along its axis will have the weakest demagnetization field and corresponding magnetostatic energy if it is long and thin.

3.2.2 Transverse magnetization M^x

Suppose that the magnet is magnetized only along the x-direction. It is simplest to look at the case of a circular cylinder. This is a magnetization along a diameter of the cross section. Any diameter is equivalent, so a fixed M^y would be the same problem. Taking $\mathbf{M} = M^x \hat{\mathbf{x}}$, this will generate a surface charge distribution on the curved surface of the cylinder,

$$\sigma_M(\theta) = \mathbf{M} \cdot \hat{\mathbf{n}} = M^x \cos\theta, \tag{3.40}$$

where θ is the angular position of a point on the surface, measured from the x-axis.

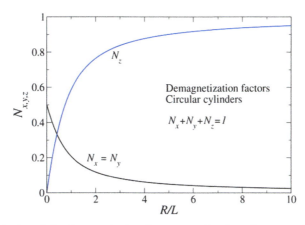

Figure 3.1. The behavior of the demagnetization factors as a function of aspect ratio for right circular cylinders, based on equations (3.38) and (3.48), for the averaged field on the cylinder axis. The point where $N_x = N_y = N_z = 1/3$ can be shown to be where $R/L = 5/12$.

This produces positive charges on one side ($x > 0$) and negative charges on the other side ($x < 0$), hence, it is easy to see that the field $\mathbf{H}_M$ will point generally towards $-\hat{\mathbf{x}}$. This now gives an integral expression for the potential, using cylindrical coordinates, with $\rho' = R$,

$$\Phi_M(\mathbf{r}) = \frac{M^x}{4\pi} \int_0^{2\pi} R\, d\theta' \int_{-\delta}^{\delta} dz' \frac{\cos\theta'}{\sqrt{\rho^2 + R^2 - 2\rho R \cos(\theta - \theta') + (z - z')^2}}. \quad (3.41)$$

There will be only small components of $\mathbf{H}_M$ along the y- and z-directions. Concentrate instead on the x-component. Performing the derivative to obtain H_M^x, using points along the x-axis, $\theta = 0$, gives

$$H_M^x(z) = -\frac{\partial \Phi_M}{\partial x} = \frac{M^x}{4\pi} \int_0^{2\pi} R\, d\theta' \int_{-\delta}^{\delta} dz' \frac{(x - R\cos\theta')\cos\theta'}{\left[\rho^2 + R^2 - 2\rho R\cos\theta' + (z - z')^2\right]^{3/2}}. \quad (3.42)$$

Evaluating on the axis of the cylinder, $x = y = \rho = 0$, leaves only one term in the numerator and simplifies the denominator. Angular integration of the factor $\cos^2\theta'$ leads to a factor of $\frac{1}{2}(2\pi)$, resulting in

$$H_M^x(z) = \frac{-M^x}{4} \int_{-\delta}^{\delta} dz' \frac{R^2}{\left[R^2 + (z - z')^2\right]^{3/2}}. \quad (3.43)$$

The integral is found by letting $z' - z = R\tan\phi$, then $dz' = R\sec^2\phi$, giving the indefinite integration,

$$\int \frac{R^2 dz'}{[R^2 + (z - z')^2]^{3/2}} = \int \frac{d\phi \sec^2\phi}{\sec^3\phi} = \int d\phi \cos\phi = \sin\phi = \frac{z' - z}{\sqrt{R^2 + (z' - z)^2}}. \quad (3.44)$$

This results in

$$H_M^x(z) = \frac{-M_x}{4}\left[\frac{\delta - z}{\sqrt{R^2 + (\delta - z)^2}} + \frac{\delta + z}{\sqrt{R^2 + (\delta + z)^2}}\right]. \quad (3.45)$$

The order of magnitude is checked by looking on the cylinder axis, $z = 0$, where

$$H_M^x(0) = \frac{-M^x}{2}\frac{\delta}{\sqrt{R^2 + \delta^2}} = \frac{-M^x}{4}\frac{L}{\sqrt{R^2 + L^2/4}} \approx \begin{cases} \dfrac{-M^x}{2} & \text{for } L \gg R \\ \dfrac{-M^x}{4}\dfrac{L}{R} & \text{for } L \ll R. \end{cases} \quad (3.46)$$

For long thin cylinders, the transverse demagnetization field just becomes half of the magnetization (in magnitude). For wide or flat cylinders (disk-shaped), the transverse demagnetization field approaches zero; the field strongly avoids entering into the side of a wide cylinder.

The average field $\overline{H_M^x}$ can be found by integration over position z. The result is

$$\overline{H_M^x} = \frac{1}{L} \int_{-\delta}^{\delta} dz \left\{ \frac{-M^x}{4} \left[\frac{\delta - z}{\sqrt{R^2 + (\delta - z)^2}} + \frac{\delta + z}{\sqrt{R^2 + (\delta + z)^2}} \right] \right\}$$

$$= \frac{-M^x}{4L} \left[-\sqrt{R^2 + (\delta - z)^2} + \sqrt{R^2 + (\delta + z)^2} \right]_{-\delta}^{\delta}$$

$$= \frac{-M^x}{2L} \left[\sqrt{R^2 + L^2} - R \right]. \tag{3.47}$$

This implies the transverse demagnetization factors (although not averaged over x),

$$N_x = N_y = \frac{1}{2L} \left[\sqrt{L^2 + R^2} - R \right]. \tag{3.48}$$

It can be stressed that in the limit of a long skinny cylinder ($R \ll L$), this demagnetization factor becomes $N_x \approx \frac{1}{2}$, and the longitudinal factor becomes $N_z \approx 0$.

Compared with the result (3.38) for N_z, the three demagnetization factors along the three Cartesian directions are seen to satisfy a well-known symmetry relation,

$$N_x + N_y + N_z = 1. \tag{3.49}$$

Considering the equivalence of any diameters across the cylinder, we have by symmetry, $N_x = N_y$. Then this relation can always be used to find N_x once N_z is known, or vice versa. Chapter 2 of the book by Jiles [3] has further discussion of demagnetization in cylindrical and other geometries. For instance, $N_x = N_y = N_z = 1/3$ for a sphere.

N_x is compared with N_z in figure 3.1, as functions of aspect ratio R/L. Notably, for long thin cylinders with $R \ll L$, the longitudinal demagnetization factor becomes $N_z \to 0$ while the transverse factors are $N_x = N_y \to \frac{1}{2}$. The state of lowest demagnetization energy E_M, equation (3.2), will be present when **M** is in the direction with the smallest demagnetization factor. For a long thin cylinder (or any needle-like object), then, the lowest magnetostatic energy corresponds to having **M** along the axis. A thin cylinder with **M** along a diameter would still be possible, but that state is of higher energy and may be unstable. The magnetization will have the energetic tendency to become and remain axial.

At the other limit, for a flat cylinder or disk, $R \gg L$, we have $N_z \to 1$ and $N_x = N_y \to 0$. The flat cylinder has a large demagnetization field if magnetized axially, and a weak demagnetization field for **M** in the plane of the disk. The states of lower magnetostatic energy will be those in which **M** remains in the plane of the disk, with very small M^z.

Generally, the greatest demagnetization effects will always take place through the shortest dimension of an object [4]. This causes the energetic preference for **M** to align with the longer dimensions of an object, if only magnetostatic energy is being considered.

3.3 Demagnetization fields in thin-film magnets

In micromagnetics simulations of the Landau–Lifshitz equations, it is important to take into account demagnetization effects. This could be for a sample of any shape. Such a system could instantaneously have a non-uniform magnetization $\mathbf{M}(\mathbf{r})$. In this book we have a particular interest in systems that are quasi-2D, such as thin-film soft magnets. Thus the discussion now is focused on magnetic systems that are effectively planar in shape, with a very small thickness L (along the z-direction) compared to the transverse dimensions of some order R in the x- and y-directions. The magnet could have some arbitrary shape in the xy-plane. We discuss here how micromagnetics is adapted [5, 6] to this quasi-3D problem. It is based on finding the Green's functions needed for quasi-2D analysis, which is described by Huang [7].

For simulations, the magnet is partitioned into computation cells. This could be achieved in different ways, but here we take tetragonal cells of size $a \times a \times L$, where a is a small displacement in the x- and y-directions and L is along z, through the thin direction of the sample. An exact solution for the demagnetization factor of rectangular prisms has been given by Aharoni [8]; here we consider an approximate analysis of a similar problem. Because this analysis is being developed to apply to simulations, it is supposed that a and L are small enough, such that the magnetization within a given cell is nearly uniform. One wants to estimate the field generated by magnetization $\mathbf{M}_i$ of cell i. That field can be considered in two regions. The first is the *interior* region within that cell. That problem is similar to the analysis performed earlier for a circular cylinder. The second region is the *exterior* region outside that cell. This is the external field problem. This field will characterize the interaction of the chosen source cell with all other cells at different points in the material. That represents the long-range dipolar interactions in the whole magnet.

A source cell is assumed to have uniform magnetization $\mathbf{M}_i$, which could be directed arbitrarily. In micromagnetics, it is commonly assumed that the magnitude of the magnetization in a cell is the saturation magnetization M_s, although this assumption could be relaxed for a more general analysis, if needed. The cells are supposed to be infinitesimal compared to the whole system. A chosen source cell is imagined to be a column of dipole density (at source points $\mathbf{r}'$), which produces the magnetic potential $\Phi_M(\mathbf{r})$ both within itself and exterior to itself. That potential will give the averaged field $\mathbf{H}_M(\mathbf{r})$ in the observer position or cell, which is then used to evolve the magnetization dynamics and evaluate magnetostatic energy.

For the thin system, each cell has surface magnetic charges on the free faces at $z = \pm \delta$, where $L = 2\delta$. These surface magnetic charges (in any cell in the system) are responsible for producing H_M^z in a chosen cell. There is a contribution from the other cells (exterior problem) and from the cell where H_M^z is being found (interior problem). A cell also contains transverse fields H_M^x, H_M^y, that are produced by the volume magnetic charge density ρ_M within the rest of the system (exterior problem), and by any magnetic charge density within the cell under consideration (interior problem). The interior problems will be considered as zero-radius limits of the general magnetic problem.

3.3.1 The longitudinal field H_M^z in a thin magnet

Start with a continuum description, where the magnetization at 2D source point (x', y') generates the field at 2D observation point (x, y). Because the system is planar, H_M^z comes only from the surface charges caused by M^z at $z' = \pm\delta$. This general case is nearly the same as what was calculated for a longitudinally magnetized circular cylinder. Start from the potential at $\mathbf{r}$ due to the sum over sources at $\mathbf{r}'$, equation (3.25) or (3.27). We use a 2D displacement notation,

$$\mathbf{r}_{xy} = (x - x', y - y'), \quad r_{xy}^2 = (x - x')^2 + (y - y')^2. \tag{3.50}$$

Then the potential is

$$\Phi_M(\mathbf{r}) = \frac{1}{4\pi} \int dx' dy' \left\{ \frac{1}{\sqrt{r_{xy}^2 + (z - \delta)^2}} - \frac{1}{\sqrt{r_{xy}^2 + (z + \delta)^2}} \right\} M^z(x', y'). \tag{3.51}$$

There is no integration over z'; this just uses the charges at the surfaces $z' = \pm\delta$. But now the (x', y')-dependence of M^z is retained. The field H_M^z can be found using the gradient with respect to z', and then averaged over z in the observer position:

$$H_M^z(x, y) = \langle H_M^z(x, y, z) \rangle = -\frac{1}{L} \int_{-\delta}^{\delta} dz \frac{d\Phi}{dz} = \frac{-1}{L}[\Phi_M(\delta) - \Phi_M(-\delta)]. \tag{3.52}$$

This gives a fairly simple result,

$$H_M^z(x, y) = \int dx' \, dy' \, \mathsf{G}^{zz}(r_{xy}) M^z(x', y'). \tag{3.53}$$

It involves the convolution of a longitudinal Green's function G^{zz} (element of a tensor) with the out-of-plane magnetization component. The Green's function depends only on the radial separation between source and observer points,

$$\mathsf{G}^{zz}(r_{xy}) = \frac{-1}{2\pi L}\left[\frac{1}{r_{xy}} - \frac{1}{\sqrt{r_{xy}^2 + L^2}}\right]. \tag{3.54}$$

Note that G^{zz} is always negative, which means it leads to the usual negative demagnetization effect through the thin direction of a 2D system. It diverges as $r_{xy} \to 0$, a behavior normally seen in Green's functions. It is good to keep in mind, however, that this divergence as r_{xy}^{-1} will be canceled out by the area element, if it is written in polar coordinates.

To check the validity of this result, apply it to a cylinder of radius R and length L with uniform M^z. Along its axis where $x = y = 0$, the integration in polar coordinates is

$$H_M^z(0, 0) = \frac{-M^z}{2\pi L} \int_0^{2\pi} d\theta' \int_0^R d\rho' \, \rho' \left[\frac{1}{\rho'} - \frac{1}{\sqrt{\rho'^2 + L^2}}\right]$$

$$= \frac{-M^z}{L}\left[L + R - \sqrt{L^2 + R^2}\right], \tag{3.55}$$

which recovers the same demagnetization factor N_z found earlier in (3.38).

When applied in simulations with discrete cells, the point $r_{xy} \to 0$ needs to be treated with care. That limit is the self-interaction or the interior field problem. Also, the computation cells are squares, yet this Green's function has circular symmetry. These problems need to be addressed if this G^{zz} is to be applied in simulations.

If $r_{xy} = 0$, the Green's function represents the field in a cell caused by the magnetization in that same cell. Thus, it makes sense to avoid a zero radius by averaging G^{zz} over a circle of radius r_0 whose area is the same as the cell area, a^2. To be specific, this averaged $G^{zz}(0)$ is found by taking $\rho = 0$, but summing over (x', y') in a circle of radius r_0 found as

$$\pi r_0^2 = a^2 \implies r_0 = \sqrt{\frac{1}{\pi}} a. \tag{3.56}$$

This circle is shown around the central cell in figure 3.2. The average is

$$\mathsf{G}_0^{zz} \equiv \langle \mathsf{G}^{zz}(0) \rangle = \frac{1}{a^2} \int_0^{r_0} \rho' d\rho' \frac{-1}{L} \left[\frac{1}{\rho'} - \frac{1}{\sqrt{\rho'^2 + L^2}} \right] = \frac{-1}{La^2} \left[L + r_0 - \sqrt{L^2 + r_0^2} \right]. \tag{3.57}$$

G_0^{zz} found this way is the same as $-N_z$ for a circular cylinder of length L and radius $r_0 = a/\sqrt{\pi}$. This will be applied to an area element $dx'\, dy'$ of size a^2, hence the factor a^2 will be canceled out. The rest is the longitudinal demagnetization factor already encountered. This can be used to redefine $\mathsf{G}^{zz}(0)$.

A similar averaging can be applied for other cells close to the source. This helps to eliminate some roughness due to using square cross section cells for a circularly symmetric function. For example, relative to one cell, there are four neighboring cells at distance a and another four cells at a distance $\sqrt{2}\, a$. The Green's function for these radii can be smoothed out by averaging between radii chosen by area mapping.

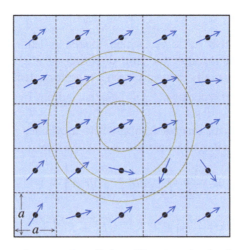

Figure 3.2. The grid of $a \times a$ micromagnetics cells for a 2D system, showing the circles of radii r_0, r_1, r_2 for averaging of the Green's function $\mathsf{G}^{zz}(r_{xy})$ near the origin.

For instance, define the radius r_1 of a circle whose area equals that of a central cell plus the four first neighbors,

$$\pi r_1^2 = a^2 + 4a^2 \implies r_1 = \sqrt{\frac{5}{\pi}}\, a. \tag{3.58}$$

Define the next smoothing radius r_2, using a circle containing the central cell, the first four neighbors and the four second neighbors,

$$\pi r_2^2 = a^2 + 4a^2 + 4a^2 \implies r_2 = \sqrt{\frac{9}{\pi}}\, a. \tag{3.59}$$

The radii r_1 and r_2 can be seen in figure 3.2. Then at radius $r_{xy} = a$, the Green function is replaced by its value averaged between r_0 and r_1, which is found to be

$$\mathsf{G}^{zz}(a) \longrightarrow \langle \mathsf{G}^{zz} \rangle_{r_0}^{r_1} = \frac{1}{4La^2}\left[\sqrt{L^2 + r_1^2} - \sqrt{L^2 + r_0^2} - (r_1 - r_0)\right]. \tag{3.60}$$

A similar expression is applied to replace the Green function at radius $r_{xy} = \sqrt{2}\,a$,

$$\mathsf{G}^{zz}(\sqrt{2}\,a) \longrightarrow \langle \mathsf{G}^{zz} \rangle_{r_1}^{r_2} = \frac{1}{4La^2}\left[\sqrt{L^2 + r_2^2} - \sqrt{L^2 + r_1^2} - (r_2 - r_1)\right]. \tag{3.61}$$

Then with this correction near the origin, the finite-element calculation of the longitudinal demagnetization field proceeds from a sum over source cells at locations (x_i, y_i), as a discrete convolution,

$$H_{\mathrm{M}}^z(x, y) = \sum_i \mathsf{G}^{zz}(x - x_i, y - y_i) \cdot M^z(x_i, y_i). \tag{3.62}$$

This gives the averaged value in a cell. In actual practice, this might be evaluated using a fast Fourier transform (FFT), to obtain the most speed, a process explained later in this chapter.

3.3.2 The transverse field (H_{M}^x, H_{M}^y) in a thin magnet

The transverse field within the plane of the magnet could be found from the effective charge density ρ_{M}, equation (3.26). Instead, it is found first here from the superposition of dipole fields from the source cells, equation (3.14). The total field $\Phi_{\mathrm{M}}(\mathbf{r})$ is a superposition of a dipole field from each layer (at fixed z') in a source cell. One supposes that a dipole element is centered in the center of the source cell, at height z'. Thus, expression (3.14) is applied to obtain the potential,

$$\Phi_{\mathrm{M}}(\mathbf{r}) = \int d^3 r' \frac{\mathbf{r} - \mathbf{r}'}{4\pi |\mathbf{r} - \mathbf{r}'|^3} \cdot \mathbf{M}(\mathbf{r}') = \int dx'dy'dz' \frac{(x - x')M^x + (y - y')M^y}{4\pi \left[r_{xy}^2 + (z - z')^2\right]^{3/2}}. \tag{3.63}$$

The integration over (x', y') is left to become the sum over source cells. The source point (x', y') for this cell is set to the center of the cell, with M^x and M^y taken as constants within a chosen cell. After integration over z', an average will be taken over the observation height z.

Doing the integration over z' gives (see the integral in (3.44)):

$$\Phi_M(x, y, z) = \int dx' dy' \left[\frac{(z'-z)}{4\pi r_{xy}^2 \sqrt{r_{xy}^2 + (z-z')^2}} \right]_{-\delta}^{\delta} \mathbf{r}_{xy} \cdot \mathbf{M}(x', y')$$

$$= \int \frac{dx' dy'}{4\pi r_{xy}^2} \left[\frac{\delta - z}{\sqrt{r_{xy}^2 + (\delta - z)^2}} + \frac{\delta + z}{\sqrt{r_{xy}^2 + (\delta + z)^2}} \right] \mathbf{r}_{xy} \cdot \mathbf{M}(x', y'). \quad (3.64)$$

The notation $\mathbf{r}_{xy} = (x - x', y - y')$ implies only an in-plane displacement. Now one can find the transverse field components, taking an averaging over z. The order in which this is performed makes no difference. Take first the average over z, defined by

$$\Phi_M(x, y) = \langle \Phi_M(x, y, z) \rangle = \frac{1}{L} \int_{-\delta}^{\delta} dz \, \Phi_M(x, y, z). \quad (3.65)$$

This gives

$$\Phi_M(x, y) = \int \frac{dx' dy'}{4\pi L r_{xy}^2} \left[-\sqrt{r_{xy}^2 + (z - \delta)^2} + \sqrt{r_{xy}^2 + (\delta + z)^2} \right]_{-\delta}^{\delta} \mathbf{r}_{xy} \cdot \mathbf{M}(x', y')$$

$$= \int \frac{dx' dy'}{2\pi L r_{xy}^2} \left[\sqrt{r_{xy}^2 + L^2} - r_{xy} \right] \mathbf{r}_{xy} \cdot \mathbf{M}(x', y'). \quad (3.66)$$

Note that this determines the vector Green's function to give the potential from the 2D source $\mathbf{M}(x', y')$. It acts on a vector source, hence it is the vector Green's function $\mathbf{g}(r_{xy})$, whose direction is radially outward from the source point. This 2D Green's function can be written:

$$\mathbf{g}(r_{xy}) = \frac{1}{2\pi L} \left(\sqrt{1 + \frac{L^2}{r_{xy}^2}} - 1 \right) \frac{\mathbf{r}_{xy}}{|\mathbf{r}_{xy}|}. \quad (3.67)$$

With that definition, the expression for the potential it produces averaged in a cell is

$$\Phi_M(x, y) = \int dx' dy' \, \mathbf{g}(r_{xy}) \cdot \mathbf{M}(x', y'). \quad (3.68)$$

The last factor in $\mathbf{g}$ is a radial unit vector. Then this Green's function has only a radial component, depending only on radius,

$$g_r(r_{xy}) = \frac{1}{2\pi L} \left(\sqrt{1 + \frac{L^2}{r_{xy}^2}} - 1 \right). \quad (3.69)$$

For the 2D problem, this must be the negative radial gradient of the Green's function g_0 that produces Φ from the volume charge density ρ, as in 3D equations

(3.12) and (3.16). So it is interesting to find the effective g_0 associated with this **g**. It is obtained from an indefinite integration,

$$g_0(r) = -\int dr \; g_r(r) = \frac{-1}{2\pi L} \int dr \left[\sqrt{1 + \frac{L^2}{r^2}} - 1 \right]. \tag{3.70}$$

This is aided by using

$$\sinh \phi = \frac{L}{r}, \quad \sqrt{1 + \frac{L^2}{r^2}} = \cosh \phi, \quad \cosh \phi \, d\phi = \frac{-L}{r^2} dr = \frac{-\sinh^2 \phi}{L} dr. \tag{3.71}$$

Then the integral with the square root is

$$-\int dr \sqrt{1 + \frac{L^2}{r^2}} = L \int d\phi \, \frac{\cosh^2 \phi}{\sinh^2 \phi} = L \int d\phi (1 + \operatorname{csch}^2 \phi) = L(\phi - \coth \phi)$$

$$= L \sinh^{-1} \frac{L}{r} - \sqrt{r^2 + L^2}. \tag{3.72}$$

Combining all terms produces

$$g_0(r_{xy}) = \frac{1}{2\pi L} \left(L \sinh^{-1} \frac{L}{r_{xy}} - \sqrt{r_{xy}^2 + L^2} + r_{xy} \right). \tag{3.73}$$

This is a well-known expression for the effective 2D in-plane Green function to be applied on $\rho(x', y')$, for finding the magnetic potential $\Phi_M(x, y)$. It acts as

$$\Phi_M(x, y) = \int dx' dy' \; g_0(x - x', y - y') \rho(x', y'). \tag{3.74}$$

Now the gradient of Φ_M in equation (3.66) can be taken to obtain the averaged field. To achieve this, recall that $r_{xy}^2 = (x - x')^2 + (y - y')^2$. Find first $H_M^x = -\partial \Phi_M / \partial x$, then H_M^y will be obtained by switching the xy-indices.

$$H_M^x = \int \frac{dx' dy'}{2\pi L} \left\{ \frac{\left[r_{xy} - \sqrt{r_{xy}^2 + L^2} \right] M^x}{r_{xy}^2} - (\mathbf{r}_{xy} \cdot \mathbf{M}) \frac{2(x - x')}{r_{xy}^4} \left[r_{xy} - \sqrt{r_{xy}^2 + L^2} \right] \right.$$

$$\left. + \frac{(\mathbf{r}_{xy} \cdot \mathbf{M})(x - x')}{r_{xy}^2} \left[\frac{1}{r_{xy}} - \frac{1}{\sqrt{r_{xy}^2 + L^2}} \right] \right\}. \tag{3.75}$$

This has contributions from both M^x and M^y. There is nothing coming from M^z, because of the assumption of a thin magnet. The part proportional to M^x depends on $G^{xx} M^x$, where

$$G^{xx} = \frac{r_{xy} - \sqrt{r_{xy}^2 + L^2}}{2\pi L \, r_{xy}^4} \left[r_{xy}^2 - 2(x - x')^2 - (x - x')^2 \frac{r_{xy}}{\sqrt{r_{xy}^2 + L^2}} \right]. \tag{3.76}$$

This can be rearranged in various ways, one of which is

$$G^{xx} = \frac{\sqrt{r_{xy}^2 + L^2} - r_{xy}}{2\pi L r_{xy}^4} \left\{ (x - x')^2 \left[1 + \frac{r_{xy}}{\sqrt{r_{xy}^2 + L^2}} \right] - (y - y')^2 \right\}. \quad (3.77)$$

Similarly, there is a term proportional to M_y depending on $G^{xy} M^y$, where

$$G^{xy} = \frac{\sqrt{r_{xy}^2 + L^2} - r_{xy}}{2\pi L r_{xy}^4} \left\{ (x - x')(y - y') \left[2 + \frac{r_{xy}}{\sqrt{r_{xy}^2 + L^2}} \right] \right\}. \quad (3.78)$$

Then these can be applied to give the x-component of the demagnetization field as

$$H_M^x(x, y) = \int dx' dy' [G^{xx} M^x(x', y') + G^{xy} M^y(x', y')]. \quad (3.79)$$

There is a similar expression to give H_M^y.

The expressions above for G^{xx} and G^{xy} might not be the best for numerical evaluations, because of the difference of two factors, $\sqrt{r_{xy}^2 + L^2} - r_{xy}$, each of which could become large. This could give a *subtractive cancellation* error, especially when r_{xy} is larger than L. If one is concerned about that, then it is good to realize the problem can be avoided by rewriting the prefactor with the help of the identity,

$$\left(\sqrt{r_{xy}^2 + L^2} - r_{xy} \right) \left(\sqrt{r_{xy}^2 + L^2} + r_{xy} \right) = L^2. \quad (3.80)$$

Then the leading factor on G^{xx} is rewritten using

$$\frac{\sqrt{r_{xy}^2 + L^2} - r_{xy}}{2\pi L r_{xy}^4} \rightarrow \frac{L}{2\pi r_{xy}^4 \left(\sqrt{r_{xy}^2 + L^2} + r_{xy} \right)}. \quad (3.81)$$

This new expression only involves addition, which is a more stable computational operation due to its lesser relative error. Using this leads to the simpler expressions,

$$G^{xx} = \frac{L}{2\pi r_{xy}^4} \left\{ \frac{(x - x')^2}{\sqrt{r_{xy}^2 + L^2}} - \frac{(y - y')^2}{\sqrt{r_{xy}^2 + L^2} + r_{xy}} \right\} \quad (3.82a)$$

$$G^{xy} = \frac{L}{2\pi r_{xy}^4} \left\{ \frac{1}{\sqrt{r_{xy}^2 + L^2}} + \frac{1}{\sqrt{r_{xy}^2 + L^2} + r_{xy}} \right\} (x - x')(y - y'). \quad (3.82b)$$

But note: see (3.114a) below for a correction to this that includes the self-demagnetization effect, which has been left out here.

> **Exercise 3.1.** Starting from the vector Green's function **g** in (3.67), take its appropriate gradients as in (3.21) to arrive at the 2D G^{xx} and G^{xy}, verifying equations (3.82a) and (3.82b).

3.3.3 Using ρ_M to obtain (H_M^x, H_M^y) in a thin magnet

If we go back to the expression (3.74) that gives the magnetic potential,

$$\Phi_M(x, y) = \int dx'dy'\ g_0(x - x', y - y')\rho(x', y'), \qquad (3.83)$$

then the transverse demagnetization field can be found by taking its in-plane gradient, which only acts on g_0 inside the integrand and gives back **g**:

$$\mathbf{H}_M = -\int dx'dy'\ \nabla g_0(x - x', y - y')\rho(x', y')$$

$$= \int dx'dy'\ \mathbf{g}(x - x', y - y')\rho(x', y'). \qquad (3.84)$$

The vector Green's function is the same as that defined in (3.67). Either it can be used to obtain Φ_M from **M** or to obtain $\mathbf{H}_M$ from ρ_M. Again, that is assuming that one is willing to go through the extra step of generating ρ_M from **M**.

3.3.4 The total field (H_M^x, H_M^y, H_M^z) in a thin magnet

These results show that there is a matrix Green's function that produces all the components of the demagnetization field, H_M^α, $\alpha = x, y, z$. It is understood that the fields are those averaged over the z-coordinate in a cell.

$$H_M^\alpha(x, y) = \int dx'dy' \sum_{\beta=x,y} G^{\alpha\beta}(x - x', y - y')M^\beta(x', y'). \qquad (3.85)$$

This expression applies to the full 3D field, when used with the 3D magnetization (on a 2D grid) and G^{zz} that was derived for the longitudinal magnetization component. The other remaining elements of the Green's matrix are obtained by swapping the xy-indices,

$$G^{yy} = \frac{L}{2\pi r_{xy}^4}\left\{\frac{(y-y')^2}{\sqrt{r_{xy}^2 + L^2}} - \frac{(x-x')^2}{\sqrt{r_{xy}^2 + L^2} + r_{xy}}\right\} \qquad (3.86a)$$

$$G^{yx} = G^{xy}. \qquad (3.86b)$$

Interestingly, these components of **G** reduce to those appropriate to give the usual far field of a point dipole, as viewed only within the xy-plane. Expansion for $r_{xy} \gg L$ leads to

$$G^{\alpha\beta}(\mathbf{r}_{xy}) = \frac{L}{4\pi r_{xy}^5} \begin{pmatrix} 2\tilde{x}^2 - \tilde{y}^2 & 3\tilde{x}\tilde{y} & 0 \\ 3\tilde{x}\tilde{y} & 2\tilde{y}^2 - \tilde{x}^2 & 0 \\ 0 & 0 & -r_{xy}^2 \end{pmatrix}. \qquad (3.87)$$

Tildes indicate the difference of source and observer points, i.e. $\tilde{x} = x - x'$, $\tilde{y} = y - y'$. Compare with the full 3D field of a point dipole $\mathbf{p} = (p^x, p^y, p^z)$ at the origin, from (3.18),

$$\mathbf{H} = \begin{pmatrix} H^x \\ H^y \\ H^z \end{pmatrix} = \frac{1}{4\pi r^5} \begin{pmatrix} 2x^2 - y^2 - z^2 & 3xy & 3xz \\ 3xy & 2y^2 - x^2 - z^2 & 3yz \\ 3xz & 3yz & 2z^2 - x^2 - y^2 \end{pmatrix} \begin{pmatrix} p^x \\ p^y \\ p^z \end{pmatrix}. \qquad (3.88)$$

In this latter expression, putting $z = 0$ for a thin magnet then recovers the large-r_{xy} limit of the 3D matrix $G^{\alpha\beta}$, accounting for the fact that $\mathbf{p}$ will be replaced by $L\, dx'dy'\, \mathbf{M}$.

This last result is good for the field outside a source cell. It appears divergent in the limit $r_{xy} \to 0$, and an averaging procedure over a circle with the same area as the central cell also diverges. This analysis has not correctly accounted for self-demagnetization of the chosen source cell. Roughly, this is an extra field of

$$\mathbf{H}_{local} = -N_x(M^x\hat{\mathbf{x}} + M^y\hat{\mathbf{y}}). \qquad (3.89)$$

Nominally N_x is a number near 1/2, but it could be smaller than this, taking, for example, the value expected for the transverse demagnetization of long thin circular cylinders as discussed earlier, equation (3.48). In the following paragraphs the transverse self-demagnetization of an individual cell of square cross section is analyzed.

3.3.5 Transverse self demagnetization—calculation of $H_M^x(0)$

Suppose a chosen cell of size $a \times a \times L$ is magnetized with a uniform value of M^x. Look at the calculation of the demagnetization field within this particular cell, and in particular its average value.

For coordinates, while z is in the range $-\delta \leqslant z \leqslant \delta$ where $L = 2\delta$, x (and y) will be taken in a symmetric range $-\Delta \leqslant x \leqslant \Delta$, where $a = 2\Delta$. Then at the faces $x = \pm\Delta$, the uniform M^x generates surface charge densities $\sigma_M = \pm M^x$. The potential at an observer point $\mathbf{r}$, due to only the surface charge density $\sigma_M = +M_x$ on the face at $x = +\Delta$, is

$$\Phi^+(\mathbf{r}) = \frac{M^x}{4\pi} \int_{-\Delta}^{\Delta} dy' \int_{-\delta}^{\delta} dz' \frac{1}{\sqrt{(x-\Delta)^2 + (y-y')^2 + (z-z')^2}}. \qquad (3.90)$$

There is another term coming from the negative charge density on the left face at $x = -\Delta$. When the H_M^x field is found and averaged over the x-axis, there is a result like that in (3.52),

$$\overline{H_M^x} = -\frac{1}{a}\int_{-\Delta}^{\Delta} dx\, \frac{\partial \Phi}{\partial x} = \frac{-1}{a}(\Phi_M(\Delta) - \Phi_M(-\Delta)). \tag{3.91}$$

Then the field averaged over the x-axis, including both positive and negative surface charges, is:

$$\overline{H_M^x} = \frac{-2M^x}{4\pi a} \int_{-\Delta}^{\Delta} dy' \int_{-\delta}^{\delta} dz' \left\{ \frac{1}{\sqrt{(y-y')^2 + (z-z')^2}} \right.$$

$$\left. - \frac{1}{\sqrt{a^2 + (y-y')^2 + (z-z')^2}} \right\} \tag{3.92}$$

This is just the double of the contribution from one face's charge. The integrals can be performed exactly, see the short paper by Aharoni [8], however, the result is rather complicated. For the purpose here, we are most interested in the case where $\Delta \ll \delta$, that of a long thin cylinder of square cross section.

An approximate analysis can be made. First, find the field without any averaging. From the positive charges, there is

$$H_M^{x+}(\mathbf{r}) = -\frac{\partial \Phi_M}{\partial x} = \frac{M^x}{4\pi} \int_{-\Delta}^{\Delta} dy' \int_{-\delta}^{\delta} dz' \frac{x - \Delta}{\left[(x-\Delta)^2 + (y-y')^2 + (z-z')^2\right]^{3/2}}. \tag{3.93}$$

The integration over z' was performed in equation (3.44). Using that result here, with the effective $R^2 \equiv (x-\Delta)^2 + (y-y')^2$, there results

$$H_M^{x+}(\mathbf{r}) = \frac{M^x(x-\Delta)}{4\pi} \int_{-\Delta}^{\Delta} dy' \left[\frac{z' - z}{R^2\sqrt{R^2 + (z'-z)^2}} \right]_{z'=-\delta}^{z'=\delta}$$

$$= \frac{M^x(x-\Delta)}{4\pi} \int_{-\Delta}^{\Delta} \frac{dy'}{R^2} \left[\frac{\delta - z}{\sqrt{R^2 + (\delta-z)^2}} + \frac{\delta + z}{\sqrt{R^2 + (\delta+z)^2}} \right]. \tag{3.94}$$

Now find the field in the middle cross section of the cylinder, $z = 0$. Further, consider the long thin limit, $L \gg a$, which means also $\delta \gg R$. Then this contribution is

$$H_M^{x+}(x, y) \approx \frac{M^x(x-\Delta)}{2\pi} \int_{-\Delta}^{\Delta} \frac{dy'}{(x-\Delta)^2 + (y'-y)^2}. \tag{3.95}$$

The integral is an inverse tangent. Use the fact that $x < \Delta$ for any point inside the system:

$$H_M^{x+}(x, y) \approx \frac{M^x}{2\pi} \frac{x-\Delta}{|x-\Delta|} \left[\tan^{-1} \frac{y'-y}{|x-\Delta|} \right]_{-\Delta}^{\Delta} = \frac{-M^x}{2\pi} \left[\tan^{-1} \frac{\Delta - y}{\Delta - x} + \tan^{-1} \frac{\Delta + y}{\Delta - x} \right]. \tag{3.96}$$

That is the contribution from the positive charge at $x = +\Delta$. The contribution from the negative charge at $x = -\Delta$ is similar, but with $\Delta \to -\Delta$ inside R, and with opposite sign:

$$H_M^{x-}(x, y) \approx \frac{-M^x}{2\pi} \frac{x + \Delta}{|x + \Delta|} \left[\tan^{-1} \frac{y' - y}{|x + \Delta|}\right]_{-\Delta}^{\Delta} = \frac{-M^x}{2\pi} \left[\tan^{-1} \frac{\Delta - y}{\Delta + x} + \tan^{-1} \frac{\Delta + y}{\Delta + x}\right]. \tag{3.97}$$

So the total in the long thin cylinder approximation is

$$H_M^x(x, y) \approx \frac{-M^x}{2\pi} \left(\tan^{-1} \frac{\Delta - y}{\Delta - x} + \tan^{-1} \frac{\Delta + y}{\Delta - x} + \tan^{-1} \frac{\Delta - y}{\Delta + x} + \tan^{-1} \frac{\Delta + y}{\Delta + x}\right). \tag{3.98}$$

At the center of the cell ($x = y = 0$), the inverse tangents are all $\frac{\pi}{4}$, which gives the field,

$$H_M^x(0) \approx -\frac{1}{2} M^x. \tag{3.99}$$

So the demagnetization factor for this point is $N_x = \frac{1}{2}$, as we would expect. Near any corner, say, $x = y = \Delta - \epsilon$ with $\epsilon \ll \Delta$, there is, surprisingly, the same value:

$$H_M^x(\Delta, \Delta) \approx \frac{-M^x}{2\pi} \left(\tan^{-1} \frac{\epsilon}{\epsilon} + \tan^{-1} \frac{2\Delta}{\epsilon} + \tan^{-1} \frac{\epsilon}{2\Delta} + \tan^{-1} \frac{2\Delta}{2\Delta}\right) \approx -\frac{1}{2} M^x. \tag{3.100}$$

Trying instead a point along the x-axis, say, $x = \Delta - \epsilon$, $y = 0$, one has

$$H_M^x(\Delta, 0) \approx \frac{-M^x}{2\pi} \left(\tan^{-1} \frac{\Delta}{\epsilon} + \tan^{-1} \frac{\Delta}{\epsilon} + \tan^{-1} \frac{\Delta}{2\Delta} + \tan^{-1} \frac{\Delta}{2\Delta}\right) = -\left(\frac{1}{2} + 0.148\right) M^x. \tag{3.101}$$

On the other hand, at the point $x = 0$, $y = \Delta - \epsilon$:

$$H_M^x(0, \Delta) \approx \frac{-M^x}{2\pi} \left(\tan^{-1} \frac{\epsilon}{\Delta} + \tan^{-1} \frac{2\Delta}{\Delta} + \tan^{-1} \frac{\epsilon}{\Delta} + \tan^{-1} \frac{2\Delta}{\Delta}\right) = -\left(\frac{1}{2} - 0.148\right) M^x. \tag{3.102}$$

This suggests that there is only little variation within the cross section. Figure 3.3 displays how the field varies within the cell versus x, at different y, all with $z = 0$ in the long thin approximation. Then one can expect that the demagnetization factor when averaging over different points, is to fair approximation,

$$N_x \approx \frac{1}{2}. \tag{3.103}$$

This suggest that the matrix $G^{\alpha\beta}(r_{xy})$ needs to include a value at zero radius of this order for the diagonal terms,

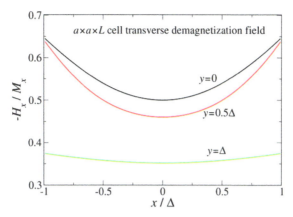

Figure 3.3. The behavior of the transverse demagnetization field for a rectangular cell of dimensions $a \times a \times L$, with $a \ll L$, as a function of position in the cross section. Note that $a = 2\Delta$, and all curves show the field at the center cross section of the cell, $z = 0$.

$$\mathsf{G}^{xx}(0) = \mathsf{G}^{yy}(0) \approx -\frac{1}{2}. \qquad (3.104)$$

Next we consider just a slight modification from this limiting value.

3.3.6 About the transverse demagnetization for computations

In actual application in calculations using computation cells of size $a \times a \times L$ for thin magnets, the transverse self-demagnetization effect must be included. We know that $N_x \approx \frac{1}{2}$ for long thin cells. But what if the cells are not so long and thin? As a slight improvement on that, one can use as a reasonable approximation the transverse demagnetization factor found for a circular cylinder. Instead of simply using $N_x = \frac{1}{2}$, a reasonable improvement is to apply

$$N_x = \frac{1}{2L}\left(\sqrt{L^2 + r_0^2} - r_0\right). \qquad (3.105)$$

The radius r_0 can be chosen by assuming a circle with the same area as the square, as was done earlier for averaging of $\mathsf{G}^{zz}(0)$,

$$A = a^2 = \pi r_0^2, \quad \Longrightarrow \quad r_0 = \frac{a}{\sqrt{\pi}}. \qquad (3.106)$$

It should be good also for the transverse demagnetization. It means we take

$$\mathsf{G}^{xx}(0) = \mathsf{G}^{yy}(0) = -N_x = \frac{-1}{2L}\left(\sqrt{L^2 + \frac{a^2}{\pi}} - \frac{a}{\sqrt{\pi}}\right). \qquad (3.107)$$

For calculations, it is important to avoid subtraction cancellations, if possible. Another way to write the transverse demagnetization factor is by eliminating the subtraction:

$$N_x = \frac{1}{2L}\left(\sqrt{L^2 + r_0^2} - r_0\right) \times \frac{\sqrt{L^2 + r_0^2} + r_0}{\sqrt{L^2 + r_0^2} + r_0} = \frac{1}{2}\frac{L}{\sqrt{L^2 + r_0^2} + r_0}. \quad (3.108)$$

Then inserting the effective radius $r_0 = a/\sqrt{\pi}$ gives the formula to use in calculations:

$$\mathsf{G}^{xx}(0) = \mathsf{G}^{yy}(0) = -N_x = \frac{-\frac{1}{2}\sqrt{\pi}\, L}{\sqrt{\pi L^2 + a^2} + a}. \quad (3.109)$$

Similarly, the longitudinal factor at the origin can be written in an alternative way:

$$N_z = \frac{1}{L}\left(L + r_0 - \sqrt{L^2 + r_0^2}\right) = \frac{2r_0}{L + r_0 + \sqrt{L^2 + r_0^2}}. \quad (3.110)$$

Then with the effective radius, the actual form in the calculations can be

$$\mathsf{G}^{zz}(0) = -N_z = \frac{-2a}{\sqrt{\pi}L + a + \sqrt{\pi L^2 + a^2}}. \quad (3.111)$$

3.3.7 Example: applying G^{xx} to a disk with uniform M^x

To check that G^{xx} works correctly, here it is applied to a circular disk of radius R, thickness L, with uniform $\mathbf{M} = M^x\hat{\mathbf{x}}$. Looking at the central point in the disk, $x = y = 0$, applying (3.77) with a transformation to polar coordinates leads to

$$H_M^x(0) = \int d^2r'\, \mathsf{G}^{xx}\, M^x = \int d^2r'\, \frac{\sqrt{r_{xy}^2 + L^2} - r_{xy}}{2\pi L\, r_{xy}^4}\left\{(x')^2\left[1 + \frac{r_{xy}}{\sqrt{r_{xy}^2 + L^2}}\right] - (y')^2\right\} M^x$$

$$= \frac{M^x}{2\pi L}\int_0^{2\pi} d\theta' \int_0^R r'dr'\, \frac{\sqrt{r'^2 + L^2} - r'}{r'^2}\left\{\cos^2\theta'\left[1 + \frac{r'}{\sqrt{r'^2 + L^2}}\right] - \sin^2\theta'\right\}$$

$$= \frac{M^x}{2L}\int_0^R dr'\, \frac{\sqrt{r'^2 + L^2} - r'}{\sqrt{r'^2 + L^2}} = \frac{M^x}{2L}\int_0^R dr'\left[1 - \frac{r'}{\sqrt{r'^2 + L^2}}\right]. \quad (3.112)$$

The angular integrations were evaluated based on $\langle\cos^2\theta'\rangle = \langle\sin^2\theta'\rangle = 1/2$, which led to a cancellation. It is interesting to notice at this point that the radial integrand is actually not singular at $r' \to 0$. Finally the radial integration gives a somewhat familiar looking result,

$$H_M^x(0) = \frac{M^x}{2L}\left(L + R - \sqrt{L^2 + R^2}\right). \quad (3.113)$$

However, this leads to a demagnetization factor N_x with the *wrong sign*. Also, the factor linear in L should not be present inside the parenthesis. This result is incorrect!

Although the function used for G^{xx} in (3.77) or in (3.82a) is not singular when combined with the area element d^2r', it does not include the self-demagnetization of a chosen cell. In the limit of very long thin cells the self-demagnetization factor is $N_x = 1/2$, which is what should be included in this continuum limit. Thus, the way to correct G^{xx} is to add a delta function contribution at $r_{xy} = 0$ that represents the self-demagnetization. This means the continuum G^{xx} and G^{yy} should be written

$$G^{xx} = \frac{L}{2\pi r_{xy}^4}\left\{\frac{(x-x')^2}{\sqrt{r_{xy}^2+L^2}} - \frac{(y-y')^2}{\sqrt{r_{xy}^2+L^2}+r_{xy}}\right\} - \frac{1}{2}\delta(r_{xy}) \qquad (3.114a)$$

$$G^{yy} = \frac{L}{2\pi r_{xy}^4}\left\{\frac{(y-y')^2}{\sqrt{r_{xy}^2+L^2}} - \frac{(x-x')^2}{\sqrt{r_{xy}^2+L^2}+r_{xy}}\right\} - \frac{1}{2}\delta(r_{xy}). \qquad (3.114b)$$

No delta function is needed in $G^{xy} = G^{yx}$. With this modification, our result for $H_M^x(0)$ will be shifted by the value $-M^x/2$, leading to

$$H_M^x(0) = -\frac{M^x}{2L}\left(\sqrt{L^2+R^2} - R\right). \qquad (3.115)$$

This now has the correct sign and it agrees exactly with the result found for the transverse demagnetization factor of a cylinder, (3.48). Thus it is very important to take into account not only the far dipole fields, but also the self-demagnetization field, even for this simple example of a uniformly magnetized cylinder.

3.4 Use of fast Fourier transforms

The numerical calculation of $\mathbf{H}_M$ is essentially the problem of finding long-range dipolar fields. On a grid of N cells this is a slow process because at each of the N cells, the field contributions from $N-1$ other cells must be calculated and summed. The computation then requires on the order of N^2 operations. By direct summation there can be some speedup if the programmer makes a table of displacement vectors and the associated Green's function for that displacement. Even so, once a system with more than 100 sites is considered, there will be considerable slowdown. Any simulation of the magnetization dynamics needs to have $\mathbf{H}_M$ recalculated at every time step, so it is crucial to calculate it efficiently.

The use of FFTs is the basic acceleration method for long-range dipolar field calculations. It was applied to vortex states in circular ferromagnets by Sasaki and Matsubara [9], as an example. It is based ultimately on the fact that any calculation of the field can be expressed as a *convolution* of a Green's function with the source, whether that source be $\mathbf{M}(\mathbf{r})$ or $\rho_M(\mathbf{r})$. We summarize some aspects of applying FFT approaches to this problem. The FFT algorithm results in speedup because it requires on the order of $N \ln N$ operations, rather than N^2. As one element in this scheme, one also needs to address some questions concerning the boundary conditions for an open system.

For the sake of argument, suppose one wishes to find the magnetic potential Φ_M based on a known configuration of effective magnetic charge ρ_M. We know that the basic Green's function g_0 will accomplish that for us, from (3.10). Here a problem is considered in only 1D, to keep the discussion simple. It is straightforward to generalize it to 2D or 3D, and also apply these ideas to the calculation of $\mathbf{H}_M$ from the source $\mathbf{M}$. So consider evaluation of an integral in the form,

$$\Phi_M(x) = \int dx' \, g_0(x, x') \rho_M(x'). \tag{3.116}$$

The Green's function depends only on the difference of source and field points, $x - x'$. That makes this integration a *convolution integral*,

$$\Phi_M(x) = \int dx' \, g_0(x - x') \rho_M(x'). \tag{3.117}$$

A convolution integral in real space becomes a product in reciprocal (Fourier) space. To show that, transform each of $\Phi_M(x)$, $g_0(x)$ and $\rho_M(x)$ into their corresponding functions in the reciprocal space, by defining their Fourier transforms (FTs) that are functions of wave vector q. Consider first a system of infinite extent. For some arbitrary function $f(x)$, its space FT is

$$F(q) = \int_{-\infty}^{+\infty} dx \, f(x) e^{-iqx}. \tag{3.118}$$

The inverse FT is the reciprocal relation,

$$f(x) = \frac{1}{2\pi} \int_{-\infty}^{+\infty} dq \, F(q) e^{iqx}, \tag{3.119}$$

which is based on the fact that the expansion basis functions $\frac{1}{\sqrt{2\pi}} \exp(iqx)$ for different q have an orthogonality relation,

$$\frac{1}{2\pi} \int_{-\infty}^{+\infty} dx \, e^{i(q-q')x} = \delta(q - q'). \tag{3.120}$$

Usually the theory of FTs is expressed in terms of wave vector q. Press *et al* [10] have made the important point that for numerical work, it is better to use *wave number* κ, because this eliminates the normalization factors of 2π. This approach will be followed here. Wave number is defined as the reciprocal of wavelength, $\kappa = \lambda^{-1}$. The wave vector is written in terms of wave number as

$$q = 2\pi\kappa. \tag{3.121}$$

The normalized basis functions are now written $\exp(i2\pi\kappa x)$, and this gives their normalization integral,

$$\int_{-\infty}^{+\infty} dx \, e^{i2\pi(\kappa-\kappa')x} = \frac{1}{2\pi} \int_{-\infty}^{+\infty} d(2\pi x) e^{i(2\pi x)(\kappa-\kappa')} = \delta(\kappa - \kappa'). \tag{3.122}$$

Then the FT $F(\kappa)$ and the inverse relation are

$$F(\kappa) = \int_{-\infty}^{+\infty} dx\, f(x) e^{-i2\pi\kappa x} \tag{3.123a}$$

$$f(x) = \int_{-\infty}^{+\infty} d\kappa\, F(\kappa) e^{i2\pi\kappa x}. \tag{3.123b}$$

Using wave number instead of wave vector for spatial properties is exactly analogous to using frequency ν instead of angular frequency $\omega = 2\pi\nu$ for describing a time sequence.

Now consider a convolution integral between two arbitrary functions $g(x)$ and $f(x)$,

$$h(x) = \int_{-\infty}^{+\infty} dx'\, g(x-x') f(x'). \tag{3.124}$$

Inserting the Fourier representations of g and f, whose FTs are $F(\kappa)$ and $G(\kappa)$, this becomes

$$h(x) = \int dx' \int_{-\infty}^{+\infty} d\kappa\, G(\kappa) e^{i2\pi\kappa(x-x')} \int_{-\infty}^{+\infty} d\kappa'\, F(\kappa') e^{i2\pi\kappa' x'}$$

$$= \int_{-\infty}^{+\infty} d\kappa \int_{-\infty}^{+\infty} d\kappa'\, G(\kappa) F(\kappa') e^{i2\pi\kappa x} \int_{-\infty}^{+\infty} dx' e^{i2\pi(\kappa-\kappa')x'}. \tag{3.125}$$

Having interchanged the order of integrations, the integration over x' produces a delta function, which leads to

$$h(x) = \int_{-\infty}^{+\infty} d\kappa \int_{-\infty}^{+\infty} d\kappa'\, G(\kappa) F(\kappa') e^{i2\pi\kappa x} \delta(\kappa - \kappa')$$

$$= \int_{-\infty}^{+\infty} d\kappa\, G(\kappa) F(\kappa) e^{i2\pi\kappa x}. \tag{3.126}$$

Comparing with the definition for FTs (3.123a), this shows that $H(\kappa)$, the FT of $h(x)$, is just the product of the FTs of g and f:

$$H(\kappa) = G(\kappa) F(\kappa). \tag{3.127}$$

Then a convolution in real space maps into a simple product at fixed wave number in the reciprocal space. Thus, the convolution can be carried out in κ-space as a product of FTs, and then be transformed back into real space to obtain $h(x)$. These arguments are easily extended to more dimensions.

3.4.1 Finite system on a 1D grid

In any simulation, the system is of finite extent. If the 1D system length is $L = Na$, the FT integral is changed to

$$F(\kappa_m) = \int_0^L dx\, f(x) e^{-i2\pi\kappa_m x} \tag{3.128}$$

and now the possible choices for wave number are *discretized*,

$$\kappa_m = \frac{m}{L}, \quad m = 0, 1, 2 \ldots (N-1). \tag{3.129}$$

In addition to this, however, the original space data must be defined on some grid in x. A uniform grid has sites at positions $x_n = na$, $n = 0, 1, 2, 3 \ldots (N-1)$. Then the integration must be evaluated as a sum, with $dx \to a$,

$$F(\kappa_m) = a \sum_{n=0}^{N-1} f(x_n) e^{-i2\pi \kappa_m x_n} = a \sum_{n=0}^{N-1} f_n \, e^{-i\frac{2\pi}{N} mn}, \tag{3.130}$$

where the original discrete data at the cell locations are $f_n = f(x_n)$. Then the Fourier description of each original data point goes over into the Fourier series (with $d\kappa \to 1/L$),

$$f_n = f(x_n) = \frac{1}{L} \sum_{m=0}^{N-1} F(\kappa_m) e^{i2\pi \kappa_m x_n}. \tag{3.131}$$

The value of $F(\kappa_m)$ is considered a continuum quantity, and has dimensions of length. Instead, it is convenient to define the discrete FT (DFT) that is a dimensionless quantity, which is typically produced from FT subroutines,

$$F_m = \sum_{n=0}^{N-1} f_n \, e^{-i\frac{2\pi}{N} mn}. \tag{3.132}$$

Obviously the continuum and discrete versions are related by $F(\kappa_m) = aF_m$. The inversion relation is then

$$f_n = \frac{1}{N} \sum_{n=0}^{N-1} F_m \, e^{i\frac{2\pi}{N} mn}. \tag{3.133}$$

The summation would be given by the same subroutine that produces F_m, using an opposite sign in the exponential. The normalization by N must be applied after that.

The DFT and its inverse depend on a sum over the set of real space data f_n, weighted by complex factors which are powers of $w_N \equiv \exp(-i2\pi/N)$. Due to the periodicity of this factor for integer powers, the FFT algorithm could be developed to evaluate this type of sum very efficiently. In terms of the phase factor w_N, the DFT is quite simple,

$$F_m = \sum_{n=0}^{N-1} w_N^{mn} f_n. \tag{3.134}$$

The factors of w_N^{mn} define elements of a matrix. The FFT algorithm makes this multiplication by building up from smaller matrices (i.e. small N) to the size desired, usually going up in powers of 2, although other bases are possible. A construction based on powers of 2 means that there must be an underlying grid of size $N = 2^p$, where p is an integer. The reader is referred to other sources for the details of the FFT algorithm.

3.4.2 Simulation of 1D open boundary conditions

One important application is that of a finite nanomagnet with free or open boundaries. The Fourier series based on (3.133) is designed to describe any function that is periodic over the interval $\Delta x = L = Na$. Numerical FT is known to be susceptible to the artifact of *aliasing*, which means that a signal with power at only one wave number will exhibit power at multiples of that basic wave number, due to the periodicity that the FFT imposes on the description. For convolution integrals, the imposed periodicity can lead to a problem known as the 'wrap-around error', unless the FFT is applied correctly. Thus it is important to see how convolution integrals for a finite system can be evaluated using a periodic basis without the wrap-around problem. This comes down to a question about boundary conditions.

It will be possible to see how this works with the calculation of a potential Φ_M in 1D due to some given effective charge density ρ_M, via a convolution integral. For that, it is good to specify the appropriate Green's function, which can be found in the following short exercise.

Exercise 3.2. The Green's function for Laplace's equation in 1D must satisfy the differential equation,

$$-\nabla^2 g_0(x, x') = -\frac{d^2}{dx^2} g_0(x, x') = \delta(x - x'). \quad (3.135)$$

Show that the solution is

$$g_0(x, x') = -\frac{1}{2} |x - x'| + A \quad (3.136)$$

where A is an arbitrary constant that sets the zero of potential.

Note that this Green's function for 1D magnetostatics gives the potential due to an infinite planar sheet of magnetic charge, which produces uniform magnetic field perpendicularly outward from the sheet. It is sketched in figure 3.4 as a solid black curve, taking the source point $x' = 0$. For the evaluation of a convolution, that curve must be shifted to the right by the value of source position, x', and then summing $\rho_M(x') g_0(x - x')$ over all different shifts x' from 0 to L.

Now suppose a certain charge distribution $\rho_M(x)$ is given, localized over some small region within the system of length L, occupying $0 \leqslant x < L$, see figure 3.5. When g_0 and ρ_M are represented as Fourier series, which are periodic, multiple copies of the original functions are generated. This is depicted in figures 3.4 and 3.5, which show the functions in the original system and in the periodic copies that are implied by the Fourier series representations.

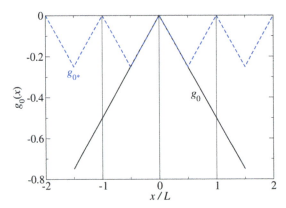

Figure 3.4. The 1D Green's function (g_0, solid black) of (3.136) with constant $A = 0$, compared with the periodic representation that would be made with a Fourier series (g_{0*}, dashed blue).

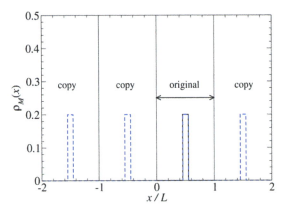

Figure 3.5. An example of a 1D magnetic charge density as in (3.140) (solid black) with $\rho_0 = 0.2$ and $l = 0.1L$, compared with the periodic representation that would be made with a Fourier series (dashed blue).

Consider the convolution as in (3.117), but integrated only over the limited range of the system,

$$\Phi_M(x) = \int_0^L dx' \, g_0(x - x')\rho_M(x'). \tag{3.137}$$

Because the Green's function is symmetric with $g_0(x - x') = g_0(x' - x)$, this is equivalent to

$$\Phi_M(x) = \int_0^L dx' \, g_0(x' - x)\rho_M(x'). \tag{3.138}$$

In this form, the interpretation of the convolution is simpler. The Green's function $g_0(x' - x)$ is obtained from $g_0(x')$ by shifting the curve to the right by a distance x. The solid black function in figure 3.4 would be shifted right by distance x, multiplied by $\rho_M(x')$, and summed over x' from 0 to L to obtain that value for $\Phi_M(x)$. Then $\Phi_M(x)$ is obtained for the whole range of x by considering different sized shifts of the Green's function.

The Fourier series representation of g_0 is periodic, rather than continuously descending away from the point $x = x'$. This is shown as the dashed blue function in figure 3.4, and represented with the symbol g_{0*}. The charge distribution also will become periodic when represented by a Fourier series. Thus the convolution using the FFT approach will not lead to the correct result. This is because the individual straight-line segments of $g_{0*}(x' - x)$ only have the extent $\Delta x' = L/2$, whereas the original system and ρ_M have the length L. Indeed, inspection shows that because $g_{0*}(x' - x)$ is forced to become periodic by the requirements of the Fourier series, it can be written as (see figure 3.4)

$$g_{0*}(x' - x) = \begin{cases} -\frac{1}{2}|x' - x| & |x' - x| \leq \frac{L}{2} \\ -\frac{L}{2} + \frac{1}{2}|x' - x| & |x' - x| > \frac{L}{2}. \end{cases} \quad (3.139)$$

This is significantly different than the correct Green's function, which will lead to incorrect results in the calculation of the demagnetization field.

Exercise 3.3. Assume that a 1D charge density is localized in a region of width $l < L$ in the center of a 1D system of length L. It can be written as

$$\rho_M(x) = \begin{cases} \rho_0 & \frac{L-l}{2} < x < \frac{L+l}{2} \\ 0 & \text{otherwise.} \end{cases} \quad (3.140)$$

Use (3.137) and the correct Green's function (3.136) from the exercise above with $A = 0$ to show that the correct potential that results is

$$\Phi_M(x) = \begin{cases} -\frac{1}{2}\rho_0 l(L - x) & 0 < x < \frac{L-l}{2} \\ -\frac{1}{2}\rho_0\left(x^2 - Lx + \frac{1}{2}Ll\right) & \frac{L-l}{2} < x < \frac{L+l}{2} \\ +\frac{1}{2}\rho_0 l(L - x) & \frac{L+l}{2} < x < L. \end{cases} \quad (3.141)$$

Also determine the magnetic field that results.

Exercise 3.4. For the charge density (3.140) localized in the system center as in the previous exercise, apply the periodic Green function (3.139) and determine how the aliasing effect changes the results for the magnetic potential and magnetic field. Compare with the results of the previous exercise.

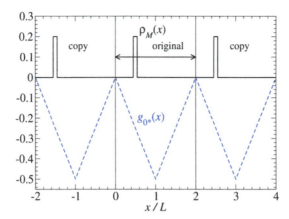

Figure 3.6. The periodic 1D Green's function (g_{0*}, blue) when embedded in a system of double length, compared with the example magnetic charge density, when also periodically repeated over distance $2L$.

For the 1D convolution using an FFT approach, the error due to wrap-around can be avoided in a simple way, by embedding the physical system into an artificial one that is twice as large. The extra places in the grid for ρ_M are packed with zeros. Now also the Green's function will be that for a system of length $2L$, but as a periodic function. Once this is done, the different periodic copies of ρ_M will not interfere with each other, see figure 3.6. Consider the convolution evaluated as in (3.138). Since each straightline segment of $g_0(x' - x)$ covers a range $\Delta x' = L$, and one considers shifting $g_0(x')$ different distances x from 0 to L, there will never be a case where one of the 'wrong segments' is overlapping the original copy of ρ_M. The convolution obtained in the Fourier representation will be the same as that from direct evaluation of the convolution integral.

For the usual FFT algorithm, the original system grid size is $N = 2^p$, where p is an integer. Then the doubled system also has size which is one higher power of 2, i.e. $2N = 2^{p+1}$. The Green's function is extended over the doubled system, length $2L = 2Na$, but the charge density is zero over the extended part of the doubled system.

> **Exercise 3.5.** By embedding the system into an artificial one twice as long, the charge density (3.140) can still be applied, together with $\rho_M = 0$ for $L < x \leqslant 2L$. Using the 1D Green's function with a period of $2L$, show that the convolution will be corrected computed when using these periodic functions, as drawn in figure 3.6.

3.4.3 Simulation of 2D and 3D open boundaries

The extension to 2D or 3D is not very different from what is explained for 1D. Again, to avoid wrap-around error, the actual physical system to be simulated can be embedded into another system on a grid twice as large in each dimension. Thus, if

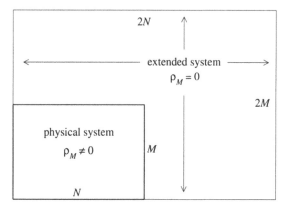

Figure 3.7. A sketch of embedding the physical system into an extended system twice as large in each coordinate, to avoid the wrap-around error. The source fields ρ_M or **M** are present only within the original physical system, not in the extended region.

the original system is to be simulated on a grid of size $N \times M$, where each is an integral power of 2, then a fictitious system of grid size $2N \times 2M$ is set up (or $2N \times 2M \times 2Z$ for 3D). One can place the original system into one corner of this fictitious extended system, as in figure 3.7. The charge density (or equivalently, the magnetization) will only be non-zero in the original system. The Green's function will be defined over the extended system. Any convolutions performed will then not involve interactions farther than distance N in one grid direction and distance M in the other direction. There will be no wrap-around error. The simulation will imitate the demagnetization field of an isolated magnetic medium, such as a nanoparticle.

In terms of the computer programming, it may be necessary to define arrays for both the original physical system and for the extended system. The FFT operations will work only on the variables in the extended system. Once $\mathbf{H}_M(\mathbf{r})$ has been calculated in the extended system, the correct section of the arrays is extracted out to describe the original physical system.

3.5 Demagnetization in a thin permalloy magnet

An example to illustrate these approaches is to consider a thin-film magnetic particle with an elliptical perimeter, see figure 3.8. The particle has been modeled by using cells of size $a \times a \times L$ with $a = 2.0$ nm and thickness $L = 2a = 4.0$ nm. The ellipse has semi-major axis $32a$ and semi-minor axis $16a$. Thus, a grid of size 32×16 just barely contains the particle. In order to apply the FFT method, the embedding system needs to have a grid at least as large as 64×32, which was used here. Figure 3.8 shows the relaxed state of lowest energy when including a combination of exchange, a weak uniaxial anisotropy in the long direction of the particle and the magnetostatic energy due to $\mathbf{H}_M$. For the plot of the $\mathbf{H}_M$ field, the values are shown at half of the grid sites within the particle, and only 1/16 of the grid sites for the region outside the particle. The magnitudes are not shown, because these are primarily of large values only very close to the particle surfaces.

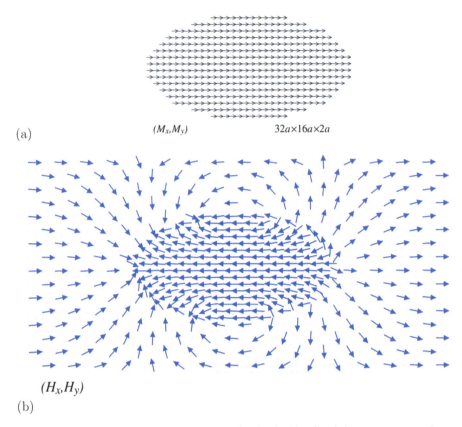

Figure 3.8. Example of an elliptical thin-film magnet simulated with cells of size parameter $a = 2$ nm and indicated semi-major axis, semi-minor axis and thickness. (a) The uniform magnetization in all of the individual cells. (b) The directions of the in-plane demagnetization field $\mathbf{H}_M$, but only on a fraction of the cells, to clarify the display. The system is embedded in an FFT grid of size 64×32.

It is helpful to mention how the different interactions have been included, especially because the cells used are not cubic for this quasi-2D problem. The micromagnetics cells have volume $v_{\text{cell}} = a^2 L$, rather than the more usual volume of a^3 for cubic cells. According to earlier discussion in chapter 2 of the exchange between micromagnetics cells, (2.28) and (2.29), or compare (2.33), the 2D inter-cell coupling is

$$J_{\text{cell}} = 2A \frac{v_{\text{cell}}}{a^2} = 2AL, \tag{3.142}$$

where A is the exchange stiffness for the material (units of Jm^{-1}). The Hamiltonian for the original magnetic system, with isotropic exchange, uniaxial energy, an external field, and magnetostatic energy, is

$$H = \int dV \left\{ A |\nabla \hat{\mathbf{m}}|^2 - K_v (\hat{\mathbf{m}} \cdot \hat{\mathbf{x}})^2 - \mu_0 \mathbf{H}_{\text{ext}} \cdot \mathbf{M} - \frac{1}{2} \mu_0 \mathbf{H}_M \cdot \mathbf{M} \right\}. \tag{3.143}$$

Recall that the 1/2 on the demagnetization term removes the double counting of dipole pair interactions. The uniaxial energy constant K_v (per unit volume) has the effect of slightly orienting $\mathbf{M}$ along $\pm\hat{\mathbf{x}}$. Now when this is transformed to the micromagnetics Hamiltonian on the grid cells, it becomes

$$H = -J_{\text{cell}} \sum_{(i,j)} \hat{\mathbf{m}}_i \cdot \hat{\mathbf{m}}_j - \sum_i v_{\text{cell}} \left[K_v(\hat{\mathbf{m}}_i \cdot \hat{\mathbf{x}})^2 + \mu_0 \left(\mathbf{H}_{\text{ext}} + \frac{1}{2}\mathbf{H}_{\text{M}} \right) \cdot \mathbf{M} \right]. \quad (3.144)$$

The exchange term has sum only over nearest neighbor cells, indicated by (i, j). The cell volume enters in the other terms when changing from volume integration to summation over cells. The saturation magnetization M_s is the natural unit in which to measure $\mathbf{H}_{\text{M}}$, as well as $\mathbf{M}$ and $\mathbf{H}_{\text{ext}}$. Typically the exchange energy is the largest energy scale in this kind of problem. Thus it makes sense to factor out J_{cell} and use it as the overall energy unit and, at the same time, define scaled external and demagnetization fields, denoted using tildes,

$$\tilde{\mathbf{H}}_{\text{ext}} \equiv \mathbf{H}_{\text{ext}}/M_s, \qquad \tilde{\mathbf{H}}_{\text{M}} \equiv \mathbf{H}_{\text{M}}/M_s. \quad (3.145)$$

One expects that the magnitude of $\tilde{\mathbf{H}}_{\text{M}}$ is unlikely to exceed unity. Then the micromagnetics Hamiltonian becomes

$$H = -J_{\text{cell}} \left[\sum_{(i,j)} \hat{\mathbf{m}}_i \cdot \hat{\mathbf{m}}_j + \sum_i \frac{v_{\text{cell}}}{J_{\text{cell}}} \left\{ K_v(\hat{\mathbf{m}}_i \cdot \hat{\mathbf{x}})^2 + \mu_0 M_s^2 \left(\tilde{\mathbf{H}}_{\text{ext}} + \frac{1}{2}\tilde{\mathbf{H}}_{\text{M}} \right) \cdot \right\} \hat{\mathbf{m}}_i \right]. \quad (3.146)$$

This suggests the definition of scaled coupling parameters for the uniaxial anisotropy and especially for the demagnetization term. For the anisotropy, define a dimensionless constant

$$\kappa_{\text{cell}} \equiv \frac{v_{\text{cell}}}{J_{\text{cell}}} K_v = \frac{K_v a^2}{2A}. \quad (3.147)$$

Although dimensionless, it does depend on the size of the micromagnetics cells being used. For the magnetic field terms, a dimensionless constant can be defined,

$$\delta_{\text{cell}} \equiv \frac{v_{\text{cell}}}{J_{\text{cell}}} \mu_0 M_s^2 = \frac{\mu_0 M_s^2 a^2}{2A}. \quad (3.148)$$

Again, this increases with the cell size. However, this last equation is usually used to define the *exchange length* λ_{ex}, which is the distance over which the exchange tends to preserve the magnetization direction (when competing against magnetostatic energy effects):

$$\lambda_{\text{ex}} \equiv \sqrt{\frac{2A}{\mu_0 M_s^2}}. \quad (3.149)$$

In addition, the competition between exchange and the uniaxial anisotropy leads to another length scale,

$$\lambda_K \equiv \sqrt{\frac{2A}{K_v}}. \tag{3.150}$$

For the purpose of simulations, one should check the cell size against these length scales. To correctly describe the spatial variation of the magnetization, it is important to have the cell size smaller than both of these lengths. K_v in many materials can be very weak and hence the length λ_K is quite large. On the other hand, the exchange length may be more important in determining cell size. For example, in permalloy, with exchange stiffness $A = 13$ pJ m^{-1} and saturation magnetization $M_s = 860$ kA m^{-1}, the exchange length is found to be $\lambda_{ex} = 5.3$ nm. The anisotropy constant is $K_v \approx 100$ J m^{-3}, leading to the anisotropy length scale $\lambda_K \approx 500$ nm, considerably larger! Although a cell parameter $a = 2.0$ nm is adequate, a better description might be obtained with an even smaller cell size, at added computational expense, however.

With the length parameter definitions, the micromagnetics Hamiltonian used can be written

$$H = -J_{cell}\left[\sum_{(i,j)}\hat{\mathbf{m}}_i \cdot \hat{\mathbf{m}}_j + \sum_i \left\{\frac{a^2}{\lambda_K^2}(\hat{\mathbf{m}}_i \cdot \hat{\mathbf{x}})^2 + \frac{a^2}{\lambda_{ex}^2}\left(\tilde{\mathbf{H}}_{ext} + \frac{1}{2}\tilde{\mathbf{H}}_M\right) \cdot \hat{\mathbf{m}}_i\right\}\right]. \tag{3.151}$$

While the Hamiltonian is needed for energy discussion, it is not as useful as the fields for actual calculations. However, from the discussion of classical spin mechanics, the effective magnetic field acting on a given cell is easily found. Those fields were used to find the equilibrium configuration in figure 3.8, by a simple method explained next.

3.5.1 Energy minimization by a spin alignment scheme

For figure 3.8 and other local minimum energy states, there are various numerical methods that could be applied to reach some approximate low energy configurations. In a minimum energy (equilibrium) configuration, the net torque on each magnetic dipole $\boldsymbol{\mu}_i$ (i.e. in a cell) in the system is zero. The only way that can be true is if each dipole is aligned with the effective magnetic induction $\mathbf{B}_i$ due to all magnetic effects. We saw in chapter 2 that a dipole has the following dynamic equation (see equations (2.61) and (2.99))

$$\frac{d\boldsymbol{\mu}_i}{dt} = \boldsymbol{\mu}_i \times \gamma \mathbf{B}_i, \tag{3.152}$$

where the effective field is

$$\mathbf{B}_i \equiv \frac{1}{\gamma}\mathcal{F}_i = -\frac{1}{\gamma}\frac{\partial H}{\partial \mathbf{S}_i} \longrightarrow -\frac{\partial H}{\partial \boldsymbol{\mu}_i}. \tag{3.153}$$

In the last step we use the fact that a magnetic dipole is $\boldsymbol{\mu}_i = \gamma \mathbf{S}_i$, where γ is the gyromagnetic ratio that converts from angular momentum to dipole moment.

For micromagnetics, each cell has a dipole of magnitude $\mu_{\text{cell}} = v_{\text{cell}} M_s$, so that $\boldsymbol{\mu}_i = \mu_{\text{cell}} \hat{\mathbf{m}}_i$. Thus, the field can be written alternatively as

$$\mathbf{B}_i = -\frac{1}{\mu_{\text{cell}}} \frac{\partial H}{\partial \hat{m}_i}. \tag{3.154}$$

Then extracting the derivatives using Hamiltonian (3.151), one has fields[1]

$$\mathbf{B}_i = \frac{J_{\text{cell}}}{\mu_{\text{cell}}} \left[\sum_{(i,j)} \hat{\mathbf{m}}_j + 2\frac{a^2}{\lambda_K^2}(\hat{\mathbf{m}}_i \cdot \hat{\mathbf{x}})\hat{\mathbf{x}} + \frac{a^2}{\lambda_{\text{ex}}^2}\left(\tilde{\mathbf{H}}_{\text{ext}} + \tilde{\mathbf{H}}_{\text{M}} \right) \cdot \hat{\mathbf{m}}_i \right]. \tag{3.155}$$

The factor in front is the unit used for magnetic induction in this formulation of micromagnetics,

$$B_0 = \frac{J_{\text{cell}}}{\mu_{\text{cell}}} = \frac{2AL}{M_s a^2 L} = \mu_0 M_s \left(\frac{\lambda_{\text{ex}}}{a}\right)^2. \tag{3.156}$$

However, to determine low energy states, we only need the direction of the local $\mathbf{B}_i$ and their magnitudes are less important.

To relax an initial state, perhaps arbitrarily chosen, into a lower energy, one can proceed iteratively as follows in this *spin alignment scheme* [11]. From the initial state, the demagnetization field $\mathbf{h}_M$ is calculated and then each $\mathbf{B}_i$ is determined. If there are N_{sites} total grid sites in the part of the system where $\mathbf{M}$ is present, one is chosen randomly, and the dipole vector $\hat{\mathbf{m}}_i$ at that cell is re-directed into the direction of its $\mathbf{B}_i$. This is then repeated for other sites, until a total of N_{sites} sites have had their $\hat{\mathbf{m}}_i$ re-directed. This would be considered one iteration step. The sites are chosen randomly to avoid producing unphysical spatial correlations. However, once a site is modified, the demagnetization field within the whole system also is modified. It would be prohibitively slow to recalculate the $\tilde{\mathbf{H}}_M$ fields every time one cell is changed. Instead, only after N_{sites} sites have been modified, the new $\tilde{\mathbf{H}}_M$ fields will be recalculated, completing one iteration step. It is not guaranteed that all N_{sites} sites will have been updated, since they are chosen randomly. However, over the course of many iterations, it is more than likely that all cells are updated, and the system does move towards the local energy minimum that is somehow nearest in phase space.

Different quantities can be calculated to determine whether the iteration should be stopped. These include the total energy and the Cartesian components of the total magnetic moment of the system. Another quantity that can be tracked is the average change in magnetic moment for all cells, from one iteration step to the next. Whatever quantity is used, the iteration will be designed to be stopped when the desired quantity does not change (or goes close to zero) relative to some desired precision.

---

[1] The factor of 1/2 on the demagnetization field $\tilde{\mathbf{H}}_M$, such as in (3.146) and (3.151), is not present here because there are two equivalent magnetostatic pair terms in the Hamiltonian that couple two chosen sites $\hat{\mathbf{m}}_i$ and $\hat{\mathbf{m}}_j$.

For the simulation in figure 3.8, the change in system magnetic moment was used, calculating the following parameter,

where
$$\Delta_m = \frac{1}{3}\Big(|\bar{m}_x - \bar{m}'_x| + |\bar{m}_y - \bar{m}'_y| + |\bar{m}_z - \bar{m}'_z|\Big), \quad (3.157)$$

$$\bar{m} = \frac{1}{N_{\text{sites}}}\sum_i \hat{\mathbf{m}}_i \quad (3.158)$$

is the average of the unit dipoles in the cells. Unprimed quantities are those in an initial state, and primed are those in the next state after one iteration. The value of Δ_m then gives a sense of the changes for each iteration. The simulation was stopped when Δ_m became smaller than a chosen limit, $\epsilon = 2.0 \times 10^{-8}$. Such a way to terminate the iteration will tend to give as good or better precision in the total system energy.

One can make modifications of this spin alignment numerical technique, such as setting the new dipole direction to be somewhere between the original direction and that of $\mathbf{B}_i$, to avoid some chance of overshooting. Typically that does not help significantly, because the true equilibrium state needs each $\hat{\mathbf{m}}_i$ parallel to its local field $\mathbf{B}_i$.

Besides spin alignment, there are other powerful techniques for studying magnetization properties, especially for dynamics and for consequences of temperature. In the following chapters approaches to spin dynamics and magnetic thermal equilibrium are developed.

Bibliography

[1] Lorrain P and Corson D 1970 *Electromagnetic Fields and Waves* 2nd edn (CA Freeman: San Fransisco)
[2] Jackson J D 1999 *Classical Electrodynamics* 3rd edn (New York: Wiley)
[3] Jiles D 1991 *Introduction to Magnetism and Magnetic Materials* (London: Chapman and Hall)
[4] Lee E W 1970 *Magnetism: An Introductory Survey* (New York: Dover)
[5] García-Cervera C J 1999 Magnetic domains and magnetic domain walls *PhD thesis* New York University
[6] García-Cervera C J, Gimbutas Z and Weinan E 2003 Accurate numerical methods for micromagnetics simulations with general geometries *J. Comput. Phys.* **184** 37
[7] Huang Z 2003 High accuracy numerical method of thin-film problems in micromagnetics *J. Comput. Math.* **21** 33
[8] Aharoni A 1998 Demagnetizing factors for rectangular ferromagnetic prisms *J. Appl. Phys.* **83** 3432
[9] Sasaki J and Matsubara F 1997 Circular phase of a two-dimensional ferromagnet with dipolar interactions *J. Phys. Soc. Japan.* **66** 2138
[10] Press W H, Teukolsky S A, Vetterling W T and Flannery B P 2007 Numerical Recipes. The Art of Scientific Computing 3rd edn (Cambridge: Cambridge University Press)
[11] Wysin G M 1996 Magnetic vortex mass in two-dimensional easy-plane magnets *Phys. Rev. B* **54** 15156

Chapter 4

Classical Monte Carlo simulation methods

The thermal properties of magnets, such as internal energy and heat capacity, depend on how the excitations are generated with increasing temperature. Magnetic properties, such as the magnetization and susceptibility, vary with temperature as different excitations appear. The Monte Carlo (MC) importance sampling method is a way to evaluate these thermal effects, according to the relative importance or statistical weight measured in phase space. In this chapter, we see different ways to apply this statistical approach to obtain equilibrium properties of magnets.

4.1 Thermal equilibrium and ergodicity

A physical system with many degrees of freedom can have its instantaneous state be described by the values of the canonical coordinates and their conjugate momenta. In classical deterministic physics the system evolves according to Hamiltonian equations of motion, and the future states can be predicted precisely from the current state. Of course, it is impossible to have perfect current knowledge of all coordinates and momenta, which leads to an uncertainty in the future behavior even in classical physics. In quantum physics there is an inherent uncertainty in the future states, even under the assumption of a perfectly known initial state, due to the statistical interpretation of quantum mechanics! Some of the aspects of this indeterminacy in the future state of any physical system come under the heading of 'chaos theory', wherein fairly simple dynamical systems can exhibit very complex future behavior. Even within this complex future behavior, however, chaos theory has been developed to show the patterns that can be present.

In statistical physics, one takes some advantage of the indeterminacy of the exact future of a system, especially for the calculation of physical measurements one would obtain by averaging results over many identical experiments. This is the idea of averaging over the *ensemble*, which means, imagining an experiment repeated many times with essentially identical systems, however, perhaps starting in somewhat different microscopic initial states, that appear equivalent from a macroscopic observation. Further, a system in thermal equilibrium can be assumed to follow the *ergodic hypothesis*, which says that given sufficient time, the system will eventually pass through all of its physically accessible microstates. The accessible microstates are states distinguished by particular coordinates and momenta, but satisfying any physical constraints such as total energy, or pressure, or particle number, and so on.

If a system is complex enough, it is likely to satisfy the ergodic principle at least to a certain extent, and this is the basic for most of equilibrium statistical mechanics. One can then say that the system passes through all of the accessible phase space, and it is assumed that each accessible microstate (described by s-coordinates q and momenta p, where s is the number of degrees of freedom) is equally probable. Statistical mechanics uses the probability of different states in the phase space to calculate averages of physically measured quantities such as total energy, heat capacity and so on. There can appear some simpler systems, however, where there is not enough mixing up of the system to truly pass through all of the allowed phase space with uniform probability. Mostly one assumes that these types of systems are singular and somehow unlikely. In a real physical system, one never has the full Hamiltonian of all possible interactions. The unknown interactions (those not accounted for in our Hamiltonian) perturb the system away from what we are calculating. It is then hoped that these unknown interactions will always be present, in sufficient strength to cause the system to seek out most of the allowed phase space, and support ergodic behavior.

4.2 Boltzmann distribution for thermal equilibrium

In the most common case of trying to reproduce classical statistical mechanics, the set of states should be selected with probabilities such that the classical Boltzmann distribution is produced. The Boltzmann distribution describes the relative frequency of the system being found in some microstate according to its total energy E_i, for a chosen absolute temperature T. A selected microstate i in the allowed phase space has a probability of occurrence p_i that is proportional the well-known *Boltzmann factor*, given by

$$p_i \propto \exp(-E_i/k_B T), \tag{4.1}$$

where k_B is Boltzmann's constant. This is a factor that comes from analysis of a system in the *canonical ensemble* for statistical mechanics, where the system is assumed to be held at constant absolute temperature T. It may be helpful to summarize its genesis, so that other related relations of statistical physics can be later derived from it. This will help to support the extension of the results to numerical simulations.

In thermal equilibrium, one assumes all allowed or accessible microstates are equally probable, that is the ergodic hypothesis. Further, in Boltzmann statistics, it is assumed that the density of particles in the system is low enough, so that quantum interference and degeneracy effects are negligible. This means, equivalently, that for each accessible microstate of a particle, there is extremely low probability of being occupied by more than one particle. Indeed, we proceed with the assumption that the average number of particles in any selected microstate is much less than one. In cases where this assumption is violated, quantum statistics must be applied.

The cases of quantum statistics for fermions (Fermi–Dirac statistics, 0 or 1 particle in each microstate) and for bosons (Bose–Einstein statistics, any number of particles in each microstate) are not considered here, because we will limit the discussion and applications to those where classical statistics can be applied.

We start with a derivation of the Boltzmann distribution for a gas of N weakly interacting particles with microstates at energies E_i. While the derivation is not completely general, it will be good to summarize a second derivation that does not depend on any assumptions about weakly interacting particles, which strongly confirms the applicability of the Boltzmann probabilities for allowed energy states.

4.2.1 Entropy and system energy

For simplicity, suppose the microstates of a particle can be explicitly counted and distinguished, as would be the case in a quantum theory with discrete eigenvalues. In any real system, a state of the system at energy E_i can be realized by a multiplicity of degenerate microstates, typically related to each other by symmetry operations that leave the energy unchanged. Therefore, suppose that G_i is the number of distinct microstates with energy E_i. G_i can be considered the density of states at energy E_i.[1] In a macroscopic system, G_i can be very large (for example, many equivalent positions or momenta). Now consider that N total particles exist in the system, and they are distributed into the accessible microstates. Note that the system quantities depend on N and on the volume V; these parameters will be suppressed on expressions. Let N_i be the number of particles that go into an energy level E_i. Due to the density of states G_i, there can be many possible ways to realize a particular macroscopic state with N particles and some total energy E. This very large number of microstates $W(E)$ helps to define the total entropy $S(E)$ by

$$S(E) = k_B \ln W(E). \tag{4.2}$$

The total system energy is a sum over the different states and their populations,

$$E = \sum_i N_i E_i \tag{4.3}$$

and total particle number is similar,

$$N = \sum_i N_i. \tag{4.4}$$

To obtain $W(E)$, it can be realized that in energy state E_i, the N_i particles can be distributed randomly among the G_i energetically equivalent microstates, and in no particular order. A particular example is shown in figure 4.1. If each particle has G_i possible choices, there would be $G_i^{N_i}$ ways to fill the microstates. However, we assume there will never be more than one particle in any of the degenerate states (this is the low density assumption). Furthermore, the particles are indistinguishable, which reduces any counting by the number of permutations, $N_i!$. Therefore, the

---
[1] If a quantum system were under consideration, G_i would be the quantum degeneracy of the state with energy eigenvalue E_i.

Figure 4.1. Sketch of an energy level with $N_i = 4$ particles distributed among its $G_i = 10$ degenerate microstates. The number of ways to realize this is $W_i = 10!/(6!\ 4!) = 210$.

number of ways to occupy this energy state is the number of ways to select N_i objects out of G_i, which is

$$W_i = \frac{G_i!}{(G_i - N_i)!\ N_i!}. \tag{4.5}$$

At low density, we assume $N_i \ll G_i$, and then this is roughly equal to

$$W_i \approx \frac{G_i^{N_i}}{N_i!}. \tag{4.6}$$

For example, in figure 4.1, with $G_i = 10$ and $N_i = 4$ (40% occupied), the exact value is $W_i = 210$ while this approximation gives $W_i \approx 417$. It turns out this will not be a significant error, especially when low occupation probability N_i/G_i holds. Try the example $G_i = 10$ with $N_i = 2$, where the approximation is much better!

Associated with the number W_i is a corresponding entropy,

$$S_i = k_B \ln W_i. \tag{4.7}$$

This applies to all other energy states, and all the factors W_i multiply together to give the approximate total number of ways to realize a macrostate of total energy E.

$$W(E) = \Pi_i W_i = \Pi_i \frac{G_i^{N_i}}{N_i!}. \tag{4.8}$$

This results in total system entropy,

$$S(E) = k_B \ln \left[\Pi_i \frac{G_i^{N_i}}{N_i!} \right] = k_B \sum_i \ln \left[\frac{G_i^{N_i}}{N_i!} \right] = k_B \sum_i S_i. \tag{4.9}$$

It is assumed that each state is only slightly populated (classical statistics), so that

$$1 < N_i \ll G_i \tag{4.10}$$

and yet, N_i is assumed to be quite large, in the limit of many particles in any macroscopic system. Then Stirling's approximation for large N_i can be applied,

$$\ln N_i! \approx N_i \ln N_i - N_i. \tag{4.11}$$

The system entropy can then be found expressed as

$$S(E) = k_B \sum_i N_i (\ln G_i - \ln N_i + 1). \tag{4.12}$$

This can define the entropy in an arbitrarily populated state. However, our goal firstly is to describe equilibrium. In that case, the equilibrium configuration will be considered as that which is most probable, that satisfies the principle of *maximum entropy*. This entropy can be maximized with respect to arbitrary variations in the N_i, under the constraints of particular values of total energy E and particle number N. Thus, one applies Lagrange's method of undetermined multipliers, and seeks the extremum of the modified entropy function,

$$\Omega(E) = S(E)/k_B + \alpha N - \beta E, \tag{4.13}$$

where α and β are Lagrange multipliers. Expressed in terms of N_i,

$$\Omega(E) = \sum_i N_i(\ln G_i - \ln N_i + 1) + \sum_i N_i(\alpha - \beta E_i). \tag{4.14}$$

The variation with respect to N_i is

$$\frac{\partial \Omega}{\partial N_i} = (\ln G_i - \ln N_i) + \alpha - \beta E_i = 0. \tag{4.15}$$

This must hold for each level, then one finds the populations of the levels,

$$N_i = G_i e^{\alpha - \beta E_i}. \tag{4.16}$$

However, α and β still must be determined from the constraints. For total particle number, there is

$$N = \sum_i N_i = \sum_i G_i e^{\alpha - \beta E_i} \implies e^\alpha = \frac{N}{Z}, \tag{4.17}$$

where the normalization factor Z is the canonical partition function:

$$Z = \sum_i G_i e^{-\beta E_i}. \tag{4.18}$$

The total system energy is found as

$$E = \sum_i N_i E_i = \sum_i G_i E_i e^{\alpha - \beta E_i} = \frac{N}{Z} \sum_i G_i E_i e^{-\beta E_i}. \tag{4.19}$$

From this one can see the definition of the statistically averaged energy of one particle,

$$\langle E_1 \rangle = \frac{1}{Z} \sum_i G_i E_i e^{-\beta E_i}, \tag{4.20}$$

which demonstrates one purpose of Z as a normalization factor for averages. To obtain the specific value of β, however, evaluate the equilibrium system entropy. One obtains from (4.12) and (4.16),

$$S(E) = k_B \frac{N}{Z} \sum_i G_i e^{-\beta E_i} \left(1 - \ln \frac{N}{Z} + \beta E_i\right) = k_B N \left(1 - \ln \frac{N}{Z}\right) + k_B \beta E. \tag{4.21}$$

In the last step the expression (4.19) for system energy was used. This shows that the equilibrium entropy increases linearly with energy. But from thermal physics we know that this is directly related to the definition of absolute temperature, which comes from the derivative of entropy:

$$\left(\frac{\partial S}{\partial E}\right)_N = \frac{1}{T}. \tag{4.22}$$

Then its application gives

$$\left(\frac{\partial S}{\partial E}\right)_N = k_B \beta = \frac{1}{T} \implies \beta = \frac{1}{k_B T}. \tag{4.23}$$

Based on these results, we have found that the population into one level at energy E_i is given by

$$N_i = N G_i \frac{e^{-\beta E_i}}{Z}. \tag{4.24}$$

From this last result, in the Boltzmann distribution, the probability for the E_i energy level to be populated, relative to any other level, is

$$p(E_i) \equiv \frac{N_i}{N} = G_i \frac{e^{-\beta E_i}}{Z}. \tag{4.25}$$

This includes the density of states at the energy E_i. Alternatively, one can state the probability to be in any of its microstates, by factoring out G_i from the expression:

$$p_i \equiv \frac{e^{-\beta E_i}}{Z}. \tag{4.26}$$

The G_i microstates all at the same energy all have probabilities equal to this value p_i. This expression is the probability of microstates for the Boltzmann distribution in thermal equilibrium. Usually the expression for the canonical partition function Z also is written as a sum over all microstates (not necessarily grouped by energy),

$$Z = \sum_{\text{states}} e^{-\beta E_{\text{state}}}. \tag{4.27}$$

Although these results were obtained by supposing particles distributed among allowed states, the range of validity is much greater, and this expression gives the canonical partition function for any classical system. The goal of simulation will be to produce a set of states in some model system of interest, that follows the distribution (4.26).

4.2.2 General derivation of Boltzmann distribution

The previous derivation assumes N weakly interacting particles, yet the Boltzmann distribution is much more general than that, with the energy in (4.26) being the system energy, regardless of the types of interactions present. This can be shown by

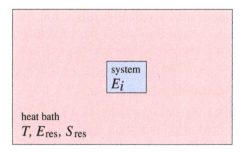

Figure 4.2. A small system embedded in and sharing energy with a large thermal reservoir or heat bath.

assuming a relatively small system in contact with a very large heat reservoir, and using entropy considerations.

We imagine that the large heat reservoir in contact with the much smaller system are isolated from outside, thus sharing a constant total energy E_0, see figure 4.2. This is in the sense of the *microcanonical ensemble*, where some degrees of freedom share a fixed total amount of energy, and temperature is a derived quantity. If the system has energy E_i, then the reservoir has energy $E_{\text{res}} = E_0 - E_i$, that is

$$E_{\text{res}} + E_i = E_0. \tag{4.28}$$

We do not try to find the configuration of maximum entropy, but rather, seek the likelihood that the system energy can take a chosen value E_i. Thus, it is assumed that the system is chosen to be in a particular microstate of energy E_i, with multiplicity $W_i = 1$. To find the likelihood of that, it is then necessary to look at the number of ways W_{res} that the reservoir can be in a state of energy E_{res}. However, this is given from the entropy function for the reservoir,

$$W_{\text{res}}(E_{\text{res}}) = \exp\{S_{\text{res}}(E_{\text{res}})/k_{\text{B}}\} = \exp\{S_{\text{res}}(E_0 - E_i)/k_{\text{B}}\}. \tag{4.29}$$

The system energy E_i is just a small value compared to E_0, assuming the reservoir is very large. Therefore make an expansion,

$$S_{\text{res}}(E_0 - E) \approx S_{\text{res}}(E_0) - \left(\frac{\partial S_{\text{res}}}{\partial E}\right) E_i + \ldots. \tag{4.30}$$

One can show that the terms that are dropped go to zero at least as fast as $1/N$. But the partial derivative of entropy with respect to energy gives the inverse absolute temperature of the reservoir. There results

$$W(E_i) = \exp\{S_{\text{res}}(E_0)/k_{\text{B}}\} \exp\{-\beta E_i\}, \tag{4.31}$$

where, as found before, $\beta = (k_{\text{B}}T)^{-1}$. This is again the Boltzmann distribution, as an unnormalized probability. It can be normalized by enforcing the constraint that all the microstate probabilities must sum to one, which leads to

$$p_i = \frac{e^{-\beta E_i}}{\sum_i e^{-\beta E_i}} = \frac{e^{-\beta E_i}}{Z}. \tag{4.32}$$

The normalization factor is the canonical partition function, with the sum over all microstates of the whole system (not single particle states). Thus, the Boltzmann distribution applies even if there are complex inter-particle interactions, such as in magnetic materials.

4.2.3 Free energy and statistical averages

The probabilities in (4.32) are those for individual microstates with specified energy E_i, when the system is held at a controlled temperature. This is the usual case for both real experiments and MC simulations, which is why the Boltzmann distribution plays such an important role.

It is interesting to compare to the probability for an individual microstate of energy E (we drop the i subscript from energy, entropy, *etc*, for simplicity) as obtained from the corresponding entropy $S = k_B \ln W$. The ergodic assumption is that each microstate is equally probable. When applied at a selected energy E (as in the microcanonical ensemble we used above), the microstate probability has to be unity divided by the number of distinct accessible states:

$$p = \frac{1}{W} = \frac{1}{\exp\{S/k_B\}} = e^{-S/k_B}. \tag{4.33}$$

This must be the same as the Boltzmann expression (4.32). Note that Z acts as a (large) normalization constant in the same way that W does. Thus it makes sense to define a new quantity called the *Helmholtz free energy* $F(T)$, which is related to Z in nearly the same way that W is related to S:

$$F = -k_B T \ln Z, \implies Z = e^{-\beta F}. \tag{4.34}$$

Now equating the two expressions (4.32) and (4.33) for p,

$$\frac{e^{-\beta E}}{Z} = \frac{1}{W} \implies e^{\beta(F-E)} = e^{-S/k_B}. \tag{4.35}$$

This shows that the Helmholtz free energy must be

$$F(T) = E - TS. \tag{4.36}$$

The free energy is shifted from the mechanical energy by an entropy effect due to temperature. One can see that the more entropy present in a state, the greater will be the effects of temperature and the so-called thermal fluctuations.

In statistical mechanics at fixed temperature, Z is the normalization factor for finding statistical averages. As well, F produces the normalization for microstate probabilities,

$$p = e^{\beta(F-E)}. \tag{4.37}$$

In theory calculations then, averages are found using this by summing over all accessible states. For the average of some observable quantity X, one uses a sum over all microstates,

$$\langle X \rangle = \sum_i p_i X_i = e^{\beta F} \sum_i e^{-\beta E_i} X_i. \tag{4.38}$$

Remembering that $e^{\beta F} = Z^{-1}$, this is normalized by the canonical partition function.

Usually it is important to find the statistical average of the energy, and then from that, other quantities such as the heat capacity C (for the whole system) or specific heat c (per particle or per unit mass). One can note that if $X = E$, the average can be found by differentiation of the partition function or the free energy, demonstrated as follows:

$$\langle E \rangle = Z^{-1} \sum_i e^{-\beta E_i} E_i = -Z^{-1} \frac{\partial}{\partial \beta} \sum_i e^{-\beta E_i} = -\frac{1}{Z} \frac{\partial Z}{\partial \beta}. \quad (4.39)$$

This can be put into the form of a logarithmic derivative,

$$\langle E \rangle = -\frac{\partial}{\partial \beta} \ln Z. \quad (4.40)$$

Finally this is equivalent to

$$\langle E \rangle = \frac{\partial}{\partial \beta}(\beta F). \quad (4.41)$$

For the heat capacity, one has

$$C = \frac{\partial}{\partial T}\langle E \rangle = \frac{d\beta}{dT} \frac{\partial}{\partial \beta}\langle E \rangle = -k_B \beta^2 \frac{\partial^2}{\partial \beta^2}(\beta F). \quad (4.42)$$

The useful identity for temperature derivative was applied,

$$\frac{\partial}{\partial T} = -k_B \beta^2 \frac{\partial}{\partial \beta}. \quad (4.43)$$

This produces specific heat at constant applied field **H**. We assume a system of fixed volume or fixed number of magnetic dipoles; pressure is not being considered here as having any effect on magnetic degrees of freedom.

One can also recover the equilibrium entropy from the free energy,

$$\langle S \rangle = \frac{\langle E \rangle - F}{T} = k_B \beta \left[\frac{\partial}{\partial \beta}(\beta F) - F \right] = k_B \beta^2 \frac{\partial F}{\partial \beta} = -\frac{\partial F}{\partial T}. \quad (4.44)$$

Note that this could have been found from the thermodynamic identity for free energy $F(T)$,

$$dF = dE - S\, dT. \quad (4.45)$$

In the practical sense, however, neither the partition function $Z(T)$ nor the free energy $F(T)$ need to be directly calculated in simulations. The same can be said of $W(E)$ and entropy $S(E)$. Instead, all that is required is a limited set of selected states that follow the desired probability distribution, usually the Boltzmann distribution (4.32). The normalization factor $Z(T)$ does not need to be known explicitly, because averages can be found from selected states that are preferentially found at lower energy as dictated by the Boltzmann factor $\exp\{-\beta E\}$. The goal in MC simulation will be to develop algorithms to evaluate these types of averages, but from a limited sample of the accessible microstates [1]. How to select that sample in an intelligent way, according to the likely frequency of microstates, is generally called *importance sampling*.

4.2.4 Averages of magnetic variables

In a magnetic model it is usual to have an interaction of the system with an externally applied field $\mathbf{H}(\mathbf{r})$. That interaction energy is described by a term in the Hamiltonian,

$$E_{\text{ext}} = -\int dV \, \mu_0 \mathbf{H}(\mathbf{r}) \cdot \mathbf{M}(\mathbf{r}) = -v_{\text{cell}} \mu_0 \sum_n \mathbf{H}_n \cdot \mathbf{M}_n. \tag{4.46}$$

The latter form accounts for the possibility of a spatially varying field at discrete cells, as would hold for micromagnetics or for a discrete spin system. It is helpful to use the cell magnetic dipole moments,

$$\boldsymbol{\mu}_n \equiv v_{\text{cell}} \mathbf{M}_n \tag{4.47}$$

and then write this interaction energy as

$$E_{\text{ext}} = -\mu_0 \sum_n \mathbf{H}_n \cdot \boldsymbol{\mu}_n. \tag{4.48}$$

This must be included in addition to any Hamiltonian that contains the internal magnetic energy of the system, E_{int}. The total Hamiltonian is $E = E_{\text{int}} + E_{\text{ext}}$. The average of the cell magnetic moments can be obtained using the canonical probabilities,

$$\langle \boldsymbol{\mu}_n \rangle = \frac{1}{Z} \sum_{\text{states}} \boldsymbol{\mu}_n \exp\{-\beta(E_{\text{int}} + E_{\text{ext}})\}, \tag{4.49}$$

where the partition function is

$$Z = \sum_{\text{states}} \exp\{-\beta(E_{\text{int}} + E_{\text{ext}})\}. \tag{4.50}$$

However, derivatives with respect to the local fields $\mathbf{H}_n$ can be used to pull out the local moments, because

$$\boldsymbol{\mu}_n = -\frac{1}{\mu_0} \frac{\partial E_{\text{ext}}}{\partial \mathbf{H}_n}. \tag{4.51}$$

Then it can be seen that the averaged value is

$$\langle \boldsymbol{\mu}_n \rangle = \frac{1}{Z} \frac{1}{\beta \mu_0} \sum_{\text{states}} \frac{\partial}{\partial \mathbf{H}_n} e^{-\beta(E_{\text{int}} + E_{\text{ext}})} = \frac{1}{\beta \mu_0} \frac{1}{Z} \frac{\partial Z}{\partial \mathbf{H}_n}. \tag{4.52}$$

This can also be put into a logarithmic form,

$$\langle \boldsymbol{\mu}_n \rangle = \frac{1}{\beta \mu_0} \frac{\partial}{\partial \mathbf{H}_n} \ln Z. \tag{4.53}$$

Using the definition of Helmholtz free energy (4.34), with $\beta F = -\ln Z$, this is

$$\langle \boldsymbol{\mu}_n \rangle = -\frac{1}{\mu_0} \frac{\partial F}{\partial \mathbf{H}_n}. \tag{4.54}$$

In the case of most interest, one might have an identical applied field $\mathbf{H} = \mathbf{H}_n$ at all of the cells, that is, a uniform applied field. If that is the case the derivative can be seen to produce the thermal average of the sum of all sites,

$$\langle \boldsymbol{\mu} \rangle = \left\langle \sum_n \boldsymbol{\mu}_n \right\rangle = -\frac{1}{\mu_0} \frac{\partial F}{\partial \mathbf{H}}. \quad (4.55)$$

This is the total magnetic moment of the whole system. Dividing by the volume of the system will give the volume-averaged magnetization,

$$\langle \mathbf{M} \rangle = \frac{1}{V} \langle \boldsymbol{\mu} \rangle. \quad (4.56)$$

Assuming there are N cells, the system volume is $V = N v_{\text{cell}}$ and the volume-averaged system magnetization is

$$\langle \mathbf{M} \rangle = \frac{1}{v_{\text{cell}}} \left\langle \frac{1}{N} \sum_n \boldsymbol{\mu}_n \right\rangle = -\frac{1}{V \mu_0} \frac{\partial F}{\partial \mathbf{H}}. \quad (4.57)$$

This could be thought of as a derivative of the free energy per unit volume, F/V. Alternatively, dividing equation (4.55) by N gives the averaged magnetic moment per grid site.

Exercise 4.1. Consider a collection of N very weakly coupled magnetic dipoles $\boldsymbol{\mu}_n$ of fixed magnitudes (i.e. Heisenberg-like spins) at temperature T within some volume V, that interact with a uniform applied field $\mathbf{H} = H\hat{\mathbf{z}}$. Assume the dipole states are a continuum defined by spherical polar angles (θ_n, ϕ_n) on a unit sphere. (a) Find an expression for the partition function Z from the definition (4.27). (b) Determine the expectation value of the energy and the specific heat per particle. (c) Find the expectation value of the magnetic moment per particle.

The next quantity of interest is the magnetic susceptibility tensor $\tilde{\chi}$, with components χ_{ab} defined by derivatives of the system's average magnetization (a, b are any of x, y, z),

$$\chi_{ab} \equiv \frac{\partial \langle M^a \rangle}{\partial H^b} = \frac{1}{V} \frac{\partial \langle \mu^a \rangle}{\partial H^b}. \quad (4.58)$$

A uniform field $\mathbf{H}$ is assumed to be applied to the system, and M^a is one component of the magnetization averaged over the whole system and statistically averaged for the given temperature (equation (4.57)). Then χ can be derived theoretically from the free energy,

$$\chi_{ab} = -\frac{1}{V \mu_0} \frac{\partial}{\partial H^b} \frac{\partial}{\partial H^a} F. \quad (4.59)$$

The order in which the derivatives are taken is irrelevant. Then the susceptibility tensor is symmetric in indices ab. It could depend on the magnitude of the applied field, but the susceptibility at zero field is usually an important limit quantity to calculate or measure.

> **Exercise 4.2.** For the collection of N very weakly coupled magnetic dipoles μ_n of fixed magnitudes at temperature T within some volume V of the previous exercise, find the components of its magnetic susceptibility tensor.

4.2.5 Averages from statistical fluctuations

The formulas derived so far for specific heat and magnetic susceptibility require knowledge of the Helmholtz free energy. In general, F is not found directly in simulations. However, both specific heat and susceptibility give indications of the rapid statistical fluctuations in the system. Specific heat indicates how the system temperature could change with response to added heat. Susceptibility indicates how the system magnetization is affected by applied magnetic field. Each of these quantities describes how easy it is for a small perturbation to change the system. Both C and χ_{ab} can be expressed in terms of average deviations, which are better for MC simulations.

One can obtain the heat capacity (at fixed **H**) without using the free energy:

$$C = \frac{\partial \langle E \rangle}{\partial T} = -k_B \beta^2 \frac{\partial \langle E \rangle}{\partial \beta}. \tag{4.60}$$

Using the average energy (4.39),

$$\frac{\partial \langle E \rangle}{\partial \beta} = \frac{\partial}{\partial \beta}\left[\frac{1}{Z}\sum_i e^{-\beta E_i} E_i\right]$$

$$= \left(\frac{\partial}{\partial \beta}\frac{1}{Z}\right)\sum_i e^{-\beta E_i} E_i + \frac{1}{Z}\left(\frac{\partial}{\partial \beta}\sum_i e^{-\beta E_i} E_i\right)$$

$$= \frac{1}{Z^2}\left(\sum_i e^{-\beta E_i} E_i\right)\left(\sum_i e^{-\beta E_i}\right) - \frac{1}{Z}\left(\sum_i e^{-\beta E_i} E_i^2\right). \tag{4.61}$$

This involves thermal averages; the first term is the square of the average energy, the second term contains the average squared energy, leading to a simple result

$$\frac{\partial \langle E \rangle}{\partial \beta} = \langle E \rangle^2 - \langle E^2 \rangle, \tag{4.62}$$

where

$$\langle E \rangle = \frac{1}{Z}\sum_i e^{-\beta E_i} E_i, \qquad \langle E^2 \rangle = \frac{1}{Z}\sum_i e^{-\beta E_i} E_i^2. \tag{4.63}$$

Then the heat capacity can be expressed as

$$C = k_B \beta^2 \left(\langle E^2 \rangle - \langle E \rangle^2\right). \tag{4.64}$$

That this is related to fluctuations can be seen by writing the average squared deviation δE of the energy away from its average value $\langle E \rangle$:

$$\langle (\delta E)^2 \rangle = \langle (E - \langle E \rangle)^2 \rangle = \langle (E^2 - 2E\langle E \rangle + \langle E \rangle^2) \rangle = \langle E^2 \rangle - \langle E \rangle^2. \quad (4.65)$$

Thus the heat capacity gives an indirect measure of energy fluctuations away from the mean. But more importantly, this shows that C can be calculated from energy fluctuations, without knowledge of the free energy function. That is a good approach to use in MC simulations.

For magnetic susceptibility, we can apply a similar approach, again bypassing the free energy. Start from an expression for the a-component of the averaged magnetic moment of the whole system, where $\boldsymbol{\mu} = \sum_n \boldsymbol{\mu}_n$,

$$\langle VM^a \rangle = \langle \mu^a \rangle = \frac{1}{Z} \sum_i e^{-\beta E_i} \mu^a. \quad (4.66)$$

Note that i labels microstates, which are summed over, and the energy includes internal and external contributions,

$$E = E_{\text{int}} + E_{\text{ext}} = E_{\text{int}} - \mu_0 \mathbf{H} \cdot \boldsymbol{\mu}. \quad (4.67)$$

Then a derivative of $-\beta E$ with respect to H^b pulls out the factor $\beta \mu_0 \mu^b$. This allows one to obtain the derivative of $\langle \mu^a \rangle$ as

$$\frac{\partial \langle \mu^a \rangle}{\partial H^b} = \frac{\partial}{\partial H^b} \left[\frac{1}{Z} \sum_i e^{-\beta E_i} \mu^a \right]$$

$$= \left(\frac{\partial}{\partial H^b} \frac{1}{Z} \right) \sum_i e^{-\beta E_i} \mu^a + \frac{1}{Z} \left(\frac{\partial}{\partial H^b} \sum_i e^{-\beta E_i} \mu^a \right)$$

$$= -\frac{\beta \mu_0}{Z^2} \left(\sum_i e^{-\beta E_i} \mu^b \right) \left(\sum_i e^{-\beta E_i} \mu^a \right) + \frac{\beta \mu_0}{Z} \left(\sum_i e^{-\beta E_i} \mu^a \mu^b \right). \quad (4.68)$$

This result contains the averages of components of $\boldsymbol{\mu}$, as well as a correlation term involving both components together. Dividing by the volume leads to the susceptibility component,

$$\chi_{ab} = \frac{1}{V} \frac{\partial \langle \mu^a \rangle}{\partial H^b} = \frac{\beta \mu_0}{V} \left(\langle \mu^a \mu^b \rangle - \langle \mu^a \rangle \langle \mu^b \rangle \right). \quad (4.69)$$

This is a useful form for calculating susceptibility, because it can be applied using averages made from a limited sample of states, as in MC simulations. Note also that the normalization factor in χ_{ab} depends on both the temperature and the system volume, or equivalently, the total number of grid sites.

Compare with the fluctuations of total magnetic moment $\boldsymbol{\mu}$ of the system away from its average value $\langle \boldsymbol{\mu} \rangle$. This requires finding the average square of the deviation $\delta \boldsymbol{\mu} = \boldsymbol{\mu} - \langle \boldsymbol{\mu} \rangle$. That is

$$\langle (\delta \boldsymbol{\mu})^2 \rangle = \langle \delta \boldsymbol{\mu} \cdot \delta \boldsymbol{\mu} \rangle = \sum_{a,b=x,y,z} \langle \delta \mu^a \delta \mu^b \rangle. \quad (4.70)$$

One of the terms here is

$$\langle \delta\mu^a \delta\mu^b \rangle = \langle (\mu^a - \langle\mu^a\rangle)(\mu^b - \langle\mu^b\rangle) \rangle = \langle \mu^a \mu^b \rangle - \langle \mu^a \rangle \langle \mu^b \rangle. \quad (4.71)$$

Therefore, each component of the susceptibility tensor is a measure of the fluctuations in the total magnetic moment of the system.

4.3 Importance sampling and the Metropolis algorithm

MC simulation is a way to take a model system through a sequence of microstates that imitates ergodicity. The basic idea is to take the system from a current state and then use an appropriate algorithm to choose a new state, and do this repeatedly, such that the whole sequence of states generated is close to that expected for thermal equilibrium, i.e. a Boltzmann distribution. A sequence with the desired equilibrium probabilities is called a *Markov chain*. It is somewhat random, and yet, there has to be an underlying structure or order, due to the particle interactions. Then this Markov chain of states can be used to calculate thermal equilibrium averages, provided that the states are selected in the correct probability distribution.

The Metropolis algorithm [2] is a simple but yet powerful way to generate a set of states that tends towards the Boltzmann distribution. It requires many iterations to do this, meaning that a long sequence of its states is to be generated and used in averages. The algorithm works in the *canonical ensemble*, which means that a temperature T is fixed. The other requirement is to know the Hamiltonian of the system, so that the system energy can be calculated for any allowed configuration. Some introductory texts on computational methods for physics [3, 4] give brief introductions to MC integration and MC simulations.

The Metropolis algorithm is a fundamental way to carry out importance sampling, selecting states essentially with a likelihood proportional to their Boltzmann factors, $\exp\{-\beta E\}$, where $\beta = (k_B T)^{-1}$. Let i label the initial microstate or configuration of the system, with energy E_i. MC works by making some change in the initial state to make a new *trial state*, labeled by j, with a new energy E_j. Physically, the system can be thought to change its energy by its weak interactions with a heat bath at temperature T. Sometimes the system will receive energy from the heat bath, and at other times the system will give energy back to the heat bath.

In the Metropolis algorithm, the trial state is either accepted or rejected based on the energy change,

$$\Delta E = E_j - E_i. \quad (4.72)$$

If the energy change is *negative*, the trial state is automatically accepted, and it becomes the new state of the system. We write for the acceptance probability,

$$p_{ij}(\Delta E) = 1, \quad \Delta E \leqslant 0. \quad (4.73)$$

On the other hand, if the energy change is *positive*, the trial state is only accepted with a probability equal to the relative Boltzmann factor,

$$p_{ij}(\Delta E) = \exp\{-\beta \Delta E\}, \quad \Delta E > 0. \quad (4.74)$$

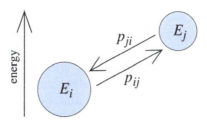

Figure 4.3. Two microstates between which the system can make transitions. If state j is higher in energy, then the Metropolis algorithm uses transition acceptance probabilities $p_{ij} = \exp\{-\beta(E_j - E_i)\}$ and $p_{ji} = 1$. This causes state i to be more highly populated, relative to state j, by the appropriate Boltzmann factor.

The subscripts ij indicate that this is the probability to make a transition from state i to state j. This result is summarized in figure 4.3, showing that transitions upward in energy have a diminished acceptance probability. In the limiting case that $\Delta E = 0$, this gives a probability of unity, and the trial state is accepted. But for an energy change large compared to $k_B T$, this probability quickly becomes exponentially small. In the Metropolis algorithm, then, negative energy changes are always accepted, while positive energy changes are only accepted sometimes, and less when the energy change is large. Note that the *acceptance probability* can be written compactly as

$$p_{ij}(\Delta E) = \min(1, \exp\{-\beta \Delta E\}). \tag{4.75}$$

This selects $p_{ij} = 1$ when $\Delta E \leq 0$ and $p_{ij} = \exp\{-\beta \Delta E\}$ when $\Delta E > 0$. Note that if the new trial state is rejected, then the system stays in its initial state. In the cases where the trial state is accepted, the system is updated into the new configuration. The total system energy is then changed by ΔE.

From a practical point of view, the decision to accept the trial state when $\Delta E > 0$ is made by comparing the Boltzmann factor $\exp\{-\beta \Delta E\}$ with a random number r uniformly distributed from 0 to 1. That is usually provided by a random number generator. The choice of random number generator is important if one is to avoid obtaining artificial periodicity in the sequence of MC states. There are many ways to take a poor random number generator and improve it, which can be found in texts on numerical analysis and computer simulations (such as Press *et al* [5]). Then for $\Delta E > 0$, the new trial state is accepted whenever

$$r < \exp\{-\beta \Delta E\}. \tag{4.76}$$

To verify that this works correctly, note that the Boltzmann factor $p_{ij} = e^{-\beta \Delta E}$ has already been fixed somewhere between 0 and 1 according to the value of ΔE. The number r is equally probable to be anywhere from 0 to 1. The larger $p_{ij}(\Delta E)$ is, the more likely r will fall below it. Conversely, when $p_{ij}(\Delta E)$ is closer to 0, it will be unlikely for r to be less than $p_{ij}(\Delta E)$. In this way the trial state will be accepted with the correct probability.

> **Exercise 4.3.** Consider a very small mass m in the space $z > 0$ above the Earth's surface with a gravitational potential energy in gravitational field g and entropy/thermal effects. Ignore its kinetic energy. Write a small computer program to implement the Metropolis algorithm for this problem. It only needs to keep track of the instantaneous altitude z of the particle, testing at each step whether some change $\Delta z = z' - z$ should be accepted or rejected. By collecting data over a long sequence of steps, one can form a histogram of the altitudes and numerically extract the probability distribution $p(z)$. Show that it obtains the expected form of a Boltzmann distribution,
>
> $$p(z) \propto \exp(-\beta m g z). \tag{4.77}$$
>
> For the histogram, the altitude could be divided into a set of bins of equal size δz, up to some maximum height. If the particle passes out of that range, the data can be ignored, and only data below that altitude will be used.

4.3.1 Averages with importance sampling

Once a sequence of equilibrium states are found in the MC simulation, those states can be used to calculate statistical averages as described in previous sections. However, the set of states is designed to be selected from the equilibrium (Boltzmann) distribution. Then, all states in the Markov chain are included into averages with equal weights. There is now no need to include a factor like $\exp\{-\beta E_i\}$, exactly because it is already accounted for by using importance sampling. Thus, the average of any quantity X is simply

$$\langle X \rangle = \frac{1}{N_S} \sum_{i=1}^{N_S} X_i, \tag{4.78}$$

where i labels a sequence of states obtained in the MC simulation, and the average is over N_S samples or states of that simulation. N_S may be the same or less than the total number of MC steps. This will give the same average as expected from the theoretical expression (4.38). Similarly, one can find quantities associated with fluctuations or deviations,

$$\langle X^2 \rangle = \frac{1}{N_S} \sum_{i=1}^{N_S} X_i^2. \tag{4.79}$$

Errors of different measurements are found the same way as in experimental science, combining $\langle X \rangle$ and $\langle X^2 \rangle$ to arrive at the standard error or standard deviation σ_X of a measurement. It is defined from the squared variation away from the mean,

$$\sigma_X^2 = \frac{1}{N_S} \sum_{i=1}^{N_S} (X_i - \langle X \rangle)^2 = \langle X^2 \rangle - \langle X \rangle^2. \tag{4.80}$$

The latter form is often used because it can be calculated on the fly (just keep updating the needed sums). But note that the sum of squares is a more stable numerical operation, but it requires saving all of the measurements before doing the sum.

Another comment is in order about σ_X for any measurement made in an MC simulation. It should be kept in mind that some of the variation found is purely statistical fluctuations due to a limited number of states being used, while another part could be due to the finite system size. For instance, it is well known that there will be natural fluctuations in average energy $\langle E \rangle$ of a system, that diminish with increasing system size as $1/N$, where N is the number of particles. If such finite size effects are present, taking more MC samples will not always lead to a reduction in the measurement error. Instead, the MC simulation truly gives a view of the physical fluctuations in a small system, which could be exactly what is interesting in some problems.

4.3.2 Why the Metropolis algorithm?

A proof is given here that the Metropolis algorithm leads to states with the correct probabilities given by the Boltzmann distribution. It is based on using the transitions between states, as produced by the algorithm. The transitions will reach an equilibrium situation as long as they satisfy the principle of *detailed balance*. This is the idea that transitions between any pair of states i and j, as in figure 4.3, should lead to a time-independent probability distribution for all microstates.

Let N_i represent the number of systems in an ensemble that are found in the microstate i. In one MC updating of the system, consider a possible transition from i into state j. This process is assumed to have a transition probability p_{ij} as given in the Metropolis algorithm, see figure 4.3. Then the product $N_i p_{ij}$ represents the number of systems that move from state i to j. Now consider also the reverse transition from state j back to state i, with reverse transition probability p_{ji}. The product $N_j p_{ji}$ gives the number of systems that move from j to i. In equilibrium, detailed balance dictates that the forward and reverse processes should balance, leaving the probability distribution unchanged:

$$N_i p_{ij} = N_j p_{ji}. \tag{4.81}$$

The MC simulation may start in a situation far from equilibrium, where one side of this expression is larger than the other, and transitions preferentially go one way. After many MC iterations or Monte Carlo steps (MCS), it is expected that equilibrium will be established and transitions between all pairs of states balance. Once that is true, (4.81) implies the relative equilibrium populations,

$$\frac{N_i}{N_j} = \frac{p_{ji}}{p_{ij}}. \tag{4.82}$$

If $E_i > E_j$, then $p_{ij} = 1$ while $p_{ji} = \exp\{-\beta(E_i - E_j)\}$. On the other hand, if $E_i < E_j$, then $p_{ij} = \exp\{-\beta(E_j - E_i)\}$ and $p_{ji} = 1$. Either way, the ratio of state populations is indeed that of the Boltzmann distribution,

$$\frac{N_i}{N_j} = \frac{\exp\{-\beta E_i\}}{\exp\{-\beta E_j\}}. \tag{4.83}$$

Obviously, the higher the energy of a state, the less likely it will be populated, in accordance with its Boltzmann factor $\exp\{-\beta E\}$.

4.3.3 Trial states for Monte Carlo simulations of magnets: single-spin updates

In MC simulation for magnets, a trial state is obtained by making some change in one or more of the magnetic moments (or spins) $\boldsymbol{\mu}_n$. In micromagnetics these are of fixed length μ. The same can be said for MC simulation of a collection of classical spins (replacing $\boldsymbol{\mu}_n$ by equivalent spin, $\gamma \mathbf{S}_n$). Then, an MC trial change needs to preserve the individual moment length. There are various ways to do this.

Consider single-spin updates. A selected $\boldsymbol{\mu}_n$ is to be changed (as a trial) to a new direction, along a unit vector $\hat{\mathbf{u}}$. This first method is a completely **uniform method**. The most unbiased way to do this is to select a new direction uniformly on the unit sphere, making all directions equally likely. Let the new direction be defined by a spherical polar angle θ (angle to the z-axis) and an azimuthal angle ϕ. The element of solid angle in spherical coordinates is either $d\Omega = \sin\theta d\theta d\phi$, or equivalently,

$$d\Omega = d(\cos\theta)d\phi. \tag{4.84}$$

The full 4π of solid angle on the sphere is obtained from

$$\int_{-1}^{+1} d(\cos\theta) \int_0^{2\pi} d\phi = (2)(2\pi) = 4\pi. \tag{4.85}$$

This means that one can select arbitrary points on the sphere with equal probabilities by choosing $\cos\theta$ uniformly between -1 and $+1$, and ϕ uniformly from 0 to 2π. These can be produced each by a call to a random number generator, which gives r in the range $0 \leqslant r \leqslant 1$. Therefore, let the new direction be selected by

$$c = \cos\theta = 2r_1 - 1, \qquad \phi = 2\pi r_2, \tag{4.86}$$

where r_1 and r_2 are two numbers from the generator. Once the angles are known, it will then also be helpful to calculate the projection of $\hat{\mathbf{u}}$ on the xy-plane,

$$s = \sin\theta = \sqrt{1 - \cos^2\theta}. \tag{4.87}$$

Then the unit vector for the new direction is calculated by

$$\hat{\mathbf{u}} = (u_x, u_y, u_z) = (s\cos\theta, s\sin\theta, c). \tag{4.88}$$

Rescaling by the conserved length μ, produces the new trial moment in the uniform method:

$$\boldsymbol{\mu}_n' = \mu \hat{\mathbf{u}}. \tag{4.89}$$

Based on the new $\boldsymbol{\mu}_n'$, the energy change ΔE can be calculated according to the system Hamiltonian (usually a local interaction), and the Metropolis scheme is used to accept or reject this trial change.

The drawback of the uniform new direction is that it may be rejected most of the time, especially at low temperature, where the magnetic fluctuations are small. Instead, a second method is to add a change to the original $\boldsymbol{\mu}_n$ and rescale back to the correct length. Call this the **increment method**. It is depicted in figure 4.4. A simple way to do this is to let the trial moment be

$$\boldsymbol{\mu}_n' = A(\boldsymbol{\mu}_n + \sigma \hat{\mathbf{u}}), \tag{4.90}$$

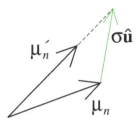

Figure 4.4. Increment method of generating a trial direction μ'_n for a magnetic moment (or spin from $\mu_n = \gamma S_n$), by adding a randomly directed unit vector $\hat{u}$ scaled by a length σ.

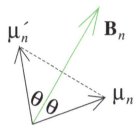

Figure 4.5. Reflection of a magnetic moment (or spin from $\mu_n = \gamma S_n$) across the effective magnetic field B_n in an over-relaxation operation.

where $\hat{u}$ is uniformly directed on the unit sphere, equation (4.88), and σ is a length parameter. Finally, A is used to rescale back to the original length μ. The parameter σ can be adjusted during the initial steps of the MC simulation to obtain a desired *acceptance rate*. The acceptance rate is the number of MC trials steps accepted divided by the number attempted. If σ is made very small, the changes in μ_n are minimal and most trials are accepted, but the system will move very slowly in phase space. Conversely, when σ is made larger, fewer trials will be accepted and the acceptance rate diminishes. If the acceptance rate is too low, there is a lot of calculation going on with little effect, and the system does not move enough through phase space. An acceptance rate around 30% to 40% will usually be good for allowing the algorithm to move adequately in the phase space for calculation of averages.

4.3.4 Over-relaxation updates

The single-spin trial states just described usually change the system energy. Another way to select new trial spin states, that are always accepted, is to move a spin in such a way that there is no energy change in the system. If $\Delta E = 0$, this is always acceptable in the Metropolis algorithm.

It should be possible to express an arbitrary local spin Hamiltonian in a form where each spin or magnetic moment interacts with an effective magnetic field B_n:

$$H = -\sum_n B_n \cdot \mu_n. \quad (4.91)$$

Even with anisotropic couplings or terms quadratic in μ_n this can be done; the field B_n may actually depend on μ_n. Then, a reflection of μ_n across the effective field B_n will preserve the angle between them, and hence the energy does not change. This is called over-relaxation, in reference to 'relaxation' being the operation of redirecting $\boldsymbol{\mu}_n$ to a point exactly along $\mathbf{B}_n$. The process is represented in figure 4.5.

To accomplish this in a program, first find the unit vector along the field direction, which is $\hat{\mathbf{b}}_n = \mathbf{B}_n/|\mathbf{B}_n|$. The component of $\boldsymbol{\mu}_n$ that is parallel to $\hat{\mathbf{b}}_n$,

$$\boldsymbol{\mu}_\| = \left(\hat{\mathbf{b}}_n \cdot \boldsymbol{\mu}_n\right)\hat{\mathbf{b}}_n \tag{4.92}$$

is unchanged by the reflection operation. The remaining part of $\boldsymbol{\mu}_n$, which is $\boldsymbol{\mu}_n - \boldsymbol{\mu}_\|$, is reversed in sign, see figure 4.5. Then using prime to indicate the new state, one moves $\boldsymbol{\mu}_n$ into

$$\boldsymbol{\mu}_n' = \boldsymbol{\mu}_\| - \left(\boldsymbol{\mu}_n - \boldsymbol{\mu}_\|\right) = 2\boldsymbol{\mu}_\| - \boldsymbol{\mu}_n. \tag{4.93}$$

This produces the following change in total magnetic moment:

$$\Delta\boldsymbol{\mu} = \boldsymbol{\mu}_n' - \boldsymbol{\mu}_n = 2\left(\boldsymbol{\mu}_\| - \boldsymbol{\mu}_n\right). \tag{4.94}$$

The over-relaxation operation can be applied to randomly selected moments in the system. Once one is moved, however, it will change the effective fields for other sites. This type of update cannot be used alone, but only in combination with other single-spin updates that change the energy, or in combination also with cluster updates, described later in this chapter.

4.4 Monte Carlo simulation of a 2D XY model

The 2D XY model (or XXZ model) is useful as an example application for the Metropolis algorithm. Consider the spin Hamiltonian (1.45) with no coupling of the z-components, so $J^z = 0$, and equal nearest neighbor couplings on x- and y-components, taking $J = J^x = J^y$:

$$H = -JS^2 \sum_{(i,j)} \left(s_i^x s_j^x + s_i^y s_j^y - 1\right) - \mu_0 \mathbf{H} \cdot \gamma S \sum_i \mathbf{s}_i. \tag{4.95}$$

The spins $\mathbf{s}_i$ are taken as dimensionless unit vectors. Their length S^2 has been factored out into the energy unit JS^2. The spins have three components, but their z-components do not enter in the Hamiltonian. Still, the z-components can be allowed to fluctuate, with an associated entropy in those fluctuations. The -1 in the exchange terms is included so that the minimum exchange energy is zero. The last term shows an interaction with a uniform external magnetic field $\mathbf{H}$, which is included here mainly for discussion of magnetic susceptibility. For the example simulations, however, $\mathbf{H}$ will be set to zero. Also recall that gyromagnetic ratio γ is present to convert angular momentum S into magnetic dipole moment at a site, $\boldsymbol{\mu}_i = \gamma \mathbf{S}_i$. This interaction term is sometimes written instead using a g-factor and the Bohr magneton, as $-(g\mu_B S)\mu_0 \mathbf{H} \cdot \mathbf{s}_i$.

When $\mathbf{H} = 0$, this XY model has full rotational symmetry in the xy-plane. Thus, there is no preferred direction for the net magnetic moment of the system within the xy-plane. In chapter 1, we noted that this symmetry allows for the presence of vortices: symmetrical configurations of spins with a flowing structure around a central core, recall figure 1.4, which shows a vortex with vorticity charge $q = +1$. There are also anti-vortices of opposite twist or vorticity charge $q = -1$. Following a circular path through angular displacement 2π around the core of an anti-vortex, the spins are seen to rotate through -2π. In a system with uniform magnetization, if a vortex is generated, there must be a corresponding anti-vortex generated, such that the total vorticity charge is conserved. These pairs affect the thermodynamics as they appear in a vortex unbinding phase transition above a critical temperature, also known as the Berezsinski–Kosterlitz–Thouless (BKT) transition. For now, we only note that this transition makes noticeable features in the specific heat and magnetic susceptibility.

4.4.1 Setting up the Monte Carlo simulation

In MC simulations where a phase transition might be present, it is important to carry out the calculations for a set of different system sizes, comparing the changes. Let us take the system to be a $L \times L$ square lattice of $N = L^2$ spins. To avoid edge effects, periodic boundary conditions are used. On each row of the grid in both directions, the last spin interacts with the first. This makes all sites equivalent and there is no edge. Every site in the square grid interacts with its four nearest neighbor sites.

For one MC run at a selected temperature T, the system is started with the spins in some appropriate initial state: we set all spins pointing in arbitrary random directions, using the algorithm (4.88) in 4.3.3. The system is then equilibrated for a number N_{skip} MCS, before any data are taken for averages. For a system with many spins or other degrees of freedom, generally, one MC step means that an *attempt* has been made to change every spin in the system by the Metropolis algorithm. One MC step corresponds to L^2 trial spin changes, and we do this on spins selected at random, to avoid any unphysical local correlations that might come from scanning through the grid. The MC step becomes the unit of 'time' for the simulation, although this is a fictitious time. For the example shown here, we used $N_{\text{skip}} = 5000$, although this may need to be selected longer if equilibrium is not established sufficiently fast. Trial new spins were selected using the increment method in section 4.3.3, equation (4.90), adding randomly oriented changes to a spin before renormalizing to unit magnitude.

4.4.2 Monte Carlo averages and errors

After equilibration, the Metropolis algorithm proceeds while data are saved for averages. This is carried out done over some number N_S of data samples. We use $N_S = 50\,000$, although now it is not uncommon to have simulations averaging over millions of MCS. Then from one such MC run, an average of the data samples X_i for some quantity is calculated as mentioned earlier,

$$\langle X \rangle = \frac{1}{N_S} \sum_{i=1}^{N_S} X_i. \tag{4.96}$$

The standard deviation could also be calculated from (4.80), however, it is not necessarily the best choice, because usually the sequence of data points X_i could be very strongly correlated. This means that the measurements X_i are not very independent, which leads to an underestimate of the error. This is because each MCS may not change the system or move it very far in phase step, especially at low temperature and also when the system is near a phase transition. Thus, one is always aware that there is a correlation time in the MCS sequence, and estimation of the errors of measurement can give trouble if the correlation time (measured in MCS) is large. One way to alleviate this effect, to some extent, is to skip some number of MCS between each data point X_i.

For estimating the error of the MC measurement, it is better to do something that gives more independent measurements, and use those independent measurements to estimate error bars. One way to achieve this is to partition the data sequence X_i into groups known as bins. For instance, the 50 000 MCS can be split into 5 bins of 10 000 MCS. The first 10 000 MCS go to the first bin, the second 10 000 to the second bin, and so on. An average is calculated within each bin. If you were calculating the average system energy $\langle E \rangle$, first you would find a result for each bin, call these E_p, $p = 1, 2, 3, 4, 5$. Then $\langle E \rangle$ comes from the average of the E_p or from the one long sequence. The result really does not depend on this partitioning. On the other hand, the error estimate is now made not from one long sequence of 50 000 (possibly highly correlated) points, but instead from the five bin averages. So we do this:

$$\langle X \rangle = \frac{1}{N_p} \sum_{p=1}^{N_p} X_p, \qquad \langle X^2 \rangle = \frac{1}{N_p} \sum_{p=1}^{N_p} X_p^2, \qquad (4.97)$$

where N_p is the number of bins (or partitions or averaged points), $N_p = 5$ in our example. Then the squared variation comes from the formula used before,

$$\sigma_X^2 = \langle X^2 \rangle - \langle X \rangle^2. \qquad (4.98)$$

The result, σ_X, represents the standard deviation one expects now from any single similar MC measurement of the quantity X. This variation could be partly due to the natural physical fluctuations due to temperature, as well as the extra statistical fluctuations introduced in the MC algorithm. However, keeping in mind that we actually made N_p measurements of X, from the N_p bins, the error given from the statistical theory of measurements is actually

$$\Delta X = \frac{\sigma_X}{\sqrt{N_p - 1}}. \qquad (4.99)$$

So the measurement from MC simulation is quoted as $\langle X \rangle \pm \Delta X$. The factor $\sqrt{N_p - 1}$ shows the slow improvement in our knowledge about X as we do more measurements. Usually it is applied as $\sqrt{N_p}$, assuming that N_p is not close to 1. Of course, the factor $N_p - 1$ gives a divergent error bar when only one measurement is made, which is just to say that a single measurement is not reliable on its own.

One can argue that splitting data into bins does not always help very much for estimate of errors. If that is the case, one can go a step further, and simply do N_p separate MC runs for one temperature, starting them from different random initial conditions, obtaining N_p much more independent estimates of X. The average and error can then be calculated from these values, just as explained for the averaging over bins. For the example here, this is the procedure used, with $N_p = 4$. This is much less likely to underestimate the error.

4.4.3 Another approach for errors of fluctuation quantities

The difficulty with error estimation tends to appear more for quantities that measure fluctuations, such as specific heat and magnetic susceptibility. If data are not strongly correlated, consider first another way to estimate their error bars from a single long sequence of data, with N_S samples.

Suppose that the sequence of energies E_i is obtained from an MC simulation. Using $X = E$ in (4.78) and (4.79), one obtains the mean value $\langle E \rangle$ and mean squared value $\langle E^2 \rangle$. These will give the heat capacity from (4.64), and then specific heat c by dividing by particle number N. But what about its error, without breaking the data into bins?

Recall that the heat capacity is related to the fluctuations in energy, as measured by the standard deviation σ_E,

$$C = k_B \beta^2 \langle (E - \langle E \rangle)^2 \rangle = k_B \beta^2 \sigma_E^2. \qquad (4.100)$$

Note that the error in the energy measurement could then be found from

$$\Delta E = \frac{\sigma_E}{\sqrt{N_S}}. \qquad (4.101)$$

Now C is dependent on the average of the following fluctuating quantity:

$$z \equiv (E - \langle E \rangle)^2. \qquad (4.102)$$

The mean value of this is

$$\langle z \rangle = \langle E^2 \rangle - \langle E \rangle^2 = \sigma_E^2. \qquad (4.103)$$

If now we want the variance in C, it is determined by the variance of z. That is obtained from the usual procedure,

$$\sigma_z^2 = \langle (z - \langle z \rangle)^2 \rangle = \left\langle [(E - \langle E \rangle)^2 - \langle E^2 \rangle + \langle E \rangle^2]^2 \right\rangle. \qquad (4.104)$$

This should be expressed in a way that can be calculated from the sequence of energy samples. Expanding and simplifying, it becomes a combination of different averages of the energy,

$$\sigma_z^2 = \left\langle [E^2 - 2E\langle E \rangle + 2\langle E \rangle^2 - \langle E^2 \rangle]^2 \right\rangle$$
$$= \langle E^4 \rangle - \langle E^2 \rangle^2 + 4\langle E \rangle [2\langle E \rangle \langle E^2 \rangle - \langle E^3 \rangle - \langle E \rangle^3]. \qquad (4.105)$$

To calculate this way, in addition to sums of E_i and E_i^2 one must also save sums for E_i^3 and E_i^4. The error in measurement of z is $\sigma_z/\sqrt{N_S}$. Then the error in the system heat capacity is

$$\Delta C = k_B \beta^2 \frac{\sigma_z}{\sqrt{N_S}}. \qquad (4.106)$$

Finally, this can be divided by N to obtain the error in the specific heat per particle.

For the magnetic properties for this model, the total magnetic moment of the system is

$$\boldsymbol{\mu} = \gamma \sum_{n=1}^{N} \mathbf{S}_n = \gamma S \sum_{n=1}^{N} \mathbf{s}_n. \quad (4.107)$$

Factor γS gives the units. The sum over all sites gives the total magnetic moment, and this is averaged over the MC states that are generated. This total $\boldsymbol{\mu}$ is then used to find susceptibility. For presentation of results, it may be convenient to show a dimensionless magnetic moment (per site),

$$\mathbf{m} = \frac{\boldsymbol{\mu}}{\gamma S N} = \frac{1}{N} \sum_{n=1}^{N} \mathbf{s}_n. \quad (4.108)$$

A component from the susceptibility tensor, equation (4.69), is obtained by putting $V = Na^2$, where a is the lattice constant of the square grid of spins. In this way, magnetization in a 2D system is magnetic dipole moment per unit area, rather than per unit volume. Then

$$\chi_{ab} = \frac{\beta \mu_0}{Na^2} \left(\langle \mu^a \mu^b \rangle - \langle \mu^a \rangle \langle \mu^b \rangle \right). \quad (4.109)$$

It is convenient to rearrange this as

$$\chi_{ab} = \chi_0 [\beta J S^2 N(\langle m_a m_b \rangle - \langle m_a \rangle \langle m_b \rangle)], \quad (4.110)$$

where the natural unit of susceptibility,

$$\chi_0 = \mu_0 \gamma^2 / J a^2 \quad (4.111)$$

multiplies a dimensionless factor (in square brackets) that is calculated in simulations. For the diagonal components, which are the most significant, χ involves the standard deviation of a magnetization component. For instance,

$$\frac{\chi_{xx}}{\chi_0} = \beta J S^2 N \left(\langle m_x^2 \rangle - \langle m_x \rangle^2 \right) = \beta J S^2 N \langle (m_x - \langle m_x \rangle)^2 \rangle = \beta J S^2 N \sigma_{m_x}^2. \quad (4.112)$$

This shows that χ_{xx} is proportional to the average of the fluctuating quantity,

$$z_{m_x} \equiv (m_x - \langle m_x \rangle)^2. \quad (4.113)$$

Then one can see that the error in χ_{xx} is found from a relation similar to that found for heat capacity. Namely, the error is determined from the variance of z_{m_x},

$$\frac{\Delta \chi_{xx}}{\chi_0} = \beta J S^2 N \frac{\sigma_{z_{m_x}}}{\sqrt{N_S}}, \quad (4.114)$$

where

$$\sigma_{z_{m_x}}^2 = \left\langle \left[m_x^2 - 2 m_x \langle m_x \rangle + 2 \langle m_x \rangle^2 - \langle m_x^2 \rangle \right]^2 \right\rangle$$

$$= \langle m_x^4 \rangle - \langle m_x^2 \rangle^2 + 4 \langle m_x \rangle \left[2 \langle m_x \rangle \langle m_x^2 \rangle - \langle m_x^3 \rangle - \langle m_x \rangle^3 \right]. \quad (4.115)$$

The same approach can be applied for the errors in χ_{yy} and χ_{zz}.

4.4.4 Spatial correlation function and susceptibility

The relation (4.112) shows that the magnetic susceptibility χ_{xx} is determined by the squared variance $\sigma_{m_x}^2$ of a corresponding component of the dimensionless magnetic moment per spin, m_x, defined in (4.108). Using those definitions, it can be expressed in a different way:

$$\sigma_{m_x}^2 = \langle (m_x - \langle m_x \rangle)^2 \rangle = \left\langle \left(\frac{1}{N} \sum_{\mathbf{n}} s_{\mathbf{n}}^x - \frac{1}{N} \sum_{\mathbf{n}} \langle s_{\mathbf{n}}^x \rangle \right)^2 \right\rangle = \left\langle \left[\frac{1}{N} \sum_{\mathbf{n}} (s_{\mathbf{n}}^x - \langle s_{\mathbf{n}}^x \rangle) \right]^2 \right\rangle. \quad (4.116)$$

The lattice indices are written in bold here to indicate that they represent the vector locations of the sites. In an XY symmetry model the average $\langle s_{\mathbf{n}}^x \rangle = 0$ for any temperature in the thermodynamic limit. These averages would not need to be identically zero in a simulation on a finite system. In any case, we include them and now rewrite the square within the average as a product of two sums, together with some further rearranging,

$$\sigma_{m_x}^2 = \left\langle \frac{1}{N^2} \sum_{\mathbf{n}} (s_{\mathbf{n}}^x - \langle s_{\mathbf{n}}^x \rangle) \sum_{\mathbf{n}'} (s_{\mathbf{n}'}^x - \langle s_{\mathbf{n}'}^x \rangle) \right\rangle$$

$$= \frac{1}{N} \sum_{\mathbf{r}} \left[\frac{1}{N} \sum_{\mathbf{n}} \langle (s_{\mathbf{n}}^x - \langle s_{\mathbf{n}}^x \rangle)(s_{\mathbf{n+r}}^x - \langle s_{\mathbf{n+r}}^x \rangle) \rangle \right] = \frac{1}{N} \sum_{\mathbf{r}} C^{xx}(\mathbf{r}), \quad (4.117)$$

where the shift $\mathbf{n}' \to \mathbf{n} + \mathbf{r}$ was applied, and we introduce the static correlation function of pairs of spins at fixed space displacement,

$$C^{xx}(\mathbf{r}) = \frac{1}{N} \sum_{\mathbf{n}} \langle (s_{\mathbf{n}}^x - \langle s_{\mathbf{n}}^x \rangle)(s_{\mathbf{n+r}}^x - \langle s_{\mathbf{n+r}}^x \rangle) \rangle. \quad (4.118)$$

The index $\mathbf{r}$ represents the vector displacement from one site to another. It applies to any number of space dimensions, with indices $\mathbf{n}$ and $\mathbf{n} + \mathbf{r}$ representing the vector locations of the lattice sites. The correlation function involves a sum over pairs of spins at a fixed displacement labeled by $\mathbf{r}$. Here in its general form the averaged values are subtracted out, although this is not necessary if one can expect those averages to be zero. Changing the sum to an integral, with total system volume V equal to Nv_{cell}, one can define the correlation functions in a continuum expression,

$$C^{xx}(\mathbf{r}) = \frac{1}{V} \int d\mathbf{n} \langle (s^x(\mathbf{n}) - \langle s^x(\mathbf{n}) \rangle)(s^x(\mathbf{n} + \mathbf{r}) - \langle s^x(\mathbf{n} + \mathbf{r}) \rangle) \rangle. \quad (4.119)$$

Obviously there are related correlation functions defined for the other spin components in the same way. Both $C^{xx}(\mathbf{r})$ and χ_{xx} can be calculated independently in a simulation, however, this shows that they must be related by

$$\frac{\chi_{xx}}{\chi_0} = \beta J S^2 \sum_{\mathbf{r}} C^{xx}(\mathbf{r}) \quad \text{or} \quad \frac{\chi_{xx}}{\chi_0} = \beta J S^2 \int \frac{d\mathbf{r}}{v_{\text{cell}}} C^{xx}(\mathbf{r}) \quad (4.120)$$

where the last form is the continuum expression. This gives a way to check the consistency of susceptibility calculations, and shows a fundamental physical relation between magnetic fluctuations and spatial correlations. Typically, $C^{xx}(\mathbf{r})$ can be expected to decay with $|\mathbf{r}|$, either as a power law at low temperature or some exponential form at high temperature. A slower decay in $C^{xx}(\mathbf{r})$ will directly translate into a larger value of the susceptibility, aside from the dependence of χ_{xx} on the factor β in (4.120).

4.4.5 XY model Monte Carlo results

Here the results of MC simulations for an $L \times L$ array of spins with $L = 8, 16, 32, 64$ are presented. As mentioned above, there were 5000 MCS used to equilibrate the system, starting from a random initial condition. Then $N_S = 50\,000$ MCS were used for averages of a single run. Finally, four such runs were performed ($N_p = 4$) to obtain four data points for the estimation of errors. A small number is being used here to show significant error bars. However, the total of 200 000 MCS is seen to be reasonably long, except for the largest system simulated here. The MC trial move is that a chosen spin is redirected by adding small increments in arbitrary directions. The lengths of those increments (σ in (4.90)) were changed during the initial equilibration until the acceptance rate fell between 10% and 40%, but σ was not allowed to exceed $\sigma = 100$.

There is no applied magnetic field. The natural energy unit being used here is JS^2. Then energies and the temperature are quoted in terms of JS^2. For all spins parallel in the xy-plane, the total energy of the Hamiltonian (4.95) is zero; this is the classical ground state. Then the total system energy is always positive.

For comparing results on different sized systems, it is best to show the energy per site, the specific heat per site, and so on, where the number of sites is $N = L^2$. Results for the system energy per site $e(T) = \langle E \rangle / N$ and specific heat per site, $c(T) = C/N$, are shown in figure 4.6. One can see that the energy is a monotonically increasing function of temperature, as needed so that c is positive. In the limit of $T \to 0$, the slope of $e(T)$ tends towards unity, then $c(T) \to k_B$ there, reflecting that fact that each classical spin has an average energy of $k_B T$ for low temperature. Obviously this is not what would be expected for a quantum spin model. At the opposite limit of very large T, because there is an upper limit for energy per site ($2JS^2$ per bond of the lattice for anti-parallel spins), there is a maximum energy, and the specific heat tends to zero for very high T.

There is very little dependence of $e(T)$ on system size, however, a greater difference can be seen in $c(T)$, which shows a distinctive peak. The peak tends to move to lower temperature and somewhat greater height as L increases. For the largest size simulated here, $L = 64$, however, the lower temperature results are not particularly reliable, as indicated by the large error bars. This is because the MC algorithm being used becomes rather inefficient at moving the system through phase space for very low temperature, or in the presence of larger fluctuations.

Consider next magnetic properties. Because of the symmetry in the xy-plane, the total system magnetic moment will be able to rotate freely in the xy-plane. Strictly speaking, the average magnetic moment, as a dimensionless quantity,

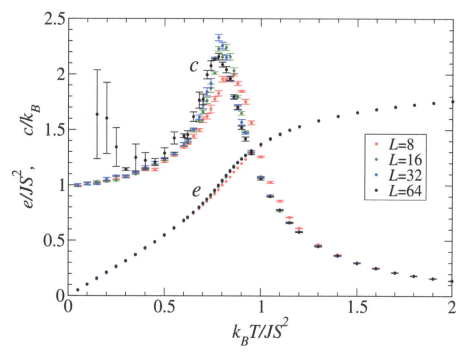

Figure 4.6. Energy and specific heat per spin as functions of temperature, for 2D XY models of indicated sizes, from MC simulations. The error bars on energy are smaller than the symbols. The specific heat values at low T for $L = 64$ (black symbols) have very large errors, indicative of a lack of efficiency in the MC algorithm.

$$\mathbf{m} = \langle \boldsymbol{\mu} \rangle / (\gamma S N) \qquad (4.121)$$

will be zero (except for statistical fluctuations). In order to look at a kind of ordering within the xy-plane, one can instead perform an averaging of the magnitude of $\boldsymbol{\mu}$, but looking only at the in-plane components. So define the averaged in-plane component,

$$m_{xy} = \frac{\mu_{xy}}{\gamma S N} \equiv \frac{1}{N} \sum_n \sqrt{(s_n^x)^2 + (s_n^y)^2} \qquad (4.122)$$

which does not average out to zero. Similarly, there will be an associated susceptibility χ_{xy}, based on the fluctuations of m_{xy}. For $L = 32$, a component m_x is compared to m_{xy} in figure 4.7. It is striking that $m_{xy}(T)$ makes a smooth increase towards the maximum possible value of 1 as $T \to 0$, while in the same temperature region $m_x(T)$ shows strong fluctuations. This region exists below the temperature where the specific heat is maximum.

In figure 4.8 $m_{xy}(T)$ is shown for different system sizes. This graph shows a striking result. For increasing L, the values of m_{xy} consistently decrease, but especially for temperatures greater than $k_B T/JS^2 \sim 0.75$, where the slope of $m_{xy}(T)$ is largest. There are rather large error bars in the results for $L = 64$. For the smaller systems, m_{xy} tends almost linearly to approach the value $m_{xy} = 1$ at zero

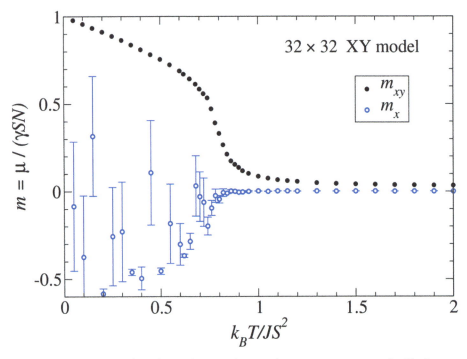

Figure 4.7. The temperature-dependence of averaged magnetic moment m_x, compared with the average magnitude of magnetic moment in the xy-plane, m_{xy}. The error bars on m_{xy} are smaller than the symbols. The large fluctuations in m_x indicate a slowing down or lack of efficiency of the MC algorithm.

temperature, when the system should fall into the ground state. The region $k_BT/JS^2 \sim 0.75$ is known to be the location of the vortex-unbinding phase transition. If one could simulate an infinite size system, theory says that m_{xy} goes to zero at a critical temperature T_c, and then remains zero for all temperatures above T_c. Only for $T < T_c$ would there be a non-zero value of m_{xy}, which acts as an *order parameter* and gives a direct indication of the state of ordering in the system, which is strong at low temperature.

The susceptibility χ_{xy} associated with fluctuations of $\mu_{xy} = \gamma SN m_{xy}$ is shown in figure 4.9. It has been calculated from

$$\chi_{xy} = \frac{\beta\mu_0}{Na^2}\left\langle\left(\mu_{xy} - \langle\mu_{xy}\rangle\right)^2\right\rangle = \chi_0\left[\beta JS^2\, N\left\langle(m_{xy} - \langle m_{xy}\rangle)^2\right\rangle\right]. \qquad (4.123)$$

In the second form, the unit χ_0 was defined in (4.111) and the factor in brackets involving m_{xy} is the calculated dimensionless quantity. This susceptibility is even more indicative of the vortex-unbinding phase transition than the changes in m_{xy}. The results for $L = 64$ have been left out here, because of the problems for low temperature on this relatively short MC simulation. For the other three sizes, there is a very strong peak in $\chi_{xy}(T)$, in the same position where the slope of $m_{xy}(T)$ is maximum. The height of the peak increases rapidly with the number of sites, $N = L^2$. This shows the very strong correlations (over the whole system) that take

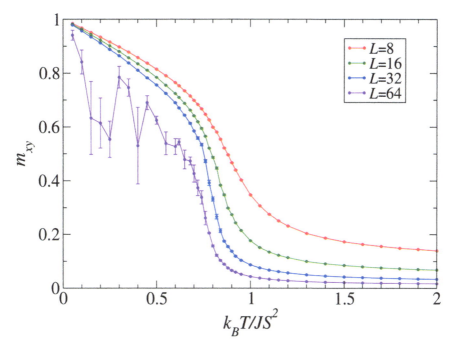

Figure 4.8. The temperature-dependence of averaged magnitude of magnetic moment in the xy-plane, m_{xy}, for the 2D XY model. The larger fluctuations for $L = 64$ indicate a slowing down or lack of efficiency of the MC algorithm to move the system in all of the accessible phase space.

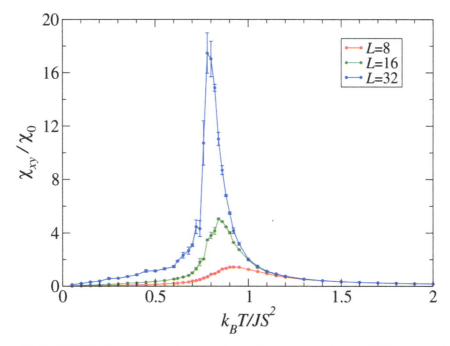

Figure 4.9. For 2D XY model, the temperature-dependence of in-plane magnetic susceptibility, χ_{xy}, calculated from fluctuations in m_{xy} using (4.123). The unit being used is $\chi_0 = \mu_0 \gamma^2 / Ja^2$.

place in this critical region, where vortex unbinding is taking place most strongly. It is also indicative of the fact that there is a correlation length in the system that tends to become as large as the system in the critical region. For an infinite sized system, the correlation length would diverge at the critical temperature.

The larger fluctuations for $L = 64$ show that there is something wrong with the MC simulation. The obvious fix is to average over more states, realizing that the average of only four numbers cannot be too reliable. However, when even larger systems are simulated, this may not remove the large fluctuations, and even more states will need to be used to obtain good results. The difficulty is that once the system begins to order, there are strong correlations between spins, and updating them by doing single-spin updates becomes very inefficient. This is especially true near a critical temperature, and as well, at low temperature. This inefficiency of MC simulation is related to a feature of critical dynamics called *critical slowing down*. Effectively, the strong correlations make it difficult for single-spin Metropolis updates to move the system efficiently through the accessible phase space. It gets stuck in regions, then averages are made over a limited part of the available phase space. This can be corrected by carrying out the updating using *cluster algorithms* and other multi-spin moves that account for strong correlations and try to move many spins together. This is discussed below.

A sense of the typical spin configurations for a low temperature ($k_BT/JS^2 = 0.10$), an intermediate temperature ($k_BT/JS^2 = 0.70$ near the transition), and a high temperature ($k_BT/JS^2 = 1.0$), are shown in figures 4.10–4.12. At low temperature, the total system magnetic moment is quite large, but can choose any random direction. It is difficult for single-spin updates to move the net magnetic moment around, however. That is really the reason for the slowing down of the MC algorithm for low T. For the intermediate temperature, vortex–anti-vortex pairs begin to appear, usually separated by only one lattice constant. With only slightly higher temperature, they can begin to unbind, exhibiting the BKT transition. Finally, at higher temperatures there are many more vortices and anti-vortices, and some even become somewhat free from any neighboring vortices or anti-vortices.

More advanced analysis of the BKT vortex unbinding transition is discussed in a later chapter. Next, cluster updating as a way to make MC more efficient, especially in critical regions, is presented.

4.5 Cluster algorithms for spin updates

In the XY model at low temperature, the spins tend to become aligned not only with their nearest neighbors, but aligned over longer distances. A similar effect also occurs near a critical temperature, where the fluctuations extend over the whole system. The distance over which spins tend to stay aligned is called the correlation length ξ. For any simulation, however, the correlation length is restricted to always be at most the size of the system being simulated. But once that happens, the MC algorithm presented so far becomes somewhat useless.

In the single-spin MC updating algorithm that attempts to change one spin at a time, only very small angular changes in in-plane spin components will be accepted

Figure 4.10. A configuration of a 16 × 16 2D XY model at $k_B T/JS^2 = 0.10$, showing the projection of spins onto the xy-plane. Arrows drawn as → (–▷) indicate positive (negative) z-components of the spins. This was generated by the single-spin update algorithm. A low temperature state is very organized, with a large in-plane magnetization, but of fluctuating direction.

if the system is strongly correlated. This means that even performing more and more MCS will not move the system very much in phase space. The system remains only in a limited region of phase space, and this effect becomes all the more worse for larger systems. Instead, one needs to use what are known as *cluster algorithms* to build linked spins that are all updated simultaneously. The idea is to change a whole set of spins all together, chosen in such a way that the net energy change is either negative, or only slightly positive. A negative energy change will be acceptable, and a positive energy change will only be acceptable according to a Boltzmann factor. Physically, the system in this situation moves in phase space in such a way that many spins change simultaneously, and a cluster algorithm tries to mimic that organized motion.

The method of building clusters depends somewhat on the interactions and types of spins. While we are most interested in Heisenberg spins having three components, it is instructive to first look at a basic cluster algorithm for Ising spins, with one component.

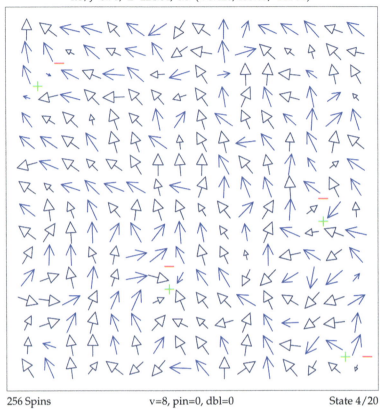

Figure 4.11. A configuration of a 16 × 16 2D XY model at $k_B T/JS^2 = 0.70$, like that in figure 4.10. Arrows drawn as → (—▷) indicate positive (negative) z-components of the spins. At this temperature the spins move more out of the xy-plane. Also, vortex–anti-vortex pairs, indicated by +/− signs, begin to appear in the BKT vortex unbinding transition.

4.5.1 Ising spins: Swendsen–Wang cluster algorithm

The Ising model [6] in two and higher dimensions has always attracted a lot of attention for being one of the simplest models to show a phase transition[2]. Although an analytic solution for some of the thermodynamic properties was found by Onsager [7], it is an ideal playground for learning about numerical simulations. To accurately study the critical properties, a cluster MC algorithm must be applied.

The Ising Hamiltonian with only nearest neighbor interactions has the Hamiltonian with ferromagnetic exchange constant J,

$$H = -JS^2 \sum_{(i,j)} (s_i s_j - 1). \qquad (4.124)$$

---

[2] In one dimension the Ising model is exactly solvable and shows no phase transition.

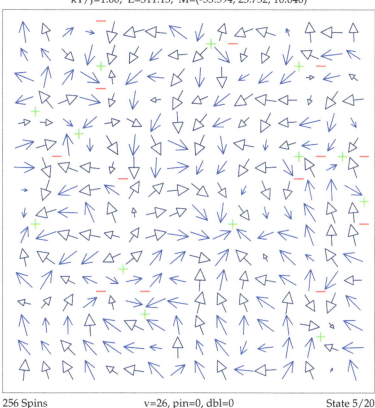

Figure 4.12. A configuration of a 16 × 16 2D XY model at $k_B T/JS^2 = 1.0$, like that in figure 4.10. Arrows drawn as →(-▷) indicate positive (negative) z-components of the spins. Besides having larger magnitude S^z-components, there are many more vortex–anti-vortex pairs being generated, in this higher temperature more disordered state. Some vortices appear to be relatively free from nearby vortices/anti-vortices.

Each spin has only two values, $s_i = \pm 1$. The spin length S is included in the exchange energy. The ground state with all spins aligned (all $s_i s_j = +1$) has zero energy. Any bond between neighboring anti-aligned spins adds energy $+2JS^2$ to the system.

The only way to modify an individual spin is to *flip it*, i.e. reverse its sign. Of course, in a 2D square lattice of spins, flipping one spin will change the energies in the four bonds to the four nearest neighbors. That could be a large energy if all the neighboring spins are aligned. A way to make it a much smaller energy is to consider whether some of the neighbors of the first spin should also be flipped. When a spin and a neighbor are flipped together, there is no change in energy in that bond. Then, for any cluster of spins connected by bonds, flipping the cluster only costs the amount of energy in the bonds on the boundary of the cluster.

In the Swendsen–Wang algorithm [8], a cluster of spins to be flipped together is built as follows. First, an initial spin s_i is selected to be in the cluster. Let us suppose

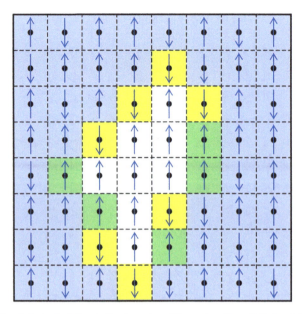

Figure 4.13. Example of cluster building in the Swendsen–Wang algorithm. Cells in white have already been included into the cluster and have already been flipped to their new states s'_j. Cells in green and yellow border the cluster and are not yet included. Green cells are now aligned with the cluster; they will certainly not be included, because they have negative energy changes on their bonds to the cluster. Yellow cells need to undergo the Metropolis test (4.129) with $\Delta E_{ij} > 0$ if no link is placed to the cluster.

it is immediately flipped, reversing its sign. Using a prime to show the new value, and unprimed to show the original value,

$$s'_i = -s_i. \tag{4.125}$$

The boundary of that one-spin 'cluster' is then its four nearest neighbors. Next, those neighbors are considered to be included into the cluster, or not, see figure 4.13. Let s_j be one of those neighbors. If it is included into the cluster, after flipping, $s'_i s'_j = s_i s_j$ and the bond ij will not be changed in energy. For every new site included into the cluster, the energy change of the included bond will be zero, because both are flipped. It is only when a site *is not included* into the cluster, that the bond's energy change will be different from zero.

When a new site is included, one imagines a link from s_i to s_j. Therefore, we calculate the bond's energy change if a link is *not made*, which is

$$\Delta E_{ij} = E_{ij}(\text{no link}) - E_{ij}(\text{link}), \tag{4.126}$$

where

$$E_{ij} = -JS^2 s_i s_j = +JS^2 s'_i s_j. \tag{4.127}$$

Of course, when no link is made, the sign of $s_i s_j$ reverses. So this bond's energy change when a link is *not made* is

$$\Delta E_{ij} = 2JS^2 s_i s_j = -2JS^2 s'_i s'_j, \qquad (4.128)$$

where $s_i s_j$ is the value of the bond before any flipping. Then, when s_i and s_j start out aligned, with $s_i s_j = +1$, if no link is placed between them, there will be a large positive energy cost as they are misaligned. On the other hand, if initially $s_i s_j = -1$, not placing a link will cause the energy on that bond to go down, as the spins will become aligned. Then the decision is made whether to put *no link* according to a Boltzmann factor, just as in the Metropolis algorithm. The probability to put *no link* on bond ij is taken as

$$p_{\text{nolink}}(\Delta E_{ij}) = \begin{cases} 1 & \Delta E_{ij} \leqslant 0 \\ \exp\{-\beta \Delta E_{ij}\} & \Delta E_{ij} > 0. \end{cases} \qquad (4.129)$$

Therefore, this is a Metropolis decision where the default action (always no energy cost) is to put a link. This probability is used to decide when not to put a link. Occasionally no link will be placed on the bond, and the cluster does not grow out in that direction. When a link is included, the newly linked spin is flipped, with new value $s'_j = -s_j$. Note that alternatively, when $\Delta E_{ij} > 0$, the probability to put a link is then

$$p_{\text{link}}(\Delta E_{ij}) = 1 - \exp\{-\beta \Delta E_{ij}\}, \qquad \Delta E_{ij} > 0. \qquad (4.130)$$

Suppose r is a uniform random number from 0 to 1, from a random number generator. Then, a link can be placed when $p_{\text{link}} > r$. Noting that $(1 - r)$ is the same as another uniform random number r', a link can be placed when

$$r' - \exp\{-\beta \Delta E_{ij}\} > 0. \qquad (4.131)$$

Otherwise, no link is placed on the bond.

An example state with a growing cluster is shown in figure 4.13. The spins in white cells are the cluster, and they have already been flipped. For simplicity, we suppose all spins in the cluster are aligned, although that would not have to be the case. Initially they were all pointing down. The cells that border the cluster are green and yellow. The green cells' spins, however, are now aligned with the spins in the cluster. It means the flip applied to the cluster sites brought the green sites' bonds to a lower energy, and then those green sites will *not* be included into the cluster. The spins in yellow cells point opposite to the cluster's spins. If no link is placed to the green cells, their bond energies will go up. These sites then need to be tested according to (4.129) with $\Delta_{ij} > 0$ to decide whether they should be linked into the cluster or not.

The decision of not putting links, governed by probability (4.129), is then iteratively applied to all bonds at the boundary of the cluster. The cluster is allowed to grow at will. Eventually it may stop growing, which is the case for higher temperature. At lower temperature, a typical cluster can grow very large, even spanning across the entire system. Once the cluster has finished growing, every spin in it has been flipped. The sum of all bonds' energy changes ΔE_{ij} will be the net

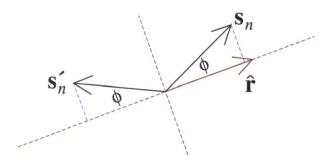

Figure 4.14. Reflection of the component of a spin along a randomly chosen axis $\hat{r}$. Note the similarity to an over-relaxation operation, figure 4.5.

energy change of the system. The energy changes where links are formed will cancel out to zero. In the end, the only energy changes are at the boundary of the cluster.

For programming of an algorithm, one needs to keep track of the sites in the cluster, which usually involves some array to mark the sites in the current cluster being built. Then, for those sites, a table of the nearest neighbors is needed, for testing for inclusion into the cluster. These are the main technical details needed to avoid attempting to add a site to the cluster more than once.

4.5.2 Heisenberg spins: the Wolff cluster algorithm

The Swendsen–Wang algorithm works on a single-spin component. It can be adapted to Heisenberg spins with three components, making what is known as the Wolff algorithm [9]. It can also be applied to the XY model, having it act to change only the xy-components of the spins.

First consider a model with isotropic Heisenberg exchange,

$$H = -JS^2 \sum_{(i,j)} (\mathbf{s}_i \cdot \mathbf{s}_j - 1) = -JS^2 \sum_{(i,j)} \left(s_i^x s_j^x + s_i^y s_j^y + s_i^z s_j^z - 1 \right). \quad (4.132)$$

The -1 is included with the dot product to set the ground state energy to zero. The couplings of x-components, taken alone, look mathematically just like the Ising Hamiltonian. Of course, the only difference is that the magnitudes of s_i^x and s_j^x are not fixed. Regardless of that, the interaction of a nearest neighbor pair will be unchanged if both have their s^x-component reversed. Stated equivalently, the pair interaction is unchanged by a reflection of both vector spins along the x-axis: $s_i^{x\prime} = -s_i^x$, $s_j^{x\prime} = -s_j^x$. However, the reflection axis need not be the x-axis. It would work just as well using the y-axis, the z-axis, or any other randomly chosen axis. This is the main idea in the Wolff algorithm. The 'flipping' of a spin in the Swendsen–Wang algorithm is effected by the 'reflection' of a spin along some random axis, see figure 4.14. Every site that will be added to a cluster being built, will be reflected along the same axis.

Therefore, to build a cluster, first choose an arbitrarily directed reflection axis, which will be some unit vector $\hat{r}$. It can be chosen uniformly on the unit sphere for the isotropic Heisenberg Hamiltonian. Then choose an initial site i randomly.

The component of $\mathbf{s}_i$ parallel to $\hat{\mathbf{r}}$, which is $s_i^r = (\hat{\mathbf{r}} \cdot \mathbf{s}_i)\hat{\mathbf{r}}$, will be reversed by the reflection operation. Therefore, $\mathbf{s}_i$ is 'flipped' by reflection into the new direction by

$$\mathbf{s}_i' = \mathbf{s}_i - 2(\hat{\mathbf{r}} \cdot \mathbf{s}_i)\hat{\mathbf{r}}. \tag{4.133}$$

This is depicted in figure 4.14. Again, if a neighbor $\mathbf{s}_j$ were also to undergo the same operation, the energy in the ij bond would not change. If site j is linked into the cluster, the energy cost is zero. If it is not linked into the cluster, there will be an energy cost. Now it is just a matter of following the same operations as for the Swendsen–Wang algorithm, testing whether to not form links to sites neighboring the ones already in the cluster.

The energy change in a bond if a link is not formed, is found analogously to the Ising problem, but using the r-components of the spins,

$$\Delta E_{ij} = E_{ij}(\text{no link}) - E_{ij}(\text{link}) = -JS^2\left(s_i^{r\prime}s_j^r - s_i^r s_j^r\right) = 2JS^2 s_i^r s_j^r. \tag{4.134}$$

The last form comes from the flipping operation on the r-component, $s_i^{r\prime} = -s_i^r$. In this way, the theory is essentially the same as that for the Swendsen–Wang algorithm for the Ising model. The cluster build proceeds again until the cluster stops growing automatically. It will stop at a smaller size for higher temperature, because the decision of 'no link' will become more preferable. The probability for deciding not to put a link comes from the same formula (4.129) used in the Swendsen–Wang algorithm.

In actual computing practice, once a spin is included into the cluster, it is immediately reflected, meaning that a spin $\mathbf{s}_i$ has already been changed to $\mathbf{s}_i'$. Typically this will be saved in terms of its Cartesian components along x, y, z. Its change due to reflection should be saved:

$$\Delta \mathbf{s}_i = \mathbf{s}_i' - \mathbf{s}_i = -2(\hat{\mathbf{r}} \cdot \mathbf{s}_i)\hat{\mathbf{r}}. \tag{4.135}$$

Then from that, the energy change in a bond ij with no link would be

$$\Delta E_{ij} = -JS^2 \Delta \mathbf{s}_i \cdot \mathbf{s}_j. \tag{4.136}$$

This will be tested for each neighbor j of the site i in the cluster. The dot product, in practice, might be calculated using the Cartesian spin components.

4.5.3 Cluster algorithms for Heisenberg spins with XXZ symmetry

Consider also an XXZ model, such as

$$H = -JS^2 \sum_{(i,j)} \left(s_i^x s_j^x + s_i^y s_j^y + \lambda s_i^z s_j^z - 1\right). \tag{4.137}$$

Here the in-plane couplings are equivalent, $J^x = J^y = J$, while the out-of-plane coupling is scaled by some dimensionless parameter λ, such that $J^z = \lambda J$. If $\lambda > 1$, the anisotropy makes the z-axis an easy axis, and the system is somewhat Ising-like. If $\lambda < 1$, the anisotropy makes the model planar or XY-like. Both of these cases can be studied with an adaption of the Swendsen–Wang or Wolff algorithms.

For $\lambda < 1$, the low temperature states are XY-like, with the spins preferring to point nearly in the xy-plane. Then, clusters can be built by using the Wolff algorithm applied only to the xy-components. A reflection axis $\hat{\mathbf{r}}$ is generated that lies within the

xy-plane. A reflection involving only xy-components will leave the term $s_i^x s_j^x + s_i^y s_j^y$ invariant if both i and j are included into the cluster. Following the same procedure as for the Heisenberg model, a cluster can be built, without paying any attention to the z-components. However, the s^z-components must still be changed by doing usual single-spin updates. Thus, a hybrid method combining cluster updates with single-spin updates will work. At low temperature, the cluster operations on the xy-components efficiently moves the system in phases space; besides, the z-components are smaller and less involved in causing critical slowing down.

For $\lambda > 1$, the low temperature states are Ising-like, with the spins preferring to point close to the $\pm\hat{z}$-directions. Then, clusters can be built using the Swendsen–Wang algorithm applied only to the z-components. For any pair ij included into the cluster, the product $s_i^z s_j^z$ is unchanged when s_i^z and s_j^z are both reversed in sign. So, clusters are built, ignoring the xy spin components, which are relatively small compared to z-components at low temperature. Again, a hybrid method combining these cluster updates with single-spin updates will be necessary to allow the xy-components also to move around in phase space. However, the xy-components will be relatively small compared to z-components, and these should be easy to change by single-spin updates; they involve energies that are small compared to $k_B T$.

4.5.4 2D XY model simulation data with cluster updates

The improvement of MC simulation using cluster updating can be seen easily by looking at some data for the 2D XY model on larger systems. Here we show some results for $L = 32, 64, 128$, using a hybrid method. The hybrid method combines cluster updates with over-relaxation and Metropolis single-spin updates.

Here an over-relaxation step is defined to be one pass through the lattice using over-relaxation applied to N randomly selected sites. Similarly, a Metropolis step is the single-spin Metropolis algorithm applied to N randomly selected sites. One Wolff cluster step is defined by letting the cluster algorithm produce clusters, until at least 1/4 of all N sites have been modified. This could be only one cluster, at low temperature, or many clusters for a higher temperature. Then, one hybrid step was taken to be a combination of an over-relaxation step, a Metropolis step, and a Wolff cluster step. For these new MC simulations, 5000 of these steps were used for equilibration, then, data were averaged from the results of four bins with 50 000 MCS per bin. The four different estimates obtained from the bins were used for estimation of error bars.

The improvement in calculation of in-plane magnetic moment per site, m_{xy}, is considerable, see the results in figure 4.15. Especially the values at lower temperature are now corrected, because the Wolff cluster moves are able to re-orient the entire system in a small number of steps. The system does not become frozen into a limited region of phase space, as long as cluster updates are being used.

Results for the associated in-plane susceptibility, χ_{xy}, are displayed in figure 4.16. This gives a really clear signal of the BKT phase transition, because it shows a peak whose height increases strongly with the system size, at a temperature slightly above the transition. For increasing system size, this peak also moves to lower temperature.

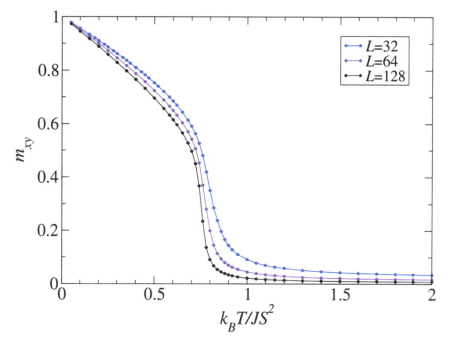

Figure 4.15. The in-plane magnetic moment per site versus scaled temperature for a 2D XY model, calculated using the hybrid cluster scheme combining cluster updates with over-relaxation and single-spin Metropolis updates. Compare with the lesser precision of single-spin update results in figure 4.8.

However, note that $\chi_{xy}(T)$ for temperatures above the location of the peak does not depend strongly on L. A method known as finite size scaling is an especially powerful analysis of $\chi_{xy}(T)$, that leads to determination of the critical temperature T_c. That is included in a more extended discussion of XY models and in the BKT transition in chapter 8.

4.6 Microcanonical Monte Carlo

MC simulations for magnetic systems can also be developed using the microcanonical ensemble, which follows some set of degrees of freedom at fixed or nearly fixed total energy. An introduction is given in chapter 15 of [3]. In the microcanonical ensemble, the controlling variables are the system energy E, the number of particles N and the volume V. These determine the total entropy $S(E, N, V)$, from which thermodynamic properties are found. In particular, the temperature is a quantity derived from the entropy by expression (4.22). Theoretically, one considers the motion in phase space on an *energy shell*, where the system energy might be selected at a value E but within some small range ΔE. A small range in energy is needed because in MC simulation it is difficult to develop a scheme that changes the system in some meaningful way without changing the energy at all.

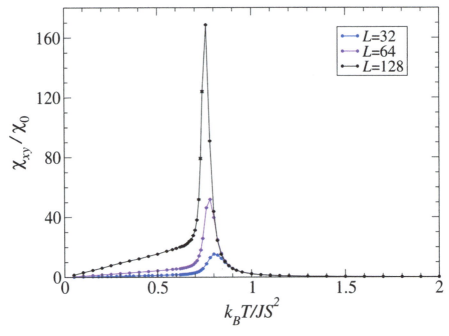

Figure 4.16. The in-plane susceptibility per site versus scaled temperature for a 2D XY model, calculated using the hybrid cluster scheme. Note that the peak increases strongly with system size while it also shifts to lower temperature. This is strongly indicative of the BKT vortex-unbinding transition.

4.6.1 Demons

To obtain small energy changes, one supposes that the system is allowed to interact weakly with another single degree of freedom called a *demon*, with reference to *Maxwell's demon*, who is supposed to have superpowers that allow him to defeat the second law of thermodynamics. The demon introduced by Michael Creutz in MC simulation [10] is allowed to take and give back small amounts of energy to the system, but the total energy of the system plus demon is strictly conserved. Gyan Bhanot and co-workers [11] also considered using many demons acting together. A demon simply moves energy around. An MC demon is not super-powerful, as it cannot acquire a negative energy. A demon runs around and exchanges energy with different spins in the system, for example. One can expect that while the system energy will only experience a small relative fluctuation, the demon's energy distribution will be wider, relative to its average energy. That distribution is then used to extract the temperature as a measured variable.

The MC rules for the demon are simple. A trial change is made in some randomly selected spin in the system, and the resulting system energy change $\Delta E = E_j - E_i$ is calculated, where i and j are the initial state and the trial state. If ΔE is negative, the change is accepted and the energy lost by the system is given to the demon. In the other case, if ΔE is positive, the change is accepted *only if* the demon has enough energy $E_d \geqslant \Delta E$ to give to the system. This is where the requirement of positive energy E_d for the demon comes in. A very large energy change might not be accepted because the demon does not have enough energy available to give. The demon is something like a banker, who cannot have a negative balance. If a move is accepted, the system's energy E is updated to its new value E' by

$$E' = E + \Delta E \tag{4.138}$$

while the demon's energy E_d is updated to its new value E_d' by

$$E_d' = E_d - \Delta E. \tag{4.139}$$

The algorithm works by using the following acceptance probability for the transition from state i to state j:

$$p_{ij}(\Delta E) = \begin{cases} 1 & \Delta E \leqslant E_d \\ 0 & \Delta E > E_d. \end{cases} \tag{4.140}$$

This process of trial states being accepted or rejected is repeated many times. Attempting to change N randomly selected spins is considered one MC step, and the simulation might be carried out for thousands of MCS. It can also be made more efficient, in terms of moving the system in phase space, by using more than one demon. Averages of desired thermodynamic properties are then made as already discussed for Metropolis and cluster MC, using the sequence of MC states that are generated. This produces averaged values as functions of total system energy E.

4.6.2 Effective temperature and equipartition

The system's energy has been held nearly constant in the simulation. Then, what is the temperature associated with the simulation? Strictly speaking, this should be found from the entropy $S(E)$ from (4.22). But it is usually not practical or even possible to count the number of available microstates, to obtain $S(E)$. Therefore, another approach is used.

Over the course of many MCS, the demon's one degree of freedom is sharing energy with all the degrees of the freedom of the system. Depending on the type of degrees of freedom, and their allowed energies, the idea of *equipartition* implies that each degree of freedom obtains the same average energy. We can assume each degree of freedom, including the demon, has an energy distribution given by the Boltzmann distribution:

$$p(E_d) = p(0)e^{-\beta E_d}. \tag{4.141}$$

This is normalized to unit area by putting $p(0) = \beta$, provided the demon energy is a continuous quantity with a value $E_d \geqslant 0$. The lowest possible demon energy is $E_d = 0$, which acts as a reference. Then, by calculating a histogram of the relative frequency of different demon energies, the temperature can be extracted. Using a logarithmic scale,

$$\ln p(E_d) = \ln p(0) - \beta E_d. \tag{4.142}$$

Therefore, the negative slope of the logarithmic histogram is β. Then the temperature is $T = (k_B \beta)^{-1}$.

One can also seek a simpler connection to the temperature, from the average demon energy. The demon has no underlying canonical coordinates or momenta. It possesses only energy. Each allowed energy state has a unit multiplicity; there is a uniform density of states for the demon, assuming that E_d has a continuous distribution from 0 to ∞. Then, the average demon energy is obtained by

$$\langle E_d \rangle = \frac{\int_0^\infty dE_d \, E_d e^{-\beta E_d}}{\int_0^\infty dE_d e^{-\beta E_d}} = k_B T. \tag{4.143}$$

In this way it is relatively easy to derive the temperature from the simulation, provided the distribution of demon energies is continuous, as it would be for interaction with Heisenberg-like spins.

A set of simulations will need to be performed to obtain thermodynamic properties over a range of temperature. In order to change the temperature, energy must be removed or added to the system, and the MC simulation run again. In molecular dynamics, where particle velocities play a major role, the energy can be scaled up or down simply by rescaling all of the velocities with a common factor. In spin MC, some other procedures can be used to allow the energy to move upward or downward. For instance, temporarily running the microcanonical MC while only accepting positive ΔE, regardless of demon energy, will allow the temperature to increase. The demon's energy can be reset to zero, and the MC restarted at the new energy scale. The reverse process, letting only negative ΔE be accepted for some duration, will allow the temperature to be reduced. Again, the demon's energy is reset to zero before restarting the microcanonical MC again.

4.6.3 Equipartition in different systems

For discussion, compare with particles in a 1D ideal gas with kinetic energy $E = \frac{1}{2}mv^2$. The equipartition theorem implies that every quadratic degree of freedom has a simple averaged value, for instance,

$$\langle E \rangle = \left\langle \frac{1}{2}mv^2 \right\rangle = \frac{\int_{-\infty}^\infty dv \, \frac{1}{2}mv^2 e^{-\frac{1}{2}mv^2\beta}}{\int_{-\infty}^\infty dv \, e^{-\frac{1}{2}mv^2\beta}} = \frac{1}{2}k_B T. \tag{4.144}$$

This results from the following standard Gaussian integrals, with parameter $\alpha = \frac{1}{2}m\beta$,

$$I_0(\alpha) = \int_{-\infty}^\infty dv \, e^{-\alpha v^2} = \sqrt{\frac{\pi}{\alpha}} \tag{4.145a}$$

$$I_2(\alpha) = \int_{-\infty}^\infty dv \, v^2 e^{-\alpha v^2} = -\frac{d}{d\alpha} I_0(\alpha) = \frac{1}{2}\sqrt{\frac{\pi}{\alpha^3}}. \tag{4.145b}$$

The factor of $\frac{1}{2}k_B T$ for every degree of freedom is half that found for the demon's average energy. This is a result of a non-unity density of states for kinetic energy E.

When a particle's average kinetic energy integral is written in terms of energy integrations, using $dv = 2dE/\sqrt{2mE}$ (the two values $\pm v$ have the same energy), the normalized probability density in energy is found to be

$$p(E) = \sqrt{\frac{\beta}{4\pi E}} e^{-\beta E}. \tag{4.146}$$

Contrast that to the demon's energy distribution. The density of states factor $1/\sqrt{E}$ explains why the average energy is $\frac{1}{2}k_B T$, in contrast to the demon's average energy $k_B T$, whose probability function has a uniform density of states.

It is interesting to also discuss equipartition for classical spins. A classical spin vector has two degrees of freedom corresponding to the two angles ϕ_n, θ_n needed to specify its direction. Only in the limit of very high temperature will spins have an average energy of $k_B T$ per spin, which is evident in figure 4.6 for the 2D XY model. At lower temperatures, a complex spin model (based on Heisenberg-like spins) may not have a simple formula for relating the average energy per spin to the temperature, especially when including anisotropies and applied fields. Usually one has resorted to numerical simulation because some basic properties such as that are unknown. Usually, there will only be a simple average energy formula in a spin model either at very high temperature or very low temperature.

4.6.4 Effective temperature in Ising models

If the spins in the model have discrete energy states, such as an Ising model or other discrete spin models, then the energies of the demon will also be discretized, rather than continuous. The probability distribution will still be that due to equipartition, (4.141), but with a different normalization constant A,

$$p(E_d) = A e^{-\beta E_d}. \tag{4.147}$$

On the other hand, the average energy will not be $k_B T$.

Suppose one considers an Ising model where each spin has bonds to z nearest neighbors, where z is called the coordination number of the lattice. For example, $z = 4$ in the 2D square lattice, $z = 6$ in the 2D triangular lattice, and so on. Any spin flip affects z bonds, and the energy change must be $\Delta E = \pm 2zJS^2$. This means the allowed demon energies are now quantized in units of $\epsilon = 2zJS^2$. They can be written as $E_d = n\epsilon$ for $n = 0, 1, 2, \ldots$. Then the normalization sum for the demon's energy probability distribution is the demon's canonical partition function,

$$Z_d = \sum_{n=0}^{\infty} e^{-\beta n \epsilon} = \sum_{n=0}^{\infty} (e^{-\beta \epsilon})^n = \frac{1}{1 - e^{-\beta \epsilon}}. \tag{4.148}$$

The sum of all probabilities is unity, then the normalization constant needed is

$$A = \frac{1}{Z_d} = 1 - e^{-\beta \epsilon}. \tag{4.149}$$

The average demon energy can be obtained in a similar way,

$$\langle E_\mathrm{d}\rangle = \frac{1}{Z_\mathrm{d}}\sum_{n=0}^{\infty} n\epsilon e^{-\beta n\epsilon} = -\frac{\partial}{\partial \beta}\ln Z_\mathrm{d} = \frac{\epsilon}{e^{\beta\epsilon}-1}. \qquad (4.150)$$

This can be inverted to solve for the resulting temperature,

$$k_\mathrm{B} T = \frac{\epsilon}{\ln\left(1 + \dfrac{\epsilon}{\langle E_\mathrm{d}\rangle}\right)}. \qquad (4.151)$$

One can see that this has the correct limit when $\epsilon \to 0$. Going over to a continuum of demon energies, it gives our previous result,

$$k_\mathrm{B} T \to \langle E_\mathrm{d}\rangle \quad \text{as } \epsilon \to 0. \qquad (4.152)$$

This constitutes a brief introduction to using demons in microcanonical MC. It is an appealing way to study thermal statistics. The lack of precise control over the temperature is its main disadvantage when compared to the other MC methods using the canonical ensemble, which fixes the temperature.

4.6.5 1D spin chain studied with microcanonical Monte Carlo

Exercise 4.4. A simple problem to test microcanonical MC with demons is a 1D chain of spins, either of three components as in the Heisenberg model, or of just one component as in the 1D Ising model. These are good for testing because the canonical partition function $Z(N, T)$ can be calculated exactly, from which average energy, specific heat, and so on, are found for comparison.

For a 1D spin chain with nearest neighbor interactions,

$$H = -J \sum_{n=1}^{N} s_n \cdot s_{n+1} \qquad (4.153)$$

set up a computer program to run the microcanonical MC algorithm. The system could have open boundaries, or periodic boundary conditions with $s_{n+1} = s_1$. Introduce a demon to exchange energy with randomly chosen sites, according to the rules presented in this section. Keep track of the demon's sequence of energies and, from that, calculate a histogram of the energies to determine its probability distribution $p(E_\mathrm{d})$. Use that to obtain the temperature. By running the simulation with different initial demon energies, the temperature can be changed. See if your simulations are able to reproduce known exact results for average energy and specific heat per particle as functions of energy. Many texts on statistical mechanics give the exact solution for the 1D Ising model, for instance, Plischke and Bergesen [12]. For Heisenberg spins there is also a well-known exact solution due to Fisher [13].

Bibliography

[1] Binder K and Heermann D 2010 *Monte Carlo Simulation in Statistical Physics: An Introduction* 5th edn (Belin: Springer)
[2] Metropolis N, Rosenbluth A W, Rosenbluth M N, Teller A H and Teller E 1953 Equation of state calculations by fast computing machines *J. Chem. Phys.* **21** 1087
[3] Gould H and Tobochnik J 1988 *Computer Simulation Methods: Applications to Physical Systems* (Reading, MA: Addison-Wesley)
[4] Koonin S E and Meredith D C 1990 *Computational Physics: FORTRAN version* (Reading, MA: Addison-Wesley)
[5] Press W H, Teukolsky S A, Vetterling W T and Flannery B P 2007 *Numerical Recipes. The Art of Scientific Computing* 3rd edn (Cambridge: Cambridge University Press)
[6] Ising E 1925 Beitrag zur Theorie des Ferromagnetismus *Z. Phys* **31** 253
[7] Onsager L 1944 Crystal statistics I. A two-dimensional model with an order-disorder transition *Phys. Rev.* **65** 117
[8] Swendsen R H and Wang J-S 1987 Nonuniversal critical dynamics in Monte Carlo simulations *Phys. Rev. Lett.* **58** 86
[9] Wolff U 1989 Collective Monte Carlo Updating for spin systems *Phys. Rev. Lett.* **62** 361
[10] Creutz M 1983 Microcanonical Monte Carlo simulation *Phys. Rev. Lett.* **50** 1411
[11] Bhanot G, Creutz M and Neuberger H 1984 Microcanonical simulation of Ising systems *Nucl. Phys.* **235** 417
[12] Plischke M and Bergersen B 1989 *Equilibrium Statistical Physics* (New York: Prentice-Hall)
[13] Fisher M E 1964 Magnetism in one-dimensional systems—the Heisenberg model for infinite spin *Am. J. Phys.* **32** 343

Chapter 5

Classical spin dynamics simulations

The theory for magnetic excitations usually requires some approximations, such as linearizations, low temperature theory, self-consistent approximations, and so on. Numerical simulations of the dynamics can show whether such theories correctly predict how the magnetization varies with time and other parameters. They also provide general ways to include the effects of temperature and finite size, which are especially important when nonlinear or topological excitations are possible. In this chapter we consider how to obtain the time-dependence of magnetic degrees of freedom from simulations. In addition, dynamic correlations are discussed as an important experimental indicator of dynamic excitations, which can be calculated via numerical simulations.

5.1 Landau–Lifshitz–Gilbert spin dynamics

The time dynamics of magnetic moments was introduced in chapter 2. Recall that the basic torque equation (2.98) for a system of spins involves the effective magnetic field $\mathcal{F}_i$, derived from the system's Hamiltonian H, and using the dot to indicate d/dt,

$$\dot{\mathbf{S}}_i = \mathbf{S}_i \times \mathcal{F}_i, \qquad \mathcal{F}_i = -\frac{\partial H}{\partial \mathbf{S}_i}. \qquad (5.1)$$

This applies in various forms for different models. For instance, the equation (2.120) determines the time development of discrete spins in a 1D chain with given exchange couplings. Continuum versions of the same equations are given in (2.142). A more general version in more than one space dimension is shown in (2.143). Generally, various forms of this Landau–Lifshitz–Gilbert (LLG) equation dictate the dynamics of a classical magnet, depending on the details for the model being used.

In earlier chapters magnets were discussed in terms of either individual spins $\mathbf{S}_n$ on a lattice, or equivalently in terms of the magnetic dipoles $\boldsymbol{\mu}_n = \gamma \mathbf{S}_n$, especially for the discretization of a magnet into cells for micromagnetics simulations. There is no strong difference between these viewpoints. Both consider magnetic moments of fixed length. In micromagnetics, however, the exchange interaction is usually isotropic, and the demagnetization fields, which are of long range, play a very important role. In this chapter, the dynamics for discrete spins $\mathbf{S}_n$ on a lattice, interacting with possibly anisotropic exchange and other energy terms in the Hamiltonian, is discussed. Demagnetization fields can also be included, if needed.

In molecular dynamics simulations for gases, liquids and other material systems, the motion and ordering of molecules is followed. From that, thermal properties, phases transitions and other features are studied. In *spin dynamics* (SD) simulations, the dynamics of discrete spins is being followed in an analogous way, from which thermodynamics, correlations, ordering and phase transitions can be studied.

To be fairly general, consider a local interaction spin Hamiltonian with anisotropic nearest neighbor exchange, an externally generated applied field and a demagnetization field. A local easy-plane anisotropy such as that in (1.52) is also included here, using the opposite sign on the coupling constant K_A. The spin Hamiltonian is initially written in the form

$$H = -J \sum_{(ij)} \left(S_i^x S_j^x + S_i^y S_j^y + \lambda S_i^z S_j^z \right)$$

$$+ \sum_i \left[K_A (S_i^z)^2 - \mu_0 \left(\mathbf{H}_{\text{ext},i} + \frac{1}{2} \mathbf{H}_{\text{M},i} \right) \cdot \gamma \mathbf{S}_i \right]. \qquad (5.2)$$

The first sum is over nearest neighbor exchange interactions. The notation (ij) indicates summing only over nearest neighbor pairs. The parameter λ characterizes a form of exchange anisotropy (XY-like) that was introduced in chapter 4, see 4.137. In some studies one may only include this exchange interaction. The second sum includes the term for local easy-plane (XY-like) anisotropy, provided $K_A > 0$, and interaction with an external field $\mathbf{H}_{\text{ext}}$ and the demagnetization field $\mathbf{H}_M$. The demagnetization field is usually included for micromagnetics simulations, whereas for many studies of theoretical spin models it might be ignored as being of lesser importance. The fields are coupled to the equivalent magnetic dipoles, $\boldsymbol{\mu}_i = \gamma \mathbf{S}_i$. Recall that the factor of 1/2 on demagnetization energy removes the double counting of dipole–dipole pair interactions in the sum.

The dynamics of spins are governed by the effective field $\mathcal{F} = \gamma \mathbf{B}$ acting on each spin, as developed earlier in (2.98) using Hamiltonian dynamics,

$$\mathcal{F}_i = -\frac{\partial H}{\partial \mathbf{S}_i} = J \sum_{(ij)} \left(S_j^x \hat{\mathbf{x}} + S_j^y \hat{\mathbf{y}} + \lambda S_j^z \hat{\mathbf{z}} \right) - 2K_A S_i^z \hat{\mathbf{z}} + \gamma \mu_0 \left(\mathbf{H}_{\text{ext},i} + \mathbf{H}_{\text{M},i} \right). \qquad (5.3)$$

In the exchange term, the sum includes only sites j that are nearest neighbors of site i. The external and demagnetization fields enter in an equivalent way, as seen earlier in (3.155). Knowing this effective field, the dynamics in the LLG equation, including damping, follows from (2.104) extended to a collection of spins,

$$\dot{\mathbf{S}}_i = \mathbf{S}_i \times \mathcal{F}_i + \frac{\alpha}{S} (\mathbf{S}_i \times \mathcal{F}_i) \times \mathbf{S}_i. \qquad (5.4)$$

Note that the undamped and damped parts can be included into one net effective field,

$$\mathbf{G}_i \equiv \mathcal{F}_i - \frac{\alpha}{S}(\mathbf{S}_i \times \mathcal{F}_i) \quad (5.5)$$

so that the dynamics is always represented as a simple torque equation:

$$\dot{\mathbf{S}}_i = \mathbf{S}_i \times \mathbf{G}_i. \quad (5.6)$$

How equation (5.4) or (5.6) is solved is somewhat dependent on the question of interest. This includes the choice of a numerical solution method, which could depend on the presence or absence of the damping parameter α. For instance, the damping parameter α can be included to relax an initial spin configuration to a local energy minimum (it might not produce the global minimum or ground state). That is not really SD, but only employing dynamics with damping as a relaxation scheme to obtain zero-temperature configurations.

On the other hand, setting $\alpha = 0$ can be used to find the dynamics for excited states and for thermal equilibrium SD at a selected finite temperature, in a kind of MC–SD hybrid simulation scheme. This scheme of using canonical (i.e. Metropolis or cluster schemes) MC to select initial states for a desired temperature, followed by SD at fixed energy, is a dynamical simulation in the microcanonical ensemble. Finally, yet another approach is possible, Langevin SD, where a non-zero value of α is combined with randomly fluctuating fields to simulate the time development, but in the canonical ensemble at fixed temperature. Each of these methods can be very productive and they will be discussed after some summary of numerical integration methods that can be used for SD.

5.1.1 Dimensionless units for spin dynamics

The LLG equations might be integrated forward in time by various algorithms. Before choosing an algorithm, it helps to choose appropriate units to be used in the simulations, such that the predominant forces or torques are of order unity. This helps to avoid underflow or overflow errors, and keeps most quantities of importance that are being calculated the same order of magnitude. There is some flexibility in the choice of computational units; only one obvious approach is given here.

Usually exchange is the strongest energy scale in the problem, then with $\mathbf{S}_i = S\mathbf{s}_i$, the factor JS can be taken out of $\mathcal{F}_i$, to make dimensionless effective field,

$$\mathbf{f}_i = \frac{\mathcal{F}_i}{JS} = \sum_{(ij)}\left(s_j^x\hat{\mathbf{x}} + s_j^y\hat{\mathbf{y}} + \lambda s_j^z\hat{\mathbf{z}}\right) - 2\kappa_A s_i^z\hat{\mathbf{z}} + \mathbf{h}_i, \quad (5.7)$$

where the dimensionless scaled anisotropy constant and local magnetic field are

$$\kappa_A \equiv \frac{K_A}{J}, \qquad \mathbf{h}_i \equiv \frac{\gamma\mu_0}{JS}(\mathbf{H}_{\text{ext},i} + \mathbf{H}_{M,i}). \quad (5.8)$$

Obviously the spins used for computation are the unit vectors $\mathbf{s}_i$. Then the factor JS taken from $\mathcal{F}_i$ scales the unit of time, leading to a dimensionless time variable,

$$\tau \equiv JSt. \tag{5.9}$$

This means that $t_u \equiv (JS)^{-1}$ is the unit of time. This gives the dimensionless LLG equation to be used in simulation, rearranged slightly from (5.4),

$$\frac{d}{d\tau}\mathbf{s}_i = \mathbf{s}_i \times [\mathbf{f}_i - \alpha(\mathbf{s}_i \times \mathbf{f}_i)]. \tag{5.10}$$

Including the damping, there is a net dimensionless field acting on $\mathbf{s}_i$,

$$\mathbf{g}_i \equiv \mathbf{f}_i - \alpha(\mathbf{s}_i \times \mathbf{f}_i) \tag{5.11}$$

and the dynamics is purely a cross product, that automatically preserves spin length,

$$\frac{d}{d\tau}\mathbf{s}_i = \mathbf{s}_i \times \mathbf{g}_i. \tag{5.12}$$

Thus, any numerical algorithm worth using has to ensure that spin length $\mathbf{s}_i \cdot \mathbf{s}_i = 1$ is conserved.

This last form (5.12) will be used to solve spin problems numerically. The general discussion of problems, however, will be performed in the original variables so that the coupling parameters J, S, K_A, *etc*, are visible in any analysis.

5.1.2 Dynamic equations for Heisenberg spins—angular variables

A special case of interest is easy-plane Heisenberg spin models with ferromagnetic (FM) exchange coupling. For theoretical investigations it is useful to consider the dynamic equations using angular variables relating to the in-plane and out-of-plane motions of the spins.

We consider an XY-like FM model with easy-plane symmetry, $0 \leqslant \lambda < 1$, without damping ($\alpha = 0$). Consider the simplest case where no applied, demagnetization, or local anisotropy fields are included. The Hamiltonian has FM exchange of nearest neighbor pairs only,

$$H = -J\sum_{(ij)}\left(S_i^x S_j^x + S_i^y S_j^y + \lambda S_i^z S_j^z\right). \tag{5.13}$$

The effective field on $\mathbf{S}_i$ due to its neighbors is

$$\mathcal{F}_i = J\sum_{(ij)}\left(S_j^x \hat{\mathbf{x}} + S_j^y \hat{\mathbf{y}} + \lambda S_j^z \hat{\mathbf{z}}\right). \tag{5.14}$$

The sum contains only sites j that are neighbors of i as indicated by (ij).

The dynamics for this or any other model follows the torque equation (5.1), written in terms of components,

$$\dot{S}_i^x = S_i^y \mathcal{F}_i^z - S_i^z \mathcal{F}_i^y \tag{5.15a}$$

$$\dot{S}_i^y = S_i^z \mathcal{F}_i^x - S_i^x \mathcal{F}_i^z \tag{5.15b}$$

$$\dot{S}_i^z = S_i^x \mathcal{F}_i^y - S_i^y \mathcal{F}_i^x. \tag{5.15c}$$

Due to the assumption of XY symmetry, it may be convenient to express a spin in planar spherical coordinates (ϕ, θ), introduced earlier in (2.42),

$$\mathbf{S}_i = (S_i^x, S_i^y, S_i^z) = S(\cos\theta_i \cos\phi_i, \cos\theta_i \sin\phi_i, \sin\theta_i). \tag{5.16}$$

A slight modification of that is to use the in-plane angle ϕ together with out-of-plane component S^z, because one can show that S^z plays the role of the momentum conjugate to ϕ. Therefore, the pair (ϕ_i, S_i^z) is sufficient to describe a spin, as

$$(S_i^x, S_i^y, S_i^z) = \left(\sqrt{S^2 - S_i^{z2}} \cos\phi_i, \sqrt{S^2 - S_i^{z2}} \sin\phi_i, S_i^z\right). \tag{5.17}$$

The in-plane angle's time derivative is found starting with

$$\phi_i = \tan^{-1}\frac{S_i^y}{S_i^x}. \tag{5.18}$$

One finds by using (5.15),

$$\dot{\phi}_i = \frac{S_i^x \dot{S}_i^y - S_i^y \dot{S}_i^x}{S_i^{x2} + S_i^{y2}} = \frac{S_i^x(S_i^z \mathcal{F}_i^x - S_i^x \mathcal{F}_i^z) - S_i^y(S_i^y \mathcal{F}_i^z - S_i^z \mathcal{F}_i^y)}{S_i^{x2} + S_i^{y2}} \tag{5.19}$$

which reduces to

$$\dot{\phi}_i = S_i^z\left(\frac{S_i^x \mathcal{F}_i^x + S_i^y \mathcal{F}_i^y}{S_i^{x2} + S_i^{y2}}\right) - \mathcal{F}_i^z. \tag{5.20}$$

Then with $S_i^{x2} + S_i^{y2} + S_i^{z2} = S^2$, the dynamic equations are generally, for any model,

$$\dot{\phi}_i = \frac{S_i^z}{\sqrt{S^2 - S_i^{z2}}}(\mathcal{F}_i^x \cos\phi_i + \mathcal{F}_i^y \sin\phi_i) - \mathcal{F}_i^z, \tag{5.21a}$$

$$\dot{S}_i^z = \sqrt{S^2 - S_i^{z2}}(\mathcal{F}_i^y \cos\phi_i - \mathcal{F}_i^x \sin\phi_i). \tag{5.21b}$$

Now apply this to the easy-plane Heisenberg model. The effective field components on site i are

$$\mathcal{F}_i^x = J\sum_{(ij)} S_j^x, \qquad \mathcal{F}_i^y = J\sum_{(ij)} S_j^y, \qquad \mathcal{F}_i^z = \lambda J\sum_{(ij)} S_j^z. \tag{5.22}$$

Then a little algebra for the in-plane component gives

$$\dot{\phi}_i = \frac{S_i^z}{\sqrt{S^2 - S_i^{z\,2}}} J \sum_{(ij)} \left(S_j^x \cos \phi_i + S_j^y \sin \phi_i \right) - \lambda J \sum_{(ij)} S_j^z$$

$$= J \sum_{(ij)} \left\{ S_i^z \sqrt{\frac{S^2 - S_j^{z\,2}}{S^2 - S_i^{z\,2}}} \left(\cos \phi_j \cos \phi_i + \sin \phi_j \sin \phi_i \right) - \lambda S_j^z \right\}$$

$$= J \sum_{(ij)} \left\{ S_i^z \sqrt{\frac{S^2 - S_j^{z\,2}}{S^2 - S_i^{z\,2}}} \cos(\phi_i - \phi_j) - \lambda S_j^z \right\}. \quad (5.23)$$

Some other algebra for the out-of-plane component gives

$$\dot{S}_i^z = \sqrt{S^2 - S_i^{z\,2}} J \sum_{(ij)} \left(S_j^y \cos \phi_i - S_j^x \sin \phi_i \right)$$

$$= J \sqrt{S^2 - S_i^{z\,2}} \sum_{(ij)} \sqrt{S^2 - S_j^{z\,2}} \left(\sin \phi_j \cos \phi_i - \cos \phi_j \sin \phi_i \right)$$

$$= -J \sqrt{S^2 - S_i^{z\,2}} \sum_{(ij)} \sqrt{S^2 - S_j^{z\,2}} \sin(\phi_i - \phi_j). \quad (5.24)$$

This would be one way to express the dynamics with only two independent coordinates per spin. It could be used as an alternative to Cartesian coordinates for numerical simulations. It is even more useful for theoretical analyses of SD, due to the clear separation of in-plane angle ϕ and out-of-plane component S^z. There would also be no difficulty to obtain the corresponding equation of motion for out-of-plane angle θ_i by the use of $S_i^z = S \sin \theta_i$.

Equations (5.23) and (5.24) are exact for the easy-plane Heisenberg model, and even apply when $\lambda = 1$, which would be the isotropic Heisenberg model, and also for $\lambda > 1$ which is a Heisenberg model with easy-axis exchange anisotropy.

5.1.3 XY model

For $0 < \lambda \leqslant 1$, the model has XY symmetry, and large out-of-plane spin components S_i^z cost extra energy. We suppose whatever excitations are present are primarily low energy excitations, slightly above the ground state energy. Therefore, it makes sense in certain circumstances (for example, low temperature dynamics) that the z-components are small, $S_i^z \ll S$. Then the square root factors simplify and the approximate equations for the easy-plane model are

$$\dot{\phi}_i \approx J \sum_{(ij)} \left\{ S_i^z \cos(\phi_i - \phi_j) - \lambda S_j^z \right\} \quad (5.25a)$$

$$\dot{S}_i^z \approx -JS^2 \sum_{(ij)} \sin(\phi_i - \phi_j). \quad (5.25b)$$

This could be a good starting point for theoretical analysis that would seek the mathematical structure of excitations such as vortices in the model. Usually a continuum approximation would be applied, which involves making assumptions about the orders of smallness for differences of nearest neighbor quantities. For an easy-plane model, the spins mostly stay near the xy-plane, and have small changes from one site to the next. Thus it is reasonable to suppose that the differences $\phi_i - \phi_j$ and the components S_i^z are of similar orders of smallness. With $\cos(\phi_i - \phi_j) \approx 1$, the leading terms in $\dot{\phi}_i$ are

$$\dot{\phi}_i \approx J \sum_{(ij)} \left(S_i^z - \lambda S_j^z \right). \tag{5.26}$$

If we specialize further to the XY model, with $\lambda = 0$, then $\dot{\phi}_i$ becomes decoupled from its neighbors and XY dynamics is determined by

$$\dot{\phi}_i \approx z_i J S_i^z \tag{5.27a}$$

$$\dot{S}_i^z \approx -JS^2 \sum_{(ij)} \sin\left(\phi_i - \phi_j\right). \tag{5.27b}$$

The factor z_i is the number of nearest neighbors of site i (local coordination number). A square lattice has $z_i = 4$, except at the boundary of the system, where it is smaller. The sum over the sines cannot be so approximated or there would be no dynamics left. It would be possible to also consider a *linearization* approximation, with $\sin(\phi_i - \phi_j) \approx \phi_i - \phi_j$, however, that could replace a bounded quantity $|\sin(\phi_i - \phi_j)| \leq 1$ by an unbounded one, so it is avoided for now. This is the approximate dynamics for the XY model for 3D spins with strong planar anisotropy.

5.1.4 Planar rotor model

The planar rotor (PR) is another model sometimes used to illustrate magnetization dynamics, with equations of motion very similar to those for the XY model. The PR model has 2D rotating masses allowed to spin around in the xy-plane[1]. The masses have angular momentum only around the z-axis, L_i, rotational inertia I and kinetic energy $L_i^2/2I$, coupled in a fashion similar to that for the XY model. Their dynamics is described by the orientation in the xy-plane, ϕ_i, of each rod or rotor. As each moves, it has an angular velocity $\omega_i = \dot{\phi}_i$. One can start from a Lagrangian, which includes a potential like that in the XY model, which tends to align the rotors with a strength given by a parameter $K > 0$. The potential mimics a FM exchange iteration of the rotors,

$$\text{PE} = -K \sum_{(ij)} \cos\left(\phi_i - \phi_j\right). \tag{5.28}$$

---

[1] Loft and DeGrand [1] conducted a study of the 2D PR dynamics but referred to it as an XY model. In this text we use the name XY model to indicate a system with three-component spins with interactions of planar anisotropy.

The Lagrangian is taken to be a difference of kinetic and potential energies,

$$\mathcal{L} = \text{KE} - \text{PE} = \sum_i \frac{1}{2} I \dot{\phi}_i^2 + K \sum_{(ij)} \cos(\phi_i - \phi_j). \quad (5.29)$$

The momenta conjugate to the angles are the angular momenta,

$$p_i = \frac{\partial \mathcal{L}}{\partial \dot{\phi}_i} = I \dot{\phi}_i = I \omega_i = L_i. \quad (5.30)$$

The Euler–Lagrange equations of motion are simple, based on

$$\frac{d}{dt}\left(\frac{\partial \mathcal{L}}{\partial \dot{\phi}_i}\right) - \frac{\partial \mathcal{L}}{\partial \phi_i} = 0. \quad (5.31)$$

This gives

$$\frac{d}{dt}\left(I\dot{\phi}_i\right) = \frac{\partial \mathcal{L}}{\partial \phi_i} = \frac{\partial}{\partial \phi_i}\left\{K \sum_{(ij)} \cos(\phi_i - \phi_j)\right\} = -K \sum_{(ij)} \sin(\phi_i - \phi_j). \quad (5.32)$$

This is a set of coupled second-order differential equations,

$$I\ddot{\phi}_i = -K \sum_{(ij)} \sin(\phi_i - \phi_j). \quad (5.33)$$

The dynamics can also be considered as coupled first-order equations for (ϕ_i, ω_i).

$$\dot{\phi}_i = \omega_i \quad (5.34a)$$

$$\dot{\omega}_i = -\frac{K}{I} \sum_{(ij)} \sin(\phi_i - \phi_j). \quad (5.34b)$$

This is mathematically the same as the XY model (when all coordination numbers z_i are the same). The parameters and the interpretations of the variables in the two models are different, however. To make the XY model equations match this, one could define a new variable there, $\omega_i = z_i J S_i^z$. This makes the first equation (5.27a) for the XY dynamics the same as the rotor, $\dot{\phi}_i = \omega_i$. Then the second equation (5.27b) becomes

$$z_i J \dot{S}_i^z = \dot{\omega}_i = -z_i (JS)^2 \sum_{(ij)} \sin(\phi_i - \phi_j). \quad (5.35)$$

One sees that the XY model will be equivalent to the rotor model if the parameters are matched using site-dependent rotational inertia,

$$z_i (JS)^2 = \frac{K}{I_i} \implies I_i = \frac{K}{z_i (JS)^2}. \quad (5.36)$$

Further, if one is solving the rotor model but imagining that it represents an XY model, then the amount of out-of-plane spin motion associated with a rotor's velocity is given by the mapping,

$$S_i^z = \frac{\omega_i}{z_i J}. \tag{5.37}$$

Faster moving rotors correspond to spins tilting more so out of the *xy*-plane. The values of z_i would only differ from the usual coordination number for the lattice at the edges of the system. Thus the PR model is equivalent to the XY model, under the approximation of small out-of-plane motions of the spins.

5.1.5 Planar rotor in Cartesian components

One can associate a planar spin for the rotor model $\mathbf{S}_i = S(\cos \phi_i, \sin \phi_i)$. Then their dynamics is simple,

$$\dot{S}_i^x = -S\dot{\phi}_i \sin \phi_i, \qquad \dot{S}_i^y = +S\dot{\phi}_i \cos \phi_i, \tag{5.38}$$

or summarize this as

$$\dot{S}_i^x = -\omega_i S_i^y, \qquad \dot{S}_i^y = +\omega_i S_i^x. \tag{5.39}$$

If ω_i were constant, it is uniform precessional motion. But ω_i changes with time, according to the applied torque,

$$\dot{\omega}_i = -\frac{K}{IS^2} \sum_{(ij)} S^2 \left(\sin \phi_i \cos \phi_j - \cos \phi_i \sin \phi_j \right) = -\frac{K}{IS^2} \sum_{(ij)} \left(S_i^y S_j^x - S_i^x S_j^y \right) \tag{5.40}$$

or more simply,

$$\dot{\omega}_i = \frac{K}{IS^2} \sum_{(ij)} \left(S_i^x S_j^y - S_i^y S_j^x \right). \tag{5.41}$$

If one is performing numerical integration of these equations, this form is good due to the lack of trigonometric evaluations. The only problem, perhaps, is that now the problem is over-specified, having three components, S_i^x, S_i^y, ω_i, when two are sufficient. The numerical integration needs to be stable enough to preserve the spin length. One sees from the three dynamic equations an effective field on each spin, with components

$$\mathcal{F}_i^x = \frac{K}{IS^2} \sum_{(ij)} S_j^x, \qquad \mathcal{F}_i^y = \frac{K}{IS^2} \sum_{(ij)} S_j^y, \qquad \mathcal{F}_i^z = -\omega_i. \tag{5.42}$$

The functions that change angular momenta, $L_i = I\omega_i$, are torques around the $\hat{z}$-axis, denoted N_i here. Thus one might also write these as the basic mechanical torques, which determine the conservative part of the dynamics,

$$N_i^{\text{mech}}(t) = \frac{K}{S^2} \sum_{(ij)} \left(S_i^x S_j^y - S_i^y S_j^x \right). \tag{5.43}$$

The dynamics in a more general case can also have other torques. In a later discussion of Langevin dynamics, one includes N^{damping} due to viscous damping and N^{random} due to random forces associated with the temperature. Considering the dynamics due to all of these torques acting together,

$$\dot{L}_i = I\dot{\omega}_i = N_i^{\text{mech}} + N_i^{\text{damping}} + N_i^{\text{random}}. \tag{5.44}$$

Because of the presence of moving masses, which are less abstract than classical spins, the rotor model is good for understanding the initial analysis for Langevin SD. More on that follows later in this chapter.

5.2 Numerical time evolution for spin dynamics

Numerical methods for the integration of first-order ordinary differential equations such as (5.12) are discussed in many textbooks, such as the classic one by Hildebrand [2]. The most important requirement of any method used is that it conserves the spin lengths. In the special case where damping $\alpha = 0$, the total system energy also must be conserved, if the external magnetic field is constant in time.

5.2.1 Fourth-order Runge–Kutta time evolution

A widely used very stable scheme for evolving (5.12) forward in time is fourth-order Runge–Kutta (RK4). It is briefly summarized here for a system with N spins, which also gives an overview of time evolution simulation regardless of the particular integration scheme. It is understood that the equations may already be in dimensionless units.

Let $y(\tau) = (\mathbf{s}_1, \mathbf{s}_2, \mathbf{s}_3, \ldots)$ be a vector of $3N$ components containing the state of all the spins at time τ. There will be some function (a subroutine in the simulation program) that evaluates the instantaneous torques $\mathbf{N}_i = \mathbf{s}_i \times \mathbf{g}_i$ on each spin in the present state. Let $f(y, \tau) = (\mathbf{N}_1, \mathbf{N}_2, \mathbf{N}_3, \ldots)$ represent the function that gives all torques in the system, when in state y. A time-dependence is included in case there is an applied field varying with time. The differential equation to be solved is represented symbolically in a standard form as

$$\frac{dy}{d\tau} = f(y, \tau). \tag{5.45}$$

The solution is to be advanced by one time step, $\Delta\tau$. This is a dimensionless number, usually quite small compared to 1. Time is discretized in time steps, so write y_n to indicate the solution $y(\tau_n)$ at the nth time step, where $\tau_n = n\Delta\tau$. One time step that produces y_{n+1} from y_n is effected by a combination of four partial steps, each of which makes an estimate to the time derivative of y_n. These time derivatives are derived from the torque function f, by the operations,

$$k_1 = f(y_n, \tau_n) \tag{5.46a}$$

$$k_2 = f\left(y_n + \frac{\Delta\tau}{2}k_1, \tau_n + \frac{\Delta\tau}{2}\right) \tag{5.46b}$$

$$k_3 = f\left(y_n + \frac{\Delta\tau}{2} k_2, \tau_n + \frac{\Delta\tau}{2}\right) \quad (5.46c)$$

$$k_4 = f\left(y_n + \Delta\tau k_3, \tau_n + \Delta\tau\right). \quad (5.46d)$$

The time variable also is updated here in half-steps, as indicated by the presence of $\tau_n + \frac{1}{2}\Delta\tau$. Then the step to advance the solution uses a combination of the intermediate torques,

$$y_{n+1} = y_n + \frac{\Delta\tau}{6}(k_1 + 2k_2 + 2k_3 + k_4). \quad (5.47)$$

The error scales with $(\Delta\tau)^5$. Details appear in many texts on numerical methods, such as those by Ceschino and Kuntzmann [3] and by Press *et al* [4]. While this is a scheme with a fixed time step, there are direct ways to monitor the error expected and then adapt the time step to be larger when errors are small and smaller when errors are larger. This takes extra programming overhead, although for RK4 it is straightforward.

In actual application, the symbol y_n may represent the $3N$ Cartesian spin components, and it is simplest to expand out the equations of motion using those components, such as that in (2.83), for every spin.

Following this procedure should conserve spin length, however, it is not guaranteed, because of round-off errors. Further, only two angles are necessary to describe a spin's direction, so using three Cartesian components makes for an overdetermined system of equations. The evolution could be developed alternatively by integrating equations for the in-plane and out-of-plane spin angles ϕ_i and θ_i as in (2.42). This sometimes has its own problems, however, because ϕ_i is undefined at the poles of the sphere. Because of that problem, and the need for evaluation of many trigonometric functions, it is usually better to avoid angular coordinates for numerical simulations of spins.

5.2.2 Adams–Bashforth–Moulton fourth-order predictor–corrector method

One drawback of RK4 is that there are many operations needed to advance by one time step. Although the programming is not difficult, the torque function $f(y, \tau)$ must be called four times per step. In predictor–corrector methods, the torque function will be called twice per step, however, its values at four previous time steps must be saved to produce the new value at time $\tau_{n+1} = \tau_n + \Delta\tau$. These methods are derived by making Taylor series of the differential equation around different points in time.

A good scheme is a fourth-order one due to Adams, Bashforth and Moulton, which uses three substeps to make one time step, advancing the solution from y_n to y_{n+1}. First is a predictor step, which gives a preliminary estimate based on one Taylor series. Then a corrector step uses the predictor's result in a Taylor series around a different point. Finally the time step is completed by a mop-up step that attempts to zero out the estimate of the errors from the predictor and corrector. The net error will be of order $(\Delta\tau)^5$. The torque function at different time steps is used, defining

$$f_n = f(y_n, \tau_n). \tag{5.48}$$

The predictor step is

$$y_p = y_n + \Delta\tau(55f_n - 59f_{n-1} + 37f_{n-2} - 9f_{n-3})/24. \tag{5.49}$$

The result of the predictor is used to estimate the torque at the next time τ_{n+1} by defining

$$f_{n+1} = f(y_p, \tau_n + \Delta\tau). \tag{5.50}$$

Then the corrector step is

$$y_c = y_n + \Delta\tau(9f_{n+1} + 19f_n - 5f_{n-1} + f_{n-2})/24. \tag{5.51}$$

Finally, the mop-up makes a slight change designed to minimize the predictor and corrector errors,

$$y_{n+1} = y_c + \frac{19}{270}(y_p - y_c). \tag{5.52}$$

This is an unevenly weighted average of the corrector and predictor results. That completes one time step. The values of y and f should be saved from the previous four times to continue to the next step. When using a predictor corrector scheme, which tends to be less stable than RK4, for example, it is important to verify that the resulting solutions do not present any numerical artifacts, due to unstable error growth. For a fixed time step approach, it is important to perform various simulations with different sized $\Delta\tau$ to ensure that the time step is small enough, and instabilities are being avoided.

A predictor–corrector method is not self-starting. Therefore, a one-step method such as RK4 can be used for the first four time steps to initiate the time evolution. Or, the initial steps can be taken by making appropriate Taylor series of the solution around the initial time. There is a very wide literature on numerical schemes that can be considered when building a simulation program.

5.3 Hybrid Monte Carlo–spin dynamics at $T > 0$

Integration of the LLG equations in the form (5.12) corresponds to a simulation at zero temperature. These equations do not include thermal fluctuations in any explicit way, which would be present for temperatures greater than zero. Here we discuss a hybrid method that determines SD numerically for a desired temperature. The simplicity of the method is reflected in the short length of its description here. The simplicity and reliability make it one of the most important practical ways to study thermalized SD.

We saw in chapter 4 that MC simulations can be carried out at nearly fixed energy in the microcanonical ensemble, using a demon to produce energy fluctuations in the system only within a thin energy shell. For SD, a very similar nearly fixed energy approach can be used, as introduced by Kawabata and co-workers [5, 6]. The approach does not require the presence of a demon, *per se*. Instead, it uses MC simulation at fixed temperature to produce initial spin states from the canonical ensemble, followed by time integration of those states at fixed energy similar to a microcanonical ensemble. Thus it is a hybrid MC–SD scheme. It was applied in the 1980s for studies of the SD in XY symmetry spin models [7] with a vortex unbinding transition. More on the characteristic dynamics in that model is presented in later chapters.

The hybrid MC–SD method is then very simple. First, use an appropriate MC scheme, single-spin Metropolis, cluster, *etc*, preferably in the canonical ensemble, from which a number of states N_{ic} will be saved as initial conditions for SD. These initial condition states taken together will have some small distribution in energy, having been selected from a canonical ensemble. It is important that they are sufficiently independent and represent a reasonable cross section of the allowed phase space for that temperature. Then, each one is integrated forward in time by equations (5.12), without any damping ($\alpha = 0$), conserving the state's initial energy. The time sequence of physically interesting quantities is saved from the time development of each initial condition. After producing the time development for N_{ic} initial conditions (this could be 100 or more), the ensemble average is found by taking averages over the different time evolutions. It is very useful for averages of *time correlation functions*, see below in this chapter for some examples.

In some of the first articles using this method [8], the ensemble consisted of only three initial states ($N_{ic} = 3$). Obviously a larger number is possible with today's machine and is certainly necessary if one is to avoid excessively large statistical errors.

The main advantage of this approach is that it combines two types of spin simulations, each of which is well understood, reliable and run independently. In actual application, it is convenient to have separate programs for the MC part and the SD part, that can be de-bugged separately. A disadvantage, more philosophical than concrete, is that it does not closely mimic what happens in a real system. A real magnetic material coupled to a thermal environment is continuously exchanging energy with its surroundings. In this MC–SD scheme, there is not a continuous heat exchange, but rather, each new initial condition undergoing time evolution is a resetting onto a different energy shell. The MC part of the algorithm allows the system to jump around to a correct distribution of energy shells. Even so, this corresponds to performing time averages over many states from the canonical ensemble, and it gives a completely reliable way to calculate time averages of physical quantities in thermal equilibrium for a chosen temperature.

5.4 Stochastic dynamics—thermal fluctuations in the planar rotor model

Next we discuss another approach for including temperature in magnetic dynamics, one that bears a greater resemblance to the physics taking place in thermal equilibrium. It is based on a stochastic equation derived from the LLG equation, that has fluctuating effects that represent the temperature. A stochastic equation of great importance in the study of molecular motion is the Langevin equation. It played an essential role in the analysis of Brownian motion by Einstein and Schmoluchowski around the year 1900. Here the extension of the Langevin equation to one for describing the dynamics of spins in the PR model in thermal equilibrium is made. To introduce the Langevin approach we consider first a system of PRs (spins with two components and rotational inertia) in thermal equilibrium, where the relation to ideal gas dynamics is closer. Later, the Langevin approach will be extended to models with three-component spins following more usual LLG dynamics.

In thermal equilibrium, the dynamics still goes on, but a system is continuously kicked by things outside it, which are in a temperature bath. Einstein and others (Langevin, Markov, *etc*) considered this problem of Brownian motion to prove the existence of atoms. The atoms or molecules are bumping larger particles and sharing their energy, at the same time, the particles' motion is slowed down by the surrounding fluid. Thus, a particle is exposed to damping and to random forces. We can suppose a similar effect takes place in the rotor or other magnetic system, although now there are viscous damping torques and random torques. The basic question, is how to describe the thermal fluctuations, the distribution of random torques, and the dynamics of the rotors (or magnetic dipoles) in their presence.

To begin, look at the dynamics of a single free rotor, not interacting with neighboring rotors. It is assumed to be exposed to a heat bath that acts on it with random and viscous torques. There is no usual mechanical torque. Still, the rotor will rotate and its direction will follow some kind of random walk. We want to consider first the correlations and fluctuations in its angular speed, $\omega(t)$.

5.4.1 Langevin equation for a free rotor in a heat bath

The equations of motion for one rotor of inertia I not coupled to others are

$$\dot{\phi} = \omega \qquad (5.53a)$$

$$I\dot{\omega} = N_s(t) - \alpha I \omega. \qquad (5.53b)$$

Here α is a damping constant with inverse time units, $N_s(t)$ is the random or stochastically fluctuating torque and $-\alpha I \omega$ is the viscous torque. This is known as the Langevin equation, especially if we had included a usual deterministic mechanical torque. Consider solving for the velocity $\omega(t)$, based on some rather simple assumptions about the random torque. For one, we suppose the random torque strength should be stronger with higher temperature. For another, it is supposed that the fluctuating torque at one time is completely unrelated to its value at an earlier time. That is the stochastic assumption. One can think there are many solutions for the velocity, starting from some initial value. We need to carry out an averaging procedure over all of the possible solutions, which correspond to different histories of the fluctuating torque.

One can write a formal solution to (5.53) by combining its homogeneous solution for no fluctuating torque, with the particular solution when the torque is present. The homogeneous equation is

$$\dot{\omega} = -\alpha\omega \tag{5.54}$$

and its solution, starting from initial velocity ω_0, is

$$\omega(t) = \omega_0 e^{-\alpha t}. \tag{5.55}$$

When the fluctuating torque is included, add some function to obtain the total solution. Make an ansatz (trial solution), letting

$$\omega(t) = (\omega_0 + u(t))e^{-\alpha t}. \tag{5.56}$$

Now one has for the time derivative,

$$\dot{\omega}(t) = (-\alpha\omega_0 - \alpha u + \dot{u})e^{-\alpha t} \tag{5.57}$$

and substituted into the dynamic equation this gives

$$I\dot{u}e^{-\alpha t} = N_s(t), \quad \Longrightarrow \quad I\dot{u} = e^{\alpha t}N_s(t). \tag{5.58}$$

This can be formally integrated, even though $N_s(t)$ is not at hand,

$$u(t) = u(0) + \frac{1}{I}\int_0^t dt'\, e^{\alpha t'} N_s(t'). \tag{5.59}$$

If one uses the boundary condition, $u(0) = 0$, the correct homogeneous solution is recovered. So the formal solution for the velocity, with an arbitrary torque function is

$$\omega(t) = \omega_0 e^{-\alpha t} + \frac{1}{I}\int_0^t dt'\, e^{-\alpha(t-t')} N_s(t'). \tag{5.60}$$

5.4.2 Planar rotator velocity autocorrelation function

Although the torque is not known, a velocity autocorrelation function can be found. The idea is that this rotor starts with initial speed ω_0 and is then affected by the random torques. But we do not know them exactly, however, we can obtain the average over all possible $N_s(t)$ (with some reasonable assumptions about their distribution). One has to assume that the averaged torque is zero, however, this does not mean that its effects are zero.

The velocity autocorrelation function is a statistical average in some ensemble, $\langle\omega(t)\omega(t)\rangle$. For large times it should converge towards a value expected from equipartition, because it is related to the averaged kinetic energy for a quadratic degree of freedom,

$$\frac{1}{2}I\langle\omega(t)\omega(t)\rangle|_{t\to\infty} \longrightarrow \frac{1}{2}k_B T, \tag{5.61}$$

where k_B is Boltzmann's constant and T is absolute temperature. Using the solution found above, one has,

$$\langle \omega(t)\omega(t)\rangle$$
$$= \left\langle e^{-2at}\left[\omega_0 + \frac{1}{I}\int_0^t dt'\, e^{at'} N_s(t')\right]\left[\omega_0 + \frac{1}{I}\int_0^t dt''\, e^{at''} N_s(t'')\right]\right\rangle$$
$$= e^{-2at}\left\{\omega_0^2 + 2\frac{\omega_0}{I}\int_0^t dt'\, e^{at'}\langle N_s(t')\rangle + \frac{1}{I^2}\int_0^t dt'\, e^{at'}\int_0^t dt''\, e^{at''}\langle N_s(t')N_s(t'')\rangle\right\}. \quad (5.62)$$

The brackets show the averaging over the torque functions. We assume there is no bias in these functions. They have an average effect of zero on the velocity. So the first average is

$$\langle N_s(t')\rangle = 0. \quad (5.63)$$

The second average involves the product of a torque function at two different times, averaged over all the possible torque functions. They are assumed to be completely independent, and stochastic, which means that the average gives zero unless the times are the same. So we take

$$\langle N_s(t')N_s(t'')\rangle = A\,\delta(t'-t''). \quad (5.64)$$

The normalization constant A will determined by equipartition. With that, one can continue to evaluate the resulting integral:

$$\int_0^t dt'\, e^{at'}\int_0^t dt''\, e^{at''} A\,\delta(t'-t'') = A\int_0^t dt'\, e^{at'}e^{at'} = A\frac{1}{2\alpha}\left(e^{2at}-1\right). \quad (5.65)$$

Finally this gives the autocorrelation function,

$$\langle \omega(t)\omega(t)\rangle = \omega_0^2 e^{-2at} + \frac{A}{2\alpha I^2}\left(1 - e^{-2at}\right). \quad (5.66)$$

The constant is determined by matching the value as $t \to \infty$ with that required by equipartition,

$$\langle \omega(t)\omega(t)\rangle_{t\to\infty} = \frac{A}{2\alpha I^2} = \frac{k_B T}{I} \implies A = 2\alpha I k_B T. \quad (5.67)$$

That means the correlations of the torques have to follow the requirement,

$$\langle N_s(t')N_s(t'')\rangle = 2\alpha I k_B T\,\delta(t'-t''). \quad (5.68)$$

This is referred to as the fluctuation-dissipation (FD) theorem, since it relates the strength of the torque fluctuations to the strength of the damping (or dissipation). The FD theorem is a mathematical statement that relates the power in thermal fluctuations to the energy being dissipated by the damping. The final velocity autocorrelation is forced to have the behavior,

$$\langle \omega(t)\omega(t) \rangle = \omega_0^2 e^{-2\alpha t} + \frac{k_B T}{I}(1 - e^{-2\alpha t}). \tag{5.69}$$

5.4.3 Planar rotor diffusion

Even the averaged position of the rotor can be determined, by integrating the other differential equation, $\dot{\phi} = \omega$, in an average sense. This is

$$\phi(t) = \phi_0 + \int_0^t dt' \left[\omega_0 e^{-\alpha t'} + \frac{1}{I} \int_0^{t'} dt'' e^{-\alpha(t'-t'')} N_s(t'') \right]$$

$$= \phi_0 + \frac{\omega_0}{\alpha}(1 - e^{-\alpha t}) + \frac{1}{I} \int_0^t dt' \int_0^{t'} dt'' e^{-\alpha(t'-t'')} N_s(t''). \tag{5.70}$$

Of course, that depends on the choice of $N_s(t)$ and so it has an infinite number of possible trajectories. Instead, look at the mean-squared displacement from the starting point, as appropriate for a random walk. This is averaged over the torque functions,

$$\langle (\phi(t) - \phi_0)^2 \rangle = \frac{\omega_0^2}{\alpha^2}(1 - e^{-\alpha t})^2$$

$$+ \frac{1}{I^2} \left\langle \int_0^t dt_1 \int_0^{t_1} dt_2 e^{-\alpha(t_1-t_2)} N_s(t_2) \int_0^t dt_3 \int_0^{t_3} dt_4 e^{-\alpha(t_3-t_4)} N_s(t_4) \right\rangle. \tag{5.71}$$

The cross term was dropped because it averages to zero. Now one can use the correlation function,

$$\langle N_s(t_2) N_s(t_4) \rangle = A\, \delta(t_2 - t_4). \tag{5.72}$$

But care is needed in the application of this. Use it to carry out the integration over t_4 first. The delta function picks out the point $t_4 = t_2$, but only if t_2 is inside the range of integration. This constraint is that $t_2 < t_3$ for a non-zero result. So the integration over only t_4 gives

$$\int_0^{t_3} dt_4 e^{-\alpha(t_3-t_4)} A\, \delta(t_2 - t_4) = \begin{cases} 0 & \text{if } t_2 > t_3 \\ A e^{-\alpha(t_3-t_2)} & \text{if } t_2 < t_3. \end{cases} \tag{5.73}$$

Next the integration over t_3 can be performed, but since it requires $t_3 > t_2$, the limits are modified to

$$\int_0^t dt_3 \to \int_{t_2}^t dt_3. \tag{5.74}$$

With that, there results

$$\int_{t_2}^t dt_3\, A e^{-\alpha(t_3-t_2)} = \frac{-A}{\alpha}(e^{-\alpha(t-t_2)} - 1). \tag{5.75}$$

The rest of the integrations are straightforward, with the integration over t_2:

$$a = \int_0^{t_1} dt_2 \, e^{-\alpha(t_1-t_2)} \frac{-A}{\alpha} \left(e^{-\alpha(t-t_2)} - 1 \right)$$

$$= -\frac{A}{\alpha} e^{-\alpha t_1} \int_0^{t_1} dt_2 \left(e^{-\alpha(t-2t_2)} - e^{\alpha t_2} \right)$$

$$= -\frac{A}{\alpha} e^{-\alpha t_1} \left[\frac{1}{2\alpha} \left(e^{-\alpha(t-2t_1)} - e^{-\alpha t} \right) - \frac{1}{\alpha} (e^{\alpha t_1} - 1) \right]$$

$$= -\frac{A}{\alpha^2} \left[\frac{1}{2} \left(e^{-\alpha(t-t_1)} - e^{-\alpha(t+t_1)} \right) - (1 - e^{-\alpha t_1}) \right]. \tag{5.76}$$

Finally there is the integration over t_1:

$$b = -\frac{A}{\alpha^2} \int_0^t dt_1 \left[\frac{e^{-\alpha t}}{2} \left(e^{\alpha t_1} - e^{-\alpha t_1} \right) - (1 - e^{-\alpha t_1}) \right]$$

$$= -\frac{A}{\alpha^2} \left[\frac{1}{2\alpha} e^{-\alpha t} \left(e^{\alpha t} - 1 + e^{-\alpha t} - 1 \right) - t - \frac{1}{\alpha} (e^{-\alpha t} - 1) \right]$$

$$= \frac{A}{\alpha^3} \left(\alpha t - \frac{3}{2} + 2e^{-\alpha t} - \frac{1}{2} e^{-2\alpha t} \right). \tag{5.77}$$

So the mean-squared displacement is found as

$$\left\langle (\phi(t) - \phi_0)^2 \right\rangle = \frac{\omega_0^2}{\alpha^2} (1 - e^{-\alpha t})^2 + \frac{2 k_B T}{I \alpha^2} \left(\alpha t - \frac{3}{2} + 2e^{-\alpha t} - \frac{1}{2} e^{-2\alpha t} \right). \tag{5.78}$$

One can look at the short time and long time behaviors. For short times $\alpha t \ll 1$, expansion of the RHS gives

$$\left\langle (\phi(t) - \phi_0)^2 \right\rangle \approx \omega_0^2 t^2. \tag{5.79}$$

Curiously, all of the constant terms and terms linear in t cancel out, and so do the temperature-dependent terms in t^2. The net result is very simple and shows a ballistic result: motion at constant angular speed. Compare at long times, where there is diffusive motion,

$$\left\langle (\phi(t) - \phi_0)^2 \right\rangle \approx \frac{\omega_0^2}{\alpha^2} + \frac{2 k_B T}{I \alpha^2} \left(\alpha t - \frac{3}{2} \right) \approx \frac{2 k_B T}{I \alpha} t. \tag{5.80}$$

At very long times the linear term dominates and the constant term does not matter. Then the root-mean-squared displacement is

$$\sqrt{\left\langle (\phi(t) - \phi_0)^2 \right\rangle} \approx \sqrt{\frac{2 k_B T}{I \alpha} t} = \sqrt{2 D t}, \tag{5.81}$$

where the diffusion constant here is

$$D = \frac{k_B T}{I\alpha}. \tag{5.82}$$

This is the Einstein relation for rotational motion. Rotor motion is diffusive for long times. The diffusion speed increases with temperature, whereas larger damping or rotational inertia slows the diffusion.

5.5 Numerical solutions of Langevin equations

Earlier in (5.34b) and in (5.43) the deterministic torques $N_i(t)$ were discussed for PRs. In general, one will consider a PR model combining those mechanical torques with the viscous and stochastic torques. Instead of a nearly free rotor, each rotor will interact with its neighbors and with the heat bath, as represented by $N_i^{\text{damping}} = -\alpha I \omega_i$ and $N_s(t)$. This leads to a set of coupled Langevin equations for PRs:

$$\dot{\phi}_i = \omega_i \tag{5.83a}$$

$$I\dot{\omega}_i = N_i(t) - \alpha I \omega_i + N_{s,i}(t). \tag{5.83b}$$

This gives a set of first-order equations that acts as a prototype for stochastic magnetics simulations. The deterministic mechanical torque is $N_i(t)$, as would be obtained from a Hamiltonian or Lagrangian without damping. We discuss the development of various numerical methods for its solution.

The main task for numerical integrations of the Langevin equations (5.83a) is how to include random torques with the correct distribution. This is usually discussed for linear motion of particles in Brownian motion with mass m, coordinate $r(t)$ and velocity $v(t) = \dot{r}$, acted on by stochastic forces $F_s(t)$. The mathematical equations for rotor motion are essentially the same, as long as the variables m, r, v, F_s are mapped over into I, ϕ, ω, N_s, respectively. For instance, the FD theorem from the rotor model becomes the following well-known expression for diffusing particles affected by stochastic forces:

$$\langle F_s(t') F_s(t'') \rangle = 2\alpha m k_B T \, \delta(t' - t''). \tag{5.84}$$

Here we follow the rotor notations but keep in mind particle diffusion as an alternative system for conceptualization. In addition, the equations will be written dropping the i subscripts, and taking ϕ, ω, *etc*, to mean the array of angular coordinates, velocities, *etc*, for a collection of coupled rotors. It simplifies subsequent equations to divide out the inertia, and define scaled torques, using f as the usual symbol for the driving functions,

$$f(t) \equiv N(t)/I, \qquad f_s(t) \equiv N_s(t)/I. \tag{5.85}$$

Then the system of Langevin equations is written as

$$\dot{\phi} = \omega \tag{5.86a}$$

$$\dot{\omega} = f(t) - \alpha \omega + f_s(t). \tag{5.86b}$$

5.5.1 Euler method for Langevin equation

With a time step Δt, define the discrete times $t_n = n\Delta t$. Then numerical integration uses quantities at discrete time steps, $\phi_n \equiv \phi(t_n), \omega_n \equiv \omega(t_n), f_n \equiv f(t_n)$, and so on. However, this cannot be done for the rapidly varying stochastic force $f_s(t)$.

Integration schemes are usually based on Taylor expansions of derivatives. Here, one must make use of integrations over time steps, and account for the stochastic variations to find changes over a time step, $\Delta\phi = \phi_{n+1} - \phi_n$ and $\Delta\omega = \omega_{n+1} - \omega_n$. The position change is

$$\Delta\phi = \int_{t_n}^{t_n+\Delta t} dt'\, \dot\phi(t') = \int_{t_n}^{t_n+\Delta t} dt'\, \omega(t') \approx \omega_n \Delta t. \tag{5.87}$$

The value of $\omega(t')$ at the initial point $t' = t_n$ was used; this is the Euler approximation. Consider the velocity change,

$$\Delta\omega = \int_{t_n}^{t_n+\Delta t} dt'\, \dot\omega(t') = \int_{t_n}^{t_n+\Delta t} dt'\, [f(t') - \alpha\omega(t') + f_s(t')]. \tag{5.88}$$

The terms with $f(t')$ and $\omega(t')$ can be approximated various ways, but in the Euler approximation their values at $t' = t_n$ are used. The fluctuating torque term is zero when averaged over many sample functions. But that does not mean a particular value of this integral is zero. There is some distribution to the value of the integral, whose average is zero. The integral is the 'stochastic velocity change' caused during the time step, labeled as

$$\Delta\omega_s = \int_{t_n}^{t_n+\Delta t} dt'\, f_s(t'), \quad \text{with} \quad \langle\Delta\omega_s\rangle = 0. \tag{5.89}$$

Although the average is zero, there is some width to the distribution of $\Delta\omega_s$. One can determine the squared variance, denoted σ_ω^2, making use of the FD theorem (5.68) for torques:

$$\sigma_\omega^2 = \langle\Delta\omega_s^2\rangle = \left\langle \int_{t_n}^{t_n+\Delta t} dt'\, f_s(t') \int_{t_n}^{t_n+\Delta t} dt''\, f_s(t'') \right\rangle$$
$$= \int_{t_n}^{t_n+\Delta t} dt' \int_{t_n}^{t_n+\Delta t} dt''\, \frac{2\alpha k_B T}{I}\delta(t'-t'') = \frac{2\alpha k_B T}{I}\Delta t. \tag{5.90}$$

This is only based on the behavior the torques must have for establishing thermal equilibrium. One step's velocity change can be expressed using the simplest approximation for the damping and conservative torque terms (using their values at time t_n),

$$\omega_{n+1} = \omega_n + (f_n - \alpha\omega_n)\Delta t + \Delta\omega_s. \tag{5.91}$$

To achieve this updating numerically, $\Delta\omega_s$ is replaced by a random number, whose average is zero and whose variance is that determined in (5.90). In computation, assume a random number generator returns a value w_n with zero mean and unit variance. Then numbers $\sigma_\omega w_n$ will have zero mean and variance σ_ω, and are used in the velocity update. A single step in the Euler–Langevin algorithm is

$$\phi_{n+1} = \phi_n + \omega_n \Delta t \qquad (5.92a)$$

$$\omega_{n+1} = \omega_n + (f_n - \alpha\omega_n)\Delta t + \sigma_\omega w_n. \qquad (5.92b)$$

This uses a stochastic velocity change,

$$\Delta\omega_s = \sigma_\omega w_n = \left[\frac{2\alpha k_B T}{I}\Delta t\right]^{1/2} w_n. \qquad (5.93)$$

This random acceleration increases with the square root of the time step. It reflects the fact that during the time step, the rotor makes a small random walk with a diffusive behavior. This random acceleration term acts together with the deterministic and damping torques to determine the net dynamics.

A note about the random number generators is needed. Some Gaussian random number generators produce numbers with unit variance, and would be good for this application. Alternatively, one could use a random number x uniformly distributed from -0.5 to $+0.5$, as Loft and DeGrand did in 1987 [1]. Then x must be scaled correctly to obtain the desired variance. For uniform random numbers x from -0.5 to $+0.5$ (a probability distribution $P(x) = 1$), the squared variance is

$$\sigma_x^2 = \int_{-0.5}^{+0.5} dx\, x^2 = 2\frac{0.5^3}{3} = \frac{1}{12}. \qquad (5.94)$$

Then to obtain the desired stochastic velocity changes, one must scale x up by $\sqrt{12}$ to obtain a unit variance, and apply them as $w = x\sqrt{12}$, or

$$\Delta\omega_s = \sigma_\omega w = \left[\frac{24\alpha k_B T}{I}\Delta t\right]^{1/2} x. \qquad (5.95)$$

This is an approach used by Loft and DeGrand [1] in early Langevin molecular dynamics simulations of the PR model in 2D.

The errors in this method are of order $(\Delta t)^2$; the scheme is a first-order method. One can improve it by going to a second-order or possibly higher-order scheme. There are different ways to do this, discussed next.

5.5.2 Developing Langevin second-order methods

One needs better approximations than first order, that include the stochastic torques correctly. Here is one way. The velocity equation can be integrated, twice, to obtain the position. For simplicity of notation, consider a time step starting at $t = 0$ and ending at $t = \Delta t$; the time origin can be shifted afterwards. First integrate to some arbitrary time inside one time step:

$$\omega(t) - \omega(0) = \int_0^t dt'\, \dot\omega = \int_0^t dt'\left[f(t') - \alpha\omega(t') + f_s(t')\right]. \qquad (5.96)$$

Then integrate again, but to the end of the time step.

$$\int_0^{\Delta t} dt\,[\omega(t) - \omega(0)] = \int_0^{\Delta t} dt \int_0^t dt'\,[f(t') - \alpha\omega(t') + f_s(t')]. \quad (5.97)$$

The LHS can be integrated exactly because $\omega = \dot\phi$. On the RHS, the deterministic and damping torques can be approximated by their values at the beginning of the step (constant values). This gives

$$\phi(\Delta t) - \phi(0) - \omega(0)\Delta t = \frac{1}{2}(\Delta t)^2[f(0) - \alpha\omega(0)] + \int_0^{\Delta t} dt \int_0^t dt'\,f_s(t'). \quad (5.98)$$

This might be improved somewhat by using average torques over the interval, or the torque at the middle of the interval, *if that can be known numerically*. The stochastic torque integral is more interesting, it gives the stochastic position change,

$$\Delta\phi_s \equiv \int_0^{\Delta t} dt \int_0^t dt'\,f_s(t'). \quad (5.99)$$

Dimensionally, $\Delta\phi_s$ is torque per inertia times time squared, which gives radians. It is the angular displacement caused by stochastic torques during Δt. Now, $\Delta\phi_s$ must be averaged over all stochastic torque functions. But that average is zero as long there is no net bias in one direction. Still there is some width to its distribution. The squared variance of $\Delta\phi_s$ can be found from integration,

$$\sigma_\phi^2 = \langle \Delta\phi_s^2 \rangle = \left\langle \int_0^{\Delta t} dt \int_0^t dt'\,f_s(t') \int_0^{\Delta t} dx \int_0^x dx'\,f_s(x') \right\rangle$$

$$= \int_0^{\Delta t} dt \int_0^t dt' \int_0^{\Delta t} dx \int_0^x dx'\,\langle f_s(t') f_s(x') \rangle$$

$$= \int_0^{\Delta t} dt \int_0^t dt' \int_0^{\Delta t} dx \int_0^x dx'\,\frac{2\alpha k_B T}{I}\delta(t' - x')$$

$$= \frac{2\alpha k_B T}{I} \int_0^{\Delta t} dt \int_0^{\Delta t} dx \int_0^t dt' \int_0^x dx'\,\delta(t' - x'). \quad (5.100)$$

This integral takes some special care. Consider especially the integrations over x' and t', and the value of

$$Z = \int_0^t dt' \int_0^x dx'\,\delta(t' - x'). \quad (5.101)$$

For the integration over x', a non-zero result requires t' between 0 and x. A diagram of the times is given in figure 5.1, supposing that $x < t$. It shows the results of the integration over x'. Then for this case, the subsequent integration over t' gives

$$Z = \int_0^t dt'\,H(x - t') = x, \quad \text{for } x < t. \quad (5.102)$$

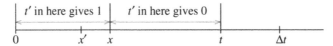

Figure 5.1. Diagram for the evaluation of integral Z in (5.101), for the case $x < t$, showing the regions where integration over x' gives different results, both summarized as a Heaviside step function, $H(x - t')$.

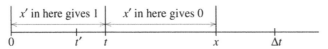

Figure 5.2. Diagram for the evaluation of integral Z in (5.101), for the case $t < x$, showing the regions where integration over t' gives different results, both summarized as a Heaviside step function, $H(t - x')$.

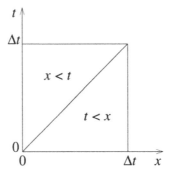

Figure 5.3. Diagram showing the region of integration for finding the integral Y in (5.106) that determines σ_ϕ^2.

The other possible case is that $t < x$, then the diagram for the integration over t' with the values of that integration shown is given in figure 5.2. For this case, the subsequent integration over x' gives

$$Z = \int_0^t dx' \, H(t - x') = t, \quad \text{for } t < x. \tag{5.103}$$

So the general result for Z is

$$Z = \int_0^t dt' \int_0^x dx' \, \delta(t' - x') = \begin{cases} x & \text{if } x < t \\ t & \text{if } t < x \end{cases} = \min(x, t). \tag{5.104}$$

This allows the remainder of the integration to be performed,

$$Y = \int_0^{\Delta t} dt \int_0^{\Delta t} dx \, \min(x, t). \tag{5.105}$$

This integration is within a unit square of the xt-plane, see figure 5.3. Half of the result comes from the lower triangle where $t < x$ and the other half comes from the upper triangle where $x < t$. So we can integrate just one of these triangles and then double that. Doubling the $t < x$ region, the integral is

$$Y = 2\int_0^{\Delta t} dx \int_0^x dt\; t$$
$$= 2\int_0^{\Delta t} dx\; \frac{1}{2}x^2 = \frac{1}{3}(\Delta t)^3. \tag{5.106}$$

Then the result for the squared variance of $\Delta\phi_s$ is

$$\sigma_\phi^2 = \left(\frac{2\alpha k_B T}{I}\right)\cdot \frac{1}{3}(\Delta t)^3. \tag{5.107}$$

This has an interesting dependence on the size of the time step. Combining with a random number generator giving a distribution of numbers w_n with zero mean and unit variance, generate random displacements $\Delta\phi_s$ by

$$\Delta\phi_s = \sigma_\phi \cdot w_n = \left[\left(\frac{2\alpha k_B T}{I}\right)\cdot \frac{1}{3}(\Delta t)^3\right]^{1/2} w_n. \tag{5.108}$$

Letting $t=0$ become t_n and $t = \Delta t$ become t_{n+1}, the twice-integrated Langevin velocity equation leads to the updated algorithm for the position:

$$\phi_{n+1} = \phi_n + \omega_n \Delta t + (f_n - \alpha\omega_n)\frac{(\Delta t)^2}{2} + \sigma_\phi \cdot w_n. \tag{5.109}$$

The reader might try to estimate the errors in the finite difference terms of the deterministic and damping torques. These are the same as in a method known as the velocity-Verlet method, which has errors of the order of Δt^3, making this a second-order method. However, we are not yet done and still need to discuss the updating of the velocity.

5.5.3 A Langevin–velocity-Verlet method

One can advance the position updating just found, and combine with a velocity updating, which could be called a 'Langevin–velocity-Verlet' algorithm. That would mean that the velocity appears explicitly in the position updating, and the deterministic forces would appear in both the position and velocity updatings. Start from the position update for one time step, but shift it forward to the next time step,

$$\phi_{n+2} = \phi_{n+1} + \omega_{n+1}\Delta t + [f_{n+1} - \alpha\omega_{n+1}]\frac{(\Delta t)^2}{2} + \sigma_\phi \cdot w_{n+1}. \tag{5.110}$$

Because this is the evolution in the next time step, there is the next random displacement term determined by w_{n+1}. Combining the two steps gives

$$\phi_{n+2} = \phi_n + (\omega_n + \omega_{n+1})\Delta t$$
$$+ [f_n + f_{n+1} - \alpha(\omega_n + \omega_{n+1})]\frac{(\Delta t)^2}{2} + \sigma_\phi \cdot (w_n + w_{n+1}). \tag{5.111}$$

Now use the symmetric difference to obtain the velocity ω_{n+1}, at the common boundary of these two intervals,

$$2\Delta t \omega_{n+1} = \phi_{n+2} - \phi_n. \tag{5.112}$$

These leads to the new velocity as

$$\omega_{n+1} = \omega_n + (f_n - \alpha\omega_n + f_{n+1} - \alpha\omega_{n+1})\frac{\Delta t}{2} + \frac{\sigma_\phi}{\Delta t} \cdot (w_n + w_{n+1}). \tag{5.113}$$

This is the stochastic generalization of the velocity-Verlet algorithm, when combined with (5.109). Because damping is velocity-dependent, ω_{n+1} appears on both sides of the equation. Isolating ω_{n+1} to the LHS, a Langevin–velocity-Verlet step is

$$\phi_{n+1} = \phi_n + \omega_n \Delta t + (f_n - \alpha\omega_n)\frac{(\Delta t)^2}{2} + \sigma_\phi \cdot w_n \tag{5.114a}$$

$$\left(1 + \frac{\alpha \Delta t}{2}\right)\omega_{n+1} = \omega_n + (f_n - \alpha\omega_n + f_{n+1})\frac{\Delta t}{2} + \frac{\sigma_\phi}{\Delta t} \cdot (w_n + w_{n+1}). \tag{5.114b}$$

The same w_n should be used in both the position and velocity update, however, the velocity update requires the generation of the next random number w_{n+1}. Then that value w_{n+1} will itself go into the subsequent position update that gives ϕ_{n+2}, and so on. It can be shown to be accurate to second order in the time step, for the position.

The time step divided into σ_ϕ is interesting. This gives the constant,

$$\frac{\sigma_\phi}{\Delta t} = \left[\left(\frac{2\alpha k_B T}{I}\right) \cdot \frac{1}{3}\Delta t\right]^{1/2} = \frac{\sigma_\omega}{\sqrt{3}}. \tag{5.115}$$

The variance σ_ω was defined for the Euler–Langevin step. So the variance in this stochastic change in velocity still depends on the square root of the time step. It causes a usual diffusive (random walk) behavior of the velocity, which is what we want. Its strength should be correctly associated with the temperature.

5.5.4 An out-of-phase second-order Langevin method

A slight variation on the first-order Euler method is considered next. It is actually a second-order method for the position *and* the velocity, if the damping is treated correctly.

The idea is to take half-steps for position updating, but whole steps for velocity updating. In the Euler method, the position is updated using the velocity at the beginning of one step. Here that is still true, but the steps are divided in half, leading to a better precision. Start at time t_n, with position ϕ_n and velocity ω_n. One step is initially sketched out as follows:

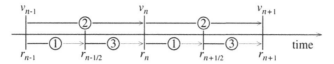

Figure 5.4. Diagram showing the substeps ①, ② and ③ made in updating of positions r_n and velocities v_n as time is evolved forward in the out-of-phase Langevin algorithm, equation (5.116).

① Position half-step: $\quad \phi_{n+\frac{1}{2}} = \phi_n + \frac{1}{2}\omega_n \Delta t, \qquad t_{n+\frac{1}{2}} = t_n + \frac{1}{2}\Delta t \qquad (5.116a)$

② Velocity full-step: $\quad \omega_{n+1} = \omega_n + f_{n+\frac{1}{2}}\Delta t - \alpha \omega_n \Delta t + \sigma_\omega w_n \qquad (5.116b)$

③ Position half-step: $\quad \phi_{n+1} = \phi_{n+\frac{1}{2}} + \frac{1}{2}\omega_{n+1}\Delta t, \qquad t_{n+1} = t_{n+\frac{1}{2}} + \frac{1}{2}\Delta t. \qquad (5.116c)$

After the position updates, the time is updated by half a step. Note also that the new velocity is produced before the final position half-step, which is responsible for the cancellation of errors in the Euler method. The process is summarized in a diagram of the time axis, figure 5.4, where circled numbers refer to the listed operations. The symbols on the diagram are the numbers that are actually calculated in the algorithm. There are velocities missing at the half-steps, because those are not calculated. Figure 5.4 shows the different substeps ①, ② and ③. Let us look at the parts ① and ③, including the error terms. For part ①, with a half time step, the Taylor expansion around time t_n is (primes indicate time derivatives)

$$\phi_{n+\frac{1}{2}} = \phi_n + \frac{\Delta t}{2}\omega_n + \frac{1}{2!}\left(\frac{\Delta t}{2}\right)^2 \omega_n' + \frac{1}{3!}\left(\frac{\Delta t}{2}\right)^3 \omega_n'' + \ldots \qquad (5.117)$$

Also write this for a reversed half time step,

$$\phi_{n-\frac{1}{2}} = \phi_n - \frac{\Delta t}{2}\omega_n + \frac{1}{2!}\left(\frac{\Delta t}{2}\right)^2 \omega_n' - \frac{1}{3!}\left(\frac{\Delta t}{2}\right)^3 \omega_n'' + \ldots \qquad (5.118)$$

Then for part ③ of the algorithm, going *into* the time t_n on the previous cycle, ϕ_n is found by solving this last equation,

$$\phi_n = \phi_{n-\frac{1}{2}} + \frac{\Delta t}{2}\omega_n - \frac{1}{2!}\left(\frac{\Delta t}{2}\right)^2 \omega_n' + \frac{1}{3!}\left(\frac{\Delta t}{2}\right)^3 \omega_n'' + \ldots \qquad (5.119)$$

Then the combination of the ③-step and the subsequent ①-step produces some cancellations (use ϕ_n from (5.119) substituted into (5.117)), giving

$$\phi_{n+\frac{1}{2}} = \phi_{n-\frac{1}{2}} + \Delta t\, \omega_n + \frac{2}{3!}\left(\frac{\Delta t}{2}\right)^3 \omega_n'' + \ldots \qquad (5.120)$$

That is the usual symmetric formula for the velocity ω_n at the mid-point of these two half-steps. As an update of the position, the error is proportional to $(\Delta t)^3$, accurate to second order in the time step. Look at the velocity update. In finding ω_{n+1} from ω_n, it uses the torque calculated at the middle of that time interval, $f_{n+\frac{1}{2}}$. This improves the precision. To see why this is so, look at the integral needed to derive this stochastic update (note that there was no stochastic term in the position update). From the Langevin equation, $\dot{\omega} = f(t) + f_s(t)$, integrated once, one obtains

$$\omega(t + \Delta t) = \omega(t) + \int_t^{t+\Delta t} dt'\, f(t') + \int_t^{t+\Delta t} dt'\, f_s(t'). \tag{5.121}$$

The last term is the stochastic push which is replaced by $\sigma_\omega w$. The other integral can be approximated in different ways, according to the time at which the force is used. Look at the expansion of that force (conservative + damping) around the *middle* of the interval— this would seem to be the best alternative to try. The mid-point time is $\bar{t} = t + \frac{\Delta t}{2}$,

$$f(t') = f(\bar{t}) + (t' - \bar{t}) \cdot f'(\bar{t}) + \frac{1}{2!}(t' - \bar{t})^2 \cdot f''(\bar{t}) + \dots. \tag{5.122}$$

With this expansion, the torque integral can be evaluated:

$$\int_t^{t+\Delta t} dt'\, f(t') = f(\bar{t}) \int_t^{t+\Delta t} dt' + f'(\bar{t}) \int_t^{t+\Delta t} dt'(t' - \bar{t}) \\ + \frac{1}{2!} f''(\bar{t}) \int_t^{t+\Delta t} dt'(t' - \bar{t})^2 + \dots. \tag{5.123}$$

But changing to the relative variable $z = t' - \bar{t}$, the integrals are of the general form

$$\int_{-\frac{\Delta t}{2}}^{+\frac{\Delta t}{2}} dz\, z^n = \left. \frac{z^{n+1}}{n+1} \right|_{-\frac{\Delta t}{2}}^{+\frac{\Delta t}{2}} = \begin{cases} \frac{2}{n+1}\left(\frac{\Delta t}{2}\right)^{n+1} & \text{for even } n \\ 0 & \text{for odd } n. \end{cases} \tag{5.124}$$

So the torque integral has only terms with odd powers of Δt:

$$\int_t^{t+\Delta t} dt'\, f(t') = \Delta t\, f(\bar{t}) + \frac{1}{24}(\Delta t)^3 f''(\bar{t}) + \dots. \tag{5.125}$$

So this gives a good velocity updating procedure, written in the index notation,

$$\omega_{n+1} = \omega_n + \Delta t \left(f_{n+\frac{1}{2}} - \alpha \omega_{n+\frac{1}{2}} \right) + \sigma_\omega w_n + \frac{1}{24}(\Delta t)^3 f''_{n+\frac{1}{2}}, \tag{5.126}$$

where damping is included explicitly and the last term is the error estimate. Surprisingly, the error is proportional to $(\Delta t)^3$, the same order as the position updating. This makes this method quite accurate.

There is only one technical detail that would limit the accuracy, which is the problem of dealing with the damping. The algorithm had to use the term $-\alpha \omega_n$, although the theory says to use $-\alpha \omega_{n+\frac{1}{2}}$. But the velocities are not being calculated at the half time steps. So, let us put in the estimate of ω at the half-step, based on the average of the end points for that interval. The expansions of ω around the midpoint $\bar{t} = t + \frac{\Delta t}{2}$ are

$$\omega(t) = \omega(\bar{t}) - \frac{\Delta t}{2}\omega'(\bar{t}) + \frac{1}{2!}\left(\frac{\Delta t}{2}\right)^2 \omega''(\bar{t}) - \frac{1}{3!}\left(\frac{\Delta t}{2}\right)^3 \omega'''(\bar{t}) + \ldots \quad (5.127a)$$

$$\omega(t + \Delta t) = \omega(\bar{t}) + \frac{\Delta t}{2}\omega'(\bar{t}) + \frac{1}{2!}\left(\frac{\Delta t}{2}\right)^2 \omega''(\bar{t}) + \frac{1}{3!}\left(\frac{\Delta t}{2}\right)^3 \omega'''(\bar{t}) + \ldots \quad (5.127b)$$

Then these added together give a good estimate of the mid-point velocity,

$$\omega(\bar{t}) = \frac{1}{2}[\omega(t) + \omega(t + \Delta t)] - \frac{1}{2!}\left(\frac{\Delta t}{2}\right)^2 \omega''(\bar{t}) + \ldots \quad (5.128)$$

Using this to write the damping $-\alpha\omega(\bar{t})$, we have in the velocity update, with index notation,

$$\omega_{n+1} = \omega_n - \frac{\alpha\Delta t}{2}[\omega_n + \omega_{n+1}] + \Delta t f_{n+\frac{1}{2}} + \sigma_\omega w_n$$
$$+ \alpha\Delta t \frac{1}{2!}\left(\frac{\Delta t}{2}\right)^2 \omega''_{n+\frac{1}{2}} + \frac{1}{24}(\Delta t)^3 f''_{n+\frac{1}{2}} \quad (5.129)$$

or just summarizing together with the error term,

$$\left(1 + \frac{\alpha\Delta t}{2}\right)\omega_{n+1} = \left(1 - \frac{\alpha\Delta t}{2}\right)\omega_n + \Delta t f_{n+\frac{1}{2}} + \sigma_\omega w_n + O(\Delta t^3). \quad (5.130)$$

In this way, the update is indeed accurate to second order in the time step, and the inclusion of the damping is totally stable and gives the correct non-forced solution.

So let us summarize this out-of-phase second-order algorithm, with stable damping and two position updates per step, but only one velocity update per step. In the original ①, ②, ③ ordering,

① $\quad \phi_{n+\frac{1}{2}} = \phi_n + \frac{\Delta t}{2}\omega_n \quad$ (5.131a)

② $\quad \left(1 + \frac{\alpha\Delta t}{2}\right)\omega_{n+1} = \left(1 - \frac{\alpha\Delta t}{2}\right)\omega_n + \Delta t f_{n+\frac{1}{2}} + \sigma_\omega w_n \quad$ (5.131b)

③ $\quad \phi_{n+1} = \phi_{n+\frac{1}{2}} + \frac{\Delta t}{2}\omega_{n+1} \quad$ (5.131c)

where each part has error of order $(\Delta t)^3$. Being simpler than the Langevin–velocity-Verlet introduced earlier, its big advantage is the evaluation of the torques in the middle of the interval for the velocity update. On the other hand, an advantage of the Langevin–velocity-Verlet is that it has stochastic terms in both the position and velocity updates, which could give it some better balance and symmetry.

Recall again that these methods can be applied to the diffusing mass problem by the translations previously mentioned, switching ϕ to position r, switching ω to linear velocity v, and inertia I to mass m.

Yet another second-order method due to Heun is discussed in the next section, where Langevin dynamics for spins is considered.

5.6 Langevin spin dynamics

In the previous section some methods for solving Langevin equations with a coordinate and a velocity (or momentum), such as rotors, were described. Now we turn to numerical schemes for spins undergoing Langevin dynamics, which do not have separated coordinates and momenta.

The deterministic dynamics for spins (with damping) is described by (5.6) or the dimensionless form (5.12). The corresponding Langevin dynamics requires including a rapidly fluctuating stochastic torque. The size of the stochastic torques will be related to the temperature and the damping constant, such that the system tends towards thermal equilibrium for the chosen temperature. As for the PR system, the FD theorem produces the relation between temperature and the relative size of the stochastic torques. The theory for this in the case of spins rather than rotors is somewhat more difficult to develop from simple principles. Therefore we will refer to the rotor problem as a prototype and map the coordinates for the rotor problem into equivalent variables for the spin problem[2].

In fact, it is more physical and more convenient for calculations to think that Langevin SD includes *random magnetic fields*, rather than torques. The dynamical equation for some selected spin exposed to a deterministic field $\mathcal{F}$ and a stochastic field $\mathcal{F}_s$ is

$$\frac{d}{dt}\mathbf{S} = \mathbf{S} \times (\mathcal{F} + \mathcal{F}_s) - \frac{\alpha}{S}\mathbf{S} \times [\mathbf{S} \times (\mathcal{F} + \mathcal{F}_s)]. \tag{5.132}$$

The first term is the free motion and the second term is the Landau–Gilbert damping, with dimensionless damping constant α. The dynamics evolves according to the superposition of the deterministic and random fields. One can also split out their contributions separately,

$$\frac{d}{dt}\mathbf{S} = \mathbf{S} \times \mathbf{G} + \mathbf{S} \times \mathbf{G}_s \tag{5.133}$$

where

$$\mathbf{G} = \mathcal{F} - \frac{\alpha}{S}(\mathbf{S} \times \mathcal{F}), \qquad \mathbf{G}_s = \mathcal{F}_s - \frac{\alpha}{S}(\mathbf{S} \times \mathcal{F}_s). \tag{5.134}$$

This shows the effective deterministic and stochastic dynamical fields acting on this spin. The stochastic fields $\mathcal{F}_s$ enter as multiplicative noise, because they appear as products with the time-dependent spins, whose solution is desired.

---

[2] The rigorous theory for Langevin SD comes from the Fokker–Planck equation of statistical physics. See advanced textbooks or the article by García-Palacios and Lázaro [9] for a good explanation.

For the stochastic fields to produce thermal equilibrium, consider the effect of their torques. For rotors, the FD theorem (5.68) gives the time–time correlation of one torque component at two different times. To map that result over to this spin problem, recall that torque is given from

$$\mathbf{N} = \mathbf{S} \times \mathcal{F}, \tag{5.135}$$

where $\mathcal{F} = \gamma \mathbf{B}$ gives effective field $\mathcal{F}$ in terms of magnetic induction $\mathbf{B}$. Further, the factor αI for a rotor maps over into αS for a spin, as one can check. For a rotor α had dimensions of time^{-1} or frequency, so that αI is angular momentum, as is the factor αS for spins (α is dimensionless in the LLG equations). Then based on this dimensional analysis, but without any rigorous proof, the FD theorem for spin degrees of freedom can be written as

$$\left\langle N_s^j(t) N_s^k(t') \right\rangle = 2\alpha S\, k_B T\, \delta_{jk}\, \delta(t - t'). \tag{5.136}$$

The indices j, k refer to any of the Cartesian coordinates. The FD theorem shows how the power of the thermal fluctuations is carried equivalently in random magnetic fields. One really wants to see a condition on those fluctuating fields, which can be implemented according to the replacement $N_s^j \to S \mathcal{F}_s^j$, giving for application with effective fields,

$$S\left\langle \mathcal{F}_s^j(t) \mathcal{F}_s^k(t') \right\rangle = 2\alpha\, k_B T\, \delta_{jk}\, \delta(t - t'). \tag{5.137}$$

Expressed more physically in terms of equivalent magnetic induction components, it is

$$\gamma^2 S \left\langle B_s^j(t) B_s^k(t') \right\rangle = 2\alpha\, k_B T\, \delta_{jk}\, \delta(t - t'). \tag{5.138}$$

All of these last three equivalent expressions apply to the fields at an individual spin.

The Langevin equation in (5.133) is a first-order differential equation where the noise is multiplicative—the second term involves a product of $\mathbf{S}$ with a net stochastic field. There is assumed to be an equation for every spin in a system. We let $y(t)$ be the array of all spins (or their components). Symbolically y obeys a differential equation in the general form,

$$\frac{dy}{dt} = f(t, y(t)) + g(t, y(t)) \cdot \mathcal{F}_s(t). \tag{5.139}$$

The function f represents the deterministic time derivative (includes damping) on the RHS of (5.133) and the function g represents the spin components that multiply the stochastic effective fields. Each is defined indirectly by comparing this with the Langevin equation. The fields f and $\mathcal{F}_s$ are vectors of $3N$ components, where N is the number of spins; g is N tensors of size 3×3. For example, from the differential equation (5.133) for $\dot{S}^x$ at a site,

$$\dot{S}^x = (S^y G^z - S^z G^y) + (S^y G_s^z - S^z G_s^y). \tag{5.140}$$

The first is the deterministic term. The stochastic term can be expanded out to show dependencies on $\mathcal{F}_s$ components,

$$(\mathbf{S} \times \mathbf{G}_s)^x = S^y \left[\mathcal{F}_s^z - \frac{\alpha}{S}(S^x \mathcal{F}_s^y - S^y \mathcal{F}_s^x) \right] - S^z \left[\mathcal{F}_s^y - \frac{\alpha}{S}(S^z \mathcal{F}_s^x - S^x \mathcal{F}_s^z) \right]$$
$$= \frac{\alpha}{S}(S^{y2} + S^{z2})\mathcal{F}_s^x + \left(-S^z - \frac{\alpha}{S}S^x S^y\right)\mathcal{F}_s^y + \left(S^y - \frac{\alpha}{S}S^x S^z\right)\mathcal{F}_s^z. \quad (5.141)$$

Thus the factors in front of $\mathcal{F}_s^x$, $\mathcal{F}_s^y$, $\mathcal{F}_s^z$ are the needed components of the function $g(t, y)$ for an x spin component's time derivative. This is shown here solely for demonstration; this expansion is not necessary for the numerical integration.

5.6.1 Second-order Heun integration scheme

An efficient method for integrating this magnetic dynamics type of equation forward in time is the second-order Heun method [9, 10]. This is in the family of predictor–corrector schemes and is rather stable. Another method that could be applied with similar precision would be second-order Runge–Kutta, however, it would be slightly less efficient for including the stochastic fields.

The predictor stage for the second-order Heun algorithm is an Euler step, which is followed by a corrector stage that is equivalent to the trapezoid rule. Each involves moving forward in time over the time step Δt, with the needed results obtained by integrating (5.139) from an initial time t_n to a final time $t_{n+1} = t_n + \Delta t$, during which the stochastic fields are acting. With notation $y_n \equiv y(t_n)$, the predictor stage produces an initial solution estimate at the end of one time step,

$$y_p = y_n + f(t_n, y_n)\Delta t + g(t_n, y_n) \cdot (\sigma_s w_n). \quad (5.142)$$

The effect of the random fields $\mathcal{F}_s$ is contained in the last term. The factor $\sigma_s w_n$ replaces the time integral of the stochastic magnetic fields. For each site of the array, there is a triple of unit variance, zero mean random numbers (w_n^x, w_n^y, w_n^z) produced by a random number generator. The physical variance σ_s needed in the stochastic fields is defined by an equilibrium average over the time step. For an individual component, say, $\mathcal{F}_s^x$, at one site, based on the discussion of an Euler step for rotors, that is found from

$$\sigma_s^2 = \left\langle \left(\int_{t_n}^{t_n+\Delta t} dt\, \mathcal{F}_s^x(t) \right)^2 \right\rangle = \int_{t_n}^{t_{n+1}} dt \int_{t_n}^{t_n+\Delta t} dt'\, \langle \mathcal{F}_s^x(t) \mathcal{F}_s^x(t') \rangle. \quad (5.143)$$

When the FD theorem (5.137) is applied to one dipole, this gives the variance of the random fields,

$$\sigma_s = \sqrt{\frac{2\alpha\, k_B T}{S}\Delta t}. \quad (5.144)$$

Thus, the individual stochastic field components, integrated over the time step, are replaced by random numbers of zero mean with the variance σ_s. One can check that the dimensions of $g(t, y)\sigma_s$ are spin dimensions (energy × time), as it should be when inserted into the predictor equation.

In the corrector stage, the points y_n and y_p are used to obtain better estimates of the slope of the solution. Their average effect is used in the trapezoid correction stage:

$$y_{n+1} = y_n + \frac{1}{2}\left[f(t_n, y_n) + f(t_{n+1}, y_p)\right]\Delta t$$

$$+ \frac{1}{2}\left[g(t_n, y_n) + g(t_{n+1}, y_p)\right] \cdot (\sigma_s w_n). \tag{5.145}$$

The error varies as $(\Delta \tau)^3$, hence it is a second-order scheme. It is important to note that the *same* random numbers w_n are used in this corrector stage as those applied in the predictor stage, for this individual time step.

In the actual implementation of this method, the functions f and g do not need to be specifically identified. Instead, it is seen that the change in any spin over a time step, $\Delta \mathbf{S} = \int dt \frac{d}{dt}\mathbf{S}$, depends linearly on $\mathcal{F}\Delta t$ from the deterministic part of the differential equation, and linearly on $\int dt\, \mathcal{F}_s(t)$ from the stochastic part. Both these contributions are small, because they depend on Δt and $\sqrt{\Delta t}$, respectively. But the stochastic contribution is replaced by random numbers of the correct variance, that is,

$$\int_{t_n}^{t_n+\Delta t} dt\, \mathcal{F}_s^x(t) \longrightarrow \sigma_s w_n^x. \tag{5.146}$$

The needed effective field integrated over time that updates this site is the deterministic/stochastic combination,

$$\mathcal{G}_p = \int_{t_n}^{t_n+\Delta t} dt(\mathcal{F} + \mathcal{F}_s) = \mathcal{F}(\mathbf{S}_n)\Delta t + \sigma_s(\Delta t)\mathbf{w}. \tag{5.147}$$

Here $\mathbf{S}_n$ refers to the state at the nth time step. The same type of combination applies in the trapezoid corrector step. Then the Euler predictor step is efficiently carried out by evaluating the combined deterministic plus stochastic field contributions, for an individual site, such as

$$\Delta \mathbf{S}_p = \mathbf{S}_n \times \left[\mathcal{G}_p - \alpha(\mathbf{S}_n \times \mathcal{G}_p)\right] \tag{5.148a}$$

$$\mathbf{S}_p = \mathbf{S}_n + \Delta \mathbf{S}_p. \tag{5.148b}$$

$\mathbf{S}_p$ is the predicted new spin position. For the corrector, the updating field at the end of the time step is calculated, using the *predicted position*, together with the *same random field*,

$$\mathcal{G}_c = \mathcal{F}(\mathbf{S}_p)\Delta t + \sigma_s(\Delta t)\mathbf{w}. \tag{5.149}$$

This leads to the corrector's estimate for the spin change,

$$\Delta \mathbf{S}_c = \mathbf{S}_p \times \left[\mathcal{G}_c - \alpha (\mathbf{S}_p \times \mathcal{G}_c) \right]. \tag{5.150}$$

Then the corrector stage gives the updated spin according to their average

$$\mathbf{S}_{n+1} = \mathbf{S}_n + \frac{1}{2}(\Delta \mathbf{S}_p + \Delta \mathbf{S}_c). \tag{5.151}$$

This algorithm does not ensure the conservation of spin length. Thus, the length of $\mathbf{S}_n$ can be rescaled to S after the step. However, in practice, when that becomes necessary, the scheme is already becoming unstable, which can happen if the net field $|\mathcal{G}|$ is greater than 1. The best way to avoid numerical instability is to make sure the time step is sufficiently small, so that $|\mathcal{G}| \ll 1$, or even better, $|\Delta \mathbf{S}| \ll 1$, to the precision desired.

The integrations requires a sequence of quasi-random numbers (the w_n stochastic fields) with a long period, so that the simulation time does not surpass the period of the random numbers. One generator called mzran13 due to Marsaglia and Zaman [11], can be implemented in the C-language for long integers, in a very short and efficient code. This generator is very simple and fast and has a period of about 2^{125}, due to the combination of two separate generators with periods of 2^{32} and 2^{95}. The reader can look for other ideas for random number generators in the text by Press et al [4], as an example.

Examples of the application of this type of stochastic LLG dynamics will be given in the chapters in part II of this book.

5.7 Dynamic correlation responses

Among the many physical quantities that depend on time in magnetization dynamics, some of the most important are the correlations of spins separated by a time interval. The correlations of spins separated in space are also of great importance. The space- and time-dependent spin correlations can be calculated in thermal equilibrium SD, and are useful for giving a detailed view of the dynamic excitations in the system. Even more interesting information can be deduced from the structures found in their time and space Fourier transforms (FTs). Here we discuss time correlations as a measure of relaxation in a system, and space–time correlations as a measure of the types and properties of excitations present. The overall Fourier-transformed space–time correlations lead to the dynamic structure functions that might be measured in spin scattering experiments. In particular, data from a discrete lattice and discrete time simulation must be converted to the corresponding quantities expected for the continuum limit theory.

5.7.1 Space Fourier transforms

For an infinite interval, 1D continuum space x, some function $f(x)$ can be expanded in Fourier space by using basis functions e^{iqx} with wave vector q, of amplitudes $A(q)$, as follows:

$$f(x) = \int_{-\infty}^{\infty} dq\, A(q)\, e^{iqx}. \tag{5.152}$$

Applying a function e^{iqx} to perform the overlap integral over all x leads to

$$\int_{-\infty}^{\infty} dx\, f(x) e^{ikx} = \int_{-\infty}^{\infty} dq\, A(q) \int_{-\infty}^{\infty} dx\, e^{i(k+q)x}. \tag{5.153}$$

But the x integration gives $2\pi\delta(k+q)$, which, after interchanging the names q and k, leads to the result for the Fourier amplitude at wave vector q,

$$A(q) = \frac{1}{2\pi} \int_{-\infty}^{\infty} dx\, f(x) e^{-iqx}. \tag{5.154}$$

This is the theoretical expression for a FT. Application in simulation needs to account for three new aspects: (1) a finite length system, (2) a finite discrete space lattice and (3) conversion from discrete sums to the equivalent continuum $A(q)$.

Recall what the function $A(q)$ really means. It represents the amplitude or weight in q-space that the original function $f(x)$ possesses near the point q. But in order to be able to find this quantity on a discrete system, it is better to realize that the product $A(q)dq$ represents the amplitude of the original function in the small interval dq of q-space.

Exercise 5.1. (a) Suppose a localized spin excitation in a 1D magnetic system has a static sinusoidal wave form for one spin component (defined on an infinite interval)

$$S^x(x) = f(x) = e^{-\alpha|x|}\cos(kx), \tag{5.155}$$

where k and α are parameters with dimensions of inverse length. Find the FT $A(q)$ for $\alpha \neq 0$. (b) What happens to the FT in the limit $\alpha \to 0$? Show how that limit is approached, by plotting $A(q)$ at various values of α. Interpret physically the shape of $A(q)$.

---

(1) Finite interval
In a system of finite length L, the FT is replaced by a Fourier series for a function that is periodic over the system length L. The basis functions are still exponentials, but there is now a discrete set of them, so that an infinite sum replaces the integral:

$$f(x) = \sum_{m=-\infty}^{+\infty} A_{q_m} e^{iq_m x}, \quad q_m \equiv \frac{2\pi}{L}m, \quad m = 0, 1, 2, 3\ldots \text{ (no limit).} \tag{5.156}$$

Here the q_m are discrete and $L = Na$ is the length of the system. Later we take the system to have N sites separated by lattice constant a. Applying another basis function to the LHS and then performing an overlap integration over the system:

$$\int_0^L dx\, f(x) e^{-iq_{m_1}x} = \sum_{m=-\infty}^{+\infty} A_{q_m} \int_0^L dx\, e^{i(-q_{m_1}+q_m)x}. \tag{5.157}$$

The x integration gives the famous Kroenecker delta function, $L\delta_{m,m_1}$, leaving one term in the sum. After switching the names q_{m_1} and q_m, it gives

$$A_{q_m} = \frac{1}{L}\int_0^L dx\, f(x) e^{-iq_m x}. \tag{5.158}$$

This is the *discrete* amplitude A_{q_m} denoted with the subscript q_m, to be distinguished from continuum quantities (different units and object) which are denoted with parenthesis (i.e. $A(q)$).

Exercise 5.2. Consider a static spin field component on a 1D magnetic chain that has a square wave form which repeats periodically over some length L,

$$S^x = f(x) = \begin{cases} +1, & 0 \leq x < \dfrac{L}{2} \\ -1, & \dfrac{L}{2} \leq x < L. \end{cases} \tag{5.159}$$

Show that the discrete Fourier amplitudes are

$$A_{q_m} = \begin{cases} \dfrac{4}{iq_m L}, & m = 1, 3, 5\ldots \\ 0, & m = 0, 2, 4\ldots. \end{cases} \tag{5.160}$$

Note that these are dimensionless numbers (i.e. the same dimension as function $f(x)$).

---

(2) Finite interval and discrete positions

If the x variable has only discrete and evenly spaced values, $x_n = na$, for $n = 1, 2, 3, \ldots N$, then the wave vectors will also have upper and lower limits and N possible values. The expansion of the function is now:

$$f_n = \sum_{m=1}^{N} A_{q_m} e^{iq_m x_n} = \sum_{m=1}^{N} A_{q_m} e^{i\frac{2\pi}{N} mn}. \tag{5.161}$$

Application of an exponential on the LHS and then summation as above gives:

$$\sum_{n=1}^{N} f_n e^{-iq_{m_1} x_n} = \sum_{m=1}^{N}\sum_{n=1}^{N} A_{q_m} e^{i\frac{2\pi}{N}(m-m_1)n}. \tag{5.162}$$

But once again, the famous summation

$$\sum_{n=1}^{N} e^{i\frac{2\pi}{N}(m-m_1)n} = N\delta_{m,m_1} \tag{5.163}$$

can be used, leading to:

$$A_{q_m} = \frac{1}{N} \sum_{n=1}^{N} f_n e^{-iq_m x_n}. \tag{5.164}$$

This is the discrete weight. It should be converted to its corresponding continuum weight, see below.

Exercise 5.3. Consider again the square wave form of the previous exercise, but for a spin field defined on discrete lattice sites, $x_n = na$, $n = 1, 2, 3 \ldots N$. The system length is $L = Na$. The square wave for some spin component is

$$S_n^x = f_n = \begin{cases} +1, & n = 1, 2, 3, \ldots \dfrac{N}{2} \\ -1, & n = \dfrac{N}{2} + 1, \ldots N. \end{cases} \tag{5.165}$$

By the summing the two geometric series that result, show that the discrete Fourier amplitudes are

$$A_{q_m} = \begin{cases} \dfrac{2}{N} \dfrac{1 + e^{-iq_m a}}{1 - e^{-iq_m a}} = \dfrac{2}{N i \tan(q_m a/2)}, & m = 1, 3, 5 \ldots \\ 0, & m = 0, 2, 4 \ldots. \end{cases} \tag{5.166}$$

Verify in an appropriate limit that the result is consistent with (5.160) from the previous exercise.

---

(3) Discrete to continuum conversion of Fourier transforms

This part is actually quite simple. If we had a *continuum* system of *finite* length L, then we would still have only discrete values of wave vectors, but with no upper or lower limit. Here we have limits, due to the discrete x-position. We also have a fixed wave vector step $dq = 2\pi/L$ between the discrete wave vectors. This means that each one can be thought to occupy a distance of $2\pi/L$ in q-space. Therefore, to obtain the equivalent continuum theory quantity $A(q)$, we should *divide* the discrete A_q by $dq = 2\pi/L$, and drop the subscript on q_m. So the corresponding continuum weight as in (5.154) is

$$A(q) = \frac{A_q}{dq} = \frac{L}{2\pi} A_q = \frac{a}{2\pi} \sum_{n=1}^{N} f_n e^{-iq x_n}. \tag{5.167}$$

This is equivalent to changing $\int dx \to a \sum_n$ in the continuum expression (5.154).

5.7.2 Space correlations

The above formula tells us how to form, for example, the space FT for each dimension of a lattice system. It is interesting that there are *no* prefactors depending on the system size. The only size-dependent factor is in the definitions of the allowed wave vectors.

Spatial correlations of spin pairs separated by some distance give information about the longer range equivalent interactions on the system. A correlation function indicates to what extent the spins tend to move together, and over what length scale. We consider a discrete correlation sum from simulation data and then transform it to the equivalent continuum quantity.

The input data for discrete sum (5.164) are actually a spin component (let us take S^x) on the lattice at time t, i.e. $f_n \to S_n^x(t)$. For now, ignore the time-dependence. Consider a space displaced correlation function (or convolution function) in 1D, defined in the continuum problem as

$$C^{xx}(r) = \frac{1}{L} \int dx \langle S^{x*}(x) S^x(x+r) \rangle. \tag{5.168}$$

This might better be referred to as an auto-correlation because it combines the same function with its shifted values. The asterisk (*) indicates complex conjugation (for generality if correlations of complex quantities are considered). This averages spin products over the system length. It is easy to generalize this to 2D or 3D, using normalization by area or volume, respectively. The angle brackets indicate an average over different runs of the same calculation, starting from different initial conditions for the same temperature and other parameters.

Exercise 5.4. Consider a convolution integral between two square pulse functions that are symmetric around the origin, defined by

$$f_1(x) = \begin{cases} A, & |x| < a \\ 0, & |x| > a. \end{cases} \quad f_2(x) = \begin{cases} B, & |x| < b \\ 0, & |x| > b. \end{cases} \tag{5.169}$$

The pulse heights are A and B; the widths are $2a$ and $2b$. Assume $a > b$. Determine the correlation function between them, defined by the generalization of (5.168),

$$C_{12}(r) = \frac{1}{L} \int dx \langle f_1(x)^* f_2(x+r) \rangle. \tag{5.170}$$

L is some large system size in which the pulse functions are embedded. A graphical solution may be helpful.

The correlation function depends on spins separated by distance $r \to x_n = na$, when discretized using $\int dx \to a \sum_m$. Its discrete-space FT is defined as follows:

$$S_{q_j}^{xx} = \frac{1}{N} \sum_{n=1}^{N} \frac{1}{N} \sum_{m=1}^{N} \langle S_m^{x*} S_{m+n}^x \rangle e^{-i q_j x_n}. \tag{5.171}$$

The discrete correlation function is the part within the angle brackets. The rest of the expression evaluates its FT. Now we can rewrite this by substituting FT expressions for S_m^{x*} and S_{m+n}^x:

$$S_m^{x*} = \sum_{k=1}^{N} A_{q_k}^* e^{-iq_k x_m}, \qquad S_{m+n}^x = \sum_{l=1}^{N} A_{q_l} e^{+iq_l(x_m + x_n)}. \qquad (5.172)$$

Then we have a sequence of trivial summations,

$$S_{q_j}^{xx} = \frac{1}{N^2} \sum_{n=1}^{N} \sum_{m=1}^{N} \sum_{k=1}^{N} \sum_{l=1}^{N} \left\langle A_{q_k}^* A_{q_l} \right\rangle e^{i\frac{2\pi}{N}(l-k)m} e^{i\frac{2\pi}{N}(l-j)n}. \qquad (5.173)$$

The summations over m and n give $N\delta_{l,k}$ and $N\delta_{l,j}$, respectively. This makes the remaining sums over j and l trivial, giving

$$S_{q_j}^{xx} = \left\langle A_{q_j}^* A_{q_j} \right\rangle = \left\langle |A_{q_j}|^2 \right\rangle. \qquad (5.174)$$

This is the discrete expression. To convert it to the equivalent continuum density, one divides as above by the unit $dq = 2\pi/L$, and drops the j-index:

$$S^{xx}(q) = \frac{S_q^{xx}}{dq} = \frac{S_q^{xx}}{2\pi/L} = \frac{L}{2\pi} \left\langle |A_q|^2 \right\rangle = \frac{L}{2\pi} \left\langle \left|\frac{2\pi}{L} A(q)\right|^2 \right\rangle = \frac{2\pi}{L} \langle |A(q)|^2 \rangle. \qquad (5.175)$$

This last form represents the easiest method to calculate $S^{xx}(q)$. FT the original data in real space, then square it, and you have the space FT of the space displaced correlation function of the original data. A normalization factor of $2\pi/L$ is needed for every space dimension that was transformed, where L is the physical length of that dimension. For example, for a $N_x \times N_y$ square lattice system with lattice constant a, the overall normalization factor to be applied is $\frac{2\pi}{N_x a} \frac{2\pi}{N_y a}$. Specifically,

$$S^{xx}(q_x, q_y) = \frac{2\pi}{N_x a} \frac{2\pi}{N_y a} \left\langle |A(q_x, q_y)|^2 \right\rangle \qquad (5.176)$$

where the 2D continuum-equivalent FT is calculated from

$$A(q_x, q_y) = \left(\frac{a}{2\pi}\right)^2 \sum_{n=1}^{N_x} \sum_{m=1}^{N_y} S_{n,m}^x e^{-iq_x na} e^{-iq_y ma}. \qquad (5.177)$$

This sum could be performed on any discrete system size, however, it could be more efficient to use a size for which these sums are evaluated by a fast FT (FFT), if all possible wave vectors are desired. Calculation for a limited number of wave vectors (and only along important directions in k-space) may not necessitate the use of an FFT scheme.

For other 2D lattices with arbitrary site locations (x_n, y_n), a more general way to express these results is to calculate

$$S^{xx}(q_x, q_y) = \frac{(2\pi)^2}{N v_{\text{cell}}} \left\langle \left| A(q_x, q_y) \right|^2 \right\rangle, \tag{5.178}$$

where each of the N unit cells containing one spin has area v_{cell} and the continuum-equivalent FT needed is

$$A(q_x, q_y) = \frac{v_{\text{cell}}}{(2\pi)^2} \sum_{n=1}^{N} S_n^x e^{-iq_x x_n} e^{-iq_y y_n}. \tag{5.179}$$

For simulations, the ensemble averaging indicated by $\langle \ \rangle$ means that $A(q_x, q_y)$ is computed for each initial condition and absolute squared before being combined with the results from other initial conditions to calculate the averages.

Exercise 5.5. Consider two localized functions $f_1(x)$ and $f_2(x)$, such as the pulse functions (5.169) in the previous example, and their correlation function defined in (5.170). Assume the functions are defined inside some system of length L which is much larger than $2a$ and $2b$. Let $S_{12,q}$ be the discrete Fourier amplitude of their correlation function, and let $F_{1,q}$ and $F_{2,q}$ be the discrete Fourier amplitudes of the two pulse functions. Show that there is a relation analogous to (5.174), namely,

$$S_{12,q} = \left\langle F_{1,q}^* F_{2,q} \right\rangle. \tag{5.180}$$

The result shows that a convolution integral in real space (the correlation function) converts into a product in the Fourier space.

5.7.3 Time correlations

Suppose some quantity $X(t)$ has been calculated from a SD simulation. X might represent the total magnetic moment along some axis, or the total energy (in SD at fixed temperature), or a density of vortices, or some other dynamically fluctuating quantity. The most common quantity, however, is a spin component, such as $S_n^x(t)$ in the previous example. A time correlation function is defined as an average over the ensemble

$$C(t) = \langle X(t_0) X(t_0 + t) \rangle. \tag{5.181}$$

This measures the correlation between X at two times separated by time interval t. Applied for thermal equilibrium, it should not depend on the starting time t_0. For the purpose of using simulation data to the fullest, it makes sense to expand the definition and extend the averaging also over t_0,

$$C(t) = \frac{1}{t_f} \int_0^{t_f} dt_0 \langle X(t_0) X(t_0 + t) \rangle. \tag{5.182}$$

The time t_f is some final time beyond which there are no more data to make the average. One can see that it could depend on the choice of t being calculated, since any simulation will have been performed over a limited total simulation time. The brackets mean that an average is taken over all particular simulation runs with the same parameters such as temperature, system size, and so on. That is the ensemble average in the simulation. $C(t)$ might be expected to generally have an exponential decay with time—at long times it will reach the equilibrium average value of $\langle X^2 \rangle$, independent of time. How it gets to that value, and how fast, could be characterized by a relaxation time scale. The relaxation may have an overall exponential decay, but that could modulate any underlying oscillatory behavior that could be due to resonant excitations. A simulation should be run long enough to average over many periods of the most important oscillations present.

5.7.4 Time Fourier transforms and correlations

For simulation, the times are discrete times t_l, and for simplicity assume they are uniformly spaced, $t_l = l\Delta t$, where $l = 0, 1, 2, \ldots N_t$. The system has been sampled at N_t equally spaced times. FT are a good way to analyze a time sequence for the oscillations present, giving a representation of different mode amplitudes as a function of frequency. In simulation the data are finite, out to final time $t_{\text{end}} = N_t \Delta t$. A Fourier series is designed to represent functions that are periodic, with a given period. While simulation data are not necessarily periodic, one can make the assumption that the finite data sequence can be periodically repeated, and then use the periodically repeated sequence to calculate time correlations. We imagine that if the sequence is expanded in a Fourier series, then the time at index $l = N_t$ is identified with time at index $l = 0$. This is the usual idea of a periodic function. As long as the primary correlations decay in less time than t_{end}, the small errors made at the end of a period will be irrelevant.

Then in the same way as for the space FT, we can expand some function f_l defined at these discrete times in discrete Fourier amplitudes A_{ω_p}, where ω_p are discrete angular frequencies. A spin component that is already space Fourier transformed is commonly used. The only difference is that the time FT uses the opposite sign in the exponent, compared to the space FT. The opposite sign comes in because the basic time-dependent mode we expand in is $e^{i(qx-\omega t)}$:

$$f_l = \sum_{p=1}^{N_t} A_{\omega_p} e^{-i\omega_p t_l}, \qquad \omega_p = \frac{2\pi}{t_{\text{end}}} p \qquad (5.183a)$$

$$A_{\omega_p} = \frac{1}{N_t} \sum_{l=1}^{N_t} f_l e^{i\omega_p t_l}. \qquad (5.183b)$$

To make it a per unit frequency interval, we need to divide by the spacing of the discrete frequency modes, which is $d\omega = 2\pi/t_{\text{end}}$. Dropping the index p, this gives

$$A(\omega) = \frac{A_\omega}{(2\pi/t_{\text{end}})} = \frac{t_{\text{end}}}{2\pi} \frac{1}{N_t} \sum_{l=1}^{N_t} f_l e^{i\omega t_l} = \frac{\Delta t}{2\pi} \sum_{l=1}^{N_t} f_l e^{i\omega t_l}. \qquad (5.184)$$

Of course, as expected, this is equivalent to a continuum FT,

$$A(\omega) = \frac{1}{2\pi} \int_0^{t_{\text{end}}} dt\, f(t) e^{i\omega t}. \tag{5.185}$$

In the same way as in real space, we can consider the correlations in time, calculating the FT of a time displaced correlation function. For a discrete correlation function, it is:

$$A_{\omega_p}^{xx} = \frac{1}{N_t} \sum_{l=1}^{N_t} \left\langle \frac{1}{N_t} \sum_{l_0=1}^{N_t} f_{l_0}^* f_{l_0+l} \right\rangle e^{i\omega_p t_l}. \tag{5.186}$$

The portion within the angle brackets is the definition of the discrete correlation function $C(t_l)$, with time displacement $t_l = l\Delta t$. But by the previous algebra applied to a space FT, this becomes:

$$A_{\omega_p}^{xx} = \left\langle A_{\omega_p}^* A_{\omega_p} \right\rangle = \left\langle |A_{\omega_p}|^2 \right\rangle. \tag{5.187}$$

Once again, converting to the corresponding continuum density by dividing by $d\omega = 2\pi/t_{\text{end}}$ and dropping the p-index, we obtain

$$A^{xx}(\omega) = \frac{t_{\text{end}}}{2\pi} \langle |A_\omega|^2 \rangle = \frac{t_{\text{end}}}{2\pi} \left\langle \left| \frac{2\pi}{t_{\text{end}}} A(\omega) \right|^2 \right\rangle = \frac{2\pi}{t_{\text{end}}} \langle |A(\omega)|^2 \rangle, \tag{5.188}$$

with $A(\omega)$ given above in both discrete and continuum forms.

5.7.5 Dynamic structure functions

A dynamic structure function $S^{xx}(q, \omega)$ combines both space and time FTs of a correlation function into one, to give a result that indicates the spatial and temporal properties of excitations in the system. We can also define a function $S^{xx}(q, t)$ that is space Fourier transformed but not yet time Fourier transformed.

Definitions and a slow way to obtain $S^{xx}(q, t)$ and $S^{xx}(q, \omega)$

First it is necessary to define a space and time displaced correlation function. In discrete form (a 1D space of N sites, length $L = Na$), with the space displacement $r_n = na$, it is

$$C^{xx}(r_n, t) = \frac{1}{N} \sum_{m=1}^{N} \frac{1}{N_t} \sum_{l=1}^{N_t} \langle S_m^x(t_l) S_{m+n}^x(t_l + t) \rangle. \tag{5.189}$$

The corresponding continuum form that might be better for theoretical analysis is

$$C^{xx}(r, t) = \frac{1}{L} \int dx\, \frac{1}{t_{\text{end}}} \int dt_0 \langle S^x(x, t_0) S^x(x + r, t_0 + t) \rangle. \tag{5.190}$$

Then performing only the discrete space FT on it gives a definition of a correlation function in wave vector and time, in discrete form,

$$S_{q_j}^{xx}(t) = \frac{1}{N} \sum_{n=1}^{N} C^{xx}(r_n, t) e^{-i q_j r_n}. \tag{5.191}$$

Inserting the definition of C^{xx} and Fourier representations (5.172) of the spin field eliminates the spatial sums, leaving the time displaced correlations of the spatial FTs at different times:

$$S_{q_j}^{xx}(t) = \frac{1}{N_t} \sum_{l_0=1}^{N_t} \left\langle A_{q_j}^{x*}(t_{l_0}) A_{q_j}^{x}(t_{l_0} + t) \right\rangle. \tag{5.192}$$

This is one way to calculate the discrete $S_{q_j}^{xx}(t)$. However, the time summation is very slow, and must be averaged over the many initial conditions. It can be followed by the discrete time FT to obtain S_{q_j,ω_p}^{xx}, which is the discrete dynamic structure function or dynamic correlation function, defined using (5.183b) as

$$S_{q_j,\omega_p}^{xx} = \frac{1}{N_t} \sum_{l=1}^{N_t} \frac{1}{N_t} \sum_{l_0=1}^{N_t} \left\langle A_{q_j}^{x*}(t_{l_0}) A_{q_j}^{x}(t_{l_0} + t_l) \right\rangle e^{i\omega_p t_l}. \tag{5.193}$$

The conversion to the equivalent continuum quantities was already described and requires dividing by a factor of $dq = 2\pi/L$ for each space dimension and a factor of $d\omega = 2\pi/t_{end}$ for the time FT, producing

$$S^{xx}(q, t) = \frac{L}{2\pi} S_q^{xx}(t) \longrightarrow \frac{1}{2\pi} \int_0^L dr \, C^{xx}(r, t) e^{-iqr}, \tag{5.194}$$

where the latter is the continuum theory expression, and for the more important dynamic structure function,

$$S^{xx}(q, \omega) = \frac{L}{2\pi} \frac{t_{end}}{2\pi} S_{q,\omega}^{xx} \longrightarrow \frac{1}{(2\pi)^2} \int_0^{t_{end}} dt \int_0^L dr \, C^{xx}(r, t) e^{-i(qr-\omega t)}. \tag{5.195}$$

Exercise 5.6. Suppose a spin field is that of an extended traveling wave in a continuum system of length L,

$$S^x(x, t) = \sin[kx - \omega(k)t], \tag{5.196}$$

where k is a particular wave vector and $\omega(k)$ is its corresponding frequency from a dispersion relation for the waves. (a) Use expression (5.33) to show that the space and time displaced correlation function is

$$C^{xx}(r, t) = \frac{1}{2} \cos[kr - \omega(k)t]. \tag{5.197}$$

Why is the amplitude equal to $\frac{1}{2}$? (b) Determine the continuum dynamic structure function $S^{xx}(q, \omega)$ that results from expression (5.38). Give a physical interpretation.

A fast way to obtain $S^{xx}(q,t)$ and $S^{xx}(q,\omega)$

Calculation of $S^{xx}(q,t)$ and $S^{xx}(q,\omega)$ can be drastically accelerated by not directly calculating the time correlations, equation (5.35). Instead, it is better to realize that the time FT of (5.35) leads to the simple result (by (5.29) and (5.30)), in discrete space and time,

$$S^{xx}_{q_j,\omega_p} = \left\langle A^*_{q_j,\omega_p} A_{q_j,\omega_p} \right\rangle = \left\langle \left| A_{q_j,\omega_p} \right|^2 \right\rangle. \tag{5.198}$$

This is a consequence of the convolution theorem: the FT of a convolution (correlation function) is the product of the FTs of the two functions in the convolution. The function A_{q_j,ω_p} is the combined space and time FT of the original data,

$$A_{q_j,\omega_p} = \frac{1}{N_t}\sum_{l=1}^{N_t} \frac{1}{N}\sum_{n=1}^{N} S^x_n(t_l) e^{-iq_j x_n} e^{i\omega_p t_l}. \tag{5.199}$$

The space sum can be performed literally in the program, using a saved table of values for the required sines and cosines coming from the exponential factors. The time summation can be performed very rapidly using a FFT routine. Then, by squaring it, you obtain the discrete structure function, $S^{xx}_{q_j,\omega_p}$. It is also very fast to apply an *inverse* time FT to it to produce the time correlation, $S^{xx}_{q_j}(t_l)$:

$$S^{xx}_{q_j}(t_l) = \sum_{p=1}^{N_t} S^{xx}_{q_j,\omega_p} e^{-i\omega_p t_l}. \tag{5.200}$$

Again the discrete times are $t_l = l\Delta t$, where $\Delta t = t_{\text{end}}/N_t$. For a typical lattice simulation, it is actually possible that a calculation of the time displaced correlation function could consume the majority of the CPU time, especially if one wants to average simulation results over many initial conditions. By this expression and (5.41), on the other hand, the calculation of $S^{xx}_{q_j}(t_l)$ is very fast. These discrete quantities are still converted to the continuum quantities, for 1D, by the expressions in (5.37) and (5.38).

Fourier transforms and autocorrelations in two space dimensions, square lattice

For two space dimensions with a $N_x \times N_y$ system, some form of SD will produce a set of discrete space- and time-dependent data, $S^x_\mathbf{r}(t_l)$, where $\mathbf{r} = (na, ma)$ are the lattice sites, and $t_l = l\Delta t$ is the time. The wave vectors are the discrete values, $\mathbf{q} = (\frac{2\pi}{L_x}i, \frac{2\pi}{L_y}j)$. The discrete and continuum equivalent space FTs (no correlations) at a fixed time are:

$$A^x_\mathbf{q}(t_l) = \frac{1}{N_x N_y} \sum_\mathbf{r} S^x_\mathbf{r}(t_l) e^{-i\mathbf{q}\cdot\mathbf{r}} \tag{5.201a}$$

$$A^x(\mathbf{q},t_l) = \left(\frac{a}{2\pi}\right)^2 \sum_\mathbf{r} S^x_\mathbf{r}(t_l) e^{-i\mathbf{q}\cdot\mathbf{r}}. \tag{5.201b}$$

Next, one can make the time FTs. In discrete and continuum forms, they are:

$$A^x_{\mathbf{q},\omega_p} = \frac{1}{N_t} \sum_{t_l} \left[\frac{1}{N_x} \frac{1}{N_y} \sum_{\mathbf{r}} S^x_{\mathbf{r}}(t_l) e^{-i\mathbf{k}\cdot\mathbf{r}} \right] e^{i\omega_p t_l}, \quad (5.202a)$$

$$A^x(\mathbf{q},\omega_p) = \left(\frac{t_{end}}{2\pi N_t}\right) \sum_{t_l} \left[\left(\frac{a}{2\pi}\right)^2 \sum_{\mathbf{r}} S^x_{\mathbf{r}}(t_l) e^{-i\mathbf{q}\cdot\mathbf{r}} \right] e^{i\omega_p t_l}. \quad (5.202b)$$

The *autocorrelation* in frequency space is just the ensemble average of the absolute square of $A^x_{\mathbf{q},\omega_p}$, with appropriate normalization factors. Find it in discrete space and time and then convert to the continuum result.

$$S^{xx}_{\mathbf{q},\omega_p} = \left\langle \left| A^x_{\mathbf{q},\omega_p} \right|^2 \right\rangle. \quad (5.203)$$

Apply conversion factors:

$$S^{xx}(\mathbf{q},\omega_p) = \left(\frac{t_{end}}{2\pi} \frac{N_x a}{2\pi} \frac{N_y a}{2\pi}\right)$$

$$S^{xx}_{\mathbf{q},\omega_p} = \left(\frac{t_{end}}{2\pi} \frac{N_x a}{2\pi} \frac{N_y a}{2\pi}\right) \left\langle \left| \frac{2\pi}{t_{end}} \frac{2\pi}{N_x a} \frac{2\pi}{N_y a} A^x(\mathbf{q},\omega_p) \right|^2 \right\rangle. \quad (5.204)$$

The simplified form is

$$S^{xx}(\mathbf{q},\omega) = \left(\frac{2\pi}{t_{end}} \frac{2\pi}{N_x a} \frac{2\pi}{N_y a}\right) \langle |A^x(\mathbf{q},\omega)|^2 \rangle. \quad (5.205)$$

This is a desired dynamic structure function. One can check that the units of $S^{xx}(\mathbf{q},\omega)$ are inverse frequency times inverse wave vector squared.

Overall normalization in programs

There are a lot of normalization factors, which can appear confusing. This can be summarized in simpler form. On the discrete space and time lattice, make a summation,

$$Q^x(\mathbf{q},\omega_p) = \sum_{t_l} \sum_{\mathbf{r}} S^x_{\mathbf{r}}(t_l) e^{-i\mathbf{q}\cdot\mathbf{r}} e^{i\frac{2\pi}{t_{end}} p t_l}. \quad (5.206)$$

Then using (5.202b) shows its square should be normalized by a simple factor to give the dynamic structure function:

$$S^{xx}(\mathbf{q},\omega_p) = \left(\frac{a}{2\pi N_x} \frac{a}{2\pi N_y} \frac{\Delta t}{2\pi N_t}\right) \langle |Q^x(\mathbf{q},\omega_p)|^2 \rangle. \quad (5.207)$$

Thus there is an obvious factor for each space or time dimension. In practice, this is calculated for each initial condition, and the ensemble averaging is then performed afterwards. The space–time FT of $S_\mathbf{r}^x(t_l)$ itself is not ensemble averaged.

5.7.6 Comparing static and dynamic correlations

The zero-time and zero-frequency limits of the dynamic correlation functions should give quantities related to the static correlation functions. Let us check this and determine the relative normalizations. We write the correlations exhibiting the actual summations that would be carried out by computation.

Static spin–spin correlations
The continuum static correlations at continuum wave vector $\mathbf{q} = (q_x, q_y) = 2\pi(i/L_x, j/L_y)$, might be calculated in MC simulations—time-dependence is unnecessary. Using sums over space points at $\mathbf{r}_{n,m} = (na, ma)$, with $L_x = N_x a, L_y = N_y a$, results in a discrete space FT,

$$A_\mathbf{q}^x = \frac{1}{N_x}\frac{1}{N_y} \sum_{n=1}^{N_x}\sum_{m=1}^{N_y} S_{n,m}^x e^{-i\mathbf{q}\cdot\mathbf{r}_{n,m}}. \tag{5.208}$$

Then the static spin–spin spatial correlations transformed to $\mathbf{q}$-space are characterized by

$$S^{xx}(\mathbf{q}) = \frac{L_x}{2\pi}\frac{L_y}{2\pi}\left\langle |A_\mathbf{q}^x|^2 \right\rangle = \frac{a}{2\pi N_x}\frac{a}{2\pi N_y}\left\langle \left|\sum_{n=1}^{N_x}\sum_{m=1}^{N_y} S_{n,m}^x e^{-i\mathbf{q}\cdot\mathbf{r}_{n,m}}\right|^2 \right\rangle. \tag{5.209}$$

Dynamic correlations at zero time
From $S_\mathbf{q}^{xx}(t)$ with $t = 0$, one obtains the discrete version of dynamic correlations at the initial time, $S_\mathbf{q}^{xx}(t=0) = \langle |A_\mathbf{q}^x|^2 \rangle$. Dividing by factors of $2\pi/L_x$ and $2\pi/L_y$ leads to the continuum result:

$$S^{xx}(\mathbf{q}, t=0) = \frac{L_x}{2\pi}\frac{L_y}{2\pi} S_\mathbf{q}^{xx}(t=0) = \frac{L_x}{2\pi}\frac{L_y}{2\pi}\left\langle |A_\mathbf{q}^x|^2 \right\rangle. \tag{5.210}$$

Then this is identical to the static correlation function that would come from MC simulations:

$$S^{xx}(\mathbf{q}, t=0) = S^{xx}(\mathbf{q}). \tag{5.211}$$

This offers a check when comparing results from MC to the more general results that would be obtained from SD simulations.

Dynamic correlations at zero frequency
The dynamic correlations at zero frequency, $S^{xx}(\mathbf{q}, \omega = 0)$, should relate to long time-averaged effects. One has

$$Q^x(\mathbf{q}, \omega_p = 0) = \sum_{t_l}\sum_{n,m} S_{n,m}^x(t_l) e^{-i\mathbf{q}\cdot\mathbf{r}_{n,m}}. \tag{5.212}$$

There will be N_t time samples, while the spins in the sum are changing with time. So the sum over time can be replaced by N_t times the time average, denoted with an overbar as $\overline{S_{n,m}^x(t)}$. Depending on the type of time-dependence for a given wave vector, this type of average could typically be very small, unless there is some kind of frozen-in order in the spins. I write $\overline{Q^x(\mathbf{q}, t)}$ as well, to indicate that it is found by using a time average of $S_{\mathbf{r}}^x(t)$:

$$\overline{Q^x(\mathbf{q}, t)} = N_t \sum_{n,m} \overline{S_{n,m}^x(t)} e^{-i\mathbf{q} \cdot \mathbf{r}_{n,m}}. \quad (5.213)$$

The dynamic correlation function becomes

$$S^{xx}(\mathbf{q}, \omega = 0) = \frac{a}{2\pi N_x} \frac{a}{2\pi N_y} \frac{t_{\text{end}}}{2\pi} \left\langle \left| \sum_{n=1}^{N_x} \sum_{m=1}^{N_y} \overline{S_{n,m}^x(t)} e^{-i\mathbf{q} \cdot \mathbf{r}_{n,m}} \right|^2 \right\rangle. \quad (5.214)$$

Except for the extra factor of $t_{\text{end}}/2\pi$, it has nearly the same normalization constants as in $S^{xx}(\mathbf{q})$. The factor of $\Delta\omega = 2\pi/t_{\text{end}}$ makes the results per unit frequency interval. However, the more significant factor is the presence of the time-averaging on $S_{n,m}^x(t)$. One can make some approximate statements about this.

Suppose the x-components have a certain time-averaged configuration (caused by vacancies or built-in disorder) with a cosine form, that is present even when averaged over different members of the ensemble. The order is characterized by a specific wave vector $\mathbf{q}_0$, so that the spins' time average is

$$\overline{S_{n,m}^x(t)} \approx w_{\mathbf{q}_0}^x \cos(\mathbf{q}_0 \cdot \mathbf{r}) = \frac{1}{2} w_{\mathbf{q}_0}^x (e^{i\mathbf{q}_0 \cdot \mathbf{r}} + e^{-i\mathbf{q}_0 \cdot \mathbf{r}}). \quad (5.215)$$

With spins dominated by this spatial structure, the zero-frequency dynamic correlations are approximated as

$$S^{xx}(\mathbf{q}, \omega = 0) \approx \frac{L_x}{2\pi} \frac{L_y}{2\pi} \frac{t_{\text{end}}}{2\pi} \left| \frac{w_{\mathbf{q}_0}^x}{2} \right|^2 (\delta_{\mathbf{q}-\mathbf{q}_0} + \delta_{\mathbf{q}+\mathbf{q}_0})^2. \quad (5.216)$$

If there is static long-range order as in a non-zero magnetization, corresponding to $\mathbf{q}_0 = 0$, then the formula gives an expected strong peak at $\mathbf{q} = 0$. If there is no zero wave vector present but a $\mathbf{q}_0 \neq 0$ wave vector is dominant, then the dynamic correlations will have a peak at that corresponding wave vector and its negative (peaks at $\mathbf{q} = \pm\mathbf{q}_0$). The amplitude $w_{\mathbf{q}_0}^x$ could at most be 1. Then comparing the peak amplitude found in numerical simulations to this upper limit can give an idea about the strength of the spatial ordering.

The expression does not take into account the effects due to more than one dominant wave vector frozen into the spin structure. For example, we could have strong $\mathbf{q}_0 = 0$ and several non-zero wave vectors $\mathbf{q}_1$, $\mathbf{q}_2$, etc, present, which would themselves interfere with each other. This result ignores possible interference terms.

Bibliography

[1] Loft R and DeGrand T A 1987 Numerical simulation and dynamics of the XY model *Phys. Rev.* B **35** 8528
[2] Hildebrand F B 1987 *Introduction to Numerical Analysis* (New York: Dover)
[3] Ceschino F and Kuntzmann J 1966 *Numerical Solution of Initial Value Problems* (New York: Prentice-Hall)
[4] Press W H, Teukolsky S A, Vetterling W T and Flannery B P 2007 *Numerical Recipes The Art of Scientific Computing* 3rd edn (Cambridge: Cambridge University Press)
[5] Kawabata C, Takeuchi M and Bishop A R 1986 Simulation of the dynamic structure factor for two-dimensional classical Heisenberg ferromagnets *J. Magn. Magn. Mater.* **54–57** 871
[6] Kawabata C, Takeuchi M and Bishop A R 1986 Monte Carlo molecular dynamics simulations for two-dimensional magnets *J. Stat. Phys.* **43** 869
[7] Mertens F G, Bishop A R, Wysin G M and Kawabata C 1987 Vortex signatures in dynamic structure factors for two-dimensional easy-plane ferromagnets *Phys. Rev. Lett.* **59** 117
[8] Mertens F G, Bishop A R, Wysin G M and Kawabata C 1989 Dynamical correlations from mobile vortices in two-dimensional easy-plane ferromagnets *Phys. Rev.* B **39** 591
[9] García-Palacios J L and Lázaro F J 1999 Langevin-dynamics study of the dynamical properties of small magnetic particles *Phys. Rev.* B **58** 14937
[10] Nowak U 2000 Thermally activated reversal in magnetic nanostructures In ed D Stauffer *Annual Reviews of Computational Physics* vol 9, page 105 (Singapore: World Scientific)
[11] Marsaglia G and Zaman A 1994 Some portable very-long-period random number generators *Comput. Phys.* **8** 117

Part II

Excitations in magnetic systems

Magnetic Excitations and Geometric Confinement
Theory and simulations
Gary Matthew Wysin

Chapter 6

Spin waves: extended but low-dimensional systems

The basic excitations in periodic but uniform systems are spin waves: magnetic ordering is not literally frozen into a magnet. Instead, the spins typically make small-amplitude oscillations around their equilibrium directions in linear excitations known as spin waves (for classical dynamics), or known as magnons when quantized. In a low-dimensional system, the geometry constrains the oscillations and affects the dependence of the frequencies on the wave vector. Surfaces or boundaries result in waves constrained to the boundaries. A discussion of these spin wave dispersion relations is developed in this chapter.

6.1 Spin waves in ferromagnetic models

Any magnetic system exhibits oscillatory motions of the spins at the atomic level, that can be excited naturally by thermal fluctuations, or by the application of some exciting stimulus such as a time-dependent applied magnetic field. The fundamental oscillatory states are called *spin waves*. An introduction to spin waves in 3D magnetic materials is given in many excellent texts [1, 2] on solid state physics. They are the linear excitations in the system, meaning their properties can be obtained from a linearization of the basic equations of motion for spin dynamics (SD) (the Landau–Lifshitz–Gilbert equations). A discussion of spin waves is important in this chapter, before going on to nonlinear excitations such as solitons and vortices in subsequent chapters. We concentrate on the properties of spin waves in reduced dimensionality, rather than 3D systems. This is a classical analysis, for classical spins. The corresponding quantum analysis is quite similar and not considered here. It leads to the corresponding quantized excitations known as *magnons*. For some models the results are very similar to the classical results, especially for the mode frequencies.

The basic spin wave property to be found is the *dispersion relation* between its wave vector and frequency. A second important property is how to describe the motion of spins geometrically, that is, the polarization properties. For an easy-plane model, that means to describe the spin motions in terms of the relative amplitudes within an easy plane and out of that easy plane, as an example. A spin wave mode is an eigenfunction of the equations of motion. The polarization properties are determined directly by the spin structure of an eigenmode.

6.1.1 Isotropic Heisenberg models

Begin with a 1D chain of spins coupled ferromagnetically and isotropically in spin space, with Hamiltonian

$$H = -J \sum_{n=1}^{N} \mathbf{S}_n \cdot \mathbf{S}_{n+1}. \tag{6.1}$$

To avoid complications with boundaries, either assume a chain of infinite length, or, one where the Nth site is coupled to the first site (periodic boundary conditions). The undamped equations of motion (5.1) are

$$\dot{\mathbf{S}}_n = \mathbf{S}_n \times \mathcal{F}_n = J\mathbf{S}_n \times (\mathbf{S}_{n-1} + \mathbf{S}_{n+1}). \tag{6.2}$$

For one of the two mentioned boundary conditions, all spins have two nearest neighbors, whose sum determines the effective field $\mathcal{F}_n$ on the RHS here. Expressed in Cartesian spin components, this is

$$\dot{S}_n^x = J \sum_{j=-1,+1} \left(S_n^y S_{n+j}^z - S_n^z S_{n+j}^y \right) \tag{6.3a}$$

$$\dot{S}_n^y = J \sum_{j=-1,+1} \left(S_n^z S_{n+j}^x - S_n^x S_{n+j}^z \right) \tag{6.3b}$$

$$\dot{S}_n^z = J \sum_{j=-1,+1} \left(S_n^x S_{n+j}^y - S_n^y S_{n+j}^x \right). \tag{6.3c}$$

The sum over index j is a way to include the two nearest neighbor terms, which assists in the generalization to higher dimensions. Next, the equations should be linearized around a ground state solution, to obtain the spin wave modes.

Because of its isotropy, the ground state for this model is highly degenerate: any states with all spins aligned is a ground state. Usually such a degeneracy is broken by some other interactions not included in the theoretical Hamiltonian. Thus, we assume a very weak symmetry breaking could be present, and it selects a particular lowest energy spin direction ($\hat{\mathbf{x}}$) for the ground state. Then all spins pointing along the x-axis define the ground state: $\mathbf{S}_n = S(1, 0, 0)$ for all n. The spin waves are obtained by assuming small-amplitude fluctuations away from the ground state direction, represented here by $\boldsymbol{\sigma}_n$, and then linearizing the equations of motion in $\boldsymbol{\sigma}_n$, which are considered to be small parameters. Then put

$$\mathbf{S}_n = S(1, 0, 0) + \boldsymbol{\sigma}_n \quad \text{with } \sigma_n^2 \ll S^2. \tag{6.4}$$

Using this, before linearization, gives the dynamics for $\boldsymbol{\sigma}_n$,

$$\dot{\sigma}_n^x = J \sum_{j=-1,+1} \left(\sigma_n^y \sigma_{n+j}^z - \sigma_n^z \sigma_{n+j}^y \right) \tag{6.5a}$$

$$\dot{\sigma}_n^y = J \sum_{j=-1,+1} \left(\sigma_n^z S_{n+j}^x - S_n^x \sigma_{n+j}^z \right) \tag{6.5b}$$

$$\dot{\sigma}_n^z = J \sum_{j=-1,+1} \left(S_n^x \sigma_{n+j}^y - \sigma_n^y S_{n+j}^x \right). \tag{6.5c}$$

With linearization, the x-components on the RHS will be approximated as S, and all terms quadratic in $\boldsymbol{\sigma}_n$ will be dropped. The linearized equations are then

$$\dot{\sigma}_n^x = 0 \tag{6.6a}$$

$$\dot{\sigma}_n^y = JS \sum_{j=-1,+1} \left(\sigma_n^z - \sigma_{n+j}^z \right) \tag{6.6b}$$

$$\dot{\sigma}_n^z = JS \sum_{j=-1,+1} \left(\sigma_{n+j}^y - \sigma_n^y \right). \tag{6.6c}$$

The equation for σ_n^x is irrelevant. On a 1D chain, summing over the two neighbors, the other equations are an eigenmode problem:

$$\dot{\sigma}_n^y = JS\left(2\sigma_n^z - \sigma_{n-1}^z - \sigma_{n+1}^z\right) \tag{6.7a}$$

$$\dot{\sigma}_n^z = -JS\left(2\sigma_n^y - \sigma_{n-1}^y - \sigma_{n+1}^y\right). \tag{6.7b}$$

This becomes a linear eigenvalue problem when seeking solutions of the form of traveling waves at angular frequency ω,

$$\boldsymbol{\sigma}_n(t) = \boldsymbol{\sigma} e^{i(qna - \omega t)}, \tag{6.8}$$

where q is a 1D wave vector, the position is $x = na$ for lattice constant a on the chain. $\boldsymbol{\sigma} = (0, \sigma^y, \sigma^z)$ is a complex vector amplitude. On the RHS of (6.7) one requires the y- and z-components of the sum over neighbors,

$$\boldsymbol{\sigma}_{n-1} + \boldsymbol{\sigma}_{n+1} = \boldsymbol{\sigma} e^{i(qna - \omega t)}(e^{-iqa} + e^{iqa}) = \boldsymbol{\sigma} e^{i(qna - \omega t)} 2 \cos qa. \tag{6.9}$$

Substitution of this traveling wave into (6.7) gives

$$-i\omega \sigma^y = JS(2 - 2\cos qa)\sigma^z \tag{6.10a}$$

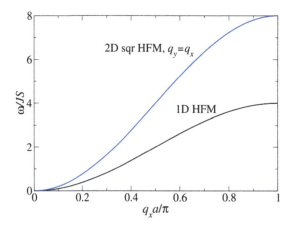

Figure 6.1. Spin wave dispersion relations for a 1D isotropic Heisenberg ferromagnetic model (labeled as 1D HFM, (6.11)) and a 2D isotropic Heisenberg ferromagnetic model on a square lattice (labeled as 2D sqr HFM, (6.18)), the latter for wave vectors along the (11)-direction. The curve (labeled as 1D HFM) also applies for the 2D square HFM model for a wave vector along the (10)-direction.

$$-i\omega\sigma^z = -JS(2 - 2\cos qa)\sigma^y. \quad (6.10b)$$

Because only these two spin components are cross-coupled, simultaneous solution gives the *dispersion relation* for a 1D ferromagnetic (FM) isotropic Heisenberg model,

$$\omega(q) = 2JS(1 - \cos qa). \quad (6.11)$$

This is the positive frequency solution, for waves traveling in the $+x$-direction. There is also a solution with the opposite sign, corresponding to waves traveling in the $-x$-direction. The result is shown in figure 6.1. With the frequency determined, the relative sizes of the y and z spin components are found to be the same, but 90° out-of-phase, regardless of wave vector: $\sigma^y = i\sigma^z$. This is reasonable, due to the isotropic couplings.

For waves on a lattice confined to positive ω, physically distinct wave vectors range from $-\pi/a$ to $+\pi/a$, which defines the first Brillouin zone. Values of q outside this range can be shifted by multiples of $2\pi/a$ back into the range. Usually one only shows the dispersion over the half-range, $0 \leqslant q \leqslant \pi/a$, as in figure 6.1. Note that at $q = \pm\pi/a$, the phase change in the waves over a displacement of one lattice site is π, equivalent to a factor of -1. This corresponds to an antiferromagnetic (AFM) variation in the spins, explaining why it is the highest energy mode in a FM model.

In a system of infinite length, the wave vector q is a continuum quantity. Then it is usually of interest to check its behavior at small qa, corresponding to long wavelengths. Expansion of the dispersion relation to leading order in qa as a small parameter, with $\cos qa \approx 1 - (qa)^2/2$, gives

$$\omega \approx JS(qa)^2. \quad (6.12)$$

This is the typical long wavelength dispersion relation in isotropic ferromagnets.

Two or more space dimensions

Most of the analysis of the isotropic Heisenberg model is unchanged when considered on a 2D or higher-dimensional lattice. However, sites of the lattice should now be identified by their position vector, **r**. The other change in the equations of motion, is that neighbor index j should be changed to a vector displacement $\mathbf{a}_j$, $j = 1, 2, 3 \ldots z$, to all the z possible nearest neighbors of a site, as was done in (1.43) and in section 2.7, where z is the coordination number of the lattice. Again supposing that spins are a slight perturbation from the ground state, as in (6.4), the linearized dynamical equations for $\sigma_\mathbf{r}(t)$ become

$$\dot{\sigma}_\mathbf{r}^x = 0 \tag{6.13a}$$

$$\dot{\sigma}_\mathbf{r}^y = JS\sum_{j=1}^{z}\left(\sigma_\mathbf{r}^z - \sigma_{\mathbf{r}+\mathbf{a}_j}^z\right) = JS\left(z\sigma_\mathbf{r}^z - \sum_{j=1}^{z}\sigma_{\mathbf{r}+\mathbf{a}_j}^z\right) \tag{6.13b}$$

$$\dot{\sigma}_\mathbf{r}^z = JS\sum_{j=1}^{z}\left(\sigma_{\mathbf{r}+\mathbf{a}_j}^y - \sigma_\mathbf{r}^y\right) = JS\left(\sum_{j=1}^{z}\sigma_{\mathbf{r}+\mathbf{a}_j}^y - z\sigma_\mathbf{r}^y\right). \tag{6.13c}$$

For a solution, again consider traveling waves, written in a general form,

$$\sigma_\mathbf{r}(t) = \sigma e^{i(\mathbf{q}\cdot\mathbf{r}-\omega t)}. \tag{6.14}$$

The mode is characterized by a wave vector **q** and frequency ω. This is quite general and to proceed further the dimension and type of lattice must be specified.

Consider a 2D square lattice. Then $z = 4$, and $\mathbf{q} = (q_x, q_y)$. The equations have sums over four neighbors at displacements $\mathbf{a}_j = \pm a\hat{\mathbf{x}}, \pm a\hat{\mathbf{y}}$, which gives

$$\sum_{j=1}^{z}\sigma_{\mathbf{r}+\mathbf{a}_j} = \sigma e^{i(\mathbf{q}\cdot\mathbf{r}-\omega t)}\left(e^{iq_x a} + e^{-iq_x a} + e^{iq_y a} + e^{-iq_y a}\right). \tag{6.15}$$

The last factor involving a sum of exponentials occurs in all spin wave problems. It is used to define a phase parameter,

$$\gamma(\mathbf{q}) \equiv \frac{1}{z}\sum_{j=1}^{z}e^{i\mathbf{q}\cdot\mathbf{a}_j} = \frac{1}{2}\left(\cos q_x a + \cos q_y a\right). \tag{6.16}$$

This leads to

$$-i\omega\sigma^y = JS(4 - 4\gamma(\mathbf{q}))\sigma^z \tag{6.17a}$$

$$-i\omega\sigma^z = -JS(4 - 4\gamma(\mathbf{q}))\sigma^y. \tag{6.17b}$$

This gives again $\sigma^y = i\sigma^z$ and a dispersion relation depending on both wave vector components,

$$\omega(\mathbf{q}) = 4JS[1 - \gamma(\mathbf{q})] = 4JS\left[1 - \frac{1}{2}\left(\cos q_x a + \cos q_y a\right)\right]. \tag{6.18}$$

> **Exercise 6.1.** With the wave vector magnitude $q = \sqrt{q_x^2 + q_y^2}$, show that in the long wavelength limit the dispersion relation for a 2D isotropic Heisenberg ferromagnet model becomes independent of direction,
>
> $$\omega(\mathbf{q}) \approx JS(qa)^2. \tag{6.19}$$

> **Exercise 6.2.** Following the examples of 1D chain and 2D square lattice Heisenberg ferromagnet models, determine the dispersion relation for a 3D simple cubic lattice. Express the result in terms of a phase function $\gamma(\mathbf{q})$. Check whether the long wavelength limit is the same as for a 1D chain and a 2D square lattice.

> **Exercise 6.3.** Develop the calculation of the dispersion relation for an isotropic Heisenberg ferromagnet on a 2D honeycomb lattice (three nearest neighbors). Express the result using a phase function $\gamma(\mathbf{q})$, and compare with the square lattice.

> **Exercise 6.4.** Develop the calculation of the dispersion relation for an isotropic Heisenberg ferromagnet on a 3D face-centered cubic lattice (12 nearest neighbors). Express the result using a phase function $\gamma(\mathbf{q})$, and compare with a simple cubic lattice.

The dispersion relation for a square lattice is plotted in figure 6.1, along two different directions in **q**-space, as a function of only $q_x a$. If the wave vector points along the (10)-direction in the lattice (the x-axis), the dispersion relation is identical to that for the 1D chain, labeled 1D HFM in figure 6.1. For **q** along the diagonal (11)-direction, the frequency is doubled in comparison.

For isotropic Heisenberg exchange, there is no particular geometric confinement effect, other than the restriction on possible q due to the lattice. Geometric confinement effects come in due to anisotropic spin couplings, such as those in XY symmetry models.

6.1.2 An isotropic ferromagnet with a weak magnetic field

Suppose the isotropic ferromagnet is subjected to a weak magnetic field **H**, which we take arbitrarily along the $\hat{\mathbf{x}}$-axis. This selects the ground state and slightly modifies the dynamics. Every site will be affected by an extra effective field that is added to $\mathcal{F}_\mathbf{r}^x$,

$$\Delta \mathcal{F}_\mathbf{r}^x = \gamma \mu_0 H^x. \qquad (6.20)$$

The extra terms produced in the spin dynamical equations are

$$\dot{S}_\mathbf{r}^y = \gamma \mu_0 H^x S_\mathbf{r}^z \qquad (6.21a)$$

$$\dot{S}_\mathbf{r}^z = -\gamma \mu_0 H^x S_\mathbf{r}^y. \qquad (6.21b)$$

There is no change in the equation, $\dot{S}_\mathbf{r}^x = 0$. The spin fluctuations now follow the modified equations,

$$\dot{\sigma}_\mathbf{r}^y = JS\left[(z + h^x)\sigma_\mathbf{r}^z - \sum_{j=1}^z \sigma_{\mathbf{r}+\mathbf{a}_j}^z\right] \qquad (6.22a)$$

$$\dot{\sigma}_\mathbf{r}^z = JS\left[\sum_{j=1}^z \sigma_{\mathbf{r}+\mathbf{a}_j}^y - (z + h^x)\sigma_\mathbf{r}^y\right]. \qquad (6.22b)$$

Here z is the coordination number, while h^x is the scaled magnetic field,

$$h^x \equiv \gamma \mu_0 H^x / JS. \qquad (6.23)$$

Applying this to the 2D square lattice with $z = 4$, making the usual traveling wave assumptions, the eigen-equations are now

$$-i\omega\sigma^y = JS[(4 + h^x) - 4\gamma(\mathbf{q})]\sigma^z \qquad (6.24a)$$

$$-i\omega\sigma^z = -JS[(4 + h^x) - 4\gamma(\mathbf{q})]\sigma^y. \qquad (6.24b)$$

These are solved with the dispersion relation for positive frequency modes given by

$$\omega = 4JS\left[1 - \gamma(\mathbf{q}) + \frac{|h^x|}{4}\right]. \qquad (6.25)$$

This is degenerate with the negative frequency solution (waves in opposite direction). Absolute value has been included on h^x, because the frequency does not depend on the direction of **H**, which selected the ground state direction.

The presence of $|h^x|$ in the dispersion relation causes there to be a gap in the spin wave spectrum at $\mathbf{q} = 0$, where $\gamma(0) = 1$. Thus, the minimum spin wave frequency is not zero, but the value of the gap,

$$\omega_{\text{gap}} = JS|h^x| = \gamma \mu_0 H^x. \qquad (6.26)$$

This is physically reasonable: the long wavelength modes will generally have all spins precessing around the applied field at this frequency, which is the Larmor frequency of an individual spin. Thus, the size of the gap is independent of the dimension of the lattice.

6.1.3 Ferromagnetic models with XY-like exchange

Next, consider an FM model with XY-symmetry nearest neighbor exchange, with the Hamiltonian written in a general way for any number of space dimensions,

$$H = -J \sum_{\mathbf{r}} \sum_{j=1}^{z} \left(S_{\mathbf{r}}^x S_{\mathbf{r}+\mathbf{a}_j}^x + S_{\mathbf{r}}^y S_{\mathbf{r}+\mathbf{a}_j}^y + \lambda S_{\mathbf{r}}^z S_{\mathbf{r}+\mathbf{a}_j}^z \right). \tag{6.27}$$

The spins occupy discrete sites $\mathbf{r}$ on a lattice. The anisotropy parameter λ is assumed to be in a range to make the xy-plane be an easy plane, $0 \leq \lambda < 1$. If λ is greater than 1, the anisotropy is of easy-axis type, and this results in a ground state of different symmetry. The choice of the ground state is essential in determination of the spin wave dispersion. Thus it is important to take care in describing the physical form of anisotropy. The effective fields that determine the dynamics are sums over the z neighbors of a site,

$$\mathcal{F}_{\mathbf{r}} = J \sum_{j=1}^{z} \left(S_{\mathbf{r}+\mathbf{a}_j}^x \hat{\mathbf{x}} + S_{\mathbf{r}+\mathbf{a}_j}^y \hat{\mathbf{y}} + \lambda S_{\mathbf{r}+\mathbf{a}_j}^z \hat{\mathbf{z}} \right). \tag{6.28}$$

With easy-plane anisotropy, the ground state is any state where all spins are aligned within the xy-plane. As for fully isotropic coupling, we assume there is some small symmetry breaking that selects a particular direction within the xy-plane as the ground state, and call that the x-axis. Then again, the spins are assumed to have small deviations $\sigma_{\mathbf{r}}$ away from the ground state, as in (6.4). It is a simple exercise to see that the equations of motion (from a slight modification of the results for isotropic exchange) for the deviations are

$$\dot{\sigma}_{\mathbf{r}}^x = 0 \tag{6.29a}$$

$$\dot{\sigma}_{\mathbf{r}}^y = JS \sum_{j=1}^{z} \left(\sigma_{\mathbf{r}}^z - \lambda \sigma_{\mathbf{r}+\mathbf{a}_j}^z \right) = JS \left(z\sigma_{\mathbf{r}}^z - \lambda \sum_{j=1}^{z} \sigma_{\mathbf{r}+\mathbf{a}_j}^z \right) \tag{6.29b}$$

$$\dot{\sigma}_{\mathbf{r}}^z = JS \sum_{j=1}^{z} \left(\sigma_{\mathbf{r}+\mathbf{a}_j}^y - \sigma_{\mathbf{r}}^y \right) = JS \left(\sum_{j=1}^{z} \sigma_{\mathbf{r}+\mathbf{a}_j}^y - z\sigma_{\mathbf{r}}^y \right). \tag{6.29c}$$

The anisotropy factor λ appears only in one place. We assume a traveling wave solution (6.14).

First consider a 1D chain, to compare with the results for the isotropic model. The equations become

$$-i\omega \sigma^y = JS(2 - \lambda \cos qa)\sigma^z \tag{6.30a}$$

$$-i\omega \sigma^z = -JS(2 - \cos qa)\sigma^y. \tag{6.30b}$$

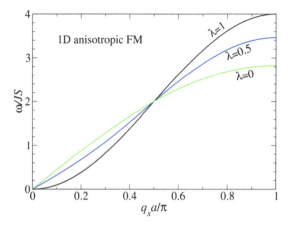

Figure 6.2. Spin wave dispersion relations (6.31) for a 1D anisotropic Heisenberg ferromagnet for different values of the anisotropy parameter, where $\lambda = 1$ gives the isotropic ferromagnet and $\lambda = 0$ gives the XY model.

The resulting dispersion relation is

$$\omega = 2JS\sqrt{(1 - \lambda \cos qa)(1 - \cos qa)}. \tag{6.31}$$

Curves of $\omega(q)$ for different λ are plotted in figure 6.2. In the limit of an XY model with $\lambda = 0$ this gives noticeably different frequencies than in the isotropic model. The other modification comes about in the relative sizes of σ^y and σ^z, which correspond to in-plane spin motions and out-of-plane spin motions (in reference to xy being the easy plane), respectively. Their sizes are found to be related by

$$i\sigma^z = \sigma^y \sqrt{\frac{1 - \cos qa}{1 - \lambda \cos qa}}. \tag{6.32}$$

This relation is plotted in figure 6.3. In particular at small qa or long wavelength, the amplitude of in-plane spin motions is considerably larger than the out-of-plane motions, as would be expected due to the easy-plane anisotropy. For larger q, however, the out-of-plane motions will dominate depending on the strength of the anisotropy. Of course, higher q is also higher energy, so that in thermal equilibrium, these higher-q modes will be less populated and the spins will still tend to stay close, on average, to the easy plane.

The dispersion relation can be expanded for small qa, using $\cos qa \approx 1 - (qa)^2/2! + (qa)^4/4!$, and the leading two terms are

$$\omega \approx 2JSqa\left[\left(\frac{1 - \lambda}{2}\right) + \left(\frac{7\lambda - 1}{24}\right)(qa)^2\right]^{1/2}. \tag{6.33}$$

With $\lambda = 1$ this exactly recovers the quadratic result (6.12) for the isotropic model, while for $\lambda = 0$, one has a linear dispersion of long wavelength spin waves in the XY chain,

$$\omega \approx \sqrt{2}\,JSqa. \tag{6.34}$$

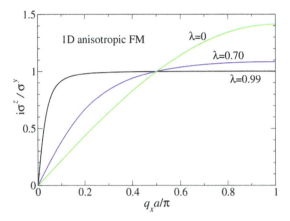

Figure 6.3. The ratio of out-of-plane to in-plane spin fluctuation amplitudes (6.32) for a 1D anisotropic Heisenberg ferromagnet, at anisotropy parameters shown, where $\lambda = 1$ gives the isotropic ferromagnet and $\lambda = 0$ gives the XY model.

While this only holds at $\lambda = 0$, still, for small values of λ and small enough qa, easy-plane symmetry does give an approximately linear dispersion relation. In fact, the crossover from approximately quadratic to linear behavior takes place where the two terms inside the square root of (6.33) are equal. This gives a crossover wave vector

$$q_c a = \sqrt{\frac{12(1-\lambda)}{7\lambda - 1}} \tag{6.35}$$

separating the two regions of different dispersion.

2D
Generalization of this model with anisotropic exchange is easy. Consider a 2D square lattice, with four nearest neighbors for every site. Following the algebra performed for the isotropic model, the dispersion relation as a function of $\mathbf{q} = (q_x, q_y)$ is found to be

$$\omega(\mathbf{q}) = 4JS\sqrt{(1 - \lambda\gamma(\mathbf{q}))(1 - \gamma(\mathbf{q}))}, \tag{6.36}$$

where $\gamma(\mathbf{q})$ is the phase factor defined in (6.16) that appears in many dispersion relations in higher dimensions. Compare the dispersion (6.31) in the 1D model, which is of this same form, with $\gamma(q) = \cos qa$. Then the relative sizes of out-of-plane spin motions compared to in-plane motions in 1D or 2D are found from

$$i\sigma^z = \sigma^y \sqrt{\frac{1 - \gamma(\mathbf{q})}{1 - \lambda\gamma(\mathbf{q})}}. \tag{6.37}$$

This indicates the restriction of spins' oscillatory motions to be more so in-plane than out-of-plane for $\lambda < 1$.

Cone state: XY symmetry with a perpendicular magnetic field
A weak magnetic field applied perpendicular to the xy-plane of symmetry in an easy-plane model ($\kappa_A = 0$, $\lambda \neq 1$) leads to an interesting situation. While the exchange makes the spins prefer the xy-plane, an H^z field will tilt the spins slightly out of the xy-plane, provided it is not too strong. The spins in this tilted state have small z spin components, however, the choice of angular direction within the xy-plane is still free. Hence, the possible ground state directions map out a cone, so this is called the cone state.

Exercise 6.5. If the applied field is too strong, the spins in the ground state will align with the applied H^z field. For the Hamiltonian (6.27) on a square lattice determine the critical H^z field where this happens and there is no more cone state.

The extra term added to the effective field is

$$\Delta \mathcal{F}_{\mathbf{r}}^z = \gamma \mu_0 H^z. \tag{6.38}$$

The extra terms produced in the spin dynamical equations affect only x- and y-components,

$$\dot{S}_{\mathbf{r}}^x = \gamma \mu_0 H^z S_{\mathbf{r}}^y \tag{6.39a}$$

$$\dot{S}_{\mathbf{r}}^y = -\gamma \mu_0 H^z S_{\mathbf{r}}^x. \tag{6.39b}$$

Consider a ground state with all spins aligned along $\mathbf{S}_{\mathbf{r}} = \mathbf{S}_0 = $ constant, satisfying $\dot{\mathbf{S}}_0 = \mathbf{S}_0 \times \mathcal{F} = 0$. The components of this equation are seen to be

$$(\mathbf{S}_0 \times \mathcal{F})^x = S_0^y(\gamma \mu_0 H^z + zJ\lambda S_0^z) - S_0^z(zJS_0^y) = 0 \tag{6.40a}$$

$$(\mathbf{S}_0 \times \mathcal{F})^y = S_0^z(zJS_0^x) - S_0^x(\gamma \mu_0 H^z + z\lambda J S_0^z) = 0 \tag{6.40b}$$

$$(\mathbf{S}_0 \times \mathcal{F})^z = S_0^x(zJS_0^y) - S_0^y(zJS_0^x) = 0. \tag{6.40c}$$

The last of these is identically true and gives nothing. The first two lead to an identical result for the tilting of spins out of the easy plane. The choice of a direction within the xy-plane is arbitrary, so take $S^y = 0$. Then one obtains the out-of-plane tilting,

$$S_0^z = \frac{\gamma \mu_0 H^z}{z(1-\lambda)J} = \frac{Sh^z}{z(1-\lambda)} \tag{6.41}$$

where $h^z = \gamma \mu_0 H^z / JS$ is the scaled field. This means that the in-plane spin component is now a little less than S,

$$S_0^x = \sqrt{S^2 - S_0^{z2}}. \tag{6.42}$$

It is clear that this cone state collapses to spins pointing on the z-axis if the field is increased to $h^z_{max} = z(1 - \lambda)$ or greater.

For the spin wave fluctuations, we consider now the variations around this ground state. Therefore let a fluctuating spin be given by

$$\mathbf{S_r} = \left(S_0^x, 0, S_0^z\right) + \sigma_\mathbf{r}. \tag{6.43}$$

The rest of the calculation is left for the following exercises.

Exercise 6.6. Using the spin fluctuations in (6.43) in the equations of motion, including the h^z field, show that the linearized equations of motion for the XY model in the cone state are

$$\dot\sigma_\mathbf{r}^x = JS_0^z \sum_{j=1}^z \left(\lambda \sigma_\mathbf{r}^y - \sigma_{\mathbf{r+a}_j}^y\right) + \gamma\mu_0 H^z \sigma_\mathbf{r}^y \tag{6.44a}$$

$$\dot\sigma_\mathbf{r}^y = JS_0^x \sum_{j=1}^z \left(\sigma_\mathbf{r}^z - \lambda\sigma_{\mathbf{r+a}_j}^z\right) + JS_0^z \sum_{j=1}^z \left(\sigma_{\mathbf{r+a}_j}^x - \lambda\sigma_\mathbf{r}^x\right) - \gamma\mu_0 H^z \sigma_\mathbf{r}^x \tag{6.44b}$$

$$\dot\sigma_\mathbf{r}^z = JS_0^x \sum_{j=1}^z \left(\sigma_{\mathbf{r+a}_j}^y - \lambda\sigma_\mathbf{r}^y\right). \tag{6.44c}$$

Exercise 6.7. Proceeding as usual with the assumption of a traveling wave such as (6.14), complete the solution and obtain a general dispersion relation in the cone state of the XY model, including the effect of the magnetic field,

$$\omega^2 = \left(zJS_0^x\right)^2 [1 - \gamma(\mathbf{q})][1 - \lambda\gamma(\mathbf{q})] + [zJS_0^z(\lambda - \gamma(\mathbf{q})) + \gamma\mu_0 H^z]^2. \tag{6.45}$$

Exercise 6.8. The overall frequency scale of the cone state is determined by coordination number z. Also, because S_0^z and S_0^x depend on H^z, it is important to use that to display the full field-dependence. Show that the dispersion relation (6.45) can be expressed in its final form as

$$\omega^2 = (1 - \gamma(\mathbf{q})) \left[(zJS)^2(1 - \lambda\gamma(\mathbf{q})) - \frac{\left(\gamma\mu_0 H^z\right)^2}{1 - \lambda}\gamma(\mathbf{q})\right]. \tag{6.46}$$

Here it is interesting that the presence of the magnetic field modifies the frequencies only slightly. However, in the limit of long wavelength with $\gamma(\mathbf{q}) \to 1$, it is seen that no gap opens up in the spectrum, unlike that found for the isotropic Heisenberg model. That is because the field is accommodated when it shifts the ground state configuration into the cone state. If one plots $\omega(\mathbf{q})$, it is found that the biggest effects due to magnetic field take place for the XY limit ($\lambda = 0$). The field depresses the frequencies near $\mathbf{q} = 0$ but augments the frequencies close to the edge of the Brillouin zone.

6.2 Spin waves in antiferromagnetic models

Spin wave modes in antiferromagnets enjoy a much richer structure than in ferromagnets, due primarily to the more complex ground state that is present. We saw for ferromagnets that the theory requires an assumption of small-amplitude oscillations relative to the ground state configuration. In an antiferromagnet, the nearest neighbor exchange interactions tend to favor anti-alignment of spins. Depending on the lattice, this can lead to some interesting ground state structures. Or, in some cases it may be impossible to make all neighboring spins anti-aligned, then the system is said to have *geometric frustration*, and possibly a non-unique or difficult to determine ground state. The discussion here is made for classical antiferromagnets; in quantum antiferromagnets there are even more interesting properties.

Consider here a general model with AFM exchange, including an easy-plane exchange anisotropy parameter λ and also a single ion (or local) easy-plane anisotropy K_A. The Hamiltonian is

$$H = J \sum_{\mathbf{r}} \sum_{j=1}^{z} \left(S_{\mathbf{r}}^x S_{\mathbf{r}+\mathbf{a}_j}^x + S_{\mathbf{r}}^y S_{\mathbf{r}+\mathbf{a}_j}^y + \lambda S_{\mathbf{r}}^z S_{\mathbf{r}+\mathbf{a}_j}^z \right) + K_A \sum_{\mathbf{r}} S_{\mathbf{r}}^{z\,2}. \tag{6.47}$$

The exchange constant J is positive, making aligned spins to be of highest energy, and anti-aligned of lowest energy. The last term is the sum of single ion anisotropy, of easy-plane type when $K_A > 0$. Both the exchange and single ion anisotropies here are taken to make the xy-plane the easy plane.

For a 1D chain, each site has two nearest neighbors, and the ground state can be arranged by letting spins alternate in direction from one site to the next. Any direction within the xy-plane will minimize the total energy; some symmetry breaking would select a particular direction, which is chosen as the $\hat{\mathbf{x}}$-axis. Other lattices in higher dimensions that are bipartite, such as square, honeycomb and cubic, can also be partitioned into two *sublattices*, denoted A and B. Any site on the A-sublattice only has nearest neighbor sites on the B-sublattice, and any site on the B-sublattice only has nearest neighbors on the A-sublattice. This means the ground state can be formed by having all spins on the A-sublattice be aligned along $\hat{\mathbf{x}}$, and all spins on the B-sublattice aligned along $-\hat{\mathbf{x}}$. This configuration would totally satisfy all the AFM exchange interactions, putting each nearest neighbor pair into its lowest energy arrangement. For now we consider only bipartite lattices. This excludes a triangular lattice from the current discussion (it has three sublattices).

Then, the ground state on the bipartite lattices has A-sublattice spins, $\mathbf{S_r}(A) = S(1, 0, 0)$, and B-sublattice spins $\mathbf{S_r}(B) = S(-1, 0, 0)$. The spin waves generate small oscillatory deviations away from this, denoted $\mathbf{A_r}$ and $\mathbf{B_r}$, so that the spins are written as

$$\mathbf{S_r}(A) = S(+1, 0, 0) + \mathbf{A_r} \qquad \mathbf{r} \in (A - \text{sublattice}) \qquad (6.48a)$$

$$\mathbf{S_r}(B) = S(-1, 0, 0) + \mathbf{B_r} \qquad \mathbf{r} \in (B - \text{sublattice}). \qquad (6.48b)$$

The deviations are assumed to be small: $\mathbf{A_r}^2 \ll S^2$, $\mathbf{B_r}^2 \ll S^2$. The SD equations of motion are obtained from

$$\dot{\mathbf{S}}_\mathbf{r} = \mathbf{S_r} \times \mathcal{F}_\mathbf{r}, \qquad (6.49)$$

where the effective fields are found to be

$$\mathcal{F}_\mathbf{r} = -\frac{\partial H}{\partial \mathbf{S_r}} = -J \sum_{j=1}^{z} \left(S^x_{\mathbf{r}+\mathbf{a}_j}\hat{\mathbf{x}} + S^y_{\mathbf{r}+\mathbf{a}_j}\hat{\mathbf{y}} + \lambda S^z_{\mathbf{r}+\mathbf{a}_j}\hat{\mathbf{z}} \right) - 2K_A(\mathbf{S_r} \cdot \hat{\mathbf{z}})\hat{\mathbf{z}}. \qquad (6.50)$$

The dynamics equations expressed in components are

$$\dot{S}^x_\mathbf{r} = -J \sum_{j=1}^{z} \left(\lambda S^y_\mathbf{r} S^z_{\mathbf{r}+\mathbf{a}_j} - S^z_\mathbf{r} S^y_{\mathbf{r}+\mathbf{a}_j} \right) - 2K_A S^y_\mathbf{r} S^z_\mathbf{r} \qquad (6.51a)$$

$$\dot{S}^y_\mathbf{r} = -J \sum_{j=1}^{z} \left(S^z_\mathbf{r} S^x_{\mathbf{r}+\mathbf{a}_j} - \lambda S^x_\mathbf{r} S^z_{\mathbf{r}+\mathbf{a}_j} \right) + 2K_A S^x_\mathbf{r} S^z_\mathbf{r} \qquad (6.51b)$$

$$\dot{S}^z_\mathbf{r} = -J \sum_{j=1}^{z} \left(S^x_\mathbf{r} S^y_{\mathbf{r}+\mathbf{a}_j} - S^y_\mathbf{r} S^x_{\mathbf{r}+\mathbf{a}_j} \right). \qquad (6.51c)$$

The sum over neighbors at displacements $\mathbf{a}_j$, on a bipartite lattice, always connects to the opposite sublattice. Now writing these out in terms of the spin wave deviations, keeping only the terms linear in $\mathbf{A_r}$, $\mathbf{B_r}$, one finds for the A-sublattice dynamics,

$$\dot{A}^x_\mathbf{r} = 0 \qquad (6.52a)$$

$$\dot{A}^y_\mathbf{r} = -J \sum_{j=1}^{z} \left[A^z_\mathbf{r}(-S) - \lambda(+S) B^z_{\mathbf{r}+\mathbf{a}_j} \right] + 2K_A(+S) A^z_\mathbf{r} \qquad (6.52b)$$

$$\dot{A}^z_\mathbf{r} = -J \sum_{j=1}^{z} \left[(+S) B^y_{\mathbf{r}+\mathbf{a}_j} - A^y_\mathbf{r}(-S) \right]. \qquad (6.52c)$$

Similarly, for the B-sublattice dynamics, there is

$$\dot{B}_\mathbf{r}^x = 0 \tag{6.53a}$$

$$\dot{B}_\mathbf{r}^y = -J\sum_{j=1}^{z}\left[B_\mathbf{r}^z(+S) - \lambda(-S)A_{\mathbf{r}+\mathbf{a}_j}^z\right] + 2K_A(-S)B_\mathbf{r}^z \tag{6.53b}$$

$$\dot{B}_\mathbf{r}^z = -J\sum_{j=1}^{z}\left[(-S)A_{\mathbf{r}+\mathbf{a}_j}^y - B_\mathbf{r}^y(+S)\right]. \tag{6.53c}$$

At this order, there are no x-component fluctuations. For yz-components, these can be expressed using the coordination number on the diagonal terms,

$$\dot{A}_\mathbf{r}^y = +JS\left(zA_\mathbf{r}^z + \lambda\sum_{j=1}^{z}B_{\mathbf{r}+\mathbf{a}_j}^z\right) + 2K_A S A_\mathbf{r}^z \tag{6.54a}$$

$$\dot{A}_\mathbf{r}^z = -JS\left(\sum_{j=1}^{z}B_{\mathbf{r}+\mathbf{a}_j}^y + zA_\mathbf{r}^y\right) \tag{6.54b}$$

$$\dot{B}_\mathbf{r}^y = -JS\left(zB_\mathbf{r}^z + \lambda\sum_{j=1}^{z}A_{\mathbf{r}+\mathbf{a}_j}^z\right) - 2K_A S B_\mathbf{r}^z \tag{6.54c}$$

$$\dot{B}_\mathbf{r}^z = +JS\left(\sum_{j=1}^{z}A_{\mathbf{r}+\mathbf{a}_j}^y + zB_\mathbf{r}^y\right). \tag{6.54d}$$

This is quite general for many different lattices. The assumed spin waves are traveling waves written as

$$\mathbf{A}_\mathbf{r}(t) = \mathbf{A}e^{i(\mathbf{q}\cdot\mathbf{r}-\omega t)}, \qquad \mathbf{B}_\mathbf{r}(t) = \mathbf{B}e^{i(\mathbf{q}\cdot\mathbf{r}-\omega t)}. \tag{6.55}$$

As for the FM problem, the time derivative d/dt acting on this is equivalent to multiplication by the factor, $-i\omega$.

6.2.1 A 1D anisotropic antiferromagnetic chain

For better transparency, consider a 1D AFM chain, with just two neighbors per site. The wave vector has one component q, and change position vector $\mathbf{r} \to na$. Only the equations for y, z-components remain, and these are

$$\dot{A}_n^y = +JS\left[2A_n^z + \lambda\left(B_{n-1}^z + B_{n+1}^z\right)\right] + 2K_A S A_n^z \tag{6.56a}$$

$$\dot{A}_n^z = -JS\left[2A_n^y + \left(B_{n-1}^y + B_{n+1}^y\right)\right] \tag{6.56b}$$

$$\dot{B}_n^y = -JS\Big[2B_n^z + \lambda\big(A_{n-1}^z + A_{n+1}^z\big)\Big] - 2K_A S B_n^z \qquad (6.56c)$$

$$\dot{B}_n^z = +JS\Big[2B_n^y + \big(A_{n-1}^y + A_{n+1}^y\big)\Big]. \qquad (6.56d)$$

Traveling waves for the 1D chain are expressed simply as

$$\mathbf{A}_n(t) = \mathbf{A}e^{i(qna-\omega t)}, \qquad \mathbf{B}_n(t) = \mathbf{B}e^{i(qna-\omega t)}. \qquad (6.57)$$

This can be substituted into (6.56), and one sees that the sums over neighbors always produce the factor,

$$\gamma(q) = \frac{1}{2}(e^{iqa} + e^{-iqa}) = \cos qa. \qquad (6.58)$$

With time derivative replaced by $-i\omega$, the four equations become an eigenvalue problem for ω. The solution is left as an exercise.

Exercise 6.9. Show that setting the determinant of equations (6.56) to zero, after inserting the traveling waves (6.57), leads to the dispersion relations,

$$\omega_\pm(q) = 2JS\Big[\big(1 + \kappa_A - \lambda\gamma^2(q)\big) \pm (1 + \kappa_A - \lambda)\gamma(q)\Big]^{1/2}. \qquad (6.59)$$

This depends on easy-plane anisotropy parameter λ and a dimensionless single ion anisotropy parameter,

$$\kappa_A = K_A/J. \qquad (6.60)$$

Allowing for solutions with positive and negative ω, the relation (6.59) gives four different solutions. Those with $\omega > 0$ correspond to waves moving towards positive x and those with $\omega < 0$ are waves moving towards negative x. Ignoring that degeneracy leaves two branches (choice of $\pm$ on second term) with positive frequency, which are referred to as optic (upper sign) and acoustic (lower sign) branches, due to the types of spin motions they involve. Note how the single ion anisotropy κ_A enters as a kind of overall stiffness (it appears in both terms), while the exchange anisotropy factor λ combines differently with $\gamma(q)$ in the two terms. If $(1 + \kappa_A - \lambda) = 0$, however, this zeros out the second term in the radical and there is only one dispersion branch.

The acoustic spin wave branch (lower, $-$ root on second term) corresponds to modes with a linear dispersion such as $\omega = cq$ at long wavelengths ($q \to 0$), where c is the acoustic group velocity (speed of 'sound' for spin waves). The optical branch (upper, $+$ root on second term) corresponds to modes with a non-zero frequency at $q = 0$. In the case of the isotropic classical antiferromagnet, with $\lambda = 1$, $\kappa_A = 0$, there is only one dispersion relation (for waves towards positive x), as plotted in figure 6.4,

$$\omega = 2JS \sin qa. \qquad (6.61)$$

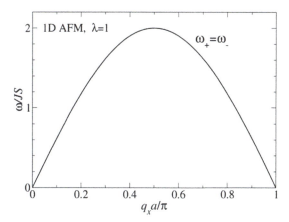

Figure 6.4. Spin wave dispersion relation for a 1D isotropic Heisenberg antiferromagnet with $\lambda = 1$, $\kappa_A = 0$. The two branches of the dispersion (6.59) become degenerate in this case.

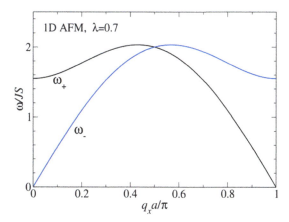

Figure 6.5. Spin wave dispersion relations for a 1D anisotropic Heisenberg antiferromagnet with $\lambda = 0.7$, $\kappa_A = 0$. The optic (+) and acoustic (−) branches of the dispersion (6.59) are now distinct, and symmetric about $qa = \pi/2$.

This has an acoustic form as $q \to 0$, with velocity $c = 2JSa$. There is no distinct optical branch, although once q reaches $\pi/2a$, the frequency is maximum and the spin motions are of a different form than near $q \approx 0$.

Once anisotropy is included into the model, the optic and acoustic branches become distinct, see figure 6.5. Furthermore, one can see that the frequency structure flips when one passes the wave vector $q = \pi/a$. To be specific, the optic branch tends linearly toward zero frequency as q approaches π/a, while the acoustic branch tends to a non-zero value at the same point. In a certain sense, the distinction of the two branches is a matter of your point of view in wave vector space.

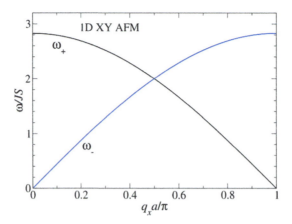

Figure 6.6. Spin wave dispersion relations for a 1D XY antiferromagnet with $\lambda = 0.0$, $\kappa_A = 0$. The optic (+) and acoustic (−) branches of the dispersion (6.59) switch character with q.

In the case of a classical XY antiferromagnet, with $\lambda = 0$, and non-zero κ_A, both branches are present as in figure 6.6. They have the following frequencies,

$$\frac{\omega_\pm}{2JS} = \sqrt{1 + \kappa_A}(1 \pm \cos qa)^{1/2} = \sqrt{2(1 + \kappa_A)} \begin{cases} \cos(qa/2) & \text{optic branch} \\ \sin(qa/2) & \text{acoustic branch}. \end{cases} \quad (6.62)$$

While the acoustic branch starting at $q = 0$ has an increasing frequency with increasing q, the optic branch starts at its highest frequency at $q = 0$, and moves towards lower ω with increasing q.

It is interesting to see the spin structure associated with the two branches. The equations for the fluctuating spin components, eliminating either z-components or y-components, lead to

$$\left[\omega^2 + (2JS)^2\left(\lambda\gamma^2(q) - \kappa_A - 1\right)\right]A^y = (2JS)^2(1 + \kappa_A - \lambda)\gamma(q)B^y \quad (6.63a)$$

$$\left[\omega^2 + (2JS)^2\left(\lambda\gamma^2(q) - \kappa_A - 1\right)\right]A^z = -(2JS)^2(1 + \kappa_A - \lambda)\gamma(q)B^z. \quad (6.63b)$$

When ω^2 from the dispersion relation (6.59) is inserted, there results for the anisotropic model,

$$A^y = \pm B^y, \qquad A^z = \mp B^z, \quad (6.64)$$

where the upper signs hold for optic modes and lower signs for acoustic modes. These show that the two sublattices have equal spin fluctuation magnitudes, with different phases for the two branches. Note that in an isotropic model, the optic and acoustic branches become degenerate and these relations are replaced by ones reflecting that degeneracy.

For optic modes, the in-plane fluctuations ($A^y = B^y$) move in-phase, while out-of-plane fluctuations ($A^z = -B^z$) are 180° out-of-phase. For acoustic modes, the in-plane fluctuations are out-of-phase while the out-of-plane fluctuations are

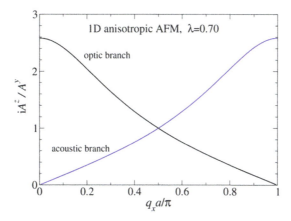

Figure 6.7. The ratio of out-of-plane to in-plane spin fluctuation amplitudes (6.65) for a 1D anisotropic antiferromagnet, with anisotropy parameter $\lambda = 0.70$, showing the different behaviors for optic and acoustic branches.

in-phase. This is a more complicated situation than the usual distinction of optic and acoustic modes in lattice waves in a solid. These relations are essential in determining how the fluctuations appear in different components of the dynamic structure functions such as $S^{xx}(k, \omega)$ and $S^{zz}(k, \omega)$.

The other important property to discuss is the relative amplitudes of in-plane fluctuations compared to out-of-plane. This can be obtained quickly from the second equation in (6.56), combining with $A^y = \pm B^y$. One obtains

$$i\omega A^z = 2JS[1 \pm \gamma(q)]A^y. \tag{6.65}$$

This is plotted in figure 6.7 for a chain with $\lambda = 0.70$. The relation (6.65) contains a somewhat complicated dependence on q, that is simpler to analyze in the particular case of the XY model, with $\lambda = 0$. Inserting the dispersion relation for that particular model, this becomes for optic modes,

$$iA^z = A^y\sqrt{\frac{1 + \cos qa}{1 + \kappa_A}} = A^y\sqrt{\frac{2}{1 + \kappa_A}} \cos\frac{qa}{2} \tag{6.66}$$

and for acoustic modes it is

$$iA^z = A^y\sqrt{\frac{1 - \cos qa}{1 + \kappa_A}} = A^y\sqrt{\frac{2}{1 + \kappa_A}} \sin\frac{qa}{2}. \tag{6.67}$$

Then, in acoustic modes the in-plane components (A^y) dominate strongly at long wavelengths. For optic modes, the in-plane components become dominant when q approaches π/a.

6.2.2 2D anisotropic antiferromagnetic model—square lattice

The generalization of the above result to a square lattice is simple. As the waves can travel along two axes, the factor $\gamma(\mathbf{q})$ is now given by expression (6.16). Together with this, the coordination number $z = 4$ comes in. Reworking the eigenvalue problem from the equations of motion (6.54) for $\mathbf{A_r}, \mathbf{B_r}$, produces a similar result for the frequencies,

$$\omega_{\pm}(\mathbf{q}) = 4JS\left[\left(1 + \frac{\kappa_A}{2} - \lambda\gamma^2(\mathbf{q})\right) \pm \left(1 + \frac{\kappa_A}{2} - \lambda\right)\gamma(\mathbf{q})\right]^{1/2}. \quad (6.68)$$

Compared to ω for a 1D chain, the overall factor out front is twice as large, due to twice the number of nearest neighbors making a stiffer oscillatory response. The single ion anisotropy, relative to exchange anisotropy, is half as strong as in a chain. Other than this, the general form of the dispersion relation is very similar to that in the 1D chain, but with $\gamma(\mathbf{q})$ as appropriate for the square lattice. The system again presents separate optic and acoustic branches, with properties like those in the AFM chain.

6.2.3 2D anisotropic antiferromagnetic model—hexagonal lattice

The anisotropic AFM model on a hexagonal lattice can also be solved by the same approach, because it is again a bipartite lattice. There are two sublattices A and B, where every A-site has three neighbors on the B-sublattice, and every B-site has three neighbors on the A-sublattice. Then, the coordination number is $z = 3$. A sketch showing the sublattices and displacements to the nearest neighbors was presented in figure 2.4. The displacements to the neighbors of an A-site that connect to B-sites are

$$\mathbf{a} = a\hat{\mathbf{x}}, \quad \mathbf{b} = a\left(-\frac{1}{2}\hat{\mathbf{x}} + \frac{\sqrt{3}}{2}\hat{\mathbf{y}}\right), \quad \mathbf{c} = -(\mathbf{a} + \mathbf{b}) = a\left(-\frac{1}{2}\hat{\mathbf{x}} - \frac{\sqrt{3}}{2}\hat{\mathbf{y}}\right). \quad (6.69)$$

For a B-site, the negatives of these connect to the nearest neighbors on the A-sublattice. In the notation of this chapter, the neighbors of an A-site are at displacements $\mathbf{a}_1 = \mathbf{a}$, $\mathbf{a}_2 = \mathbf{b}$, $\mathbf{a}_3 = \mathbf{c}$, while those of a B-site are at displacements $\mathbf{a}_1 = -\mathbf{a}$, $\mathbf{a}_2 = -\mathbf{b}$, $\mathbf{a}_3 = -\mathbf{c}$.

The ground state is still a state (6.48) where the spins on one sublattice are all aligned with each other but anti-parallel to the spins on the other sublattice. Assuming the traveling waves (6.55), the equations of motion for the yz-components of the fluctuations are still given by expressions (6.54). The primary modification here occurs in the summations over the neighbors. For instance, in the equation for $\dot{A}_\mathbf{r}^y$, there is the sum

$$\sum_{j=1}^{z} B_{\mathbf{r}+\mathbf{a}_j}^z = B^z e^{i(\mathbf{q}\cdot\mathbf{r} - \omega t)}(e^{i\mathbf{q}\cdot\mathbf{a}} + e^{i\mathbf{q}\cdot\mathbf{b}} + e^{i\mathbf{q}\cdot\mathbf{c}}). \quad (6.70)$$

The last factor is that needed to define the phase factor for this lattice,

$$\gamma(\mathbf{q}) = \frac{1}{3}(e^{i\mathbf{q}\cdot\mathbf{a}} + e^{i\mathbf{q}\cdot\mathbf{b}} + e^{i\mathbf{q}\cdot\mathbf{c}}) = \frac{1}{3}\left[e^{iq_x a} + e^{-iq_x \frac{a}{2}}\left(e^{iq_y \frac{\sqrt{3}a}{2}} + e^{-iq_y \frac{\sqrt{3}a}{2}}\right)\right]. \quad (6.71)$$

On the other hand, in the equation for $\dot{B}_r^y$, the displacements to the neighbors on the A-sublattice are reversed, and there is the term

$$\sum_{j=1}^{z} A_{r+a_j}^z = A^z e^{i(q \cdot r - \omega t)}(e^{-i q \cdot a} + e^{-i q \cdot b} + e^{-i q \cdot c}). \tag{6.72}$$

This can be seen to produce a factor of $\gamma^*(q)$ in the dynamic equations. They become

$$-i\omega A^y = +3JS(A^z + \lambda\gamma(q)B^z) + 2K_A S A^z \tag{6.73a}$$

$$-i\omega A^z = -3JS(A^y + \gamma(q)B^y) \tag{6.73b}$$

$$-i\omega B^y = -3JS(B^z + \lambda\gamma^*(q)A^z) - 2K_A S B^z \tag{6.73c}$$

$$-i\omega B^z = +3JS(B^y + \gamma^*(q)A^y). \tag{6.73d}$$

An additional application of $-i\omega$ to any of these can be used to combine them and eliminate two of the unknowns. Doing so on the first and third equations, we arrive at

$$\left[\omega^2 + (3JS)^2\left(\lambda|\gamma|^2 - 1 - \frac{2\kappa_A}{3}\right)\right]A^y = (3JS)^2\left(1 + \frac{2\kappa_A}{3} - \lambda\right)\gamma B^y \tag{6.74a}$$

$$\left[\omega^2 + (3JS)^2\left(\lambda|\gamma|^2 - 1 - \frac{2\kappa_A}{3}\right)\right]B^y = (3JS)^2\left(1 + \frac{2\kappa_A}{3} - \lambda\right)\gamma^* A^y. \tag{6.74b}$$

These then can be combined to produce the dispersion relation with a structure similar to that for the square lattice antiferromagnet,

$$\omega_\pm(q) = 3JS\left[\left(1 + \frac{2\kappa_A}{3} - \lambda|\gamma(q)|^2\right) \pm \left(1 + \frac{2\kappa_A}{3} - \lambda\right)|\gamma(q)|\right]^{1/2}. \tag{6.75}$$

As expected there are two branches, optic and acoustic, corresponding to the upper and lower signs. While $\gamma(q)$ is a complex function, only its absolute square enters in the frequency, which is

$$|\gamma(q)|^2 = \frac{1}{9}\left[1 + 4\cos q_y \frac{\sqrt{3}a}{2}\left(\cos q_y \frac{\sqrt{3}a}{2} + \cos q_x \frac{3a}{2}\right)\right]. \tag{6.76}$$

It is interesting to check the long wavelength behavior in the dispersion relation. To that end, expand $|\gamma(q)|^2$ for small q_x, q_y,

$$|\gamma(q)|^2 \approx \frac{1}{9}\left\{1 + 4\left[1 - \frac{1}{2}\left(\frac{\sqrt{3}q_y a}{2}\right)^2\right]\left[1 - \frac{1}{2}\left(\frac{\sqrt{3}q_y a}{2}\right)^2 + 1 - \frac{1}{2}\left(\frac{3q_x a}{2}\right)^2\right]\right\}. \tag{6.77}$$

This simplifies to

$$|\gamma(\mathbf{q})|^2 = 1 - \frac{1}{2}\left(q_x^2 + q_y^2\right)a^2 = 1 - \frac{1}{2}\mathbf{q}^2 a^2. \tag{6.78}$$

Using this in the dispersion relation, we have

$$\frac{\omega_\pm(\mathbf{q})}{3JS} \approx \left[\left(1 + \frac{2\kappa_A}{3} - \lambda\left(1 - \frac{1}{2}\mathbf{q}^2 a^2\right)\right) \pm \left(1 + \frac{2\kappa_A}{3} - \lambda\right)\left(1 - \frac{1}{4}\mathbf{q}^2 a^2\right)\right]^{1/2}. \tag{6.79}$$

This can be analyzed at some particular limits. For instance, for the isotropic AFM model, putting $\lambda = 1$, $\kappa_A = 0$ makes the optic and acoustic branches degenerate, and

$$\omega(\mathbf{q}) \approx \frac{3JS}{\sqrt{2}} qa. \tag{6.80}$$

That indicates a group velocity of $c = 3JSa/\sqrt{2}$. This is slower than in the square lattice isotropic antiferromagnet, where $c = 2\sqrt{2}JSa$, as can be found from its dispersion relation (6.68). Thus the lattice structure does imply certain changes in the spin wave spectrum.

For a more general case with anisotropy present, the long wavelength optical dispersion relation becomes

$$\omega_+ \approx 3JS\left[2\left(1 + \frac{2\kappa_A}{3} - \lambda\right) - \frac{1}{4}\left(1 + \frac{2\kappa_A}{3} - 3\lambda\right)\mathbf{q}^2 a^2\right]^{1/2}. \tag{6.81}$$

The initial frequency at $q = 0$ is directly related to the combination of the single ion and exchange anisotropies. Note that the parameter $(1 - \lambda)$ gives a sense of the strength of exchange anisotropy. The acoustic modes are found to have the approximate dispersion,

$$\omega_- \approx \frac{3}{2}JS\left(1 + \frac{2\kappa_A}{3} + \lambda\right)^{1/2} qa, \tag{6.82}$$

which displays the dependence of the group velocity on the anisotropy parameters.

6.3 Dynamic correlations of spin wave fluctuations

Spin waves can be detected in materials by scattering experiments, typically with neutrons whose spin-1/2 can interact with spins in the magnet. The dynamic correlation functions discussed in chapter 5 are relevant to these scattering experiments. Depending on the details of the scattering, the cross section is proportional to different components of $S^{\alpha\beta}(\mathbf{q}, \omega)$, where α, β refer to Cartesian components.

6.3.1 Ferromagnets

In an idealized situation, consider the dynamic correlation functions $S^{yy}(\mathbf{q}, \omega)$ and $S^{zz}(\mathbf{q}, \omega)$ when a 2D ferromagnet has a single spin wave mode propagating in it, with wave vector **k** and frequency $\omega(\mathbf{k})$ as given by the system's dispersion relation. Because we assumed the fluctuations are around a uniform state with all $S_\mathbf{r}^x = S$, there are no fluctuations in x spin components to be considered in the linearized theory. The yz-components of the time-dependent spins are vibrating according to

$$S_\mathbf{r}^{y,z}(t) = \sigma^{y,z} e^{i(\mathbf{k}\cdot\mathbf{r}-\omega(\mathbf{k})t)}. \tag{6.83}$$

The dynamic correlations can be found by using the space and time Fourier transforms (FTs) of these. In the general case, the time is a continuous variable. For a simulation, we suppose the time ranges from 0 to the final time t_{end}. Then one should calculate the generalization of (5.202b) to continuous time, changing the summation over time to an integration, where $t_{\text{end}}/N_t \to dt$. Also the discrete frequencies $2\pi p/t_{\text{end}}$ are replaced by frequency variable ω for simplicity. Then the space–time FT on the y spin components is taken as

$$S^y(\mathbf{q}, \omega) = \frac{1}{2\pi} \int_0^{t_{\text{end}}} dt \left[\left(\frac{a}{2\pi}\right)^2 \sum_\mathbf{r} S_\mathbf{r}^y(t) e^{-i\mathbf{q}\cdot\mathbf{r}}\right] e^{i\omega t}. \tag{6.84}$$

The part in square brackets is the space FT, which is

$$S^y(\mathbf{q}, t) = \left(\frac{a}{2\pi}\right)^2 \sum_\mathbf{r} S_\mathbf{r}^y(t) e^{-i\mathbf{q}\cdot\mathbf{r}}$$

$$= \left(\frac{a}{2\pi}\right)^2 \sum_\mathbf{r} \sigma^y e^{i(\mathbf{k}\cdot\mathbf{r}-\omega(\mathbf{k})t)} e^{-i\mathbf{q}\cdot\mathbf{r}} = \frac{N_x N_y a^2}{4\pi^2} \sigma^y \delta_{\mathbf{q},\mathbf{k}} e^{-i\omega(\mathbf{k})t}. \tag{6.85}$$

The summation over all $N_x N_y$ spin sites produces the Kronecker delta $\delta_{\mathbf{q},\mathbf{k}}$, due to an equal contribution from every site if $\mathbf{q} = \mathbf{k}$, and zero otherwise, due to destructive interference of non-equal wave vectors. Note that $N_x N_y a^2$ is a large factor equal to the area of the system. Then the time FT of this result is

$$S^y(\mathbf{q}, \omega) = \frac{1}{2\pi} \int_0^{t_{\text{end}}} dt\, S^y(\mathbf{q}, t) e^{i\omega t} = \frac{N_x N_y a^2}{(2\pi)^3} \sigma^y \delta_{\mathbf{q},\mathbf{k}} \left[\int_0^{t_{\text{end}}} dt\, e^{-i\omega(\mathbf{k})t} e^{i\omega t}\right]. \tag{6.86}$$

The time integration in square brackets gives a function

$$f(\omega) = \int_0^{t_{\text{end}}} dt\, e^{i(\omega-\omega(\mathbf{k}))t} = \frac{e^{i(\omega-\omega(\mathbf{k}))t_{\text{end}}} - 1}{i(\omega - \omega(\mathbf{k}))}. \tag{6.87}$$

The continuum dynamic structure function is obtained from a square of $S^y(\mathbf{q}, \omega)$ with appropriate normalization factors as in (5.205),

$$S^{yy}(\mathbf{q}, \omega) = \frac{2\pi}{t_{\text{end}}} \frac{2\pi}{N_x a} \frac{2\pi}{N_y a} |S^x(\mathbf{q}, \omega)|^2 = \frac{1}{2\pi t_{\text{end}}} \frac{N_x a}{2\pi} \frac{N_y a}{2\pi} |\sigma^y|^2 \delta_{\mathbf{q},\mathbf{k}} |f(\omega)|^2. \tag{6.88}$$

Magnetic Excitations and Geometric Confinement

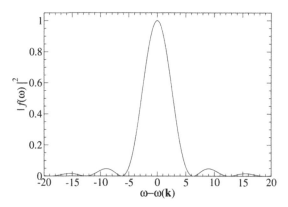

Figure 6.8. Behavior of the function $|f(\omega)|^2$ as in (6.89), for the value $t_{end} = 1$. The peak height will be proportional to t_{end}^2, and the width of the central maximum is about $2\pi/t_{end}$.

The absolute square of $f(\omega)$ needed here is seen to be

$$|f(\omega)|^2 = \left| \frac{\sin\left(\frac{1}{2}(\omega - \omega(\mathbf{k}))t_{end}\right)}{\frac{1}{2}(\omega - \omega(\mathbf{k}))t_{end}} \right|^2 t_{end}^2. \tag{6.89}$$

This is strongly peaked around the point $\omega = \omega(\mathbf{k})$, see figure 6.8, with a peak height equal to t_{end}^2, and a width on the order of $\Delta\omega \approx 2\pi/t_{end}$. The same type of integral appears in quantum theory for the development of time-dependent perturbation theory and Fermi's golden rule [3]. Changing to a new variable $u = \frac{1}{2}[\omega - \omega(\mathbf{k})]t_{end}$, an integration over all frequencies gives the total area under the peak

$$\int_{-\infty}^{+\infty} d\omega \, |f(\omega)|^2 = 2t_{end} \int_{-\infty}^{+\infty} du \, \frac{\sin^2 u}{u^2} = 2\pi t_{end}. \tag{6.90}$$

This indicates that for large t_{end} the function $|f(\omega)|^2$ is proportional to a Dirac delta function:

$$\lim_{t_{end} \to \infty} |f(\omega)|^2 = 2\pi t_{end} \, \delta[\omega - \omega(\mathbf{k})]. \tag{6.91}$$

Then the dynamic structure function that results is

$$S^{yy}(\mathbf{q}, \omega) = \frac{N_x a}{2\pi} \frac{N_y a}{2\pi} |\sigma^y|^2 \, \delta_{\mathbf{q},\mathbf{k}} \, \delta[\omega - \omega(\mathbf{k})]. \tag{6.92}$$

This is a rather simple and highly physical result. It shows that the presence of a pure spin wave mode in the system will give a very strong peak in a dynamic structure function, exactly at a frequency given by the dispersion relation, $\omega_{peak} = \omega(\mathbf{k})$. The strength of the peak is determined by the size of the system and the squared spin wave amplitude. One can expect a similar result also for the dynamic correlation function for the z-components.

Going away from the above idealization, a real system at finite temperature will have a statistical distribution of spin wave modes present. Each mode will tend to give a term such as that in (6.92), however, due to nonlinearities, the different modes can interfere and scatter amongst themselves. This will give them a finite length scale and also a finite time scale over which a mode stays coherent. Then, instead of an extremely sharp response at one frequency, each mode will be significantly broadened. Therefore, in a real system, the spin wave response will have peaks that are not sharp delta functions, but instead broad responses with widths that depend on the temperature. Even so, analysis of $S^{yy}(\mathbf{q}, \omega)$ and the other components will contain valuable information about the spin wave spectrum.

For ferromagnets with XY symmetry, our choice that spins were making small fluctuations about the x-direction was arbitrary. Due to the symmetry, any axis in the xy-plane is equivalent. Then, the dynamic structure functions $S^{yy}(\mathbf{q}, \omega)$ and $S^{xx}(\mathbf{q}, \omega)$ will be essentially the same, when averaged over many states. The correlations $S^{zz}(\mathbf{q}, \omega)$ will naturally be different, unless one considers a fully isotropic FM model, where we had $\sigma^y = i\sigma^z$.

6.3.2 Antiferromagnets

In an antiferromagnet, the spins on different sublattices may be moving 180° out-of-phase, depending on the spin components under consideration, and whether an optic mode or an acoustic mode is under consideration. The relation was summarized in (6.64) for a 1D AFM chain, however, it would apply equally well on a 2D square lattice.

Consider the idealized case of an individual optical mode of wave vector k on a 1D chain. The S^y-fluctuations on the two sublattices move in-phase, while the S^z-fluctuations are out-of-phase. We can write

$$S_n^y(t) = A^y\, e^{i(kna - \omega_+(k)t)} \tag{6.93a}$$

$$S_n^z(t) = A^z\, e^{i\pi n}\, e^{i(kna + \omega_+(k)t)} = A^z\, e^{i(ka+\pi)n}\, e^{-i\omega_+(k)t}. \tag{6.93b}$$

The FT and its square leads to the corresponding dynamic structure functions.

Exercise 6.10. Use the fluctuations (6.93) for *optical modes* and show that the dynamic structure functions that result are

$$S^{yy}(q, \omega) = \frac{Na}{2\pi}\, |A^y|^2\, \delta_{q,k}\, \delta[\omega - \omega_+(k)] \tag{6.94a}$$

$$S^{zz}(q, \omega) = \frac{Na}{2\pi}\, |A^z|^2\, \delta_{q,k+\frac{\pi}{a}}\, \delta[\omega - \omega_+(k)]. \tag{6.94b}$$

For optic modes, the in-plane fluctuations appear at $q = k$, the optic mode frequency, while the out-of-plane fluctuations appear at $q = k + \frac{\pi}{a}$. Shifting that by $-\frac{2\pi}{a}$ back into the Brillouin zone, it is equivalent to $q = k - \frac{\pi}{a}$. Here we assumed a mode propagating towards $+x$. Had we assumed a mode moving towards $-x$, it would appear in the dynamic correlations at $q^* = -k + \frac{\pi}{a}$. For modes with small k, the out-of-plane fluctuations then appear in $S^{zz}(q, \omega)$ at q^*, near the edge of the Brillouin zone, while the in-plane fluctuations appear in $S^{yy}(q, \omega)$ at $q = k$.

For an acoustic mode, the S^y-fluctuations on the two sublattice move out-of-phase, while the S^z-fluctuations are in-phase. They are expressed as

$$S_n^y(t) = A^y\, e^{i\pi n}\, e^{i(kna + \omega_-(k)t)} = A^y\, e^{i(ka+\pi)n}\, e^{-i\omega_-(k)t} \tag{6.95a}$$

$$S_n^z(t) = A^z\, e^{i(kna - \omega_-(k)t)}. \tag{6.95b}$$

The dynamic structure functions now are interchanged in structure compared to optic modes.

Exercise 6.11. Use the fluctuations (6.95) for **acoustic modes** and show that the dynamic structure functions that result are

$$S^{yy}(q, \omega) = \frac{Na}{2\pi} |A^y|^2\, \delta_{q, k+\frac{\pi}{a}}\, \delta[\omega - \omega_-(k)] \tag{6.96a}$$

$$S^{zz}(q, \omega) = \frac{Na}{2\pi} |A^z|^2\, \delta_{q, k}\, \delta[\omega - \omega_-(k)]. \tag{6.96b}$$

This last exercise shows that the in-plane fluctuations appear at $q = k + \frac{\pi}{a}$ in $S^{yy}(q, \omega)$, for the forward moving mode, or at $q^* = -k + \frac{\pi}{a}$ for the reverse direction mode. The out-of-plane fluctuations will appear at $q = k$ in $S^{zz}(q, \omega)$.

The assumption that spins were making small oscillations about the x-direction was arbitrary; the spin wave properties would be the same if the y- or z-directions had been used for this isotropic model. The dynamic correlation $S^{xx}(q, \omega)$ would be essentially the same as $S^{yy}(q, \omega)$ once one averages over many states in a thermal equilibrium situation. Furthermore, the infinitely narrow peaks will be broadened due to thermal fluctuations breaking up the coherence of the modes, just as in ferromagnets.

The result above also will extend to a 2D antiferromagnet on a bipartite lattice. Consider a square lattice, where the sites' locations are $\mathbf{r} = (n, m)a$. One can write for an *optic mode* the spin fluctuations,

$$S_\mathbf{r}^y(t) = A^y\, e^{i(k_x na + k_y ma - \omega_+(\mathbf{k})t)} \tag{6.97a}$$

$$S_\mathbf{r}^z(t) = A^z\, e^{i\pi n}\, e^{i\pi m}\, e^{i[k_x na + k_y ma + \omega_+(\mathbf{k})t]} = A^z\, e^{i(k_x a+\pi)n}\, e^{i(k_y a+\pi)m}\, e^{-i\omega_+(\mathbf{k})t}. \tag{6.97b}$$

Then one can see that the out-of-plane fluctuations will appear at the wave vector $\mathbf{q} = \mathbf{k} + \mathbf{b}$, where the vector $\mathbf{b} = (\frac{\pi}{a}, \frac{\pi}{a})$ is the corner of the Brillouin zone. A mode going in the opposite direction will produce intensity at $q^* = -k + b$. The in-plane fluctuations, of course, will still appear at $q = k$. It is clear that for an acoustic mode the relations are switched, and in-plane fluctuations will appear at $q = k + b$ (or at q^* for the mode in the opposite direction), and out-of-plane fluctuations at $q = k$.

6.4 Nonlinear spin waves—ferromagnets

Linearization is not always necessary for finding spin wave solutions. In the case of an isotropic FM chain, there exist traveling wave magnetic excitations that do not require the linearization approximation. They are known as nonlinear spin waves.

For the equations of motion (6.3), these solutions are found by looking for precessional motions of the spins around one axis (really, any direction could be chosen as the precessional axis). Choosing the x-axis, look for a solution with the following form:

$$S_n^x(t) = S\cos\theta = \text{constant} \tag{6.98a}$$

$$S_n^y(t) = S\sin\theta \cos(qna - \omega t) \tag{6.98b}$$

$$S_n^z(t) = S\sin\theta \sin(qna - \omega t). \tag{6.98c}$$

The angle θ is a constant angle that is the deviation of the spins away from the x-axis. As this wave propagates in the positive direction down the chain, the spins precess around the x-axis. The relation between wave vector q and frequency ω is to be determined. A short bit of algebra shows that this is a valid solution. It is helpful to define the phase or precessional angle at site n,

$$\phi_n = qna - \omega t. \tag{6.99}$$

Using the differential equations (6.3), one can check the time derivatives:

$$\dot{S}_n^x = J\{S_n^y(S_{n-1}^z + S_{n+1}^z) - S_n^z(S_{n-1}^y + S_{n+1}^y)\}$$
$$= JS^2 \sin^2\theta \{\cos\phi_n[\sin(\phi_n - qa) + \sin(\phi_n + qa)]$$
$$- \sin\phi_n[\cos(\phi_n - qa) + \cos(\phi_n + qa)]\}$$
$$= JS^2 \sin^2\theta \{\sin(\phi_n - qa - \phi_n) + \sin(\phi_n + qa - \phi_n)\} = 0. \tag{6.100}$$

Therefore the constant values of S_n^x are ensured. For the y-components, we have

$$\dot{S}_n^y = J\{S_n^z(S_{n-1}^x + S_{n+1}^x) - S_n^x(S_{n-1}^z + S_{n+1}^z)\}$$
$$= JS^2 \sin\theta \cos\theta \{2\sin\phi_n - [\sin(\phi_n - qa) + \sin(\phi_n + qa)]\}$$
$$= 2JS^2 \sin\theta \cos\theta \sin\phi_n(1 - \cos qa)$$
$$= 2JS\cos\theta(1 - \cos qa)S_n^z. \tag{6.101}$$

A similar result holds for the z-components,

$$\begin{aligned}
\dot{S}_n^z &= J\{S_n^x(S_{n-1}^y + S_{n+1}^y) - S_n^y(S_{n-1}^x + S_{n+1}^x)\} \\
&= JS^2 \sin\theta \cos\theta \{\cos(\phi_n - qa) + \cos(\phi_n + qa) - 2\cos\phi_n\} \\
&= -2JS^2 \sin\theta \cos\theta \cos\phi_n(1 - \cos qa) \\
&= -2JS \cos\theta(1 - \cos qa)S_n^y.
\end{aligned} \qquad (6.102)$$

Comparing with the time derivatives obtained from the assumed solution (6.98),

$$\dot{S}_n^y = \omega S \sin\theta \sin\phi_n = \omega S_n^z \qquad (6.103a)$$

$$\dot{S}_n^z = -\omega S \sin\theta \cos\phi_n = -\omega S_n^y \qquad (6.103b)$$

then the solution is valid, provided the frequency is given by a nonlinear dispersion relation,

$$\omega = 2JS \cos\theta(1 - \cos qa). \qquad (6.104)$$

This is a very interesting relation, because of the dependence on the deviation angle θ. At very small θ with $\cos\theta \approx 1$, the linear spin wave dispersion relation for an isotropic FM chain is recovered, see figure 6.9. Furthermore, the amplitude of the spin fluctuations is given by $\sin\theta \approx \theta$. Going away from this limit, the maximum possible amplitude will be when $\sin\theta$ approaches 1, but then $\cos\theta \to 0$ in the same limit. These large-amplitude nonlinear oscillations take place at very low frequency. In a situation of thermal equilibrium, one might expect a high population of the these large-amplitude states, if their energy scales proportional to their frequency. However, a high population of large-amplitude fluctuations is counterintuitive and at odds with the usual interpretation of mode amplitudes in equilibrium.

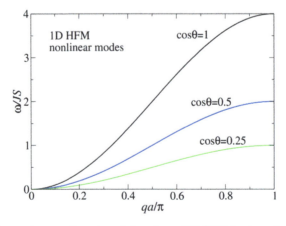

Figure 6.9. The nonlinear spin wave dispersion relations for an FM 1D Heisenberg chain, for three choices of the deviation angle θ, see the spin wave mode (6.98) and dispersion relation (6.104).

One can assign a phase velocity v to these waves by

$$v = \frac{\omega}{q} = 2JS \cos\theta \frac{1 - \cos qa}{q} = 4JS \cos\theta \frac{\sin^2\left(\frac{1}{2}qa\right)}{q}. \qquad (6.105)$$

In a continuum limit of long wavelength modes, $qa \ll 1$, this gives $v \approx JSqa^2 \cos\theta$, revealing a speed that depends strongly on wave vector and deviation angle. Looked at from a different point of view, this shows how the speed determines the approximate wave vector,

$$q \approx \frac{v}{JSa^2} \sec\theta \qquad (6.106)$$

and the frequency,

$$\omega = qv \approx \frac{v^2}{JSa^2} \sec\theta. \qquad (6.107)$$

Then the oscillating spin components depend on the precession angle, which is written as a continuum expression

$$\phi_n \to qx - \omega t = \frac{v \sec\theta}{JSa^2}(x - vt). \qquad (6.108)$$

Looked at this way, there is a family of solutions that can be selected according to speed and deviation angle.

It is useful to calculate the energy density in a nonlinear spin wave mode. Relative to the aligned ground state, the energy in a single bond is $\Delta E_n = -J\mathbf{S}_n \cdot \mathbf{S}_{n+1} + JS^2$. Inserting a mode's spin components, this is

$$\Delta E_n = -JS^2\{\cos^2\theta + \sin^2\theta[\cos\phi_n \cos(\phi_n + qa) + \sin\phi_n \sin(\phi_n + qa)] - 1\}$$
$$= JS^2 \sin^2\theta (1 - \cos qa). \qquad (6.109)$$

Every bond carries this same energy. Note that it is not proportional to the mode frequency. This result shows that the most populated modes will be those with small θ, and with the preference for long wavelengths. This is what one would typically expect, that the small-amplitude modes will have the highest statistical weight.

Bibliography

[1] Kittel C 2005 *Introduction to Solid State Physics* 8th edn (New York: Wiley)
[2] Ashcroft N W and Mermin N D 1976 *Solid State Phys.* (New York: Holt, Reinhart and Winston)
[3] Schiff L I 1968 *Quantum Mechanics* 3rd edn (New York: McGraw-Hill)

Chapter 7

Solitons in magnetic chains

A chain of magnetic ions can be considered as a model for quasi-1D magnets, where exchange interactions in a 3D crystal are highly anisotropic. As a result, the dynamics is closer to that of a 1D system than of 3D. The combination of low-dimensionality and nonlinearity leads to soliton excitations: concentrated regions of magnetic energy that behave somewhat like particles, with a kind of topological charge. In this chapter a summary is given of how 1D magnets generate localized topological excitations. The properties of these solitons depend on whether the system is ferromagnetic (FM) or antiferromagnetic (AFM), and on details of the anisotropy. The role of an applied magnetic field, and the relation to domain walls is considered. The notion of soliton stability is introduced.

7.1 Nonlinear excitations: solitons in FM magnetic chains

In the previous chapter spin waves were presented as the excitations found in magnets when the equations of motion are linearized around some ground state. Linearization requires the assumption of small-amplitude deviations. However, magnets can also have large-amplitude deviations away from a ground state, that include the effects of *nonlinear* interactions in the equations of motion. In fact, there are types of magnetic excitations that are impossible to produce starting from a ground state, by making continuous deformations in the spin structure. A soliton is one of these kinds of excitations. Solitons are large-amplitude wave motions where nonlinearity competes with dispersion and possibly also with dissipation [1], and they can be present in many condensed matter systems [2], including magnetic ones. We will see that in order to generate a soliton from the ground state, it is actually necessary to form a pair: a soliton and its anti-particle or anti-soliton. This is known as a topological constraint; it is derived from geometry. This is very reminiscent of particle generation in the presence of conservation laws such as charge, where an electron might be generated out of a vacuum only with the simultaneous generation of a positron, so that total charge is conserved. In the case of magnetic solitons, there is a conserved topological charge, rather than some electric charge.

In a quasi-1D magnetic material, the exchange interactions are strongest along one spatial direction, leading to strong interactions that can be essentially isolated along chains. An important historical example is the compound $CsNiF_3$, whose Ni^{2+} ions possess angular momentum $S = 1$ which are the magnetically active species. Babel [3] and Steiner and co-workers [4] were able to determine various crystal properties via neutron diffraction. The ions also are subjected to a crystal field of the neighboring fluorines that induces a local easy-plane anisotropy interaction. Thus, a Hamiltonian for a chain in this type of compound can be written

$$H = \sum_n \left[-J(\mathbf{S}_n \cdot \mathbf{S}_{n+1} - S^2) + K_A S_n^{z\,2} - g\mu_B B^x (S_n^x - S) \right]. \tag{7.1}$$

In addition to isotropic FM exchange, the single ion anisotropy K_A is included, as well as an applied magnetic induction along the x-axis. The exchange term has a constant JS^2 added per site, and the field term has $g\mu_B B^x S$ added per site, to give the ground state zero energy. We should assume that B^x is weak compared to the exchange and anisotropy. As in the problems with spin waves present, B^x is used to break the symmetry for spins in the easy plane (the xy-plane). By breaking that symmetry, it selects a ground state direction (along x). Although the spin value is not very large, one proceeds by assuming that classical spin dynamics (SD) can be applied. For $CsNiF_3$, the coupling parameters, when measured in temperature units, are $JS^2 = 23.6$ K, $K_A S^2 = 4.5$ K (for the classical model)[1]. Note that the intra-chain Ni–Ni distance is $a = 0.26$ nm, while the chain-to-chain distance is 0.62 nm. This accounts for the quasi-1D behavior dominated by the exchange interactions. However, at very low temperatures (less than $T_N = 2.65$ K), there is a phase transition to a 3D AFM phase, which is caused by the weak inter-chain couplings. Thus, we are considering temperatures higher than that, where 1D ordering is present.

The model has XY symmetry which is only weakly broken by the presence of B^x. Steiner and co-workers [5] used neutron scattering to study the SD in a field. Due to K_A, the spins prefer to lie in the xy-plane. Rotations around the z-direction only cost a small energy in the region where the spins point against the magnetic induction. A soliton excitation is a twisting of the spins around the z-axis, over a well-defined length scale or width. An example of the structure of a magnetic soliton for this model is shown in figure 7.1. The width is determined by the competition between exchange and the magnetic induction. The center of the soliton will be the point where the spins point against B^x. In the regions far from the center of the soliton, the spins rotate back to their ground state direction along B^x. These nonlinear excitations in magnets are better called *kinks*, due to the appearance of the spin profile versus position on the chain. Also, the term soliton is reserved for robust nonlinear excitations that preserve their structure upon collisions with other solitons. For the magnetic kinks under consideration in this model, they are not found to

---

[1] If considering a quantum model the value is $K_A S^2 = 9.0$ K. This is renormalized by half when mapped into an equivalent classical problem.

have this property in any strong sense. Thus, they are only approximate *solitary waves*.

The main theory for magnetic kinks comes from a continuum limit theory, applied to the above Hamiltonian, much of it developed by Mikeska [6]. Let us first obtain the discrete equations of motion, and from that, go over to a continuum description. The discrete effective fields that determine the dynamics as in (2.98) are

$$\mathcal{F}_n = J(\mathbf{S}_{n-1} + \mathbf{S}_{n+1}) + g\mu_B B^x \hat{\mathbf{x}} - 2K_A S_n^z \hat{\mathbf{z}}. \tag{7.2}$$

The equations of motion are the usual ones,

$$\dot{\mathbf{S}}_n = \mathbf{S}_n \times \mathcal{F}_n. \tag{7.3}$$

It is convenient to write the equations by expressing the spins in planar spherical coordinates, (2.42), where each spin has an in-plane angle ϕ_n (measured to the x-axis) and an out-of-plane angle θ_n (measured to the xy-plane). In terms of those angles, the discrete equations of motion are found as in (5.19),

$$(JS)^{-1}\dot{\phi}_n \cos\theta_n = \sin\theta_n \Big[\cos\theta_{n+1}\cos(\phi_{n+1} - \phi_n) + \cos\theta_{n-1}\cos(\phi_{n-1} - \phi_n)\Big]$$
$$- \cos\theta_n(\sin\theta_{n+1} + \sin\theta_{n-1}) + (2K_A/J)\cos\theta_n \sin\theta_n$$
$$+ (g\mu_B B^x/JS)\sin\theta_n \cos\phi_n \tag{7.4a}$$

$$(JS)^{-1}\dot{\theta}_n = \cos\theta_{n+1}\sin(\phi_{n+1} - \phi_n) + \cos\theta_{n-1}\sin(\phi_{n-1} - \phi_n)$$
$$- (g\mu_B B^x/JS)\sin\phi_n. \tag{7.4b}$$

On the RHS of these equations, the parameters that determine the easy-plane character and the magnetic field strength, relative to exchange, are

$$\alpha \equiv 2K_A/J, \qquad \beta \equiv g\mu_B B^x/JS. \tag{7.5}$$

To obtain a continuum limit, let index n be replaced by position $z = na$ along the chain, and use the leading terms in Taylor series for objects such as θ_{n+1}, and expand small arguments of the trigonometric functions. For example, consider the equation for $\dot{\theta}_n$. With lattice constant a, and using subscripts to denote partial derivatives, one needs

$$\cos\theta_{n+1} \to \cos\left(\theta + a\theta_z + \frac{1}{2}a^2\theta_{zz}\right)$$
$$= \cos\theta \cos\left(a\theta_z + \frac{1}{2}a^2\theta_{zz}\right) - \sin\theta \sin\left(a\theta_z + \frac{1}{2}a^2\theta_{zz}\right)$$
$$\approx \cos\theta\left[1 - \frac{1}{2}\left(a\theta_z + \frac{1}{2}a^2\theta_{zz}\right)^2\right] - \sin\theta\left(a\theta_z + \frac{1}{2}a^2\theta_{zz}\right). \tag{7.6}$$

This must multiply

$$\sin(\phi_{n+1} - \phi_n) \approx a\phi_z + \frac{1}{2}a^2\phi_{zz} \tag{7.7}$$

and then that combination must be combined with another similar term but with $a \to -a$. Keeping only terms to second order in derivatives, one obtains

$$\cos\theta_{n+1}\sin(\phi_{n+1} - \phi_n) + \cos\theta_{n-1}\sin(\phi_{n-1} - \phi_n) \to a^2(\phi_{zz}\cos\theta - 2\phi_z\theta_z\sin\theta). \quad (7.8)$$

Continuing this way for all the terms, the continuum equations that are found are

$$(JS)^{-1}\dot\phi\cos\theta = -a^2\theta_{zz} + \left(\alpha - a^2\phi_z^2\right)\sin\theta\cos\theta + \beta\sin\theta\cos\phi \quad (7.9a)$$

$$(JS)^{-1}\dot\theta = a^2(\phi_{zz}\cos\theta - 2\phi_z\theta_z\sin\theta) - \beta\sin\phi. \quad (7.9b)$$

For further analysis it is convenient to define new units for distance and time, to bring the equations to their simplest form. Dividing through by anisotropy parameter α, one can define dimensionless length and time parameters,

$$\xi \equiv \sqrt{\frac{2K_A}{Ja^2}}\, z, \qquad \tau \equiv (2K_A S)t. \quad (7.10)$$

Then the equations in these units are simplified to

$$\phi_\tau \cos\theta = -\theta_{\xi\xi} + \left(1 - \phi_\xi^2\right)\sin\theta\cos\theta + b\sin\theta\cos\phi \quad (7.11a)$$

$$\theta_\tau = \phi_{\xi\xi}\cos\theta - 2\theta_\xi\phi_\xi\sin\theta - b\sin\phi, \quad (7.11b)$$

where the magnetic field parameter is

$$b \equiv \frac{\beta}{\alpha} = \frac{g\mu_B B^x}{2K_A S}. \quad (7.12)$$

In these units, only the parameter b is needed to parameterize the solutions.

Equations (7.11) are valid for slow spatial variations and arbitrary magnetic field. They can be brought to a more interesting but approximate form if the magnetic field is weak, and the easy-plane anisotropy is strong. To that end, suppose that b, θ and all space derivatives are of similar orders of smallness, and $\sin\theta \approx \theta$. Keeping only the leading terms with this approximation, the equations become [6]

$$\phi_\tau = \theta \quad (7.13a)$$

$$\theta_\tau = \phi_{\xi\xi} - b\sin\phi. \quad (7.13b)$$

Note that we have kept an important nonlinear term involving $\sin\phi$. That angle is not assumed to be small! One can see that these combine into a single second-order partial differential equation, which is known as the *sine-Gordon* (sG) equation:

$$\phi_{\xi\xi} - \phi_{\tau\tau} = b\sin\phi. \quad (7.14)$$

This sG limit of the magnetic equations is the regime in which solitary waves can appear. Because of the approximations, one cannot expect the perfect sG behavior

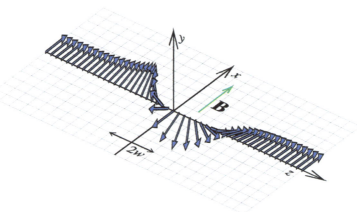

Figure 7.1. A sketch of the spins' rotation versus position in a stationary sG magnetic kink for $q = +1$, equation (7.19). The double width of (7.23), inversely proportional to $\sqrt{B^x/J}$, is indicated. The twist is right-handed; a kink with $q = -1$ will be left-handed.

for magnetic kinks. However, some of their most important properties can be discussed. Note that the LHS involves the usual wave operator. The RHS involves a type of force due to the effective potential produced by the applied magnetic field.

Due to the wave operator on the LHS, the sG equation (7.14) has traveling wave solutions. It is possible to solve for these solutions first by seeking the static solutions of the form $\phi(\xi)$, and then modifying those static solutions to obtain one depending as $\phi(\xi - u\tau)$, where u is a velocity in the dimensionless units. The determination of the static solution is an exercise in integrating the differential equation twice. The first integral is found with the help of multiplication by ϕ_ξ, which acts as an integrating factor that brings both sides to perfect differentials:

$$\phi_\xi \phi_{\xi\xi} = \phi_\xi b \sin \phi \tag{7.15a}$$

$$\frac{d}{d\xi}\left(\frac{1}{2}\phi_\xi^2\right) = \frac{d}{d\xi}(-b \cos \phi). \tag{7.15b}$$

As a boundary condition, suppose $\phi \to 0$, $\phi_x \to 0$, as $\xi \to -\infty$. Then the first integral here is

$$\frac{d\phi}{d\xi} = \sqrt{2b(1 - \cos \phi)}. \tag{7.16}$$

This is now in separated form, and can be integrated a second time. Starting from

$$\int \frac{d\phi}{\sqrt{2(1 - \cos \phi)}} = \int \sqrt{b}\, d\xi \tag{7.17}$$

then a sequence of trigonometric substitutions, starting with $1 - \cos \phi = 2 \sin^2 \frac{\phi}{2}$, eventually leads to a result

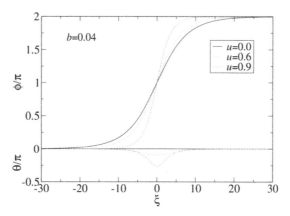

Figure 7.2. Examples of in-plane (ϕ, upper curves) and out-of-plane (θ, lower curves) sG soliton profiles ($q = +1$) in (7.22) and (7.26) for magnetic field $b = 0.04$, at indicated velocities in units of maximum speed c.

$$\ln\left(\tan\frac{\phi}{4}\right) = \sqrt{b}\,\xi. \tag{7.18}$$

One could also add an arbitrary constant to the RHS to effect a shift in the center position of the resulting soliton. Then, a static sG soliton solution centered at the origin is found,

$$\phi_{sG}(\xi) = 4q\tan^{-1}\left\{\exp\left[\sqrt{b}\,\xi\right]\right\}. \tag{7.19}$$

A charge-like parameter $q = \pm 1$ has been included. For $q = +1$, one can see (figure 7.2) that as $\xi \to +\infty$, the angle $\phi \to 2\pi$. This is called the soliton (or magnetic kink) solution. For $q = -1$, an equally valid solution is achieved; it is called an anti-soliton (or magnetic anti-kink). Normally we might simply say soliton (or kinks) when referring to both of these in a general sense. Thus, an sG soliton is a solution where the in-plane spin angle rotates through $\pm 2\pi$ from one end of the chain to the other. However, most of the rotation takes place over a small region of width determined by $1/\sqrt{b}$. At weak magnetic field, the soliton width will be large. If b is large, the $\pm 2\pi$ rotation would take place over a short distance, and spatial derivatives would become large, violating the assumptions of smallness for derivatives. So it is important to keep in mind the limited range of validity for this type of solution if applied to the magnetic problem. Also note, there is no way to build the soliton shape from a superposition of small-amplitude perturbations of the ground state. It is an inherently nonlinear excitation.

It is possible to boost the static soliton solutions and give them a velocity u, assuming a solution of the form $\phi(\xi - u\tau)$ in (7.14). On the LHS this produces an extra factor $1 - u^2$, via the time derivatives. The sG equation then is in the form

$$\phi_{\xi\xi} = \frac{b}{1 - u^2}\sin\phi. \tag{7.20}$$

The velocity factor is grouped with the magnetic field factor, leading to a relativistic length contraction of the soliton width! We can define the relativistic factor just as in Einstein's theory,

$$\gamma \equiv \frac{1}{\sqrt{1-u^2}}. \tag{7.21}$$

The moving soliton/anti-soliton solution is then

$$\phi_{sG}(\xi, \tau) = 4q \tan^{-1}\left\{\exp\left[\gamma\sqrt{b}(\xi - u\tau)\right]\right\}. \tag{7.22}$$

This displays the most important feature of a solitary wave. The shape of the profile is preserved as the wave moves. Although this is a single-soliton solution, there exist multi-soliton solutions that truly demonstrate the ability of sG solitons to pass through each other and eventually recover their original structure. Thus, they attract a lot of interest, with this particle-like behavior.

In terms of the original coordinates, the soliton width is

$$w = \frac{1}{\gamma\sqrt{b}}\sqrt{\frac{Ja^2}{2K_A}} = \sqrt{\frac{JS}{g\mu_B B^x}} \frac{a}{\gamma}. \tag{7.23}$$

This reflects that the width results as a competition between exchange and the magnetic induction. For the moving soliton, the maximum speed is $u = 1$. It means there is a speed of the light in the problem, which is the ratio of length unit to time unit,

$$c = \frac{\sqrt{Ja^2/2K_A}}{(2K_A S)^{-1}} = \sqrt{2JS^2 K_A}\, a. \tag{7.24}$$

This speed acts as the upper limit of soliton speed. Indeed, it is essentially a speed for linearized spin waves in the system. If one looks for small-amplitude traveling wave solutions of the sG equation at frequency ω, wave vector k, it is easy to show that their dispersion relation in physical units is

$$\omega^2 = 2K_A g\mu_B B^x S + c^2 k^2. \tag{7.25}$$

The first factor is a gap frequency that is determined by the applied field. Thus, the upper limit for soliton speed is related to the spin wave spectrum.

The out-of-plane structure in a sG soliton is given from the time derivative of ϕ. Therefore, in the static structure, the out-of-plane angle is $\theta = 0$. If the soliton is moving, we obtain

$$\theta_{sG} = \phi_\tau = -2q\gamma\sqrt{b}\, u\, \text{sech}\left[\gamma\sqrt{b}(\xi - u\tau)\right]. \tag{7.26}$$

This is a small pulse at the soliton center, whose amplitude grows with speed. Whether θ_{sG} is positive or negative is determined by the sign of qu. Some examples of the soliton structure, both in-plane and out-of-plane, are shown in figure 7.2 for $b = 0.04$. For this sG limit to be valid in the magnetic problem, we see that $2\gamma\sqrt{b}\, u \ll 1$ is required.

Based on the derivation of the continuum limit used above, one can also evaluate the soliton energy, using the associated continuum Hamiltonian. That Hamiltonian is found with the techniques in chapter 2 to be

$$H = \int \frac{dz}{a} \left\{ \frac{Ja^2}{2} \left(\frac{d\mathbf{S}}{dz} \right)^2 + K_A (S^z)^2 + g\mu_B (S - S^x) \right\}. \quad (7.27)$$

The extra magnetic field term is used to set the ground state energy to zero with all $S^x = S$. This can also be written in the dimensionless coordinates, and using the angular description of the spins,

$$H = \epsilon_0 \int d\xi \left\{ \frac{1}{2} \left(\theta_\xi^2 + \phi_\xi^2 \cos^2 \theta \right) + \frac{1}{2} \sin^2 \theta + b(1 - \cos\theta \cos\phi) \right\}, \quad (7.28)$$

where the energy unit is

$$\epsilon_0 \equiv \sqrt{2K_A J}\, S^2. \quad (7.29)$$

This is the full continuum magnetic Hamiltonian. In the sG limit, we considered θ, b, and space derivatives as similar small orders of magnitude. If one does that here as well, and using $\theta = \phi_\tau$, then the sG limit of this Hamiltonian is obtained:

$$H_{sG} = \epsilon_0 \int d\xi \left\{ \frac{1}{2} \phi_\xi^2 + \frac{1}{2} \phi_\tau^2 + b(1 - \cos\phi) \right\}. \quad (7.30)$$

One sees the first two contributions are exchange and a kinetic energy; the last term is the potential of the applied field. It is clear these are the only terms needed in a Hamiltonian to obtain the sG equation (7.14) as its dynamical equation. Using the sG solution (7.22) one finds the needed quantities,

$$\cos\phi_{sG} = 1 - 2\,\text{sech}^2\left[\gamma\sqrt{b}(\xi - u\tau)\right] \quad (7.31a)$$

$$\phi_{sG,\xi} = 2q\gamma\sqrt{b}\,\text{sech}\left[\gamma\sqrt{b}(\xi - u\tau)\right]. \quad (7.31b)$$

Using the sG-limit Hamiltonian, the only integrals that appear are of the form,

$$\int_{-\infty}^{+\infty} d\xi\ \text{sech}^2\left[\gamma\sqrt{b}(\xi - u\tau)\right] = \frac{1}{\gamma\sqrt{b}} \tanh\left[\gamma\sqrt{b}(\xi - u\tau)\right]\Big|_{-\infty}^{+\infty} = \frac{2}{\gamma\sqrt{b}}. \quad (7.32)$$

Then the energy of the sG-kink solution is found to be

$$E_{sG} = 8\epsilon_0 \sqrt{b}\,\gamma. \quad (7.33)$$

An sG soliton has a rest energy, and its total energy increases in proportion to the relativistic factor when in motion.

7.2 Ferromagnetic sine-Gordon kink instability

With certain approximations we arrived at the sG-kink magnetic structure. We know that this solution has limitations of applicability. Besides that, it is important to look at the stability properties of nonlinear excitations. Even if a nonlinear excitation is present in the system, small perturbations could appear dynamically on top of that profile. The spectrum of small excitations on top of the soliton as a background tells us about the stability of the soliton.

We are considering the stability of the sG-kink solution (7.19), but for the full magnetic equations (7.11). The kink is an exact solution there only when static, $u = 0$, which is the case considered. Assume the total spin angles are superpositions of the static sG-kink plus small fluctuations $\tilde{\phi}, \tilde{\theta} \ll 1$:

$$\phi = \phi_{sG} + \tilde{\phi}, \qquad \theta = \theta_{sG} + \tilde{\theta}. \tag{7.34}$$

Of course, the static solution has $\theta_{sG} = 0$. Using these in (7.11), and linearizing in the fluctuations, leads to these equations,

$$\tilde{\phi}_\tau = -\tilde{\theta}_{\xi\xi} + \left(1 - \phi_{sG\,\xi}^2 + b\cos\phi_{sG}\right)\tilde{\theta} \tag{7.35a}$$

$$\tilde{\theta}_\tau = \tilde{\phi}_{\xi\xi} - b\cos\phi_{sG}\,\tilde{\phi}. \tag{7.35b}$$

Using the sG-kink generates a pair of dynamical equations, with $x \equiv \sqrt{b}\,\xi$,

$$\tilde{\phi}_\tau = -\tilde{\theta}_{\xi\xi} + (1 + b - 6b\,\text{sech}^2 x)\tilde{\theta} \equiv L_1 \cdot \tilde{\theta} \tag{7.36a}$$

$$\tilde{\theta}_\tau = \tilde{\phi}_{\xi\xi} - (1 - 2\,\text{sech}^2 x)\tilde{\phi} \equiv -L_2 \cdot \tilde{\phi}. \tag{7.36b}$$

This shows that the in-plane and out-of-plane fluctuations are cross-coupled. The differential operators L_1 and L_2 characterize this coupling.

Generally we expect that the fluctuations should be able to oscillate with some frequency $\tilde{\omega}$, with a time variation $\exp(-i\tilde{\omega}\tau)$. This is, after all, essentially a spin wave problem, but on the background of a soliton. But the in-plane and out-of-plane components form a vector, and we seek the oscillations of this vector. In particular, the question arises as to whether there are unstable modes of oscillation. In terms of the differential operators above, the dynamic equations can be placed in matrix form,

$$\frac{\partial}{\partial \tau}\begin{pmatrix}\tilde{\phi}\\ \tilde{\theta}\end{pmatrix} = \begin{pmatrix}0 & L_1\\ -L_2 & 0\end{pmatrix}\begin{pmatrix}\tilde{\phi}\\ \tilde{\theta}\end{pmatrix}. \tag{7.37}$$

Now one can suppose that the fluctuations are free to vibrate in any allowed way. One particular choice could be any eigenfunction of the operators L_1 and L_2. Suppose we have eigenvalues labeled as λ_1 and λ_2, respectively. Note that with $\partial/\partial\tau = -i\omega$, the square of this matrix operation is a diagonal operator,

$$\omega^2\begin{pmatrix}\tilde{\phi}\\ \tilde{\theta}\end{pmatrix} = \begin{pmatrix}L_1 L_2 & 0\\ 0 & L_2 L_1\end{pmatrix}\begin{pmatrix}\tilde{\phi}\\ \tilde{\theta}\end{pmatrix}. \tag{7.38}$$

Looked at this way, one sees that the eigenspectrum of either L_1 or L_2 is relevant to the soliton's stability. If either one of these operators has a negative eigenvalue in its spectrum, that could be excited and indicate an instability of the soliton. It indicates instability, because it would imply a negative value of ω^2. This type of analysis was performed independently by Kumar [7] and by Magyari and Thomas [8], from slightly different approaches.

The operator L_2 actually has a bound state of the form $\psi_2 = \text{sech } x$ with a zero eigenvalue, where $x = \sqrt{b}\,\xi$, as can be seen from differentiation:

$$\frac{\partial^2}{\partial x^2} \text{sech } x = \frac{\partial}{\partial x}(-\tanh x \text{ sech } x) = (1 - 2\text{ sech}^2 x)\text{sech } x. \tag{7.39}$$

Then

$$L_2 \cdot \text{sech } x = 0. \tag{7.40}$$

This function $\psi_2 = \text{sech } x$ is known as the translational mode of the soliton. It costs zero energy for it to change position, hence there is a mode in its spin wave spectrum that carries out this translation for zero energy cost. There is no instability implied by this mode. One can see that $\psi_2 = \text{sech } x$ is a function proportional to the space derivative of ϕ_{sG}, as would be expected for the translational mode.

For the operator L_1, there are two bound states. The first is

$$\psi_{1a} = \text{sech } x \tanh x, \qquad L_1 \cdot \psi_{1a} = \psi_{1a}. \tag{7.41}$$

The eigenvalue $\lambda_{1a} = 1$ is always positive, hence this does not imply any instability. There is also a second bound state:

$$\psi_{1b} = \text{sech}^2 x, \qquad L_1 \cdot \psi_{1b} = (1 - 3b)\psi_{1b}. \tag{7.42}$$

Its eigenvalue is $\lambda_{1b} = 1 - 3b$. This is importantly different. This eigenvalue can become negative, and destabilize the soliton, if

$$b > \frac{1}{3}. \tag{7.43}$$

In this range of magnetic field, there is a structural instability of the soliton. It can deform according to the shape of this mode, and then lower the total energy. What it implies, then, is that it will be impossible for a magnetic kink to maintain the sG profile, once the magnetic field exceeds this limiting value. This shows the outer limit of applicability of the sG theory for the magnetic chain.

7.2.1 The Liebmann *et al* ansatz

The dynamic effect of this soliton instability presents itself in the propagation properties of magnetic kinks. For the magnetic model on a lattice, the kinks do not propagate at the speed expected, based on the sG profile for a given velocity. Instead, the instability discussed above causes them to travel at slower speeds. Indeed, for larger values of b, a kink expected to travel with speed $+u$ can even end up propagating in the opposite direction! These effects can be seen in numerical simulations, using the sG soliton profile as an initial condition. There is also a way to see this effect using an ansatz due to Liebmann, Schöbinger and Hackenbracht [9], which is an assumed variational form of the solution, that may not be perfect, but

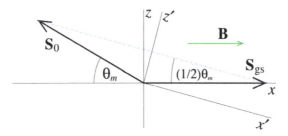

Figure 7.3. A depiction of the tilted plane (dashed blue line) where the spins rotate in a kink in the Liebmann *et al* ansatz, equation (7.48a), for parameter θ_m. The ground state spin direction $\mathbf{S}_{gs}$ is along x, parallel to $\mathbf{B}$, while the spin $\mathbf{S}_0$ at the kink center makes angle θ_m to the xy-plane, partially opposing the applied field.

carries the essential dynamic structure. An ansatz can be used to find the energy as a function of the assumed velocity. The procedure works by evaluating the Lagrangian or an effective energy functional of the system. It is a variational approach, such as that used in quantum mechanics to obtain upper bounds on the ground state energy of a system.

In the Liebmann *et al* ansatz [9] for a moving soliton, a certain structure of the spins **S** along the chain is assumed, which allows for them to tilt out of the xy-plane near the center of the kink, and then go back to the ground state direction along $\hat{x}$ at the ends of the chain. At the center of the kink, the spins tilt out-of-plane at some maximum angle θ_m, which acts as a variational parameter of the ansatz. The unit spins **s** are assumed to map out a circular path on the unit sphere, whose plane is tilted at an angle $\frac{1}{2}\theta_m$ relative to the xy-plane. This tilted circle has a central axis z' also tilted at the angle $\frac{1}{2}\theta_m$ away from $\hat{z}$ leaning towards the $\hat{x}$-axis, see figure 7.3. On the defined circular path, the spins rotate through an azimuthal angle given by the sG profile, but with a variational width parameter w:

$$\phi_L = \tan^{-1}\left\{\exp\left[(\xi - u\tau)/w\right]\right\}. \tag{7.44}$$

The unit spins can be expressed most easily in the tilted coordinate system $x'y'z'$, which is rotated from xyz by angle $\frac{1}{2}\theta_m$ around the y-axis, with the transformation:

$$s^x = s^{x'}\cos\frac{1}{2}\theta_m + s^{z'}\sin\frac{1}{2}\theta_m \tag{7.45a}$$

$$s^y = s^{y'} \tag{7.45b}$$

$$s^z = -s^{x'}\sin\frac{1}{2}\theta_m + s^{z'}\cos\frac{1}{2}\theta_m. \tag{7.45c}$$

The ansatz in the tilted coordinates corresponds to the spins confined on a circle of radius

$$c \equiv \cos\frac{1}{2}\theta_m \tag{7.46}$$

at a constant value of

$$z' = s \equiv \sin\frac{1}{2}\theta_m. \tag{7.47}$$

Then the ansatz is expressed as

$$s^{x'} = c \cos \phi_L \tag{7.48a}$$

$$s^{y'} = c \sin \phi_L \tag{7.48b}$$

$$s^{z'} = s. \tag{7.48c}$$

By design, the spin length $c^2 + s^2 = 1$ is conserved to unity, and it contains the essential geometric structure of a kink. The transformation back to the original coordinates gives

$$s^x = c^2 \cos \phi_L + s^2 \tag{7.49a}$$

$$s^y = c \sin \phi_L \tag{7.49b}$$

$$s^z = cs(1 - \cos \phi_L). \tag{7.49c}$$

The parameters θ_m and w should be chosen to make the Lagrangian of the system an extremum, because this is a dynamic problem (the velocity not equal to zero implies a kinetic term). The Lagrangian is obtained from the Hamiltonian by including the sum over the kinetic term $\dot\phi S^z$ for each site:

$$L[S] = K[S] - H[S], \tag{7.50}$$

where the kinetic integral (recall that ϕ_n and S_n^z are conjugate variables) is

$$K[S] = \sum_n \dot\phi_n S_n^z \to S \int \frac{dz}{a} \dot\phi \sin\theta. \tag{7.51}$$

Transforming to the coordinates ξ and τ, this is seen to be

$$K[S] = \epsilon_0 \int d\xi \, \phi_\tau \sin\theta. \tag{7.52}$$

In these coordinates, the evaluation is complicated. However, note that the time derivative is equivalent to $-u$ times the space derivative, so this is

$$K[S] = -u\epsilon_0 \int d\xi \, \frac{d\phi}{d\xi} s^z. \tag{7.53}$$

There is no problem to evaluate this in the primed coordinates, where $s^{z'} = s =$ constant. Then it becomes a perfect differential, and the result is very simple:

$$K[S] = -u\epsilon_0 \, 2\pi s = uP. \tag{7.54}$$

It is interesting to note that this is proportional to the momentum for the spin field,

$$P = -\int d\xi \, \frac{d\phi}{d\xi} s^z. \tag{7.55}$$

One needs also to calculate the energy H for the ansatz, using the continuum magnetic expression (7.28). However, it will again be simplest to evaluate it in the primed spin components. The result found by Liebmann *et al* [9] is

$$H[\mathbf{S}] = 4\epsilon_0 c^2 \left[\frac{1}{w} + \left(b + \frac{2}{3} s^2 \right) w \right]. \tag{7.56}$$

Then it is possible to find the extremum of $L = K - H$ with respect to independent variations in the width w, still letting c and s be free parameters determined by θ_m.

$$\frac{\partial L}{\partial w} = -4\epsilon_0 c^2 \left[-\frac{1}{w^2} + b + \frac{2}{3} s^2 \right] = 0. \tag{7.57}$$

This determines the width as

$$w(\theta_m) = \left[b + \frac{2}{3} \sin^2\left(\frac{\theta_m}{2} \right) \right]^{-1/2}. \tag{7.58}$$

The energy $E = H$ becomes

$$E(\theta_m) = 8\epsilon_0 c^2 w^{-1}. \tag{7.59}$$

Then also the Lagrangian should be extremized independently with respect to the angle θ_m. One has some algebra,

$$\frac{\partial L}{\partial \theta_m} = -\pi u c + 4cs \left(\frac{1}{w} + bw \right) - \frac{4w}{3}(2cs)(c^2 - s^2) = 0. \tag{7.60}$$

A simplification and use of the result for w leads to the kink velocity:

$$u(\theta_m) = \frac{8s}{\pi w} \left(1 - \frac{1}{3} c^2 w^2 \right). \tag{7.61}$$

Note that small positive values of velocity u correspond to negative values of θ_m when the field is weak, see figure 7.4 for typical behaviors of $u(\theta_m)$ and $w(\theta_m)$. These results for w, E and u all use θ_m as a parameter. Elimination of θ_m between the relations then can lead to, most importantly, the energy as a function of velocity, $E(u)$, with w also eliminated. That is plotted in figure 7.5, for $b = 0.04, 0.16, 0.33$. These curves are somewhat surprising, in that there can be two values of energy for a given velocity. Note that small positive u here maps to small negative θ_m, at weak field b.

Looking at the relation $u(\theta_m)$, the quadratic dependence on w on the RHS indicates that there are two values of θ_m for a chosen velocity. This is a significant finding. The ansatz correctly predicts the presence of two different types of forward-moving magnetic kinks, or two energy branches. Typically for a chosen velocity $u > 0$, there is a lower energy solution with a smaller $|\theta_m|$ and a higher energy

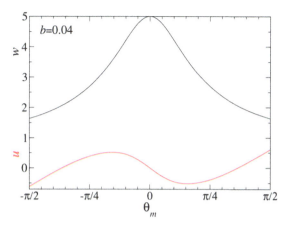

Figure 7.4. Magnetic kink width w, upper curve, equation (7.58), and velocity u, lower curve, equation (7.61), as functions of the tilting angle θ_m in the Liebmann *et al* ansatz at field strength $b = 0.04$.

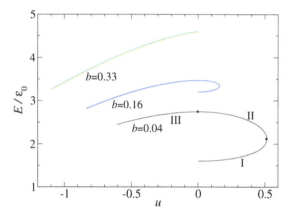

Figure 7.5. Magnetic kink energy–velocity dispersion relations from the Liebmann *et al* ansatz for indicated magnetic fields. Branch I is forward motion and lowest energy, branch II is forward motion at the higher energy and branch III is backwards motion. These curves were generated using θ_m from 0 to $-\pi/2$.

solution with a larger value of $|\theta_m|$. These have been called branch I and branch II types of kinks. They both move to the right (negative values of θ_m were used to make figure 7.5), at the same speed, but the energies and structures differ. The branch II kink has both a higher energy and a larger magnitude of out-of-plane tilting $|\theta_m|$ at the kink center. It might be considered a kink with an extra amount of deformation.

The branch II kink even exists with non-zero θ_m but zero velocity. From the expression (7.61), the velocity is zero when $s = 0 = \theta_m$ (branch I static kink) or when $cw = \sqrt{3}$ (branch II static kink). Using this in the result for $w(\theta_m)$, the branch II static kink has an out-of-plane angle given by

$$s^2 = 1/3 - b. \tag{7.62}$$

This shows the existence of branch II only up to magnetic field strength $b = 1/3$. Beyond that, there is only branch III, which is a set of backwards moving kinks.

They are backwards, because the direction of motion is opposite that expected for the sign of θ_m. In figure 7.5, we used negative values of θ_m that would correspond to positive velocity for sG type kinks, see equation (7.26). Furthermore, the branch III kinks have negative mass! They reduce their energy by moving faster. All of this curious behavior can be attributed to the magnetic soliton instability, that also comes into full force at $b = 1/3$. The $E(u)$ behavior has been verified by numerical simulations of kink dynamics. Thus, a structural instability in a magnetic kink must be taken into account for its significant modification of the excitation dynamics.

7.3 π-kinks in ferromagnetic chains

An easy-plane FM chain with no applied magnetic field supports a kind of pulse soliton, as found by Mikeska [10], where the spins rotate only through π. If in addition, there is a weak uniaxial (Ising-like) local anisotropy within the easy plane, the π pulse soliton, found by Sklyanin [11], is still present. It is important to study these for their intrinsic value, as well as the fact that they relate closely to the nearly π solitons that appear in chains with AFM interactions. Their structure is also reminiscent of the Liebmann *et al* ansatz.

7.3.1 Easy-plane ferromagnetic chain

We consider the easy-plane FM chain with continuum Hamiltonian (7.28), with field $b = 0$, and the corresponding equations of motion (7.11) can be written

$$\phi_\tau \cos\theta = -\theta_{\xi\xi} + \left(1 - \phi_\xi^2\right)\sin\theta\cos\theta \qquad (7.63a)$$

$$\theta_\tau \cos\theta = \phi_{\xi\xi}\cos^2\theta - 2\theta_\xi\phi_\xi\sin\theta\cos\theta \qquad (7.63b)$$

The RHS of (7.63b) is a perfect differential, see the related derivation of (2.149b), so that assuming a solution depending on $(\xi - u\tau)$, one has

$$-u\frac{\mathrm{d}}{\mathrm{d}\xi}\sin\theta = \frac{\mathrm{d}}{\mathrm{d}\xi}\left(\phi_\xi\cos^2\theta\right). \qquad (7.64)$$

We take a boundary condition that spins are in the xy-plane at $\xi \to \pm\infty$, or equivalently, $\theta(\pm\infty) = 0$, as well as a uniform state where the space derivatives are zero at the far ends of the chain. Then this is integrated and no boundary constant appears,

$$\phi_\xi = -\frac{u\sin\theta}{\cos^2\theta}. \qquad (7.65)$$

This result is used to eliminate ϕ. Consider this in the following exercises which produce the pulse soliton solutions.

Exercise 7.1. a) Use the result (7.65) found for ϕ_ξ and apply it in the out-of-plane dynamic equation (7.63b), showing that the differential equation governing θ becomes

$$\theta_{\xi\xi} = \sin\theta\cos\theta - u^2\frac{\sin\theta}{\cos^3\theta}. \tag{7.66}$$

b) Bring the LHS to an exact differential, and integrating from $\xi = -\infty$ to an arbitrary point on the chain, using the boundary condition $\theta(-\infty) = 0$, show that the first integral is

$$\theta_\xi^2 = \tan^2\theta\left(1 - u^2 - \sin^2\theta\right). \tag{7.67}$$

Note how this result places a velocity-dependent restriction on the out-of-plane angle, because the RHS must remain positive for all ξ. A faster pulse soliton must have a smaller magnitude out-of-plane angle.

Exercise 7.2. The result in the previous exercise, (7.67), can be integrated to obtain the pulse soliton structure. Define a relativistic factor, $\gamma = (1 - u^2)^{-1/2}$, and show that the pulse soliton solutions are of the form

$$\sin\theta = -q\gamma^{-1}\,\text{sech}\left[\gamma^{-1}(\xi - u\tau)\right] \tag{7.68a}$$

$$\phi = \phi_0 + q\tan^{-1}\left\{u^{-1}\sinh\left[\gamma^{-1}(\xi - u\tau)\right]\right\}. \tag{7.68b}$$

The soliton/anti-soliton solutions correspond to $q = +1$, $q = -1$, respectively, from the two roots of (7.67). The constant ϕ_0 is an arbitrary in-plane angle at the pulse center. Although ϕ has a singular behavior when $u \to 0$, the Cartesian spin components are well behaved in that limit.

The pulse soliton solution derived in these exercises has unusual but very interesting behavior and structure. Setting the arbitrary azimuthal constant $\phi_0 = 0$, the Cartesian spin components of unit spins are found to be

$$s^x = u\,\text{sech}\left[\gamma^{-1}(\xi - u\tau)\right] \tag{7.69a}$$

$$s^y = q\tanh\left[\gamma^{-1}(\xi - u\tau)\right] \tag{7.69b}$$

$$s^z = -q\gamma^{-1}\,\text{sech}\left[\gamma^{-1}(\xi - u\tau)\right]. \tag{7.69c}$$

One can see in the limit $u = 0$, the spins rotate only in the yz-plane. This is the static limit of the pulse soliton. For non-zero u, there will be a lesser maximum out-of-plane angle θ_m, measured at the pulse center, where $s^y(0) = 0$. One has the relations at the pulse center:

$$s^x(0) = u = \cos\theta_m, \qquad s^z(0) = -q\gamma^{-1} = \sin\theta_m. \tag{7.70}$$

Thus, the faster pulses have smaller magnitude out-of-plane angles.

Consider the structure in a spin frame $x'y'z'$, obtained from xyz by rotating through θ_m around the y-axis. The transformation of coordinates needed is

$$s^{x'} = s^x \cos\theta_m + s^z \sin\theta_m \quad = \text{sech}\left[\gamma^{-1}(\xi - u\tau)\right] \quad (7.71a)$$

$$s^{y'} = s^y \quad = q \tanh\left[\gamma^{-1}(\xi - u\tau)\right] \quad (7.71b)$$

$$s^{z'} = -s^x \sin\theta_m + s^z \cos\theta_m \quad = 0. \quad (7.71c)$$

This then shows a sG structure within this tilted plane, just as in the FM Liebmann ansatz, but with only a π rotation:

$$\phi' = \tan^{-1}\left(\frac{s^{y'}}{s^{x'}}\right) = \frac{\pi}{2} + 2q\tan^{-1}\left\{\exp\left[\gamma^{-1}(\xi - u\tau)\right]\right\}. \quad (7.72)$$

Evaluation of the Hamiltonian (7.28) for a pulse soliton leads to the unusual result,

$$E_{\text{pulse}} = 2\epsilon_0(1 - u^2)^{1/2}. \quad (7.73)$$

The total energy *decreases* with increasing speed. Using $u = \cos\theta_m$, this means a pulse soliton with small tilt angle has a high speed but a low energy. Conversely, one with a larger tile angle is moving slowly and has high energy. This is a rather strange behavior, which might be partly explained if one investigated the momentum P of this excitation, that can be defined from the kinetic integral $K = uP$, see the following section.

Note that these results can be obtained from an ansatz very similar to the Liebmann *et al* ansatz, using the obvious spin rotation in the plane tilted at angle θ_m to the xy-plane.

7.3.2 Easy-plane ferromagnetic chain with in-plane Ising symmetry

Referring to the spin rotation in a pulse soliton, as sketched in figure 7.6, one could imagine that this type of solution will not be strongly modified in the presence of a uniaxial interaction within the xy-plane. Remember, the sketch was made under the assumption of an integration constant ϕ_0 in (7.68) being set to zero. With some other

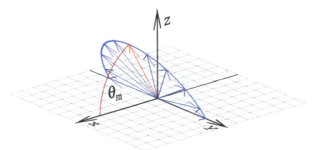

Figure 7.6. Sketch of the spin rotation in a π pulse soliton in an easy-plane FM chain, equation (7.69a), with out-of-plane tilting $\theta_m = 60°$, velocity $u = \cos\theta_m$ and charge $q = -1$.

choice of ϕ_0 the solution would be rotated around the z-axis. A particular choice of ϕ_0 can be forced by the introduction of a uniaxial (Ising-like) term in the Hamiltonian. An interaction that forces $\phi_0 = 0$ solutions, with spins along $\pm\hat{y}$ at the far ends of the chain, can be added to the discrete Hamiltonian,

$$H_D = -D \sum_n \left(S_n^y\right)^2. \tag{7.74}$$

The total continuum limit Hamiltonian becomes

$$H = \epsilon_0 \int d\xi \left\{ \frac{1}{2}\left(\theta_\xi^2 + \phi_\xi^2 \cos^2 \theta\right) + \frac{1}{2}\sin^2 \theta - \frac{1}{2}d \cos^2 \theta \sin^2 \phi \right\} \tag{7.75}$$

where $d \equiv D/K_A$ is the uniaxial interaction divided by the easy-plane interaction strength (both are positive numbers). Recall that K_A has been absorbed into the length and energy units.

One can then seek the pulse soliton solutions from the new equations of motion. Another method is via an ansatz approach that reflects the structure in figure 7.6. While an ansatz method is normally expected to be approximate, in this special case it succeeds to give the exact pulse soliton solution, because the extra uniaxial term does not modify the geometry or symmetry of the exact solution found in its absence.

The ansatz is rather simple. We assume a sG variation of the spins, with width w, in a plane (coordinates $x'y'z'$) tilted at angle θ_m to the xy-plane. A rotation around $\hat{y}$ is used. In this tilted plane, the assumed structure is like that in (7.71),

$$s^{x'} = \sin \phi_{sG} = \text{sech}[(\xi - u\tau)/w] \tag{7.76a}$$

$$s^{y'} = \cos \phi_{sG} = q \tanh[(\xi - u\tau)/w] \tag{7.76b}$$

$$s^{z'} = 0. \tag{7.76c}$$

The azimuthal angle used here, ϕ_{sG}, and hyperbolic functions, are based on (7.72) for ϕ'. This can then be rotated back to the original xyz-coordinates, performing the inverse of the transform in (7.71):

$$s^x = s^{x'} \cos \theta_m - s^{z'} \sin \theta_m \;\; = \cos \theta_m \text{ sech}[(\xi - u\tau)/w] \tag{7.77a}$$

$$s^y = s^{y'} \;\; = q \tanh[(\xi - u\tau)/w] \tag{7.77b}$$

$$s^z = s^{x'} \sin \theta_m + s^{z'} \cos \theta_m \;\; = \sin \theta_m \text{ sech}[(\xi - u\tau)/w]. \tag{7.77c}$$

This is the ansatz; note it is identical in structure to the pulse soliton found above. The system Lagrangian must be made extremum with respect to the choice of w and θ_m, to complete the solution.

For the Lagrangian $L = K - H$, one needs the kinetic integral, which is

$$K = \epsilon_0 \int d\xi \, s^z \, \phi_\tau = -\epsilon_0 u \int d\xi \, s^z \, \phi_\xi. \tag{7.78}$$

The next steps of the calculation are carried out in the following exercises.

Exercise 7.3. Show that the kinetic integral (7.78) for the pulse soliton ansatz (7.77) is

$$K = -2qu\theta_m\epsilon_0. \tag{7.79}$$

Note that with $K = uP$, this also determines the canonical soliton momentum, $P = -2q\theta_m\epsilon_0$.

Exercise 7.4. Show that the Hamiltonian (7.75) for the pulse soliton ansatz (7.77) is

$$H = \left[w(\sin^2\theta_m + 2d) + w^{-1}\right]\epsilon_0. \tag{7.80}$$

With the results for K and E, the Lagrangian is $L(w, \theta_m) = K - H$, and it is simple to make it stationary with respect to independent variations in width w and angle θ_m. The results describe the pulse soliton with the addition of the uniaxial term, using θ_m as the controlling parameter:

$$w = \left(\sin^2\theta_m + 2d\right)^{-1/2} \tag{7.81a}$$

$$E = 2\epsilon_0/w \tag{7.81b}$$

$$qu = w\sin\theta_m\cos\theta_m. \tag{7.81c}$$

One can verify that in the limit of in-plane symmetry, $d = 0$, this indeed recovers the correct pulse soliton of (7.69) and correct energy (7.73). For anisotropy $d = 0.01$,

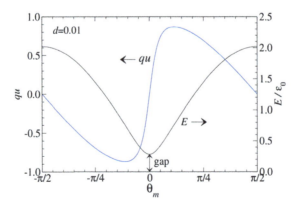

Figure 7.7. Kink velocity $u(\theta_m)$ and energy $E(\theta_m)$ (see (7.81)) for π-kinks in the presence of uniaxial anisotropy $d = 0.01$ within the easy plane from the ansatz. This leads to double-valued $E(u)$ such as those in figure 7.8. Note also the maximum speed below 1 and the energy gap at $\theta_m = 0$.

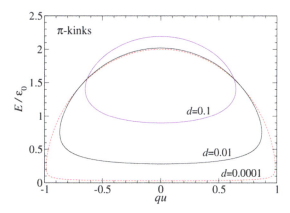

Figure 7.8. $E(u)$ dispersion relations, equation (7.81), for π-kinks in the presence of uniaxial anisotropy d within the easy plane. The maximum energy corresponds to tilting angle $\theta_m = \pm\pi/2$; for the minimum energy the tilt is $\theta_m = 0$.

graphs of $qu(\theta_m)$ and $E(\theta_m)$ are shown in figure 7.7. The presence of q implies inversion of the results for anti-kinks compared to kinks. It can be noticed that for a given choice of qu, there are two possible values of tilt angle θ_m, which then implies two values of E.

There are several interesting consequences of having non-zero symmetry breaking parameter d. One effect is to shift the minimum soliton energy, opening a gap in the spectrum for $\theta_m = 0$. Note the strange behavior of the $E(u)$ dispersion relation, as seen in some examples in figure 7.8. As determined from the behavior in figure 7.7, the energy $E(u)$ exhibits an upper energy branch and a lower energy branch. Also, one sees that the top speed of the excitations is reduced below unity. In addition to these effects, at $\theta_m = 0$ the soliton will now have a finite width given by $w_{\min} = (2d)^{-1/2}$, rather than a divergent width. In the limit that $d \to 0$, only the top branch of the energy dispersion remains; the lower branch becomes of infinitesimal weight (the range of θ_m corresponding to that branch will become insignificant). Thus we find the magnetic π-kinks possess a number of curious properties not usually seen in typical particle-like excitations.

7.4 Magnetic kinks in antiferromagnetic chains

1D easy-plane AFM chains offer some similarities to FM chains and, at the same time, there are considerable differences. With an applied magnetic field perpendicular to the chain, they support magnetic kink excitations, of two different types, depending on the plane in which the spins tend to move. These are referred to as XY kinks and YZ kinks. They could equivalently be called in-plane kinks and out-of-plane kinks, where the easy plane is the xy-plane. There is also a critical field at which the static energies of these different kinks becomes equal. This is rather different than the generation of backwards moving kinks in the FM system. The analysis is again carried out in a continuum limit. Due to the need for two sublattices, that derivation is quite different from the mathematics for the FM model.

The Hamiltonian for a chain of spins is similar to that for the FM model, but with the opposite sign on the exchange terms (J is positive here):

$$H = \sum_n \left[J\mathbf{S}_n \cdot \mathbf{S}_{n+1} + K_A S_n^{z\,2} - g\mu_B B^x S_n^x \right]. \tag{7.82}$$

A compound for which this Hamiltonian applies in a low temperature regime (below 20 K) is $(CH_3)_4NMnCl_3$ (tetramethyl ammonium manganese chloride), also known as TMMC. The Mn^{2+} ions have spin $S = 5/2$, which helps considerably in the applicability of classical spin mechanics. The structure of the compound leads to quasi-1D coupling behavior. The energy scales are $JS^2 = 85$ K, $K_A S^2 = 2$ K and $g = 2.0$.

Our intention is to start from an appropriate ground state, and discuss then the structure and properties of the linearized spin wave excitations and the nonlinear kink excitations. Again, the kink structure is approximately described by sG solitons. Their stability in the full magnetic model can be analyzed as was done for FM kinks. Finally, an ansatz can be made to describe both the XY and YZ kinks as being members of a single energy–velocity dispersion curve.

7.4.1 Antiferromagnetic chain ground state

The ground state of an AFM chain is important for determination of spin wave dispersion. Its energy is important as a reference energy from which to measure the energy of the kinks. The presence of an applied magnetic field affects the ground state. The analysis needs to assume two sublattices: odd sites along the chain comprise the A-sublattice and even sites form the B-sublattice. In a ground state, all sites on A have spin vector $\mathbf{S}_A$, and those on B have spins $\mathbf{S}_B$. Due to the presence of applied field B^x, both sublattices' spins tilt towards the x-direction, however, that effect competes with exchange that tries to keep neighboring sites anti-parallel. This results in the two sublattices' spins being nearly perpendicular to $\mathbf{B}$. A way to make an energy minimizing structure, with spins in the xy-plane, is

$$\mathbf{S}_A = S(\sin\phi_0, +\cos\phi_0, 0) \tag{7.83a}$$

$$\mathbf{S}_B = S(\sin\phi_0, -\cos\phi_0, 0), \tag{7.83b}$$

where ϕ_0 is a constant angle to be determined. The energy per spin for this configuration is

$$e = E/N = JS^2\left(\sin^2\phi_0 - \cos^2\phi_0\right) - g\mu_B B^x S \sin\phi_0. \tag{7.84}$$

This is minimized by making it stationary with respect to variations in ϕ_0,

$$\frac{de}{d\phi_0} = 4JS^2 \sin\phi_0 \cos\phi_0 - g\mu_B B^x S \cos\phi_0 = 0. \tag{7.85}$$

There are two degenerate solutions,

$$\sin\phi_0 = \frac{g\mu_B B^x}{4JS} \equiv \frac{1}{4}\beta, \quad \cos\phi_0 = \pm\sqrt{1 - \frac{1}{16}\beta^2}. \tag{7.86}$$

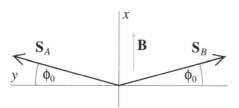

Figure 7.9. The sublattice spin directions in the spin-flop ground state (7.83) for an AFM easy-plane chain in a magnetic field. The canting angle ϕ_0 is given in (7.86).

In terms of the usual in-plane angles ϕ_n to the x-axis, these states are

$$\phi_n = \begin{cases} \pm\left(-\dfrac{\pi}{2} + \phi_0\right) & n \in A \\ \pm\left(+\dfrac{\pi}{2} - \phi_0\right) & n \in B. \end{cases} \quad (7.87)$$

These are called spin-flop states, for the canted structure of the two sublattices, see figure 7.9. The angle ϕ_0 is a small tilting of spins away from the $\pm\hat{y}$-axes on the two sublattices. Because the spins lie in the easy plane, this is the overall ground state. It is two-fold degenerate, considering that the A and B spin directions can be interchanged. The two ground state unit spin directions can be summarized as

$$\mathbf{g}_\pm = (\sin\phi_0, \pm\cos\phi_0, 0). \quad (7.88)$$

Then the two ground states are

$$1.\ \mathbf{S}_n = \begin{cases} S\mathbf{g}_+ & n \in A \\ S\mathbf{g}_- & n \in B, \end{cases} \quad 2.\ \mathbf{S}_n = \begin{cases} S\mathbf{g}_- & n \in A \\ S\mathbf{g}_+ & n \in B. \end{cases} \quad (7.89)$$

The energy per site for either of these is found to be

$$e_0 = -JS^2\left(1 + \frac{1}{8}\beta^2\right). \quad (7.90)$$

It is interesting to look also at another local minimum state, where the spins point out of the easy plane:

$$\mathbf{S}_A = S(\sin\theta, 0, +\cos\theta) \quad (7.91a)$$

$$\mathbf{S}_B = S(\sin\theta, 0, -\cos\theta). \quad (7.91b)$$

The energy per site is now

$$e = JS^2(\sin^2\theta - \cos^2\theta) + K_A S^2 \cos^2\theta - g\mu_B B^x S \sin\theta. \quad (7.92)$$

To become stationary we need

$$\frac{de}{d\theta} = 4JS^2 \sin\theta\cos\theta - 2K_A S^2 \sin\theta\cos\theta - g\mu_B B^x S \cos\theta = 0. \quad (7.93)$$

With anisotropy parameter $\alpha = 2K_A/J$, the solutions are again two-fold degenerate,

$$\sin\theta_1 = \frac{g\mu_B B^x}{(4J - 2K_A)S} \equiv \frac{\beta}{4-\alpha}, \qquad \cos\theta_1 = \pm\sqrt{1 - \frac{\beta^2}{(4-\alpha)^2}}. \qquad (7.94)$$

The energy of these states is

$$e_1 = -JS^2\left(1 + \frac{\beta^2}{2(4-\alpha)} - \frac{\alpha}{2}\right). \qquad (7.95)$$

A little inspection shows that this is higher energy than the spin-flop state, e_0. It is, however, a local minimum.

7.4.2 Spin waves

Starting from the spin-flop ground state, we can obtain the spin wave spectrum by assuming small perturbations ($\tilde{\phi}_n, \tilde{\theta}_n$) about that state. If ($\phi_n, \theta_n$) are planar spherical angles, then the fluctuating state is described by

$$\phi_n = \begin{cases} -\dfrac{\pi}{2} + \phi_0 + \tilde{\phi}_n & n \in A \\ +\dfrac{\pi}{2} - \phi_0 + \tilde{\phi}_n & n \in B \end{cases} \qquad (7.96a)$$

$$\theta_n = \tilde{\theta}_n, \quad \forall\, n. \qquad (7.96b)$$

Then, the discrete equations of motion are like those for the FM chain, equation (7.4), but with the sign of J reversed. After being linearized, and using $\sin\phi_0 = \beta/4$, they become

$$(JS)^{-1}\dot{\tilde{\phi}}_n = \tilde{\theta}_{n-1} + \tilde{\theta}_{n+1} + (2+\alpha)\tilde{\theta}_n \qquad (7.97a)$$

$$(JS)^{-1}\dot{\tilde{\theta}}_n = \left(1 - \frac{1}{8}\beta^2\right)(\tilde{\phi}_{n-1} + \tilde{\phi}_{n+1}) - 2\tilde{\phi}_n. \qquad (7.97b)$$

By assuming traveling waves varying as $\exp\{i(kna - \omega t)\}$ as in chapter 6, with wave vector k, the dispersion relation is easily found to be

$$\omega^2 = (2JS)^2\left[1 - \left(1 - \frac{1}{8}\beta^2\right)\cos ka\right]\left(1 + \frac{1}{2}\alpha + \cos ka\right). \qquad (7.98)$$

This dispersion relation is plotted for different values of scaled magnetic field strength β in figure 7.10. Although there is only one branch, the limits near $ka = 0$ and $ka = \pi$ (zone boundaries) are of interest, because they correspond to different manners of oscillation. Both limits have a quadratic dependence,

$$\omega^2 \approx \begin{cases} (2JS)^2\left[\dfrac{1}{4}\beta^2 + (ka)^2\right], & \text{near } ka = 0 \\ (2JS)^2\left[\alpha + (\pi - ka)^2\right], & \text{near } ka = \pi. \end{cases} \qquad (7.99)$$

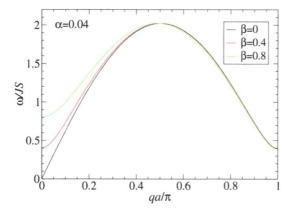

Figure 7.10. The AFM chain spin wave dispersion relation (7.98) for anisotropy $\alpha = 0.04$ and magnetic fields below ($\beta = 0$), equal to ($\beta = 0.4$) and above ($\beta = 0.8$) the critical field $\beta_c = 2\sqrt{\alpha}$.

Note that near $ka = 0$, the frequency is close to $\omega \approx JS\beta$ and the relative amplitudes of in-plane and out-of-plane motion are related by

$$-i\beta\tilde{\phi} \approx (4 + \alpha)\tilde{\theta}. \quad (7.100)$$

The field β is weak as is the anisotropy parameter α, so in this part of the dispersion curve the in-plane amplitude is much stronger, $\tilde{\phi} \gg \tilde{\theta}$, showing that this is predominantly an in-plane oscillation. At the other zone boundary ka near π, the frequency is close to $\omega \approx 2JS\sqrt{\alpha}$ and the relative amplitudes have the relation

$$-2i\tilde{\phi} \approx \sqrt{\alpha}\tilde{\theta}. \quad (7.101)$$

With α a small parameter, now the out-of-plane component has the greater amplitude, $\tilde{\theta}$. These two limiting modes of oscillation obtain the same frequency when $\alpha = \frac{1}{4}\beta^2$, which relation defines a critical magnetic field:

$$\beta_c = 2\sqrt{\alpha}, \quad \text{or} \quad g\mu_B B_c = \sqrt{8K_A JS^2}. \quad (7.102)$$

This quick analysis of the spin waves has implications for the nonlinear kinks in the system. It turns out that also the XY and YZ kinks acquire equal static energies at this same critical field.

7.4.3 Continuum limit dynamics in an antiferromagnetic chain

A continuum limit for the 1D AFM chain can be obtained as was done by Flüggen and Mikeska [12] using four angles in spherical polar coordinates, with slightly different expression on the two sublattices:

$$\mathbf{S}_{A,B} = \pm S\bigl(\sin(\Theta \pm \theta)\cos(\Phi \pm \phi),\ \sin(\Theta \pm \theta)\sin(\Phi \pm \phi),\ \cos(\Theta \pm \theta)\bigr). \quad (7.103)$$

The + signs give the A-sublattice and the − signs give the B-sublattice. The upper case Θ, Φ will be considered the large angles, while lower case θ, ϕ are smaller angles. Slow spatial variations in Θ, Φ are assumed, and at the same time, $\theta \ll 1$, $\phi \ll 1$.

Let time t be measured in units of $(JS)^{-1}$ and length z in units of the lattice constant a, defining dimensionless variables for the AFM problem,

$$\xi \equiv z/a, \qquad \tau \equiv JS\, t. \tag{7.104}$$

Then keeping the leading terms in the equations of motion leads to

$$\Theta_\tau = 4\phi \sin\theta + \beta \sin\Phi \tag{7.105a}$$

$$\Phi_\tau = -4\theta \csc\Theta - \alpha\theta \sin\Theta + \beta \cot\Theta \cos\Phi \tag{7.105b}$$

$$\theta_\tau = -(4\theta\phi + 2\Theta_\xi\Phi_\xi)\cos\Theta - \Phi_{\xi\xi}\sin\Theta + \beta\phi\cos\Phi \tag{7.105c}$$

$$\phi_\tau = \left(4\theta^2 \csc^2\Theta - 4\phi^2 - \Phi_\xi^2\right)\cos\Theta + \Theta_{\xi\xi}\csc\Theta + \alpha\left(1 - \frac{1}{2}\theta^2\right)\cos\Theta$$
$$- \beta\left(\phi \cot\Theta \sin\Phi + \theta \csc^2\Theta \cos\Phi\right). \tag{7.105d}$$

Assuming weak easy-plane anisotropy with $\alpha \ll 4$, the small angles can be eliminated, first using the equations for Θ_τ and Φ_τ,

$$\theta = \frac{1}{4}\left(\beta \cos\Theta \cos\Phi - \Phi_\tau \sin\Theta\right) \tag{7.106a}$$

$$\phi = \frac{1}{4}\csc\Theta\left(\Theta_\tau - \beta \sin\Phi\right). \tag{7.106b}$$

Then these can be used in the other two equations, leaving only the large angles, with wave-like operators acting,

$$\Phi_{\xi\xi} - \frac{1}{4}\Phi_{\tau\tau} + 2\left(\Theta_\xi\Phi_\xi - \frac{1}{4}\Theta_\tau\Phi_\tau\right)\cot\Theta = -\frac{1}{4}\beta^2 \sin\Phi \cos\Phi + \frac{1}{2}\beta\Theta_\tau \cos\Phi \tag{7.107a}$$

$$\Theta_{\xi\xi} - \frac{1}{4}\Theta_{\tau\tau} - \left(\Phi_\xi^2 - \frac{1}{4}\Phi_\tau^2\right)\frac{1}{2}\sin 2\Theta = \left(\frac{1}{4}\beta^2 \cos^2\Phi - \alpha\right)\frac{1}{2}\sin 2\Theta$$
$$- \frac{1}{2}\beta\Phi_\tau \cos\Phi \sin^2\Theta. \tag{7.107b}$$

7.4.4 XY antiferromagnetic kinks

Now consider the structure of an XY kink. The spins in this excitation tend to stay near the xy-plane but rotate through π, starting from a ground state and rotating back through the xy-plane to a ground state. Statically, (7.107b) has an exact solution with constant $\Theta = \pi/2$. Then, (7.107a) becomes a static sG equation for the larger in-plane angle, parameterized by a new variable,

$$\Psi = 2\Phi \mp \pi. \tag{7.108}$$

Either choice of sign on π is valid, which leads to multiple solutions. The resulting static sG equation is

$$\Psi_{\xi\xi} = \frac{1}{4}\beta^2 \sin\Psi. \tag{7.109}$$

For a dynamic sG equation, Mikeska noticed that small deviations of Θ away from $\pi/2$ should be allowed, by writing

$$\Theta = \frac{\pi}{2} - \Theta_s \qquad (7.110)$$

where $\Theta_s \ll 1$ is assumed. Then including dynamics, the in-plane angle is now governed by

$$\Psi_{\xi\xi} - \frac{1}{4}\Psi_{\tau\tau} = \frac{1}{4}\beta^2 \sin\Psi, \qquad (7.111)$$

With $\frac{1}{2}\sin 2\Theta = \sin\Theta_s \approx \Theta_s$, the small out-of-plane variations are determined by

$$\left(\Theta_{s,zz} - \frac{1}{4}\ddot{\Theta}_s\right) + \left(\Phi_z^2 - \frac{1}{4}\dot{\Phi}^2 + \frac{1}{4}\beta^2\cos^2\Phi - \alpha\right)\Theta_s = \frac{1}{2}\beta\dot{\Phi}\cos\Phi. \qquad (7.112)$$

Comparing with the soliton solution found for the FM chain, we can see here that an XY kink ($q = +1$) or anti-kink ($q = -1$) moving at dimensionless velocity u is

$$\Psi_{xy} = 4q\tan^{-1}\exp\left\{\frac{1}{2}\gamma\beta(\xi - u\tau)\right\}, \qquad \gamma = \left(1 - \frac{1}{4}u^2\right)^{-1/2}. \qquad (7.113)$$

One can see here the limiting speed of spin waves is $c = 2$ for these units, which acts as the limiting soliton velocity[2]. Transformed to the original in-plane angle $\Phi = (\Psi \pm \pi)/2$, this gives

$$\Phi_{xy} = \pm\frac{\pi}{2} + 2q\tan^{-1}\left\{\exp\left[\frac{1}{2}\gamma\beta(\xi - u\tau)\right]\right\}. \qquad (7.114)$$

Note that this $\pm\pi$ rotation of the spins takes them from one spin-flop ground state to the other degenerate spin-flop ground state (sublattices A and B interchanged), see figure 7.11. There are two types of kinks and two types of anti-kinks, according to the sign on the leading factor of $\frac{\pi}{2}$. Those choices are also restricted by the boundary conditions. At low magnetic field $\beta^2 \ll 4\alpha$, the associated out-of-plane variations are given approximately by

$$\Theta_{s,xy} \approx -\frac{\beta^2}{4\alpha}\gamma qu\ \mathrm{sech}^2\left[\frac{1}{2}\gamma\beta(\xi - u\tau)\right]. \qquad (7.115)$$

This is somewhat similar to the out-of-plane structure for the FM solitons. Note the dependence on the product qu.

For the XY kink energy, one should evaluate it using the AFM continuum limit Hamiltonian. Using the same approach as that for finding the equations of motion, with $\beta = g\mu_B B^x/JS$, $\alpha = 2K_A/J$, that Hamiltonian is found to be

$$H = \frac{1}{2}JS^2 \int d\xi \Big\{\Theta_\xi^2 + 4\dot{\theta}^2 + \left(\Phi_\xi^2 + 4\dot{\phi}^2\right)\sin^2\Theta \\
+ \alpha\left(\cos^2\Theta - \dot{\theta}^2\cos 2\Theta\right) - 2\beta\left(\dot{\theta}\cos\Theta\cos\Phi - \dot{\phi}\sin\Theta\sin\Phi\right)\Big\}. \qquad (7.116)$$

---

[2] The limiting speed is $c = 2JSa$ in physical units.

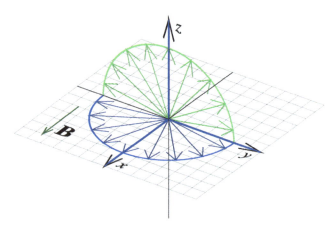

Figure 7.11. Sketch indicative of the rotation of the spins on one sublattice in AFM kinks. The ground state directions are near $\mathbf{S} \approx \pm S\hat{\mathbf{y}}$, almost perpendicular to the applied field. In an XY kink (blue), the spins rotate from one ground state to the other close to the xy-plane. In a YZ kink (green), the variation follows close to the yz-plane. The canting angle is very small and is not shown here.

Using this, the dynamic XY sG kink has a very simple energy dependence, determined only by the applied field,

$$E_{xy} = JS^2 \gamma \beta = \gamma g \mu_B B^x S. \tag{7.117}$$

Small corrections for higher field and velocity, scaled by $1/\alpha$ have been dropped. This is the energy relative to the ground state energy.

7.4.5 YZ antiferromagnetic kinks

There is another kink solution of the equations of motion, where the spins rotate predominantly up out of the easy plane in going from one ground state to the other. In this case, the rotation takes place with spins staying close to the yz-plane.

Equation (7.107a) has an exact solution, $\Phi = \Phi_{yz} = \pi/2$, which defines the yz-plane. Then (7.107b) becomes

$$\Theta_{\xi\xi} - \frac{1}{4}\Theta_{\tau\tau} = -\frac{\alpha}{2}\sin 2\Theta. \tag{7.118}$$

Changing to a new variable $\Psi = 2\Theta \mp \pi$, this is again an sG equation, but for the rotation within the yz-plane:

$$\Psi_{\xi\xi} - \frac{1}{4}\Psi_{\tau\tau} = \alpha \sin \Psi. \tag{7.119}$$

One can see that this results in sG kink/anti-kink solutions such as that in (7.113) and (7.114) but with $\frac{1}{2}\beta$ replaced by $\sqrt{\alpha}$,

$$\Theta_{yz} = \pm\frac{\pi}{2} + 2q \tan^{-1}\left\{\exp\left[\gamma\sqrt{\alpha}(\xi - u\tau)\right]\right\}. \tag{7.120}$$

This structure is indicated approximately in figure 7.11, however, the canting of the ground state is not represented, which is assumed to be very small.

One should also check the smaller angles, which were assumed to be small. Using (7.106), we obtain

$$\theta_{yz} = 0 \tag{7.121a}$$

$$\phi_{yz} = \frac{1}{4}\{\gamma\sqrt{\alpha}\ qu\ \text{sech}\ x + \beta\}\coth x, \qquad x = \gamma\sqrt{\alpha}(\xi - u\tau). \tag{7.121b}$$

One sees that there is a divergence in ϕ at the kink center. This is a mathematical anomaly due to the use of polar spherical coordinates, and the fact that the azimuthal angle of a spin is not defined when it points towards one of the poles. It produces a cusp, especially in S^x, making this coordinate system inadequate for the YZ kinks, especially if a formula is needed as initial conditions for simulations.

A way to correct this problem is to use spherical coordinates where the x-axis is the polar axis, using new x-polar spherical coordinate angles as

$$\mathbf{S}_{A,B} = \pm S\bigl(\cos(\Theta \pm \theta),\ \sin(\Theta \pm \theta)\cos(\Phi \pm \phi),\ \sin(\Theta \pm \theta)\sin(\Phi \pm \phi)\bigr). \tag{7.122}$$

This transformation does not affect the isotropic exchange terms in the dynamics, but it does modify the easy-plane term and the field term. The equations of motion are now

$$\Theta_\tau = 4\phi\sin\theta + \alpha\left(\phi\sin\Theta\cos 2\Phi + \frac{1}{2}\theta\cos\Theta\sin 2\Phi\right) \tag{7.123a}$$

$$\Phi_\tau = -4\theta\csc\Theta + \alpha\bigl(\theta\sin\Theta\sin^2\Phi - \phi\cos\Theta\sin 2\Phi\bigr) - \beta \tag{7.123b}$$

$$\theta_\tau = -\bigl(4\theta\phi + 2\Theta_\xi\Phi_\xi\bigr)\cos\Theta - \Phi_{\xi\xi}\sin\Theta + \frac{1}{2}\alpha\sin\Theta\sin 2\Phi \tag{7.123c}$$

$$\phi_\tau = \bigl(4\theta^2\csc^2\Theta - 4\phi^2 - \Phi_\xi^2\bigr)\cos\Theta + \Theta_{\xi\xi}\csc\Theta - \alpha\sin\Theta\sin 2\Phi. \tag{7.123d}$$

Now under the assumption of weak anisotropy, $\alpha \ll 4$, the small angles can be eliminated with

$$\theta = -\frac{1}{4}(\beta + \Phi_\tau)\sin\Theta \tag{7.124a}$$

$$\phi = \frac{1}{4}\Theta_\tau\csc\Theta. \tag{7.124b}$$

The resulting dynamic equations for the larger angles again have wave-like operators,

$$\Phi_{\xi\xi} - \frac{1}{4}\Phi_{\tau\tau} + 2\left(\Theta_\xi\Phi_\xi - \frac{1}{4}\Theta_\tau\Phi_\tau\right)\cot\Theta = \frac{1}{2}\alpha\sin 2\Phi + \frac{1}{2}\beta\Theta_\tau\cot\Theta \tag{7.125a}$$

$$\Theta_{\xi\xi} - \frac{1}{4}\Theta_{\tau\tau} - \left(\Phi_\xi^2 - \frac{1}{4}\phi_\tau^2\right)\frac{1}{2}\sin 2\Theta = \left(\alpha\sin^2\Phi - \frac{1}{4}\beta^2 - \frac{1}{2}\beta\Phi_\tau\right)\frac{1}{2}\sin 2\Theta. \tag{7.125b}$$

In these coordinates, the yz-plane is defined by $\Theta = \frac{1}{2}\pi$, $\theta = 0$. One can see that the choice $\Theta = \Theta_{yz} = \frac{1}{2}\pi$ solves (7.125b), then (7.125a) becomes a sG equation for twice the large angle within the yz-plane:

$$(2\Phi)_{\xi\xi} - \frac{1}{4}(2\Phi)_{\tau\tau} = \alpha \sin 2\Phi. \tag{7.126}$$

A first type of YZ kink/anti-kink solution is then seen to be

$$\Phi_{yz} = 2q \tan^{-1} \exp x, \qquad \Theta_{yz} = \frac{1}{2}\pi \tag{7.127a}$$

$$\theta_{yz} = -\frac{1}{4}(\beta - \gamma\sqrt{\alpha}\, qu \operatorname{sech} x), \qquad \phi_{yz} = 0. \tag{7.127b}$$

The argument x is defined in (7.121b). There is another type of solution which starts at $\Phi_{yz} = \pi$ for $x \to -\infty$. It simply has a shift on the larger azimuthal angle,

$$\Phi_{yz} = \pi + 2q \tan^{-1} \exp x. \tag{7.128}$$

Thus there are again a total four types of YZ kinks. These solutions are smooth profiles, good for use as initial conditions in simulations. The use of x-polar coordinates has helped to improve the solution.

In x-polar coordinates one can write out the Hamiltonian; again the isotropic exchange terms are no different than in z-polar coordinates. Only the applied field and anisotropy contributions are modified. The continuum result is

$$H = \frac{1}{2}JS^2 \int d\xi \Big\{ \Theta_\xi^2 + 4\theta^2 + \left(\Phi_\xi^2 + 4\phi^2\right)\sin^2\Theta - 2\beta \sin\Theta \sin\theta$$
$$+ \alpha\Big[\left(\sin^2\Theta + \theta^2 \cos 2\Theta\right)\left(\sin^2\Phi + \phi^2 \cos\Phi\right) + 2\theta\phi \sin 2\Theta \sin 2\Phi\Big]\Big\}. \tag{7.129}$$

Applying this to the YZ kink solution, the energy is found to be

$$E_{yz} = 2JS^2\gamma\sqrt{\alpha}\left[1 - \frac{1}{32}\beta^2\left(1 - \frac{1}{4}u^2\right)\right] \approx 2\sqrt{2JK_A}\, S^2\gamma. \tag{7.130}$$

Here one sees that the XY kink energy is less than E_{yz} if $\beta < 2\sqrt{\alpha}$. The two energies become equal at the critical field, $\beta_c = 2\sqrt{\alpha}$. This crossover effect also has implications for the stability of these excitations. As can be expected, the existence of a lower energy state can make it possible for one of the kink solutions to destabilize via a structural instability, just as was found for FM kinks in 1D chains. This can be investigated by looking at small fluctuations around these solutions. Later, a type of ansatz can be considered for the kink structures, which can be used to estimate the energies and stabilities.

7.4.6 Antiferromagnetic YZ kink stability analysis

The analytic calculation of kink stability can be done most easily for YZ kinks. The approach is the same as that used in the 1D FM chain. We start from the YZ sG kink solution, and assume small perturbations around it. If the original solution is described by x-polar spherical angles Φ_{yz}, $\phi_{yz} = 0$, $\Theta_{yz} = \pi/2$ and θ_{yz} as given earlier in (7.127), then the assumed fluctuations are possible in all these coordinates, denoted as $\tilde{\Phi}$, $\tilde{\phi}$, $\tilde{\Theta}$ and $\tilde{\theta}$. We write the perturbed fields as

$$\Phi = \Phi_{yz} + \tilde{\Phi}, \qquad \Theta = \Theta_{yz} + \tilde{\Theta} \qquad (7.131a)$$

$$\theta = \theta_{yz} + \tilde{\theta}, \qquad \phi = \phi_{yz} + \tilde{\phi}. \qquad (7.131b)$$

It is suggested as an exercise to show that for $\alpha \ll 4$, linearizing the equations of motion in these small fluctuations, and eliminating the smaller angles, $\tilde{\theta}$ and $\tilde{\phi}$, leads to *decoupled* stability equations for the larger angles,

$$-\tilde{\Theta}_{\xi\xi} + \left(\Phi_{xy\,\xi}^2 - 4\theta_{yz}^2 + \alpha \sin^2 \Phi_{yz}\right)\tilde{\Theta} = \frac{1}{4}\tilde{\Theta}_{\tau\tau} \qquad (7.132a)$$

$$-\tilde{\Phi}_{\xi\xi} - \left(\alpha \cos 2\Phi_{yz}\right)\tilde{\Phi} = \frac{1}{4}\tilde{\Phi}_{\tau\tau}. \qquad (7.132b)$$

These are typical 1D eigenfunction and related eigenvalue problems, once we assume time-dependencies such as $\tilde{\Theta} \sim \exp(-i\tilde{\omega}_1 \tau)$, $\tilde{\Phi} \sim \exp(-i\tilde{\omega}_2 \tau)$. Using the kink solution, which determines the potentials in these equations, they can be expressed in standard forms, for moving kinks,

$$-\gamma^2 \tilde{\Theta}_{xx} + \left(1 - 2\,\mathrm{sech}^2 x - \frac{\gamma \beta q u}{2\sqrt{\alpha}}\,\mathrm{sech}\,x\right)\tilde{\Theta} = \lambda_1 \tilde{\Theta} \qquad (7.133a)$$

$$-\gamma^2 \tilde{\Phi}_{xx} + \left(1 - 2\,\mathrm{sech}^2 x\right)\tilde{\Phi} = \lambda_2 \tilde{\Phi}. \qquad (7.133b)$$

The argument in these functions is $x \equiv \gamma\sqrt{\alpha}(\xi - u\tau)$. The eigenvalues λ_1 and λ_2 are related to the frequencies by

$$\tilde{\omega}_1^2 = 4\alpha(\lambda_1 - 1) + \beta^2 \qquad (7.134a)$$

$$\tilde{\omega}_2^2 = 4\alpha\lambda_2. \qquad (7.134b)$$

Static YZ kinks
In the case of a **static** YZ kink with $u = 0$, $\gamma = 1$, the two eigen-equations are identical. The potential $V = 1 - 2\,\mathrm{sech}^2 x$ has a bound state solution of the form $\psi = \mathrm{sech}\,x$, with eigenvalue $\lambda = 0$. Note that the function $\psi = \mathrm{sech}\,x$ is related to the space derivative of Φ_{yz}. This function and eigenvalue appeared in the FM kink stability problem, see (7.36b) and (7.40). Note, however, that its effect is different when used for the $\tilde{\Theta}$-fluctuations compared to the $\tilde{\Phi}$-fluctuations. For the $\tilde{\Phi}$ equation, $\lambda_2 = 0$ implies $\tilde{\omega}_2 = 0$, for any parameter values. This zero-frequency mode is the zero-energy translational mode of the kink, relating to the freedom it has to shift position at no energy cost. There is no instability in this mode. For the $\tilde{\Theta}$ equation, however, a mode with $\lambda_1 = 0$ implies that the associated eigenfrequency is

$$\tilde{\omega}_1 = \sqrt{\beta^2 - 4\alpha}. \qquad (7.135)$$

Now, for a weak magnetic field the frequency is imaginary, leading to a structural instability of the kink. This instability mode corresponds to having the spins tilt away from the yz-plane. On the other hand, for fields $\beta > 2\sqrt{\alpha} \equiv \beta_c$, a real frequency results, and there is no instability. Therefore, the kink becomes unstable when $\beta < \beta_c$, i.e. when the field crosses below the critical field. Of course, this is the same point at which the static XY kink and YZ kink have the same energy. This structural instability in YZ kinks is also an energetic instability, due to the possibility of the YZ kink to modify its structure via spin tilting that deforms it eventually into the lower energy XY kink. Although we have not given a similar analysis of the XY kink stability, one might expect that conversely, it would also become unstable once $\beta > \beta_c$.

Moving YZ kinks

For a **moving** YZ kink, the velocity u and the charge q affect the stability, as can be seen by the presence of qu in (7.133a) for the $\tilde{\Theta}$-fluctuations. A positive value of qu will make the potential in (7.133a) deeper, which will then cause the eigenvalue λ_1 to move below zero. With the requirement that $\tilde{\omega}_1^2 > 0$, for a moving kink the stability criterion is

$$\beta^2 > 4\alpha(1 - \lambda_1). \tag{7.136}$$

With $\lambda_1 < 0$, a $q = +1$ kink moving in the positive z-direction requires a field *larger than* $\beta_c = 2\sqrt{\alpha}$ to be stable. On the other hand, if the same kink is moving in the negative z-direction ($u < 0$), the potential becomes shallower, and the eigenvalue λ_1 will become positive. As long as $0 \leqslant \lambda_1 < 1$, the kink can still be stable, and a *smaller field* is required to accomplish that. The kink moving in the negative direction will tend to be more stable for a given field. This means that even when $\beta < \beta_c$, there can be stable YZ kinks, as long as they are moving in the negative direction. These statements apply in the reversed sense to anti-kinks ($q = -1$). An anti-kink moving in the positive direction would be more stable than those moving in the negative direction. For fields $\beta < \beta_c$, the only stable anti-kink should be those moving in the positive direction.

These results show an interesting handedness or chirality to the interactions. The sense of the rotation of the spins along the chain is directly coupled to their motion and dynamic properties. One can summarize the results by saying that the product qu acts as a chirality parameter for these excitations. Positive chirality parameter tends to destabilize a YZ kink or anti-kink. Negative chirality parameter may help to stabilize a YZ kink or anti-kink, but this statement requires more knowledge of the effect of qu on the eigenvalue λ_1.

Perturbation analysis for moving kinks

The effect of non-zero velocity can be estimated quantitatively using first-order perturbation theory for the Schrödinger-like equation (7.133a) governing $\tilde{\Theta}$. The modification to the original potential $V_0 = 1 - 2\,\text{sech}^2 x$ (and to an effective Hamiltonian H_0 for this problem) keeping only term linear in u is

$$V_1 = \frac{-\beta qu}{2\sqrt{\alpha}} \,\text{sech}\, x. \tag{7.137}$$

Then with $\psi_0 = \text{sech } x$ being the exact solution without this perturbation, the rules of first-order perturbation theory as found in most texts on quantum mechanics give the leading correction to the eigenvalue $\lambda_1 = 0$,

$$\Delta\lambda_1 = \frac{\langle\psi_0|V_1|\psi_0\rangle}{\langle\psi_0|\psi_0\rangle} = \frac{-\beta q u}{2\sqrt{\alpha}} \frac{\int_{-\infty}^{+\infty} dx \text{ sech}^3 x}{\int_{-\infty}^{+\infty} dx \text{ sech}^2 x} = \frac{-\beta q u}{2\sqrt{\alpha}} \frac{\pi/2}{2} = \frac{-\pi\beta q u}{8\sqrt{\alpha}}. \qquad (7.138)$$

Then to linear order in u, this is the perturbed eigenvalue, $\lambda_1 = \Delta\lambda_1$. Combining with the stability result (7.136) leads to a quantitative constraint on the velocity needed for stability:

$$q\frac{u}{c} < \frac{\beta^2 - 4\alpha}{\pi\beta\sqrt{\alpha}} = \frac{2}{\pi}\frac{\beta^2 - \beta_c^2}{\beta\beta_c}. \qquad (7.139)$$

Here the value $c = 2$ for the units used is being applied. This confirms the qualitative statements made earlier. If $\beta > \beta_c$, kinks with $q = +1$ can be stable moving in the negative direction for any speed (up to c), but in the positive direction only up to a velocity limit given by the RHS. Anti-kinks with $q = -1$, however, can move at any speed in the positive direction, but only up to the speed limit given by the RHS when moving in the negative direction. Similarly, if $\beta < \beta_c$, this result also shows the uni-directional negative motion of kinks and positive motion of anti-kinks. These quantitative results are changed only slightly if one goes to the next higher order of perturbation theory.

Exercise 7.5. Carry out the calculations leading to the dynamic fluctuations, equations (7.132). Using the YZ kink solutions, confirm the relations (7.134) between the eigenfrequencies and the corresponding eigenvalues.

Exercise 7.6. Perform a similar stability analysis for XY kinks, assuming small perturbations $\tilde{\Phi}$, $\tilde{\Theta}$, $\tilde{\phi}$, $\tilde{\theta}$ about the sG XY kink solutions that were found earlier. After eliminating the smaller angles $\tilde{\phi}$, $\tilde{\theta}$, show that the resulting dynamic equations for $\tilde{\Phi}$, $\tilde{\Theta}$, do not decouple, leading to a situation more similar to that found for FM kink stability.

7.4.7 Antiferromagnetic kink ansatz analysis

The kink stability analysis can be augmented with an ansatz that encompasses both XY and YZ kink structures. An ansatz can give some idea about the energy–velocity relationship and kink effective mass. By itself, it will not determine the kink stability, which must be achieved ultimately through numerical simulations.

The basic idea of the ansatz is simple, and similar to that for the π pulse solitons in the FM chain. The unit spins s_A on the A-sublattice rotate in a plane tilted at some angle θ_A to the xy-plane; on the B-sublattice unit spins s_B move in another plane tilted at angle θ_B. The rotation on each sublattice has to go from one of the spin-flop ground states to the other one. In addition, the spatial variation of the angle in these tilted planes is of the sG form with a width parameter w. The Lagrangian of the system is then found as a function of θ_A, θ_B and w, and minimized with respect to independent variations in all three. The main complication is to ensure that the profile connects to the spin-flop ground states at the far ends of the chain.

On the A-sublattice, the spin trajectory is the intersection of the unit sphere with the plane that passes through the two ground state points $\mathbf{g}_\pm = (\sin\phi_0, \pm\cos\phi_0, 0)$ and tilted at angle θ_A from the xy-plane. At the kink center, the highest point on this trajectory is $\mathbf{s}_0 = (\cos\theta_m, 0, \sin\theta_m)$, where θ_m is a maximum out-of-plane spin angle. Suppose the trajectory starts at one chain end $\mathbf{g}_-$, passes to $\mathbf{s}_0$ at kink center and then ends at $\mathbf{g}_+$ at the opposite chain end. From figure 7.12 that plane's tilt angle and the center spin out-of-plane angle are related by

$$\tan\theta_A = \frac{\sin\theta_m}{\cos\theta_m - \sin\phi_0}. \tag{7.140}$$

Note that $\sin\phi_0 = \tfrac{1}{4}\beta$ and it is seen that in the rotated $x'y'z'$ frame all the spins have the same z'-component, given by

$$s^{z'} = -\tfrac{1}{4}\beta\sin\theta_A. \tag{7.141}$$

For the $y'z'$-components one can use FM and AFM vectors $\mathbf{M}$ and $\mathbf{N}$ in the $x'y'$-plane as an aid, letting the A-sublattice spins be written as a combination,

$$\mathbf{s}_A = f_A(x)(\mathbf{M} + \mathbf{N}) + s^{z'}\hat{\mathbf{z}}'. \tag{7.142}$$

The factor $f_A(x)$ will be chosen to ensure unit spin length. One might suppose a similar expression for the B-sublattice, with the sign of $\mathbf{N}$ reversed and another normalization factor $f_B(x)$. Then the ansatz made by Wysin [13] is to assume circular motions for both $\mathbf{M}$ and $\mathbf{N}$ of different centers and radii, from which the $x'y'$-components of $\mathbf{s}_A$ are derived. These circular paths are sketched in figure 7.13. Vector $\mathbf{M}$ traces out a circle of

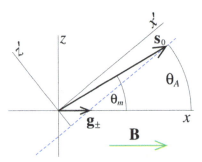

Figure 7.12. Notation and vectors for the AFM kink ansatz on the A-sublattice. $\mathbf{s}_0$ is the spin at the kink center, $\mathbf{g}_\pm$ shows the projection of the ground states at the chain ends. The diagram is made for $\theta_m = 30°$ and a large value $\beta = 1$ to clarify the geometry, which gives $\theta_A = 39°$ from (7.140).

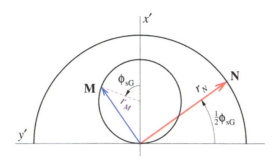

Figure 7.13. Sketch of the paths in the tilted $x'y'$-plane for the **M** and **N** vectors in the AFM ansatz. The radii are $r_M = \frac{1}{8}\beta \cos \theta_A$ and $r_N = \cos \phi_0 = (1 - \frac{1}{16}\beta^2)^{1/2}$. The diagram is made for $\theta_m = 30°$ and a large value $\beta = 1$, which gives $\theta_A = 39°$ from (7.140). The size of the circle for **M** is exaggerated by a factor of 4 relative to that for **N**.

radius $r_M = \frac{1}{8}\beta \cos \theta_A$ centered at $(x', y') = (r_M, 0)$. Vector **N** traces out a semicircle of radius $r_N = \cos \phi_0 = (1 - \frac{1}{16}\beta^2)^{1/2}$ centered at the origin. Within the tilted plane, suppose there is a sG profile assumed for an in-plane angle,

$$\phi_{sG}(x) = 4 \tan^{-1}\{\exp x\}, \qquad x = (\xi - u\tau)/w. \tag{7.143}$$

Then the ansatz for the FM and AFM vectors is

$$M_{x'} = r_M(1 + \cos \phi_{sG}) = 2r_M \tanh^2 x \tag{7.144a}$$

$$M_{y'} = r_M \sin \phi_{sG} = -2r_M \tanh x \, \text{sech} \, x \tag{7.144b}$$

$$N_{x'} = r_N \sin \frac{1}{2}\phi_{sG} = r_N \, \text{sech} \, x \tag{7.144c}$$

$$N_{y'} = -r_N \cos \frac{1}{2}\phi_{sG} = r_N \tanh x. \tag{7.144d}$$

This is depicted in figure 7.13. To guarantee unit spin length requires

$$s_A^2 = f_A^2(x)(\mathbf{M} + \mathbf{N})^2 + s^{z'2} = 1. \tag{7.145}$$

It is convenient to use that to define the radius in the $x'y'$-plane,

$$r_A^2 \equiv 1 - s^{z'2} = 1 - \frac{1}{16}\beta^2 \sin^2 \theta_A. \tag{7.146}$$

This leads to the spatially dependent normalization factor,

$$f_A(x) = \frac{r_A}{\sqrt{\cos^2 \phi_0 + \frac{1}{16}\beta^2 \cos^2 \theta_A \tanh^2 x}} = \frac{r_A}{\sqrt{r_A^2 - \frac{1}{16}\beta^2 \cos^2 \theta_A \, \text{sech}^2 x}}. \tag{7.147}$$

With these definitions, the ansatz on the A-sublattice is defined,

$$s_A^{x'} = f_A(x)\left(M^{x'} + N^{x'}\right) \tag{7.148a}$$

$$s_A^{y'} = f_A(x)\left(M^{y'} + N^{y'}\right) \tag{7.148b}$$

$$s_A^{z'} = -\frac{1}{4}\beta \sin\theta_A. \tag{7.148c}$$

The same type of structure is assumed for the B-sublattice, but at a different tilt angle θ_B, and with the sign on $\mathbf{N}$ reversed, and another normalization factor f_B defined similar to f_A in (7.147). Using double primes to indicate the B-sublattice tilted plane, the ansatz there uses the same profile $\phi_{sG}(x)$, and

$$s_B^{x''} = f_B(x)\left(M^{x''} - N^{x''}\right) \tag{7.149a}$$

$$s_B^{y''} = f_B(x)\left(M^{y''} - N^{y''}\right) \tag{7.149b}$$

$$s_B^{z''} = -\frac{1}{4}\beta \sin\theta_B. \tag{7.149c}$$

In order to use the ansatz, the spins should be rotated back into the original xyz-coordinates. This is effected by the same transformation as applied in the discussion of FM pulse solitons, namely, for either sublattice,

$$s^x = s^{x'} \cos\theta_{A,B} - s^{z'} \sin\theta_{A,B} \tag{7.150a}$$

$$s^y = s^{y'} \tag{7.150b}$$

$$s^z = s^{x'} \sin\theta_{A,B} + s^{z'} \cos\theta_{A,B}. \tag{7.150c}$$

The Lagrangian
The rest of the ansatz calculation is straightforward but tedious. It rests on evaluation of the continuum limit Lagrangian of the system. Although each sublattice occupies only half of the sites, it is useful to imagine both spin fields existing together on all sites, and write out the Hamiltonian, kinetic integral and Lagrangian with appropriate factors of 1/2 to cancel any double counting. A short exercise leads to the continuum Lagrangian as $L = K - H$ with general expressions for the kinetic term and Hamiltonian,

$$K = \frac{1}{2}JS^2 \int d\xi \left\{ s_A^z \frac{d}{d\tau}\left(\tan^{-1}\frac{s_A^y}{s_A^x}\right) + s_B^z \frac{d}{d\tau}\left(\tan^{-1}\frac{s_B^y}{s_B^x}\right) \right\} \tag{7.151a}$$

$$H = \frac{1}{2}JS^2 \int d\xi \left\{ 2\mathbf{s}_A \cdot \mathbf{s}_B - \frac{\partial \mathbf{s}_A}{\partial \xi} \cdot \frac{\partial \mathbf{s}_B}{\partial \xi} + \alpha\left(s_A^{z\,2} + s_B^{z\,2}\right) - \beta\left(s_A^x + s_B^x\right) \right\}. \tag{7.151b}$$

Recall that time and space units are those defined in (7.104), and the easy-plane anisotropy α and scaled magnetic field β are those as defined in the FM chain, equation (7.5). For practical purposes it is sufficient to evaluate both K and H to second order in small parameters $\sqrt{\alpha}$ and β; higher terms are not necessary (this is an approximate analysis, after all). The tilt angles are also expected to be nearly the same. Then a difference angle is assumed to be small:

$$\Delta \equiv \theta_B - \theta_A \ll 1. \tag{7.152}$$

The kinetic term needs careful treatment. It can be simplified somewhat by shifting the z spin components: $s_A^z \to s_A^z - 1$, and $s_B^z \to s_B^z + 1$, which changes K by an irrelevant additive constant, and removes unphysical step functions at $\theta_A = \pi/2$ and $\theta_B = -\pi/2$. The results are summarized in the following exercises.

Exercise 7.7. Show that the kinetic term (7.151a) in the AFM ansatz (7.148) and (7.149) can be expressed in the following form:

$$K(u, \theta_A, \theta_B) = JS^2 u \left[(\theta_B - \theta_A) + \frac{\pi}{8}\beta(\sin\theta_A + \sin\theta_B) \right] = uP. \tag{7.153}$$

Using $\Delta = \theta_B - \theta_A \ll 1$, then show that the kink momentum P can be expressed as

$$P(\theta_A, \Delta)/JS^2 \approx \frac{\pi}{4}\beta \sin\theta_A + \left(1 + \frac{\pi}{8}\beta \cos\theta_A\right)\Delta = p_0(\theta_A) + p_1(\theta_A)\Delta. \tag{7.154}$$

Note how K and P do not depend on the kink width, because they are indirect measures of the area covered by the spin profiles on the unit sphere.

Exercise 7.8. Show that up to second order in small parameters $\sqrt{\alpha}$ and β, the Hamiltonian (7.151b) in the AFM ansatz (7.148) and (7.149) is evaluated as

$$H(\theta_A, \Delta)/JS^2 = w^{-1}F(\theta_A, \Delta) + wG(\theta_A, \Delta), \tag{7.155}$$

where

$$F(\theta_A, \Delta) = \left(r_N^2 - \frac{1}{48}\beta^2 \cos^2\theta_A\right) - \left(\frac{\pi}{16}\beta r_N \sin\theta_A\right)\Delta - \frac{1}{6}\Delta^2 \tag{7.156a}$$

$$G(\theta_A, \Delta) = \left(\alpha r_N^2 \sin^2\theta_A + \frac{1}{4}\beta^2 \cos^2\theta_A\right) + \frac{1}{2}\sin 2\theta_A\left(\alpha r_N^2 - \frac{1}{4}\beta^2\right)\Delta + \Delta^2. \tag{7.156b}$$

For the purpose of simplified notation, these are used to define new functions f_0, f_1, f_2, and g_0, g_1, g_2, implicitly, by

$$F(\theta_A, \Delta) \equiv f_0 + f_1 \Delta + \frac{1}{2}f_2 \Delta^2 \tag{7.157a}$$

$$G(\theta_A, \Delta) \equiv g_0 + g_1 \Delta + \frac{1}{2}g_2 \Delta^2. \tag{7.157b}$$

Kink extremum solution
The resulting Lagrangian is first made stationary with respect to changes in w. Then $\partial L/\partial w = 0$ leads to some simple results,

$$w(\theta_A, \Delta) = \sqrt{F/G} \tag{7.158a}$$

$$E(\theta_A, \Delta) = 2JS^2\sqrt{FG} \approx JS^2\left(e_0 + e_1\Delta + \frac{1}{2}e_2\Delta^2\right) \tag{7.158b}$$

$$L(\theta_A, \Delta) = uP - E. \tag{7.158c}$$

The leading terms in the energy are shown, defined via functions of θ_A alone,

$$e_0(\theta_A) = 2\sqrt{f_0 g_0} \tag{7.159a}$$

$$e_1(\theta_A) = \frac{1}{2}(f_1/f_0 + g_1/g_0)e_0 \tag{7.159b}$$

$$e_2(\theta_A) = \frac{1}{2}\left[(f_2/f_0 + g_2/g_0) - \frac{1}{2}(f_1/f_0 - g_1/g_0)^2\right]e_0. \tag{7.159c}$$

From that, L is now made stationary with respect to variations in Δ, leaving θ_A to act as the independent parameter of the ansatz. Then $\partial L/\partial \Delta = 0$ leads to

$$\Delta(\theta_A) = (up_1 - e_1)/e_2 \tag{7.160a}$$

$$L(\theta_A) = L_0 + L_1 u + L_2 u^2, \tag{7.160b}$$

where the components needed to express L are functions of θ_A,

$$L_0(\theta_A) = \frac{1}{2}e_1^2/e_2 - e_0 \tag{7.161a}$$

$$L_1(\theta_A) = p_0 - p_1 e_1/e_2 \tag{7.161b}$$

$$L_2(\theta_A) = \frac{1}{2}p_1^2/e_2. \tag{7.161c}$$

This indirectly defines $L(\theta_A)$. Minimization with respect to θ_A then determines the kink velocity from a quadratic equation,

$$L_0' + L_1'u + L_2'u^2 = 0. \tag{7.162}$$

Primes indicate $d/d\theta_A$. The quadratic has two possible solutions,

$$u_\pm = \frac{1}{2L_2'}\left[-L_1' \pm \sqrt{L_1'^2 - 4L_0'L_2'}\right]. \tag{7.163}$$

After a value of u is determined, it should be verified with (7.160a) that $\Delta \ll 1$, as assumed. Assuming that is acceptable, an input value of θ_A then determines $u_\pm$, the width w, momentum P and energy E from expressions listed above. The ansatz solution is fully determined with θ_A as a control parameter, and α and β as fixed parameters. The kink energy–velocity dispersion $E(u)$ is found by eliminating θ_A numerically. This defines the general solution. The results are found from here numerically, and have been tested well against numerical simulations of kink motion by Wysin [13]. The reader is referred to the PhD thesis by Wysin [13] for further results and details of the full ansatz.

7.4.8 XY kink limit: $\theta_A = \theta_B$

For XY kinks, the three-parameter ansatz can be simplified to two parameters, which is easier to analyze, by forcing the tilt angles to be the same, $\theta_A = \theta_B$, see figure 7.14(a). This is a reasonable assumption. It is exact for the stationary XY kink, and one expects $\Delta = \theta_A - \theta_B$ in the full ansatz to be a small parameter for non-zero velocity kinks. This simplification should work best for $u/c \ll 1$. Then, setting $\Delta = 0$, with the notation,

$$s \equiv \sin \theta_A \tag{7.164}$$

one has the reduced ansatz results,

$$P(s) = JS^2 \frac{\pi}{4}\beta s \tag{7.165a}$$

$$F(s) = r_N^2 - \frac{1}{48}\beta^2(1 - s^2) \tag{7.165b}$$

$$G(s) = \frac{1}{4}\beta^2 + \left(\alpha r_N^2 - \frac{1}{4}\beta^2\right)s^2. \tag{7.165c}$$

The width and energy are still given by the expressions (7.158) for the full ansatz. It is then possible to find a result for the velocity by a ratio of derivatives, and the energy,

$$u(s) = \frac{\partial E/\partial s}{\partial P/\partial s} = \frac{4}{\pi}\left(1 - \frac{1}{12}\beta^2\right)^{1/2}\left(\frac{4\alpha}{\beta^2} - 1\right)s \tag{7.166a}$$

$$E_{xy}(s) = JS^2\beta\left(1 - \frac{1}{12}\beta^2\right)^{1/2}\left[1 + \frac{1}{2}\left(\frac{4\alpha}{\beta^2} - 1\right)s^2\right]. \tag{7.166b}$$

At the limit $u = 0$, which corresponds to $s = 0$, the energy is

$$E_{xy}^0 = JS^2 \beta\left(1 - \frac{1}{12}\beta^2\right)^{1/2} \tag{7.167}$$

consistent with the result (7.117) obtained earlier; some extra terms in β^2 have been kept but are small. However, an interesting new feature appears here. The energy either increases or decreases with s^2, depending on whether β is smaller than or greater than the critical field $\beta_c = 2\sqrt{\alpha}$. This is a change in the sign of the second derivative of $E(s)$. It implies a sign change in the kink effective mass; see the following exercise.

Exercise 7.9. Assuming XY kinks of low velocity, eliminate s between the relations for $E(s)$ and $u(s)$ and show that the energy–velocity dispersion relation is approximated by

$$E_{xy}(u) \approx E_{xy}^0 + \frac{1}{2}m^*u^2 \qquad (7.168)$$

where the effective mass is

$$m^* = \frac{(\pi\beta/4)^2}{\left(1 - \frac{1}{12}\beta^2\right)(4\alpha - \beta^2)} E_{xy}^0. \qquad (7.169)$$

Then, the XY kink effective mass is positive for $\beta < \beta_c$ and becomes negative once the applied field surpasses the critical field. $E_{xy}(u)$ changes from parabolic upward to parabolic downward at the critical field, a result found also by Flüggen and Mikeska [12] via another ansatz. This does not prove an instability but it is certainly suggestive of a change in dynamical behavior, somewhat like that in the kinks in the easy-plane FM chain at $b = 1/3$. Note also that the relationship between the signs of s and u also reverses at $\beta = \beta_c$.

7.4.9 YZ kink limit: $\theta_A + \theta_B = \pi$

In the static YZ kink the spins on each sublattice move in-plane close to $\theta = \pi/2$. In the moving YZ kink, the spins obtain an extra tilting towards or away from the magnetic field. Therefore, the expected structure of a YZ kink can be reduced to a description with a single angle parameter ϵ, which determines the tilt angles of the planes by

$$\theta_A = \frac{1}{2}\pi - \epsilon, \qquad \theta_B = \frac{1}{2}\pi + \epsilon. \qquad (7.170)$$

A YZ structure like this is shown in figure 7.14(b). Although ϵ at first may seem the same as small angle θ, up to a sign, this is not the case. The tilt angle of the plane in which the spins move is not the same as $\pi/2 \pm \theta$. Under this assumption the momentum and other functions that determine the Lagrangian are found to be reduced to

$$P(\epsilon) = JS^2\left[2\pi + 2\epsilon + \frac{1}{4}\pi\beta \cos \epsilon\right] \qquad (7.171a)$$

$$F(\epsilon) = r_N\left(r_N - \frac{1}{8}\pi\beta\right)s - \left(\frac{2}{3} - \frac{7}{48}\beta^2\right)s^2 \qquad (7.171b)$$

$$G(\epsilon) = \alpha r_N\left(r_N - \frac{1}{8}\pi\beta\right)s + \left(4r_N^2 - \alpha\right)s^2, \qquad (7.171c)$$

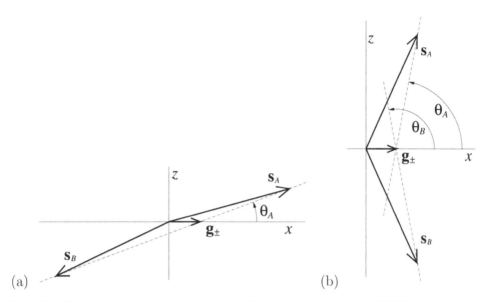

Figure 7.14. Diagrams showing the spins on the two sublattices at the center of an AFM kink in the ansatz. (a) XY kink limit with $\theta_A = \theta_B$. (b) YZ kink limit with $\theta_A + \theta_B = \pi$. The dashed blue lines indicate the tilted planes in which the spins move and $\mathbf{g}_\pm$ represents the ground state directions.

where now we use the notation

$$s \equiv \sin \epsilon. \tag{7.172}$$

These are results assuming that $\sqrt{\alpha}$, β and ϵ are all small parameters. Making the Lagrangian an extremum with respect to variations in w and ϵ then leads to the velocity and energy results,

$$u(s) = 2\sqrt{\alpha}\, r_N^2 \left[-\frac{\pi\beta}{16 r_N} + \left(\frac{2}{\alpha} - \frac{5}{6 r_N^2} \right) s \right] \tag{7.173a}$$

$$E_{yz}(s) = 2JS^2\sqrt{\alpha}\, r_N^2 \left[1 - \frac{\pi\beta}{8 r_N} s + \left(\frac{2}{\alpha} - \frac{5}{6 r_N^2} \right) s^2 \right]. \tag{7.173b}$$

One sees that $s = 0$ does not correspond to $u = 0$, and this is because the static kink solution rediscovered here has a finite canting of the spins on the two sublattices. For a zero velocity kink, the required canting angle at kink center is approximately

$$\epsilon(u = 0) = \frac{\pi\beta}{16 r_N} \left(\frac{2}{\alpha} - \frac{5}{6 r_N^2} \right)^{-1} \approx \frac{\pi}{32} \alpha\beta. \tag{7.174}$$

We used $r_N \approx 1$ and $\alpha \ll 1$ to arrive at this result. Note again that ϵ is not the same as the small angle $\pm\theta$, as can be seen when looking at the vertical situation when $\theta_A = \theta_B = \frac{\pi}{2}$, where $\epsilon = 0$ but $\theta = -\frac{1}{4}\beta$.

For the static YZ kink, using $\epsilon(u = 0)$ leads to the static energy,

$$E_{yz}^0 = 2JS^2\sqrt{\alpha}\left(1 - \frac{1}{16}\beta^2\right), \tag{7.175}$$

which is consistent with but slightly lower than the sG result found earlier. One can also check the energy–velocity relationship at small velocity, to extract an effective mass. A short calculation leads to the results,

$$E_{yz}(u) = E_{yz}^0 + \frac{1}{2}m^*u^2 \tag{7.176a}$$

$$m^* = \frac{JS^2}{(JSa)^2}\left[\sqrt{\alpha}\,r_N^2\left(\frac{2}{\alpha} - \frac{5}{6r_N^2}\right)\right]^{-1} \approx \frac{1}{Ja^2}\frac{\sqrt{\alpha}}{2r_N^2}. \tag{7.176b}$$

This mass is equivalent to the sG effective mass. The result of the ansatz for YZ kinks does not lead to much of anything unexpected relative to the sG solution. There is no change of sign in m^* for reasonable parameter values; it is always positive as long as $\alpha < \frac{12}{5}r_N^2$, which will always be true for materials with weak easy-plane anisotropy. However, keep in mind that the earlier stability analysis still holds true: YZ kinks have a minimum velocity necessary for dynamic stability. This effect is not evaluated in the ansatz variational approach.

7.4.10 Energy–velocity relations from the simplified antiferromagnetic ansatz results

For XY kinks, we can let s in (7.166) range from -1 to $+1$, acting as a parameter, and effectively eliminate s from the relation between E_{xy} and u. Because of the limited range of s, there is a corresponding limited range of u, within the ansatz assumption. For low values of magnetic field strength β, the two-parameter ansatz is rather poor at reproducing the more accurate results of the full three-parameter ansatz. However, as β approaches closer to β_c, the two-parameter ansatz becomes much more reliable and gives results the same as the three-parameter ansatz. Thus we show some results for $\beta = 0.30, 0.37, 0.43$ in figure 7.15.

The same procedure can be applied to the YZ kinks, using (7.173). These results are also shown in figure 7.15. It turns out there is very little change in $E_{yz}(u)$ with varying β. For $E_{xy}(u)$, however, there are strong changes as β crosses from below β_c to above β_c. For the case shown, with $\alpha = 0.04$, one has $\beta_c = 2\sqrt{\alpha} = 0.40$. At $\beta = 0.30, 0.37$, or any value below β_c, XY kinks have positive effective mass and the XY kink dispersion falls below the lower velocity part of the YZ kink dispersion. In that range, the higher energy YZ kink is unstable. This is obtained from the earlier stability analysis and also from numerical simulations. There is also a very limited range of possible velocities for XY kinks (using the finite range of $s = \sin\theta_A$). Further, the XY curve for $\beta = 0.30$ is very approximate here; in the full three-parameter ansatz, it actually connects to the YZ dispersion curve.

At the critical point, $\beta = \beta_c$, the XY kink dispersion contracts to a single point (at $u = 0$) on the $E(u)$ diagram. This implies only a static XY kink is possible

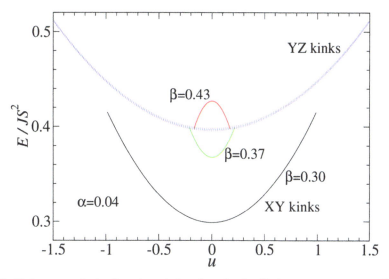

Figure 7.15. Kink energy–velocity dispersion relations from the simplified two-parameter ansatz for XY and YZ kinks, at $\alpha = 0.04$. The YZ results, from (7.173), are shown as dotted curves and there is little dependence on β. The XY results, from (7.166), are solid curves for the indicated field strengths.

when $\beta = \beta_c$. That is an intriguing elaboration of a kink instability. Of course, at this critical limit, the static XY and YZ kink energies become degenerate. At $\beta = \beta_c$, a wide range of YZ kink velocities is still possible. The assumption $\theta_A = \theta_B$ for XY kinks and $\theta_A + \theta_B = \pi$ for YZ kinks is exactly true at $\beta = \beta_c$ when the full three-parameter ansatz is used.

For any $\beta > \beta_c$, the XY kink effective mass becomes negative, as can be seen in the downward parabolic $E_{xy}(u)$ for $\beta = 0.43$ in figure 7.15. Now the XY kink dispersion again acquires a finite velocity range, however, simulations indicate that both XY and YZ kinks are stable. Even so, the XY kinks with negative effective mass are similar to the kinks in a FM chain above its critical field. For a chosen out-of-plane spin component at the kink center, they move in the opposite direction that would be expected for sG magnetic kinks. Then, the XY kinks become 'backwards moving' once $\beta > \beta_c$.

The dispersion relations are symmetrical curves close to parabolic in shape. We have reviewed the ansatz solutions for a particular sign of $q = +1$, and when performed carefully, the full three-parameter ansatz actually selects only a limited region of these curves [13]. However, once the opposite sign of $q = -1$ is allowed, one will populate both sides of the dispersion curves and the situation becomes fully symmetrical for both directions of motion along the chain.

Bibliography

[1] Scott A C, Chu F Y F and McLaughlin D W 1973 The soliton: a new concept in Applied Science *Proc. IEEE* **61** 1443

[2] Bishop A R, Krumhansl J A and Trullinger S E 1980 Solitons in condensed matter physics: a paradigm *Physica* D **1** 1
[3] Babel D 1969 Die Kristallstrukturen der hexagonalen Fluorperorvskite *Z. Anorg. Allg. Chem.* **369** 117
[4] Steiner M and Dachs H 1974 The magnetic phase diagram of $CsNiF_3$ as determined by neutron diffraction *Solid State Commun.* **14** 841
[5] Steiner M, Kakurai K and Kjems K J 1983 Experimental study of the spin dynamics in the 1-D-ferromagnet with planar anisotropy, $CsNiF_3$, in an external magnetic field *Z. Phys.* B **53** 117
[6] Mikeska H J 1978 Solitons in a one-dimensional magnet with an easy plane *J. Phys. C: Solid State Phys.* **11** L29
[7] Kumar P 1982 Soliton instability in a one-dimensional magnet *Phys. Rev.* B **25** 483
[8] Magyari E and Thomas H 1982 Kink instability in planar ferromagnets *Phys. Rev.* B **25** 531
[9] Liebmann R, Schöbinger M and Hackenbracht D 1983 Extended soliton band in easy-plane ferromagnets *J. Phys. C: Solid State Phys.* **16** L633
[10] Mikeska H J 1981 Solitons in one-dimensional magnets *J. Appl. Phys.* **52** 1950
[11] Sklyanin E K 1979 On complete integrability of the Landau–Lifshitz equation *Sov. Phys.— Dokl.* **24** 107
[12] Flüggen N and Mikeska H J 1983 On the nonlinear dynamics of the easy-plane antiferromagnetic chain in an external magnetic field *Solid State Commun.* **48** 293
[13] Wysin G M 1985 Classical kink dynamics and quantum thermodynamics in easy-plane magnetic chains with an applied magnetic field *PhD thesis* Cornell University https://www.phys.ksu.edu/personal/wysin/phdthesis/index.html

IOP Publishing

Magnetic Excitations and Geometric Confinement
Theory and simulations
Gary Matthew Wysin

Chapter 8

Vortices in layered or 2D ferromagnets

Many magnetic crystals have an anisotropic structure that is layered and, hence, the magnetic interactions have a 2D anisotropy. Then, the dynamics is close to that of independent planes of magnetic ions, i.e. 2D magnets. Depending on the details of the magnetic interactions, different types of topological excitations are possible, known both as solitons and vortices. In this chapter the structure of the two types of vortices in ferromagnets is presented. The unbinding of vortices in particle–anti-particle pairs produces the Berezinskii–Kosterlitz–Thouless (BKT) topological phase transition. The transition can be studied using Monte Carlo (MC) and spin dynamics (SD) simulations. An ideal gas of vortices is discussed as a model for their contribution to the thermodynamics. Some simulations of the dynamic correlations in the ferromagnetic (FM) XY model are presented to show the changes in dynamic correlations with temperature and wave vector.

8.1 A 2D ferromagnet with easy-plane exchange anisotropy

In this chapter we consider a model of three-component spins occupying the sites of some 2D lattice, with a FM nearest neighbor interaction. The model applies to materials such as K_2CuF_4 [1, 2], where the magnetic ions occupy well-defined layers in a crystal, separated by larger spacing than the intra-layer nearest neighbor distance. Then, the spin interactions are quasi-2D; for the most part we ignore the inter-layer couplings. It can be kept in mind that although the inter-layer couplings are small, at low enough temperature they can be considered important, where they lead to 3D ordering. The model is considered above any 3D ordering temperature. Much of the analysis can be performed assuming a square lattice; the analysis and certain results have some dependencies on the particular lattice, however, the basic physics effects are the same.

A model Hamiltonian is assumed with XXZ symmetry as in (1.45), and defining the in-plane exchange constant as $J^x = J$, and the out-of-plane exchange constant as

$$J^z = \lambda J^x, \quad 0 \leqslant \lambda < 1, \tag{8.1}$$

the Hamiltonian, with only nearest neighbor interactions, is taken as

$$H = -J \sum_{(i,j)} \left(S_i^x S_j^x + S_i^y S_j^y + \lambda S_i^z S_j^z - S^2 \right). \tag{8.2}$$

The sum is over nearest neighbor pairs of sites (i, j), which could occupy a square, triangular, hexagonal, or other 2D lattice. The last factor of S^2 forces the ground state energy to be zero. The usual XY model is the limit $\lambda = 0$. At $\lambda \to 1$ the interaction becomes isotropic; this isotropic Heisenberg limit is outside the discussion in this chapter. It is important for the considerations here to include at least an infinitesimal XY or easy-plane anisotropy. By including the anisotropy parameter λ, the properties of vortices and the BKT vortex-unbinding transition can be found to change with anisotropy strength. Besides, the exchange interactions in real quasi-2D materials may be very close to isotropic, with only weak easy-plane character ($\lambda \approx 0.99$ for K_2CuF_4 [1, 2]). This model does not include the plane rotor (PR) model (two-component spins). By using three-component spins, the model has true SD without the need for introducing an inertia as in the rotor model. In this chapter only FM coupling is assumed, and the demagnetization field is neglected.

The Hamiltonian (8.2) has already been discussed concerning MC simulations (equation (4.137)) and the discrete SD equations of motion, for the in-plane angle ϕ_i and out-of-plane spin component S_i^z, were found in (5.23) and (5.24). Using $S_i^z = S \sin \theta_i$ in planar spherical coordinates, the dynamics can alternatively be set in terms of the out-of-plane angles θ_i. This Hamiltonian is a combination of an isotropic part ($-J \mathbf{S}_i \cdot \mathbf{S}_j$ pair interactions) and an anisotropy term,

$$H_{\text{ani}} = +J\delta \sum_{(i,j)} S_i^z S_j^z, \quad \text{where } \delta \equiv 1 - \lambda. \tag{8.3}$$

For much of the analysis, a continuum description is useful. The continuum isotropic Hamiltonian has been developed in chapter 2, see in particular (2.57) for an isotropic chain, (2.28) for a square lattice and (2.33) for arbitrary lattices. We suppose that the original discrete system is defined on a square lattice. Letting $S_i^z \to S^z(\mathbf{r})$ and $S_j^z \to S^z(\mathbf{r} + \mathbf{a}_j)$, it is straightforward to expand this anisotropic part and arrive at the total continuum limit Hamiltonian that was derived from interactions on a square lattice (four nearest neighbors $\mathbf{a}_j = \pm a\hat{\mathbf{x}}, \pm a\hat{\mathbf{y}}$):

$$H = \frac{1}{2} JS^2 \int d^2r \left\{ (1 - \delta \cos^2 \theta) |\nabla \theta|^2 + \cos^2 \theta |\nabla \phi|^2 + 4\delta \sin^2 \theta \right\}. \tag{8.4}$$

In this chapter lengths such as r are assumed to be in units of the lattice constant a.

Exercise 8.1. Verify that Hamiltonian (8.4) results from taking a continuum limit of (8.2) on a square lattice. The isotropic part is the 1D to 2D generalization of (2.57), which should be included with the anisotropic part from (8.3).

A Hamiltonian similar to (8.4) was developed and the equations of motion analyzed by Hikami and Tsuneto [1], but with half the strength on the term $4\delta \sin^2 \theta$. Either by the techniques of chapter 2 or by the continuum limits of (5.23) and (5.24), one arrives at the dynamical equations (see also the work of Gouvêa et al [3] and Wysin et al [4]),

$$(JS)^{-1}\dot{\phi} \cos \theta = \left[2\delta - \frac{1}{2}(\delta |\nabla\theta|^2 + |\nabla\phi|^2) \right] \sin 2\theta - \left(1 - \delta \cos^2 \theta \right) \nabla^2 \theta \quad (8.5a)$$

$$(JS)^{-1}\dot{\theta} = \cos \theta \, \nabla^2 \phi - 2 \sin \theta \nabla\theta \cdot \nabla\phi. \quad (8.5b)$$

Exercise 8.2. Verify that the dynamical equations (8.5) result from the continuum Hamiltonian (8.4), for instance, by using functional derivatives (2.129) as described in chapter 2, and then finding the dynamics from (2.103).

Takeno and Homma [5] also arrived at equivalent dynamic equations, using different notation, but mainly analyzed them for a 1D model. Nikiforov and Sonin [6] considered a similar Hamiltonian but with the anisotropy inserted in a local interaction form, $+J\delta(S_i^z)^2$. This does not make a strong difference in the vortex structures that result or the basic physics, beyond some rescaling of parameters. The first step in the subsequent analysis will be to describe the kinds of vortex solutions, both static and dynamic, that are possible from these equations.

After describing the basic vortex structures (the spin waves present were described in chapter 6), the role played by vortices in the BKT topological transition will be described. Dynamics of vortices can be seen in the dynamic correlations above the BKT transition temperature, although this response is mixed in with a similar response due to spin waves. There is also considerable interest in the particular interaction between an individual vortex and the surrounding bath of spin waves.

8.2 In-plane and out-of-plane vortices

One can seek the static vortex solutions of (8.5), putting $\dot{\phi} = \dot{\theta} = 0$. Then (8.5b) has solutions of a form where $\phi = \phi(\varphi)$ depends only on azimuthal position coordinate φ and $\theta = \theta(r)$ depends only on radial position coordinate r, both measured relative to an arbitrary center, which is taken as the origin for now. This means $\nabla \phi \cdot \nabla \theta = 0$, and then (8.5b) becomes a 2D Laplace equation for the in-plane angle:

$$\nabla^2 \phi = 0. \tag{8.6}$$

Using either circular coordinates (r, φ) or Cartesian coordinates (x, y), the in-plane angle for a vortex solution is

$$\phi = q\varphi + \phi_0 = q \tan^{-1}\left(\frac{y}{x}\right) + \phi_0. \tag{8.7}$$

$q = \pm 1, \pm 2, \ldots$ is the *vorticity charge* or vorticity for short, and ϕ_0 is a constant of integration or *phase angle* that gives the direction of the spin field on the positive x-axis. For example, with $\phi_0 = \pi/2$, the spins point in the counterclockwise sense (for $q = +1$) as one moves along a path around the origin. The solutions are called vortices ($q = +1, +2, \ldots$) or anti-vortices ($q = -1, -2, \ldots$) according to the sign of q. A vortex solution has a gradient given by

$$\nabla \phi = \frac{1}{r}\frac{\partial \phi}{\partial \varphi}\hat{\varphi} = \frac{q}{r}\hat{\varphi}, \tag{8.8}$$

where the azimuthal unit vector is

$$\hat{\varphi} = \hat{z} \times \hat{r} = -\hat{x}\sin\varphi + \hat{y}\cos\varphi = (-\sin\varphi, \cos\varphi). \tag{8.9}$$

At the origin there is a singularity; it might be referred to as the *vortex core*. The field $\nabla \phi$ is analogous to the magnetic induction around an infinitely thin current-carrying wire[1]. There is a corresponding Ampere's law for the total vorticity charge enclosed within a path surrounding the vortex core:

$$\oint \nabla \phi \cdot d\ell = 2\pi q. \tag{8.10}$$

Although the in-plane angle for an individual vortex satisfies a 2D Laplace equation, the gradient can be rotated 90° to the radial direction by forming the object $\nabla \phi \times \hat{z} = (q/r)\hat{r}$. This is like the 2D electric field of a line of charge. This implies a 2D Poisson equation for that situation, corresponding to Gauss's Law,

$$\nabla \cdot (\nabla \phi \times \hat{z}) = 2\pi q\, \delta(\mathbf{r}). \tag{8.11}$$

Also with the identity,

$$\nabla \cdot (\nabla \phi \times \hat{z}) = \hat{z} \cdot (\nabla \times \nabla \phi) \tag{8.12}$$

(or via Stokes' theorem), one can see that an Ampere's law equation holds,

$$\nabla \times \nabla \phi = 2\pi q\, \delta(\mathbf{r})\hat{z}. \tag{8.13}$$

Although evaluated here for an individual vortex, one can see by superposition that (8.10) applies to any number of enclosed vortices, if q on the right-hand side is the

---

[1] A real current-carrying wire has no true singularity in its field because the current must be spread out over some cross-section. In the same way, there is no true singularity in the spin field of the original model on a lattice. The singularity appeared only as a result of the continuum limit from a discrete system.

algebraic sum of their vorticity charges. If the path of integration encloses a vortex and the opposite signed anti-vortex, this integral will be zero. This already shows that the vorticity charge follows a topological conservation law. Vortices can only be created into a system if the total vorticity generated through any process does not change. Every vortex created (in a closed system) must be balanced by a corresponding anti-vortex. In an open system, vortices or anti-vortices can enter or leave at the edges, and total vorticity charge is not conserved.

8.2.1 In-plane vortices

The vortex type is determined by the structure of the out-of-plane angle. Equation (8.5a) in the static limit has been found both from analytics [1, 5] and numerics [3, 4] to have two types of solutions. The first is the *planar vortex* or *in-plane vortex* solution, where all spins remain in the xy-plane:

$$\theta_{\rm ip} = 0. \tag{8.14}$$

In this case, the only contribution to the energy is from that associated with $|\nabla \phi|^2$,

$$E_{\rm ip} = \frac{1}{2}JS^2 \int {\rm d}^2r \, |\nabla\phi|^2 = \frac{1}{2}JS^2 \int_0^{2\pi} {\rm d}\varphi \int_{r_0}^R r \, {\rm d}r \, \frac{q^2}{r^2} = \pi JS^2 q^2 \ln \frac{R}{r_0}. \tag{8.15}$$

To make the result finite, a short-distance cutoff r_0, of the order of a lattice spacing, is applied. In addition, the large radius integration is cutoff at some upper limit R, which is of the order of the system size. The individual, perfect vortex, therefore, would have a weakly divergent energy in an infinite system. Estimates of the single in-plane vortex energy on a square lattice ([7], figure 6) in circular systems of varying radius R, give the same mathematical form as (8.15), but with π replaced by $\tilde{\pi} = 3.06$ and the cutoff distance $r_0 = 0.24a$, where a is the lattice spacing. In practice, the more important energetic result would be the total energy of a vortex–anti-vortex (VA) pair. It is an interesting exercise to show that the energy of a VA pair, separated by some distance R, takes a similar form[2], see the derivation of (9.118).

8.2.2 Out-of-plane vortices

There is a second type of vortex solution, the *out-of-plane vortex*, with non-zero θ that is maximum at the vortex core position, and decays away exponentially far away. Consider (8.5a) at small radius, where by symmetry only one of the values $\theta \approx p\pi/2$ is assumed, with $p = \pm 1$ being the *core polarization*. Expanding as $\theta \approx p\pi/2 + \epsilon$, with $\epsilon \ll 1$, and keeping the leading terms in ϵ, one obtains

$$-\left(4\delta - \frac{q^2}{r^2}\right)\epsilon - \left(\frac{{\rm d}^2\epsilon}{{\rm d}r^2} + \frac{1}{r}\frac{{\rm d}\epsilon}{{\rm d}r}\right) = 0. \tag{8.16}$$

Terms that were cubic in ϵ have been dropped. This allows for a power law solution of the form $\epsilon = r^n$. For small enough r, the term involving 4δ can be ignored relative

---

[2] Refer to a pair of long straight wires separated by distance R and carrying electric currents in opposite directions. The energy stored in their magnetic field is also finite and depends on $\ln R$.

to the others, which leads to the power being $n = |q|$. For large r, the asymptotics requires $\theta \to 0$ and the solution needs to obey the leading terms,

$$4\delta \cdot \theta - (1-\delta)\left(\frac{d^2\theta}{dr^2} + \frac{1}{r}\frac{d\theta}{dr}\right) = 0. \tag{8.17}$$

This has an exponential behavior with the exponent determined by the ratio $4\delta/(1-\delta)$. This gives the length scale of the far field, r_v, which is known as the *vortex core radius*,

$$r_v \equiv \frac{1}{2}\sqrt{\frac{1-\delta}{\delta}} = \frac{1}{2}\sqrt{\frac{\lambda}{1-\lambda}}. \tag{8.18}$$

Then, Gouvêa *et al* [3] determined the asymptotic form for the out-of-plane structure in an out-of-plane vortex, in which we include the core polarization $p = \pm 1$ and the possibility of $|q| \geq 1$,

$$\theta_{\text{oop}} \approx p\left[\frac{\pi}{2} - A_0\left(\frac{r}{r_v}\right)^{|q|}\right], \quad r \to 0 \tag{8.19a}$$

$$\theta_{\text{oop}} \approx pA_\infty \sqrt{\frac{r_v}{r}} e^{-r/r_v}, \quad r \to \infty. \tag{8.19b}$$

Parameters $A_0 = 3\pi/10$ and $A_\infty = \pi e/5$ are constants that are determined by matching these two solutions at the radius $r = r_v$. This is an approximate analysis, as there is no analytic solution known for all r. For the compound K_2CuF_4 with $\lambda \approx 0.99$, the core radius is $r_v \approx 5$, in lattice constants. That means the discreteness effects of the underlying lattice may not be too strong, and the continuum description should be acceptable. Note, however, there is not a corresponding length scale for the in-plane vortex solution.

The energy of the out-of-plane vortex can be evaluated approximately from this asymptotic solution. In [3] it is estimated that the out-of-plane vortex energy E_{oop} is greater than E_{ip} when $\lambda \ll 0.8$; conversely, E_{oop} is lower than E_{ip} for values of λ closer to 1. This is an indication of their ranges of stability, which is discussed below in section 8.3.

8.2.3 Discrete lattice vortex solutions

Static in-plane or out-of-plane vortex solutions on a lattice can be found numerically. They may be close to the approximate analytic solutions already described, such as (8.7) for the in-plane angle, and (8.19) for the out-of-plane component if $\lambda > \lambda_c$. These become modified slightly on a grid. To obtain the correct solution on a grid, start by setting the in-plane angles ϕ_n according to (8.7), for the vortex centered in the system. For instance, if an in-plane vortex is assumed, one can show from the equations of motion such as (2.103) that the static in-plane angles must satisfy

$$\sum_{(j)} \sin(\phi_i - \phi_j) = 0. \tag{8.20}$$

This must hold at all sites i, summing over the nearest neighbors j, indicated by (j). Most sites have z nearest neighbors, where z is the coordination number of the lattice. The sites at the boundary of the system have a few fewer neighbors. The equation is analogous to $\nabla^2 \phi = 0$ for the continuum problem. In the core region, the discrete and continuum problems give slight differences in the structure.

The discrete structure can be calculated numerically by the spin alignment technique described earlier in section 3.5.1. To do so, one uses the effective field $\mathcal{F}_i$ that acts on the spin at site i, as defined in (5.3). Referring also to (5.4), the static vortex structure is obtained when each spin is parallel to its field $\mathcal{F}_i$, leading to

$$\dot{\mathbf{S}}_i = \mathbf{S}_i \times \left[\mathcal{F}_i - \frac{\alpha}{S}(\mathbf{S}_i \times \mathcal{F}_i)\right] = 0. \tag{8.21}$$

The spin alignment algorithm iteratively sets each unit spin to be in the direction of its current $\mathcal{F}_i$, starting from some initial state. To be completely general, it is best to initiate the system also with some small out-of-plane spin components, breaking the planar symmetry numerically. This allows for the generation of an organized out-of-plane structure for the cases where that is the most stable vortex. Conversely, if the in-plane vortex is the more stable one, any initial out-of-plane structure will decay away under the iteration. The process is repeated until the structural changes are less than some desired precision. The stopping criterion could be based on the overall changes in spin components, or on the changes in the total energy, or both, see section 3.5.1. Indeed, any method that relaxes into the locally lowest energy configuration is acceptable.

Profiles of $S^z(r)$ as obtained by spin alignment relaxation are given in figure 8.1 for some different values of $\lambda > \lambda_c$, on a circular system of radius $R = 15a$ using a square lattice grid (lattice constant a). A strong demagnetization boundary condition was used in the spin alignment relaxation, setting the spins on the circular edge of the system to remain in the xy-plane and follow the circular boundary. The approximate solution (8.19a) for $r < r_v$ is shown as solid curves; the corresponding solution (8.19b) for $r > r_v$ is indicated with dotted curves. Note that the point $r = r_v$ always gives the value $\theta = \pi/5$ for the analytic solution, marked with an arrow on the plot. The comparison to the approximate analytic solution (8.19) is reasonable but not particularly good, probably because of the linearization used to arrive at that solution. In addition, the asymptotic solutions do not take into account some rearrangement of the in-plane angles in the core region on the discrete lattice.

8.3 Vortex instability

One can note that for $\lambda = 0$ (the XY limit), the vortex core radius for an out-of-plane vortex becomes $r_v = 0$. This would imply that there is no far field asymptotic region. The validity of the continuum limit to obtain that solution also might be questionable. One can note that $\lambda = 0.80$ gives $r_v = 1$, and $\lambda = 2/3$ gives $r_v = 1/\sqrt{2}$. One might expect that a vortex core radius near these values would be the limit where the continuum description must be strongly corrected by lattice discreteness effects. In fact, these effects cause an instability of both vortex types.

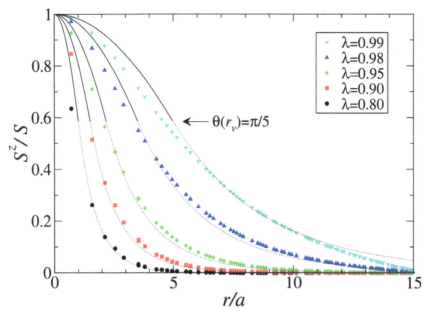

Figure 8.1. Out-of-plane vortex structure as seen in the out-of-plane spin component S^z as a function of radius from the vortex center, for indicated anisotropy constants $\lambda > \lambda_c$. Symbols were obtained by the spin alignment numerical relaxation calculations for a circular system of radius $R = 15a$, discretized on a square lattice. The solid curves are asymptotic expression (8.19a) for $r < r_v$; dotted curves are asymptotic expression (8.19b) for $r > r_v$. The curves connect at $r = r_v$, indicated with the arrow. Note how the results for $\lambda = 0.99$ have been limited by the Dirichlet ($S_z(R) = 0$) strong demagnetization boundary condition that was applied.

8.3.1 Numerical simulations for stability

The vortex stability can be tested by numerical simulations for the vortices on a lattice. The approach was carried out by Wysin *et al* [3, 4], integrating the discrete equations of motion (5.4) written in terms of Cartesian coordinates, including Landau–Lifshitz–Gilbert damping and using unit length spins $\mathbf{s}_i$,

$$\dot{\mathbf{s}}_i = \mathbf{s}_i \times \mathcal{F}_i - \alpha \mathbf{s}_i \times (\mathbf{s}_i \times \mathcal{F}_i) \qquad (8.22a)$$

$$\mathcal{F}_i = -\frac{\partial H}{\partial \mathbf{S}_i} = JS \sum_{(i,j)} \left(s_j^x \hat{\mathbf{x}} + s_j^y \hat{\mathbf{y}} + \lambda s_j^z \hat{\mathbf{z}} \right). \qquad (8.22b)$$

The sum for $\mathcal{F}_i$ contains only the nearest neighbor sites j. A damping parameter $\alpha = 0.1$ can be used to allow the system to seek energetically stable configurations. The equations can be integrated forward in time by any good method, such as fourth-order Runge–Kutta (RK4), provided one checks energy conservation for $\alpha = 0$. RK4 with a time step of $\Delta t = 0.04$ in units of $(JS)^{-1}$ works fairly well. A system on a square lattice as small as 40×40, with open boundaries, is sufficient to test vortex stability.

For these simulations, the initial condition for the in-plane angle is the planar vortex structure, equation (8.7). For the initial out-of-plane component s_i^z, one can check whether the system prefers the in-plane of out-of-plane structure by starting from some non-zero values. The obvious choices are 1) set all s_i^z to the same initial small constant $z_0 \ll 1$ and 2) set the initial s_i^z to a narrow distribution of random values with zero mean, which adds some small non-zero fluctuations, breaking the symmetry. In the first choice, if the out-of-plane vortex is more stable, the system will grow the s^z profile until it becomes an out-of-plane vortex profile with core polarization p the same sign as z_0. In the second choice, if the out-of-plane vortex is more stable, it could emerge with either sign of polarization. For both methods, if the planar vortex is more stable, it will emerge. The process can be carried out for a sequence of values of λ, thus mapping out the stability region of both vortex types. One can also make simulations with an approximate out-of-plane vortex initial condition, based on the asymptotic solution (8.19). Usually the vortex stability can be decided already by observing the simulation out to an integration time of $t = 100(JS)^{-1}$.

The results of this type of study are simple, although they can depend on the final time of the integration. For $\lambda < 0.72 \pm 0.01$, the vortex structure evolves towards the planar configuration, with $s_i^z \approx 0$ for all sites, and especially for those at the vortex core. On the other hand, for $\lambda > 0.72 \pm 0.01$, a non-zero s_i^z profile emerges, with its maximum at the vortex core, and decaying away as a function of the radius, similar to the asymptotic form found above. This is the stable out-of-plane vortex. This shows that there is a *critical anisotropy constant* $\lambda_c \approx 0.72 \pm 0.01$ at which the stability of each vortex type is destroyed, and the other type becomes the stable structure. There is an uncertainty of the order of ± 0.01 for a fairly simple reason: when $\lambda \approx \lambda_c$, the time scales of the evolution become longer. Only with fairly extended integrations in this region is it possible to accurately decide which type is more stable.

These vortex stability simulations can also be done on triangular ($z = 6$ nearest neighbors) and hexagonal ($z = 3$ nearest neighbors). One finds $\lambda_c \approx 0.62$ for the triangular lattice and $\lambda_c \approx 0.86$ for the hexagonal lattice. This suggests that a greater coordination number (more interactions) leads to greater stability of the planar vortex. This can be tested by some analysis of the spin energetics in the vortex core, described next.

8.3.2 Discrete energetics of vortex core stability

An analytic stability analysis of the vortex solutions is difficult to do in the continuum limit, because of the singularity at the vortex core. The continuum evaluation of vortex energy is not well defined and requires a cutoff. However, the energy should have a very specific value on the original lattice being considered. This leads one to consider an analysis of the vortex structure and energy in the core region, on the lattice as carried out in [8].

The approach is the following. Consider a vortex centered in one cell on a square lattice, see 8.2 for the notation. The in-plane spin structure is taken to follow (8.7),

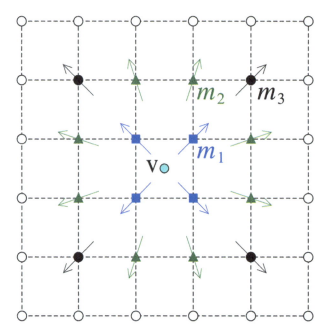

Figure 8.2. Notation for the spins sites with different out-of-plane spin components s^z around a $q = 1$ vortex centered at point v. There are four sites (blue squares) at radius $r_1 = \sqrt{1/2}$ with $s^z = m_1$, eight sites (green triangles) at radius $r_2 = \sqrt{5/2}$ with $s^z = m_2$ and four sites (black dots) at radius $r_3 = \sqrt{9/2}$ with $s^z = m_3$. In the outer surroundings $s^z = 0$ is assumed (gray dots). The in-plane spin directions follow (8.7) with $\phi_0 = 0$.

with the vortex at the origin, and with $q = 1$ and $\phi_0 = 0$. The out-of-plane spin components s_i^z are set to discrete values m_1, m_2, m_3, etc, symmetrically around the vortex center, according to increasing radius of the sites from the vortex center. All sites at the same radius are assigned the same value of m. Assuming this circularly symmetric arrangement, one can evaluate the Hamiltonian and write out the vortex energy-dependence on m_j, $j = 1, 2, 3...$, where j labels the radii r_j. A minimization of the total energy so obtained on the original lattice can reveal the energetic stability limits.

This procedure can be carried out first letting only the sites nearest to the core have non-zero out-of-plane component m_1, and all others $m_j = 0$. The four spins nearest the central core occupy at radius $r_1 = \sqrt{1/2}$ and have in-plane angles 45°, 135°, 225°, 315°, and their in-plane components are reduced by the factor $\cos\theta = (1 - m_1^2)^{1/2}$ due to their out-of-plane tilting. The differences of their in-plane angles are $\pm 90°$, so those four exchange bonds contribute an amount $4JS^2(1 - \lambda m_1^2)$ to the energy (above the ground state). There are also eight bonds connecting outward to the next eight sites at radius $r_2 = \sqrt{(3/2)^2 + (1/2)^2} = \sqrt{5/2}$, and their in-plane spin angles are $\tan^{-1}(1/3)$, $\tan^{-1}(3)$, etc. Then the interactions of the four central sites with the next eight always involve the same absolute difference in in-plane angle, whose cosine is

$$\cos(\phi_1 - \phi_2) = \frac{\mathbf{r}_1 \cdot \mathbf{r}_2}{r_1 r_2} = \frac{\frac{1}{2} \cdot \frac{3}{2} + \frac{1}{2} \cdot \frac{1}{2}}{\frac{1}{\sqrt{2}} \cdot \frac{\sqrt{5}}{\sqrt{2}}} = \frac{2}{\sqrt{5}}. \quad (8.23)$$

There are a total of eight bonds connecting r_1-sites to r_2-sites, with their exchange energy being $8JS^2(1 - \frac{2}{\sqrt{5}}\sqrt{1 - m_1^2})$. The total energy in the first 12 bonds in the core region is then

$$E_{\text{core}} = -4JS^2\left(\lambda m_1^2 + \frac{4}{\sqrt{5}}\sqrt{1 - m_1^2} - 3\right). \quad (8.24)$$

The energies in bonds farther out from the core are ignored. One can see that for small m_1, there are two competing terms, and an expansion to quadratic order gives

$$E_{\text{core}} \approx -4JS^2\left[\frac{4}{\sqrt{5}} + \left(\lambda - \frac{2}{\sqrt{5}}\right)m_1^2 - 3\right]. \quad (8.25)$$

An in-plane vortex has all $m_j = 0$. This shows that the energy of an in-plane vortex will be reduced with non-zero m_1, provided that $\lambda > \frac{2}{\sqrt{5}}$. Therefore, the in-plane vortex will become unstable towards formation of an out-of-plane vortex at a critical anisotropy parameter $\lambda_c = \frac{2}{\sqrt{5}} \approx 0.894$, in this initial approximation.

The energy E_{core} in (8.24) can be analyzed more generally for arbitrary $m_1 < 1$ by finding the extrema, according to

$$\frac{\partial E_{\text{core}}}{\partial m_1} = -8JS^2 m_1 \left[\lambda - \frac{2}{\sqrt{5}} \frac{1}{\sqrt{1 - m_1^2}}\right] = 0. \quad (8.26)$$

This has two solutions. The in-plane solution with $m_1 = 0$ exists for any λ, and has energy in these 12 bonds,

$$E_{\text{ip}} = 4JS^2(3 - 2\lambda_c), \quad \lambda_c = \frac{2}{\sqrt{5}}. \quad (8.27)$$

The out-of-plane solution exists only for $\lambda > \lambda_c$, with the out-of-plane component given by

$$m_1 = \sqrt{1 - (\lambda_c/\lambda)^2}. \quad (8.28)$$

This is zero at $\lambda = \lambda_c$ and then grows for $\lambda > \lambda_c$. The out-of-plane vortex energy in this approximation is

$$E_{\text{oop}} = 4JS^2(3 - \lambda - \lambda_c^2/\lambda), \quad \text{only for } \lambda > \lambda_c. \quad (8.29)$$

When the out-of-plane vortex is possible, E_{oop} is less than E_{ip}. Although both vortex types are possible for $\lambda > \lambda_c$, this explains why any small fluctuation will cause the in-plane structure to deform into an out-of-plane structure when $\lambda > \lambda_c$. This simple analysis shows that the energetics of the core region on a discrete lattice is responsible for the crossover from in-plane vortices for $\lambda < \lambda_c$ to out-of-plane vortices for $\lambda > \lambda_c$.

The calculation can be made more precise by considering sets of spins with non-zero m at larger radii. In addition to including four sites at radius $r_1 1/\sqrt{2}$ with $m_1 \neq 0$, the additional eight sites at radius $r_2 = \sqrt{5/2}$ with $m_2 \neq 0$ are included. These latter sites interact with another four sites (with $m_3 = 0$) at a third radius $r_3 = \sqrt{9/2}$. Including the total of 32 bonds among these 16 sites, the next approximation for the core exchange energy is

$$E_{\text{core}} = -4JS^2 \left[\lambda(m_1^2 + m_2^2) + \frac{4}{\sqrt{5}}\sqrt{1-m_1^2}\sqrt{1-m_2^2} \right.$$
$$\left. + \frac{4}{\sqrt{5}}\left(1 + \frac{4}{\sqrt{13}}\right)\sqrt{1-m_2^2} + \frac{4}{5}(1 - m_2^2) - 8 \right]. \quad (8.30)$$

The first term is the coupling of s^z-components; the remaining terms are the in-plane components. The variations with respect to m_1 or m_2 both are zero for an allowed solution:

$$\frac{\partial E_{\text{core}}}{\partial m_1} = -8\lambda(m_1 + m_2) + \frac{16}{\sqrt{5}} \frac{m_1}{\sqrt{1-m_1^2}} \sqrt{1-m_2^2} = 0 \quad (8.31a)$$

$$\frac{\partial E_{\text{core}}}{\partial m_2} = -8\lambda(m_1 + m_2) + \frac{16}{\sqrt{5}} \frac{m_2}{\sqrt{1-m_2^2}} \left[\sqrt{1-m_1^2} + \left(1 + \frac{4}{\sqrt{13}}\right) \right]$$
$$+ \frac{32}{5} m_2 = 0. \quad (8.31b)$$

It may not be easy to solve this nonlinear equation for m_1, m_2, but that is not necessary for determining the critical anisotropy. One only needs to assume that $m_1 = m_2 = 0$ for the in-plane vortex, and the value of λ at which they begin to take non-zero values is λ_c. This means an expansion of these equations for small values $m_1 \ll 1$, $m_2 \ll 1$, is all that is needed. That gives a pair of homogeneous linearized equations,

$$+(A - \lambda)m_1 - \lambda m_2 = 0 \quad (8.32a)$$

$$-\lambda m_1 + (B - \lambda)m_2 = 0 \quad (8.32b)$$

$$A = \frac{2}{\sqrt{5}}, \quad B = \frac{4}{\sqrt{5}}\left(1 + \frac{1}{\sqrt{5}} + \frac{2}{\sqrt{13}}\right). \quad (8.32c)$$

A non-trivial solution (the out-of-plane vortex) begins to appear only if the determinant D of the coefficients is zero:

$$D = (A - \lambda_c)(B - \lambda_c) - \lambda_c^2 = 0. \tag{8.33}$$

This gives the critical anisotropy parameter,

$$\lambda_c = \frac{AB}{A + B} \approx 0.716. \tag{8.34}$$

This is a considerable improvement over the initial approximation, and much closer to the value $\lambda_c \approx 0.72$ found in numerical simulations. Note that the *strength* of the easy-plane critical anisotropy is

$$\delta_c = 1 - \lambda_c \approx 0.284. \tag{8.35}$$

This procedure of discrete core analysis can be taken successively to larger core regions. By using three sets of sites with non-zero m_j, the critical parameter moves down to $\lambda_c \approx 0.7044$, somewhat smaller than the initial simulations on small systems. When simulations are performed for 50 × 50 or larger, one finds that indeed λ_c is actually slightly less than this value. Zaspel and Godinez [19] found a procedure to calculate the discrete vortex energy, successively summing over larger and larger core regions. A similar procedure can be applied to the calculation of the critical anisotropy, allowing for an extrapolation to the infinite sized limit [10], where $\delta_c \to 0.29659051$ ($\lambda_c = 0.70340949$). This result means than in any square-lattice materials where $\lambda < \lambda_c$, the static vortex structure is essentially planar, with no large out-of-plane component. This discrete analysis also is able to give the relative sizes of the m_j; it produces the profile of the out-of-plane vortex, at its stability limit. That is actually a dynamic mode of oscillation associated with the crossover between the two vortex types.

This same discrete core analysis can be extended to other lattices. For the triangular lattice (six nearest neighbors) one obtains $\delta_c \to 0.38714359$ ($\lambda_c = 0.61285641$), and for the hexagonal lattice (three nearest neighbors) one has $\delta_c \to 0.16704412$ ($\lambda_c = 0.83295588$). The greater concentration of sites in the triangular lattice makes the system more continuous, leading to a wider range of stability for the out-of-plane vortices. The opposite is true for the hexagonal lattice, which more greatly favors the in-plane vortices.

Overall, the results show that the vortex instability is a discrete lattice effect, something that cannot be described in the continuum limit. This is quite different from the instability found for kinks in a 1D easy-plane ferromagnet, using the continuum limit.

Exercise 8.3. Consider a vortex centered in a cell of a hexagonal lattice, surrounded by six nearest sites at distance $r_1 = 1$ with $s^z = m_1$, and another six second nearest sites at distance $r_2 = 2$ with $s^z = m_2$. Assume the in-plane structure given in (8.7), with a $q = 1$ vortex at the origin and $\phi_0 = 0$. (a) Show that the core energy in the 24 bonds among these sites is

$$E_{\text{core}} = -6JS^2 \left[\lambda \left(m_1^2 + m_1 m_2 \right) + \frac{1}{2}\left(1 - m_1^2\right) \right.$$
$$\left. + \sqrt{1 - m_1^2}\sqrt{1 - m_2^2} + \frac{5}{\sqrt{7}}\left(1 - m_2^2\right) - 4 \right] \quad (8.36)$$

(b) Defining a constant $A_h \equiv 1 + 5/\sqrt{7}$, show that the determinant of the resulting linearized stability system goes to zero at a critical anisotropy parameter given by

$$\lambda_c = -A_h + \sqrt{A_h(A_h + 2)} \approx 0.869. \quad (8.37)$$

Simulations find a vortex instability near $\lambda \approx 0.86$, see further discussion of vortex stability issues in [8–10].

8.4 Moving in-plane and out-of-plane vortices

Vortex motion couples the in-plane angle with the out-of-plane spin components, because ϕ_i and S_i^z are canonically conjugate coordinates. This means that even an in-plane vortex will develop non-zero out-of-plane spin components if it is moving at some velocity. Here we take a brief look at how vortex motion modifies the structures, and allows one to define vortex momentum and an effective mass. Taken together, these properties describe a magnetic vortex as a particle-like object, with an interesting dynamical equation of motion for its center.

The leading correction to vortex structure, due to motion at velocity $\mathbf{v}$, can be obtained from the dynamics equations (8.5) by assuming a traveling solution of the form $\mathbf{S}(\mathbf{r} - \mathbf{v}t)$, implying that the time derivative is replaced by $-\mathbf{v} \cdot \nabla$. At the same time, one can assume small perturbations ϕ_1 and m_1 on top of the static vortex solutions ϕ_0 and $m_0 = \sin\theta_0$ corresponding to the unperturbed vortex. The velocity $\mathbf{v}$ is also considered a small parameter of the same order as ϕ_1 and m_1. This is simple for the case of in-plane vortices, where initially $m_0 = \theta_0 = 0$. The dynamic equations linearized in the perturbations are now

$$-\mathbf{v} \cdot \nabla \phi_0 \approx JS\left[\left(4\delta - |\nabla\phi_0|^2\right)\sin\theta_1 - (1 - \delta)\nabla^2\theta_1 \right] \quad (8.38a)$$

$$-\mathbf{v} \cdot \nabla \theta_1 \approx JS\, \nabla^2 \phi_1. \quad (8.38b)$$

Away from the vortex center, we used $\nabla^2 \phi_0 = 0$ from (8.11). It is difficult to obtain an exact solution even for this linearized system. That may not be too surprising, considering that we know the continuum limit does not describe the vortex core correctly, which is exactly the region where all the interesting action is. Assuming δ might be close to unity, the first equation gives a result that has a singular point at $r = |q|/2\sqrt{\delta}$,

$$\sin\theta_1 \approx \frac{-\mathbf{v}\cdot\nabla\phi_0}{JS\left(4\delta - |\nabla\phi_0|^2\right)}. \tag{8.39}$$

Using (8.8), and $\hat{\boldsymbol{\varphi}} = \hat{\mathbf{z}} \times \hat{\mathbf{r}}$, this can be expressed as

$$\sin\theta_1 \approx \frac{q(\mathbf{v}\times\mathbf{r})\cdot\hat{\mathbf{z}}}{JS\left(4\delta\, r^2 - q^2\right)}. \tag{8.40}$$

This has been mostly applied to the simplest cases, $q = \pm 1$. Note that coordinate $\mathbf{r}$ is measured from the instantaneous location of the vortex core. Along a line through the vortex core and parallel to $\mathbf{v}$, one has $\theta_1 = 0$. Crossing this line, the sign of θ_1 reverses. This field then depends on the sine of the angle between $\mathbf{v}$ and $\mathbf{r}$. As a vortex moves through a square lattice, however, one can expect that the precise description of the vortex core is not contained in this solution. There will be strong discreteness effects if the core moves too close to a lattice site, violating the assumptions made to obtain the continuum limit.

By (8.38b), the perturbation ϕ_1 in the in-plane field is determined by θ_1, but one can see that the particular solution for ϕ_1 will be of the order of $\mathbf{v}^2$, hence, it is considered negligible to leading order.

For out-of-plane vortices, a similar perturbation analysis for a moving vortex can be performed [3]. However, the corrections θ_1 and ϕ_1 due to motion are both small compared to the original unperturbed angles. A very slight deviation in the out-of-plane structure similar to that for in-plane vortices appears, and in addition, the in-plane structure obtains a slight asymmetry on the line parallel to the velocity. The asymmetry that appears due to motion can be readily demonstrated by performing numerical simulations of a pair of vortices, for both in-plane and out-of-plane vortex types [4]. The radial dependence, however, may not follow expression (8.40) very closely in the core region: there is no sign change of θ_1 near $r = |q|/2\sqrt{\delta}$.

8.5 The vortex unbinding transition

Models with XY or easy-plane symmetry have always attracted much attention because of the presence of vortices and the existence of a special kind of phase transition, known as the BKT topological phase transition [11–13], associated with the thermal generation of vortices in VA pairs. Pair creation is also called *vortex unbinding*, imagining that before a pair is created, the vortex and anti-vortex are somehow coupled tightly together (this should not be taken literally, because they do not exist until they are created). Here we take a short look at some aspects of this transition, as obtained from combinations of MC and SD simulations.

According to the Ampere's law (8.10) for the gradient of the in-plane spin angle, the total vorticity within a boundary is quantized. If no vortices cross the boundary, the total vorticity is conserved, which implies that vortex charges can be created only in VA pairs ($q = +1$ created along with $q = -1$). The pair interaction energy is shown later in (9.118) to depend on the separation of the vortex from the anti-vortex. This energy is in addition to self-energies for both the vortex and the anti-vortex, as given in (8.15). Although there is an energy cost for creation of a pair, it also increases the entropy of the system, which increases the likelihood of pair formation as the temperature is raised. We can obtain a very approximate estimate of the temperature at which VA pair formation becomes favorable, by asking what temperature is needed to make the change in free energy become negative.

A vortex could be created somewhere in a circular system of radius R, and its corresponding anti-vortex at another arbitrary location, separated by distance X_{12}. The two self-energies combined are approximately (at $\lambda = 0$, for in-plane vortices)

$$E_1 + E_2 = 2 \times \pi J S^2 \ln \frac{R}{r_0}, \tag{8.41}$$

where the short cutoff is on the order of $a/4$ in terms of the lattice spacing a (derived for square lattice). The pair interaction energy is

$$U_{\text{pair}} = 2\pi J S^2 q_1 q_2 \ln \frac{R}{X_{12}} = -2\pi J S^2 \ln \frac{R}{X_{12}}, \tag{8.42}$$

where the two vortices are of opposite sign. This is an attractive interaction that competes with the self-energies. The total energy change for creation of a pair is estimated as the sum,

$$\Delta E = E_1 + E_2 + U_{\text{pair}} = 2\pi J S^2 \ln \frac{X_{12}}{r_0}. \tag{8.43}$$

If the vortex and anti-vortex become free from each other, one can let $X_{12} \to R$, as the typical maximum separation. On the other side is the question of the entropy available for the pair creation, which is estimated very roughly by supposing that both the vortex and anti-vortex can be created almost anywhere in the system. Allowing the precision of a location to be on the order of r_0^2, the entropy change in the creation process is about

$$\Delta S \approx k_B \ln \left(\frac{R^2}{r_0^2} \times \frac{R^2}{r_0^2} \right) = 4 k_B \ln \frac{R}{r_0}. \tag{8.44}$$

Both the vortex and the anti-vortex contribute a factor within the logarithm. Then the change in Helmholtz free energy for a fixed temperature is estimated as

$$\Delta F = \Delta E - T \Delta S \approx 2\pi J S^2 \ln \frac{R}{r_0} - 4 k_B T \ln \frac{R}{r_0}. \tag{8.45}$$

The typical pair separation has been set to R. ΔF becomes negative at an estimate of the BKT unbinding temperature,

$$k_B T_c \approx \frac{\pi}{2} J S^2. \tag{8.46}$$

The particular number obtained here is not very precise, in fact, the number 0.699 replaces $\pi/2$ for the XY model ($\lambda = 0$). But the analysis shows that the entropy available to vortex pairs is sufficient to overcome their energy cost, and once the temperature is high enough, thermally generated vortices and anti-vortices are a certainty. To obtain better estimates of T_c requires other theory or we can defer to MC simulations, which have the ability to find T_c as a function of anisotropy parameter λ and other factors.

8.6 Monte Carlo simulations of the Berezinskii–Kosterlitz–Thouless transition

A hybrid cluster MC method described in section 4.5.4 can be used to produce some thermal averages for the 2D XY model. Here we show some typical results that exhibit the main features of the BKT transition; this is a wide field of study and many other results can be found in the literature. Recall that the hybrid scheme described earlier involves a combination of single-spin Metropolis steps and over-relaxation steps that modify all spin components, and Wolff cluster steps acting only on the in-plane spin components. One MC step is the combination of an effective pass through the lattice by each of these methods. We show data using four to ten bins of 30 000 to 50 000 MCS for a total of 200 000 to 300 000 MCS. One can take averages over more bins and more MCS but for many quantities this is already sufficient to see the major trends. In addition, it is important to use a range of temperatures and system sizes; the variation with system size helps to determine T_c precisely according to finite size scaling methods. The calculations have been performed on $L \times L$ square lattice systems with periodic boundary conditions.

First consider the XY model ($\lambda = 0$). The internal energy per spin ($e = E/N = \langle H \rangle/N$) in figure 8.3 starts near zero at low temperature and has a slow rise with increasing T, but exhibits nothing dramatic. The specific heat per spin ($c = C/N$) also shown in figure 8.3 does show a strong peak whose position moves to lower temperature with increasing system size. However, c must tend to $c \to 1$ at $T \to 0$ (classical low-T limit for two degrees of freedom) and it also necessarily must tend to zero at very high T, because a spin system has a maximum energy state, unlike a usual mechanical system with a quadratic kinetic energy. For the largest systems the peak is near $k_B T/JS^2 \approx 0.73$.

The in-plane magnetic moment (m_{xy}, see figure 4.15) does make a dramatic drop to values close to zero, around the same region as the peak in specific heat. This quantity acts as an *order parameter* because it better indicates the geometrical structure in the average spin configuration. At low temperatures, neighboring spins are strongly correlated over a larger length scale known as the correlation length; the whole set of spins tend to move around together if the correlation length is of the order of the system size. Because we can only simulate a finite size, that limits the maximum correlation length and the drop of m_{xy} towards zero at T_c is rounded. In principle, m_{xy} would be zero for $T > T_c$ in an infinite sized system, but finite size effects prevent that.

The vortex unbinding can be seen by measuring the number density of vortices plus anti-vortices as a function of temperature. Vorticity can be found using a discrete version of Ampere's law (8.10) around each square cell or plaquette of the

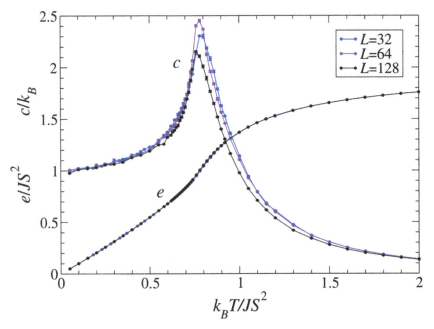

Figure 8.3. Energy and heat capacity per spin versus temperature in $L \times L$ XY models from the hybrid cluster MC simulations.

lattice. Each plaquette contains four spins and four bonds between neighboring spin pairs. Going in the counterclockwise sequence around a plaquette, one needs to form the sum of changes in in-plane spin angles, from which the vorticity within that path is found by

$$q = \frac{1}{2\pi} \sum_{\text{edge bonds}} \Delta\phi_{\text{bond}}. \qquad (8.47)$$

The symbol $\Delta\phi_{\text{bond}}$ indicates the difference of two spin angles in a bond, $\phi_{i+1} - \phi_i$, taken in the counterclockwise sense and shifted into the branch $-\pi/2 \leqslant \Delta\phi_{\text{bond}} \leqslant +\pi/2$. This way, each plaquette has only the possible values $q = 0, \pm 1$. We should also like to mention that theoretically there is also the possibility of vortices with larger integral values, such as $q = \pm 2$. They can appear especially if there are vacancies or non-magnetic impurities in the lattice. A sum around the outside of a block of four neighboring plaquettes can be used to locate these higher vorticity charges. Then once all vortices of different charges are located, an averaged vortex number density ρ can be defined by dividing the total absolute charges by the number of sites,

$$\rho = \frac{1}{N} \sum_k |q_k|, \qquad (8.48)$$

where k here labels the charges found. Note that if there are charges larger then ± 1 present they contribute with a weight proportional to the size of the charges. It is

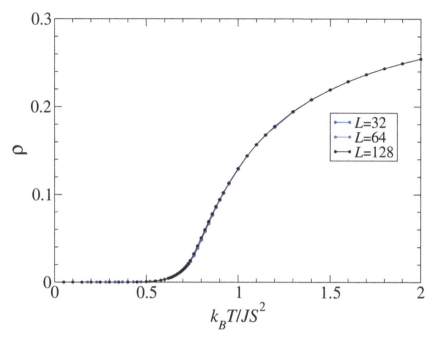

Figure 8.4. Number density of vortices (both charge signs) per unit cell versus temperature in $L \times L$ XY models from the hybrid cluster MC simulations.

possible to sketch a state where every plaquette has a unit charge, but alternating in signs. Then the maximum possible value is $\rho = 1$, however, that is an improbable higher energy state that will contribute little in thermal averages.

The resulting vortex number density $\rho(T)$ are shown in figure 8.4. The density is nearly zero for temperatures $k_B T/JS^2 < 0.5$, and starts to make a stronger rise around $0.6 < k_B T/JS^2 < 0.8$. There is very weak dependence on system size, with the initial rise in ρ occurring at slightly lower T for larger L. Ironically, the vortex density does not give a precise way to determine the critical temperature.

8.6.1 Estimations of the critical temperature—Binder's cumulant

A quantity that can sometimes indicate the precise location of T_c is Binder's fourth-order cumulant, denoted U_L. It is a quantity that indicates the shape of the probability distribution of the system magnetic moment, $\mathbf{M} = \sum_n \mathbf{S}_n$. Suppose a component M_x is being measured at high temperature. Because $\mathbf{M}$ is very random and typically of a small value, due to cancellations among the spins, one component M_x will tend to have a Gaussian distribution around a zero mean. On the other hand, at very low temperature, the whole system is in a highly correlated state, with all spins nearly parallel in the xy-plane (for models of easy-plane symmetry). The vector $\mathbf{M}$ will tend to be distributed mostly near a circle in the xy-plane of radius near NS. The corresponding distribution of a component M_x will not be Gaussian, but instead, have a minimum at $M_x = 0$ between two peaks at positive and negative values.

The moments of the spin distribution indicate its shape. Suppose one component M_x has a normalized Gaussian probability distribution for high temperature,

$$p(M_x) = \frac{1}{\sqrt{2\pi\sigma_x^2}} \exp\left\{-\frac{M_x^2}{2\sigma_x^2}\right\}. \tag{8.49}$$

The second- and fourth-order moments of this distribution are found by

$$\langle M_x^2 \rangle = \int_{-\infty}^{+\infty} dx\, p(M_x) M_x^2 = \sigma_x^2 \tag{8.50a}$$

$$\langle M_x^4 \rangle = \int_{-\infty}^{+\infty} dx\, p(M_x) M_x^4 = 3\sigma_x^4. \tag{8.50b}$$

Then for high temperature one has $\langle M_x^4 \rangle = 3\langle M_x^2 \rangle^2$. On the other hand, at very low temperature, with all spins nearly aligned, one expects $\langle M_x^4 \rangle = \langle M_x^2 \rangle^2$. For a single variable such as this, Binder's fourth-order cumulant is then defined by

$$U_L = 1 - \frac{\langle M_x^4 \rangle}{3\langle M_x^2 \rangle^2} \tag{8.51}$$

which takes the limiting values $U_L \approx 2/3$ at low temperature and $U_L \approx 0$ at high temperature, regardless of system size. Also at $T = T_c$ the spin distribution becomes independent of L and U_L is expected to have a universal value independent of L. Usually U_L is measured in a set of MC simulations with the curves for $U_L(T)$ at different L plotted together, and their common crossing point gives an estimate of T_c.

For the 2D XY model, the definition needs to be generalized to use both the x and y in-plane spin components. Each one will be Gaussian distributed with the same variance $\sigma_x = \sigma_y$ for high temperature. Then a sum $M_x^2 + M_y^2$ has the following leading order moments,

$$\langle M_x^2 + M_y^2 \rangle = \langle M_x^2 \rangle + \langle M_y^2 \rangle = \sigma_x^2 + \sigma_y^2 = 2\sigma_x^2 \tag{8.52a}$$

$$\langle (M_x^2 + M_y^2)^2 \rangle = \langle M_x^4 + 2M_x^2 M_y^2 + M_y^4 \rangle = 3\sigma_x^4 + 2\sigma_x^2\sigma_y^2 + 3\sigma_y^4 = 8\sigma_x^4. \tag{8.52b}$$

This shows that $\langle (M_x^2 + M_y^2)^2 \rangle = 2\langle M_x^2 + M_y^2 \rangle^2$ at high temperature, so it inspires the definition of the appropriate function for planar symmetry,

$$U_L = 1 - \frac{\langle (M_x^2 + M_y^2)^2 \rangle}{2\langle M_x^2 + M_y^2 \rangle^2}. \tag{8.53}$$

One sees that $U_L \approx 1/2$ at low temperature and $U_L \approx 0$ at high temperature, for any L. It is expected to have a universal value at T_c independent of L, as mentioned for the single component U_L, which is used to estimate T_c.

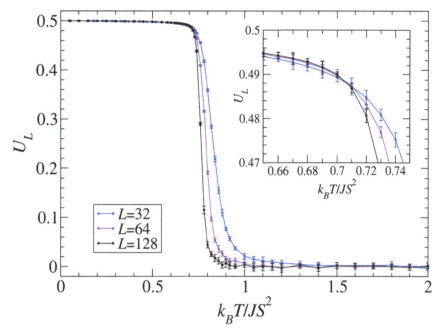

Figure 8.5. Binder's fourth-order cumulant as defined in (8.53) versus temperature in $L \times L$ XY models from the hybrid cluster MC simulations. The inset shows where the curves cross, which gives an estimate of the BKT transition temperature $k_B T_c/JS^2 \approx 0.70 - 0.71$.

A typical plot of U_L using definition (8.53) for different sizes is shown in figure 8.5. Although there is a crossing point, it is hard to locate unless the statistical noise in the data is very well under control. It does give an estimate of T_c for $\lambda = 0$ in the range $0.70 < k_B T_c/JS^2 < 0.71$. For the XY model the crossing appears too close to where $U_L \approx 1/2$, however, which means it is not the best way to find T_c.

Exercise 8.4. Check that the distribution $p(M_x)$ in (8.49) is correctly normalized, so that $\int_{-\infty}^{+\infty} \mathrm{d}x \, p(M_x) = 1$. Then verify the second- and fourth-order moments of the distribution in (8.50).

Exercise 8.5. Consider an isotropic Heisenberg model, with total magnetic moment $\mathbf{M} = (M_x, M_y, M_z)$. Show that an appropriate definition of the fourth-order cumulant that will distinguish the shape of the low and high temperature distributions of $\mathbf{M}$ is

$$U_L = 1 - \frac{3}{5} \frac{\langle \mathbf{M}^4 \rangle}{\langle \mathbf{M}^2 \rangle^2}. \tag{8.54}$$

What are the limiting values at low and high temperatures?

8.6.2 Estimations of the critical temperature—scaling of susceptibility

Binder's fourth-order cumulant can give an estimate of T_c for some models but it can require extensive calculations to obtain even two digit accuracy. A better approach has been to use the finite size scaling (FSS) of in-plane magnetic susceptibility χ_{xx}, such as that applied by Cuccoli *et al* [14] for $\lambda = 0$. In this theory, the in-plane susceptibility for temperatures near and below T_c is expected to follow a power law scaling with system size, according to the form

$$\chi_{xx} \propto L^{2-\eta}, \tag{8.55}$$

where $\eta(T)$ is the exponent for the in-plane spin–spin correlations below T_c. The exponent determines the large-distance form of decay of static spin–spin correlations. The static correlation function is $C^{xx}(\mathbf{r})$, as defined in (4.118). According to the BKT theory the spin–spin correlations are expected to be of power law form below T_c but change to a faster exponential form above T_c. To see this change in behavior, the correlation function has been plotted on a log-log scale in figure 8.6 for equally spaced temperatures from $k_B T/JS^2 = 0.1$ to $k_B T/JS^2 = 1.5$, for a 128×128 system. The expected behavior of the correlations is

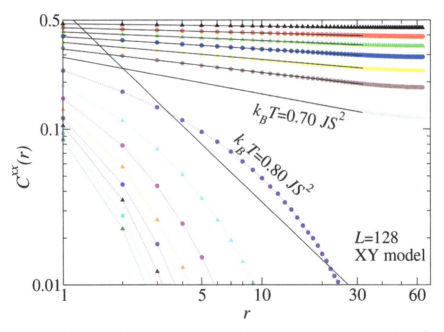

Figure 8.6. From hybrid cluster MC for 128×128 (data points), the static in-plane spin correlation function versus radius, for a sequence of evenly spaced temperatures $k_B T/JS^2 = 0.1, 0.2, \ldots 1.5$ (lowest T at top). Black lines are fits to power law decays $C^{xx} \propto r^{-\eta}$ using the data at $r \leqslant 32$, which works well only for $T < T_c$ where $k_B T_c/JS^2 \approx 0.70$. A fit is also shown for the data at $k_B T/JS^2 = 0.80$, which is poor because above T_c the correlations decay with an exponential form, see (8.56). Dotted lines on the higher temperature curves are guides for the eye, not fits.

$$C^{xx}(r) = \begin{cases} C_0 r^{-\eta} & T < T_c \\ C_1 r^{-1/2} \exp(-r/\xi) & T > T_c. \end{cases} \quad (8.56)$$

where C_0, C_1 are fitting constants and ξ is a *correlation length* in the state with unbound vortices, that also may depend on temperature. In figure 8.6 only fits to the power law form have been made, mostly for $T \leqslant T_c$, but also for one temperature above T_c. Only the data for $r \leqslant 32$ were used in the fits, because on a finite system, the correlation function is defined only to a maximum radius $L/2$, and we want to avoid the end region where finite size effects change the power law decay ($C^{xx}(r)$ must be symmetrical around $r = L/2$ in a finite system). The power law fits are very good for $T < T_c$. For $T < T_c$ the fits can be used to estimate $\eta(T)$, if desired, but a form that takes into account the limited system size is better. However, it is seen strikingly that the power law form does not work at all above the critical temperature.

Recalling the relation (4.120), it is then feasible that the exponent η appearing in the low temperature decay of $C^{xx}(\mathbf{r})$ should affect the scaling properties of χ_{xx} with system size. It is possible to take our MC results and fit χ_{xx} to the form (8.55) and extract $\eta(T)$ from the slope of $\ln(\chi_{xx})$ versus L, for each temperature simulated. However, the BKT theory determines that $\eta = 1/4$ at the critical point, regardless of the system size. Then, the FSS method to find T_c is to plot $\chi_{xx}/L^{2-\eta}$ versus temperature, for different system sizes. The resulting curves tend to cross at the universal point $\eta(T_c) = 1/4$, and even in XY systems with vacancies this crossing is a very sharp point. This type of plot is shown in figure 8.7. In order to display the details better, the error bars are not shown; the crossing is close to the point $k_B T_c/JS^2 \approx 0.700$, which is very good precision considering that the temperature increments of the MC simulation here were $k_B \Delta T/JS^2 = 0.01$ in these units. Then, for temperatures below T_c, the power law correlations of spin components, which are limited by the system size L, give one of the best ways to determine the transition temperature of the infinite sized system, T_c.

8.6.3 A measure of spin twist resistance—the helicity modulus

An easy-plane system also has another type of susceptibility, known as the helicity modulus per spin, $\Upsilon(T)$, that measures how the system reacts to imposing a slight spin twist Δ across the system along one coordinate. The helicity modulus maps over into superfluid density ρ_s according to $\Upsilon = (\hbar^2/m)\rho_s$, when the XY model is used to describe such a system with macroscopic quantum phase effects.

The spin twist would be an imposed angular displacement of the in-plane angle from one side of the system to the opposite side. It can be thought of as a generalized applied field, similar to a magnetic field. If the spins in the system are strongly correlated, as at low temperature, there will a strong resistance to a twist, and Υ is large. If the spins are more disordered and moving mostly independently, as at high temperature, there will be little resistance and Υ will be small. It can be expected that $\Upsilon(T)$ will diminish with temperature mostly in the region where vortices are being thermally produced, because the creation of VA pairs leads to weaker correlations for spins on opposite sides of the pairs.

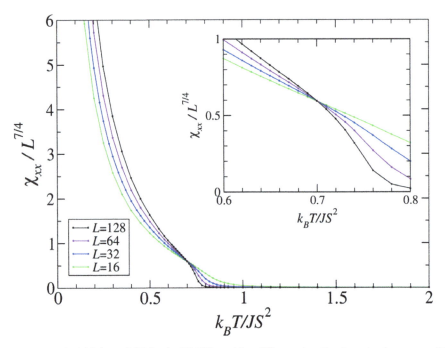

Figure 8.7. From hybrid cluster MC for the 2D XY model at different sizes $L \times L$, an in-plane susceptibility component (see the definition in (4.110)) scaled by $L^{2-\eta}$, under the assumption that $\eta(T_c) = 1/4$. The inset shows where the curves cross, which gives an estimate of the BKT transition temperature $k_B T_c/JS^2 \approx 0.70$. The error bars are suppressed to bring out the crossing point.

The definition of helicity modulus can be based on the free energy F, and finding the system's susceptibility to spin twists,

$$\Upsilon \equiv \frac{1}{N}\frac{\partial^2 F}{\partial \Delta^2}. \qquad (8.57)$$

Compare the definition of magnetic susceptibility, equation (4.59). This definition is used in the limit of infinitesimal twist Δ. From the definition of free energy (4.34), the required derivative gives for any spin Hamiltonian,

$$\Upsilon = \frac{1}{N}\left\{\left\langle \frac{\partial^2 H}{\partial \Delta^2}\right\rangle - \beta\left[\left\langle \left(\frac{\partial H}{\partial \Delta}\right)^2\right\rangle - \left\langle \frac{\partial H}{\partial \Delta}\right\rangle^2\right]\right\}, \qquad (8.58)$$

where $\beta = (k_B T)^{-1}$. A simple model that imposes a twist on bonds in the $\hat{\mathbf{x}}$-direction is to modify the in-plane interactions in the spin Hamiltonian (8.2) into a form such as

$$J\left(S_i^x S_j^x + S_i^y S_j^y\right) = JS^2 \sin\theta_i \sin\theta_j \cos(\phi_i - \phi_j)$$
$$\longrightarrow JS^2 \sin\theta_i \sin\theta_j \cos(\phi_i - \phi_j - \Delta). \qquad (8.59)$$

The extra Δ inside the cosine then shifts the minimum angular displacement in the bond to Δ. In theory, it should only be included for bonds in one space direction. One then finds the following derivatives,

$$\frac{\partial H}{\partial \Delta} = J \sum_{(i,j)} (\hat{\mathbf{e}}_{ij} \cdot \hat{\mathbf{x}}) \left(S_i^x S_j^y - S_i^y S_j^x \right) \tag{8.60a}$$

$$\frac{\partial^2 H}{\partial \Delta^2} = \frac{1}{2} J \sum_{(i,j)} \left(S_i^x S_j^x + S_i^y S_j^y \right), \tag{8.60b}$$

where $\hat{\mathbf{e}}_{ij}$ is a unit vector in the direction from i to j, that selects only the bonds along one direction. Although $\frac{\partial H}{\partial \Delta}$ may have a small average value, its fluctuations contribute to Υ. The second partial derivative of H is seen to be half of the in-plane exchange energy. Once $\Delta \to 0$ is assumed, the direction of the bonds does not matter. Then these expressions can be implemented for MC calculations with symmetrization:

$$\frac{\partial H}{\partial \Delta} = \frac{1}{2} J \sum_{(i,j)} \left(S_i^x S_j^y - S_i^y S_j^x \right). \tag{8.61}$$

This averages over all nearest neighbor bonds. Note that each term is proportional to a triple cross product, $\hat{\mathbf{z}} \cdot (\mathbf{S}_i \times \mathbf{S}_j)$.

Exercise 8.6. Suppose a system has a spin twist field Δ that shifts the in-plane angles between neighboring spins along one direction, in a form such as that in (8.59), i.e. $\phi_i - \phi_j \to \phi_i - \phi_j - \Delta$. From the definition (4.34) of free energy F, verify that (8.58) for Υ results.

According to the renormalization group theory of Kosterlitz and Thouless [13], in an infinite system the helicity modulus jumps downward from a finite value $(2/\pi)k_B T_c$ at the critical temperature to zero for higher temperatures. This is similar in behavior to the average magnetic moment. Because the value at T_c is then known (but only for $L \to \infty$), this gives a way to estimate T_c, by plotting $\Upsilon(T)$ and marking its intersection with the straight line,

$$\Upsilon = \frac{2}{\pi} k_B T. \tag{8.62}$$

Examples of the application of this approach are given in figure 8.8(a) for the XY model and in figure 8.9(a) for the PR model. For the initial calculations of $\Upsilon(T)$, averages are from 3.2×10^5 to 1.28×10^6 MCS, with shorter runs on the larger systems for practical reasons. The curves of $\Upsilon(T)$ become steeper in the critical region as L increases. The intersection of the data with the straight line of (8.62)

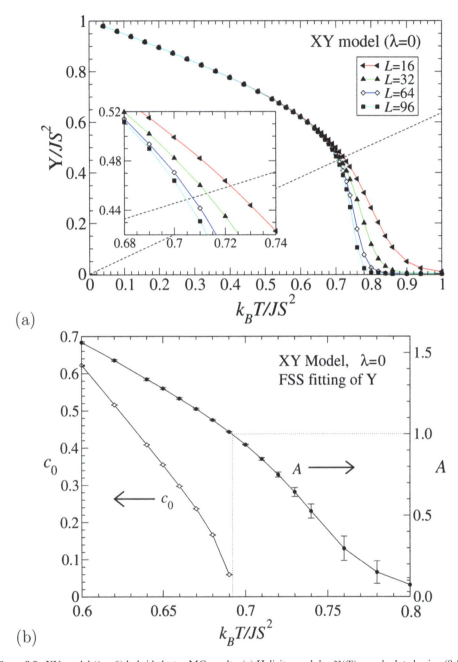

Figure 8.8. XY model ($\lambda = 0$) hybrid cluster MC results. (a) Helicity modulus $\Upsilon(T)$ as calculated using (8.58) on indicated $L \times L$ systems. The intersection of the dashed line (8.62) with the data gives an overestimate of the critical temperature, $k_B T_c/JS^2 \approx 0.705$. (b) FSS analysis of the same data by fitting to expression (8.63) for $T < T_c$ to obtain $c_0(T)$ and to expression (8.65) to obtain $A(T)$, leading to $k_B T_c/JS^2 \approx 0.692$ for the infinite system.

moves towards lower estimates of T_c for larger L. This method then always overestimates T_c, even at the largest possible L that can be simulated. At $L = 96$, it gives $k_B T_c/JS^2 \approx 0.705$ for the XY model and $k_B T_c/JS^2 \approx 0.912$ for the PR model. More data at $L = 160$ for the PR model change the estimate slightly, to $k_B T_c/JS^2 \approx 0.910$.

A finite size scaling analysis [15] can be used to improve the estimates. That theory shows that for temperatures below T_c, the helicity scales with system size according to [16]

$$\frac{\pi \Upsilon}{2k_B T} = 1 + c_0 \coth\left[2c_0 \ln(L/L_1)\right], \tag{8.63}$$

where c_0 and L_1 are fitting constants. This expression is valid only for $T < T_c$, and once the critical temperature is reached, the fitting constant c_0 goes to zero. The fits to MC with this expression also become very poor or nearly impossible once T passes above T_c. The fits are made versus different system sizes L, for a set of temperatures in the critical region. Note also that once $c_0(T)$ has been determined, then the estimate for the helicity modulus in the limit $L \to \infty$ is obtained as

$$\Upsilon_\infty(T) = \begin{cases} \dfrac{2}{\pi} k_B T [1 + c_0(T)] & T < T_c \\ 0 & T \geqslant T_c. \end{cases} \tag{8.64}$$

The helicity modulus jumps suddenly to zero at the transition.

Another scaling expression that has been widely applied for planar spin models is

$$\frac{\pi \Upsilon}{2 k_B T} = A(T)\left[1 + \frac{1}{2\ln(L/L_0)}\right], \tag{8.65}$$

where both $A(T)$ and L_0 are fitting parameters. The expression is known to be exact at $T = T_c$, where $A(T_c) = 1$ also gives a method to obtain good estimates of the critical temperature. The expression does not fit very well for temperatures away from T_c, thus the appearance of a tight fit is a good indication of being close to T_c. Note that for the FSS approach to work well may require substantial computational effort, in order to have data with low enough statistical noise.

Examples of applying these scaling methods are shown in figure 8.8(b) for the XY model and figure 8.9(b) for the PR model. For the XY model, the same initial data used for systems with $L = 16, 32, 64, 96$ and 3.2×10^5 to 1.28×10^6 MCS were used to perform the scaling analysis. One finds that $c_0(L)$ can be fit quite well by expression (8.63), giving $c_0 > 0$, until T reaches 0.70, at which point a nonlinear least squares scheme (one found in the 'fit' command of gnuplot) is unable to come to convergence. The point where $c_0 \to 0$ should be T_c, but it is not exactly located. The fitting on $A(L)$ works both above and below T_c, and gives a value $A = 1$ at a temperature slightly higher than $k_B T/JS^2 = 0.69$. Thus, the two fittings bracket the

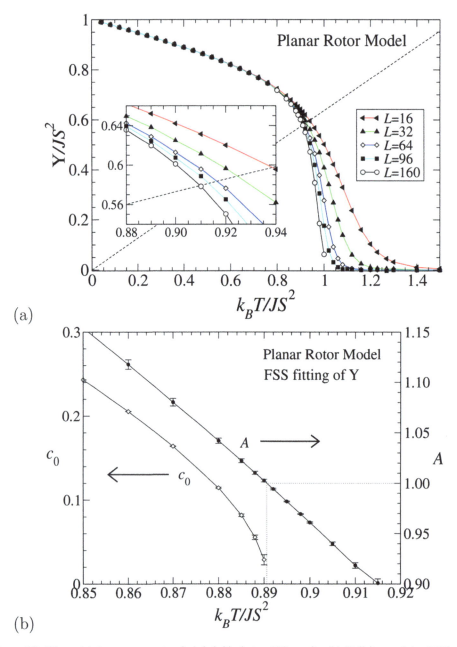

Figure 8.9. PR model (two-component spins) hybrid cluster MC results. (a) Helicity modulus $\Upsilon(T)$ as calculated using (8.58) on indicated $L \times L$ systems. The intersection of the dashed line (8.62) with the data gives an overestimate of the critical temperature, $k_B T_c / JS^2 \approx 0.912$. (b) FSS analysis of more extensive data at $L = 20, 32, 48, 64, 96$, fitting to expression (8.63) for $T < T_c$ to obtain $c_0(T)$ and to expression (8.65) to obtain $A(T)$, giving $k_B T_c / JS^2 \approx 0.891$ for the infinite system.

critical temperature, and combined they indicate $k_B T_c/JS^2 \approx 0.692$, slightly lower than that from using the susceptibility.

For the PR model, a new set of longer runs, all to 20×10^6 MCS, was made for temperatures near T_c, on systems with $L = 20, 32, 48, 64, 96$, to perform the FSS analysis. The result in figure 8.9(b) is quite consistent between where c_0 goes to zero and A goes to unity, leading to the estimate, $k_B T_c/JS^2 \approx 0.891$, rather high compared to the transition in the XY model. Of course, the extra S^z spin component present in the XY model is allowing the spins to move out of the xy-plane, causing their in-plane components to be shorter, on average. This implies a lower value of effective in-plane coupling JS_{xy}^2, hence the transition takes place at lower temperature in the XY model. One can come to the same conclusion by considerations of entropy for the two models.

One sees that the application of finite size scaling drastically improves the utility of the helicity data for estimations of T_c and extraction of the infinite sized limit.

8.6.4 Dependence of critical temperature on anisotropy and vacancies

It is interesting to consider the easy-plane model for different values of the anisotropy parameter λ, besides the case $\lambda = 0$ (the XY model). As λ is increased towards 1, the easy-plane anisotropy becomes weaker, and the spins can move more out of the xy-plane. This extra freedom of motion corresponds to greater out-of-plane fluctuations, and smaller typical in-plane spin components. The critical temperature for the XY model was found to be proportional to JS^2, however, one can think that S^2 in that expression is really the in-plane part of the spins, which now becomes smaller with weaker easy-plane anisotropy. This suggests that T_c should decrease as λ increases. This is indeed the case, as seen in the upper curve in figure 8.10, showing estimates of $T_c(\lambda)$ from finite sized scaling for model (8.2).

Another effect that has been of great interest, and which we only briefly touch on here, is the question of spin vacancies in the system. A small percentage of the lattice sites could be occupied by non-magnetic impurities, hence, those sites are not magnetically active and, for all practical purposes, behave as a missing site or spin vacancy. Although this might seem to be an unimportant perturbation to the system, even a few vacancies can have a drastic effect on both the static averages and the dynamics. In particular, simple calculations for individual vortices show that they will tend to be pinned on vacancies [17], which produce a strong pinning potential [18]. Furthermore, Zaspel et al [19] showed that a vortex formed around a vacancy will require weaker easy-plane anisotropy (smaller parameter δ) to remain in the in-plane state. While the critical anisotropy $\lambda_c \approx 0.7034$ for unpinned vortices, this changes to $\lambda_{cv} \approx 0.9545$ for a vortex pinned on a vacancy (both numbers for square lattice). The missing exchange bonds in the vicinity of the vacancy are responsible for the lower vortex energy there and the energetic preference to remain in the in-plane state with $s_z = 0$. Pereira and co-workers [20] also found that vacancies are responsible for causing oscillations of vortices and, in addition, an increasing density of vacancies further weakens their attractive potential [21] and the oscillation frequencies.

In the lower curve of figure 8.10, data are shown for $T_c(\lambda)$ with 16% of the sites randomly occupied by 'repulsive' vacancies [22], of density $\rho_{vac} = 0.16$, which were not allowed to be closer than the second nearest neighbor distance of $\sqrt{2}$. This model forces all vacancies to be surrounded with eight spin-occupied sites, which allows for each vacancy to be searched easily for the presence or absence of a pinned vorticity. In fact, this model produces pinned vorticities $q = \pm 2$ as well as the more usual $q = \pm 1$. One sees that in addition to the possibility of higher vorticity objects, the presence of vacancies lowers the BKT transition temperature. The maximum possible density of repulsive vacancies is $\rho_{vac} = 0.25$, however, because they are placed suddenly and randomly (or, *quenched*), the maximum density that can be placed is $\rho_{vac} \approx 0.1872$.

In another model of *randomly placed* vacancies [23], with no restrictions on their positions, their maximum density is not limited below unity. The hybrid cluster MC calculations on this model used averages over 4 to 64 realizations of the vacancy positions, on $L \times L$ systems with $L = 16, 32, 64, 96$ and 160. T_c could be found most precisely by fitting $C^{xx}(r)$ to power law form and locating the point where $\eta(T) = 1/4$. The results were also verified by comparison with Binder's fourth-order cumulant and the helicity modulus. For instance, figure 8.11 shows the trends in $\Upsilon(T)$ with increasing vacancy density. The intersection of the data with the straight line of (8.62) monotonically moves to lower T with increasing ρ_{vac}, exhibiting a striking trend. When many vacancies are present, the lattice becomes

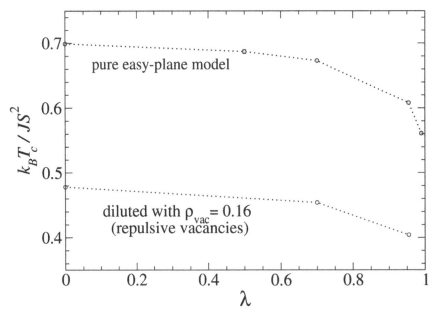

Figure 8.10. From scaling analysis of susceptibility and also checked by Binder's fourth-order cumulant, results for the BKT transition temperature in the easy-plane model (8.2) from hybrid cluster MC. The upper curve is the square lattice model, fully occupied, while the lower curve is a square lattice model with 16% of the sites occupied by repulsive vacancies. Dotted lines are guides for the eye.

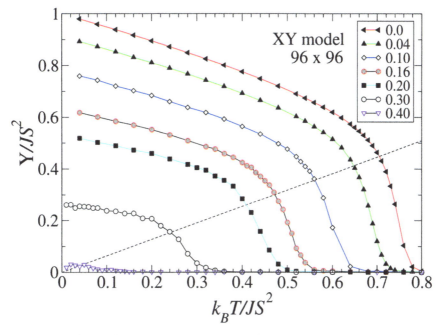

Figure 8.11. Results for helicity modulus versus temperature in the square lattice XY model with random vacancies, as calculated from expression (8.58) in hybrid cluster MC. Curves are labeled by the vacancy density ρ_{vac}. The dashed line is (8.62), whose intersection with the data gives an estimate of T_c. No transition is present for $\rho_{\text{vac}} > 0.41$, which coincides with the percolation limit.

rather disconnected, and this causes greatly weakened exchange interactions and correlations. This results in a lowering of T_c with increasing vacancy density, see figure 8.12. The effect of lower T_c with increasing ρ_{vac} is so strong that around a vacancy density of $\rho_{\text{vac}} \approx 0.41$, the transition temperature goes to zero. This vacancy density is then seen to be complementary to the site percolation threshold of about $p_c \approx 0.59$ for a square lattice, which is the minimum site density needed to have a connection across the system. Thus, the BKT transition is only present as long as the lattice is sufficiently occupied that the exchange interactions are able to percolate throughout the entire system. There must be sufficient couplings to lead to a correlation length equal to the system size at $T = T_c$. Note that even though the system may be somewhat disconnected due to all the vacancies, even so, at the critical temperature all spins in the system tend to be strongly correlated, which means that a cluster update scheme is still a necessity for accurate results. See [23] for further details.

Exercise 8.7. Consider a square lattice with a vacancy (missing spin) at the origin. Assume that a nearly planar vortex with in-plane angles from (8.7) is centered on the missing site. (a) Using the nearest out-of-plane spin components m_1 at radius $r_1 = 1$ (four sites) and m_2 at radius $r_2 = \sqrt{2}$ (four sites), estimate the core energy $E_{\text{core}}(m_1, m_2)$ for a vortex pinned on the vacancy. (b) Perform the

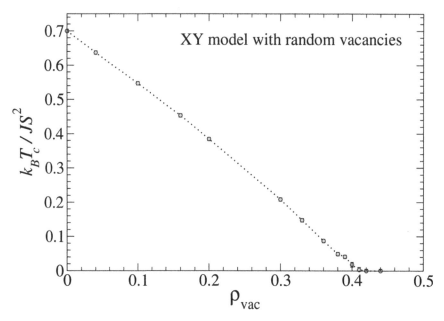

Figure 8.12. From scaling analysis of susceptibility and also checked by helicity modulus [23], results for the BKT transition temperature in the square lattice XY model (8.2) with $\lambda = 0$ from hybrid cluster MC, with randomly placed vacancies of density ρ_{vac}. The dotted line is a guide for the eye.

stability analysis in the limit $m_1 \ll 1$ and $m_2 \ll 1$, and estimate the critical anisotropy parameter λ_{cv} where the planar pinned vortex develops non-zero out-of-plane components.

8.7 Dynamic correlations in XY models

So far, we have discussed mostly the static properties due to vortices in thermal equilibrium in easy-plane models. Now we consider some aspects of the time-dependent response, especially, in the space- and time-dependent *dynamic correlation functions* or dynamic structure functions, such as $S^{xx}(\mathbf{q}, \omega)$. The basic definition and properties of $S^{xx}(\mathbf{q}, \omega)$ were introduced in section 5.7.5. The dynamic structure function gives an indication of the types of excitations in the system, in terms of indicating what frequency oscillations are present at a chosen wave vector $\mathbf{q}$. A significant intensity in $S^{xx}(\mathbf{q}, \omega)$ usually gives a direct indication of the spin wave spectrum, which appears as sharp peaks for $\mathbf{q}$ and ω related by the spin wave dispersion relation. For more localized objects such as vortices, the space Fourier transform (FT) of such a structure determines its intensity in $\mathbf{q}$-space, and its motion if present determines how intensity in $S^{xx}(\mathbf{q}, \omega)$ appears as a function of frequency.

The dynamic correlation function is the space–time FT of a space and time displaced spin pair correlation function. The definition (4.118) for the static correlations $C^{xx}(\mathbf{r})$ was generalized in (5.189) to include time displacement. We apply it here in discrete form using **n** to represent lattice sites and **r** the possible displacements, using unit spins,

$$C^{xx}(\mathbf{r}, t) = \frac{1}{N} \sum_{\mathbf{n}} \frac{1}{N_t} \sum_{t_0} \langle s_{\mathbf{n}}^x(t_0) s_{\mathbf{n}+\mathbf{r}}^x(t_0 + t) \rangle \tag{8.66}$$

N is the total number of spin sites and N_t is the number of time samples. This function averages pairs of spins at constant space and time displacements **r** and t, respectively. The sums can be changed to integrations if a continuum expression is needed. We are not subtracting out the thermal average values, as was introduced originally for the static correlations, equation (4.118). The dynamic correlation function is the space and time FT, but converted to continuum weight as explained in chapter 5 in the derivation of (5.195). The structure functions are defined per frequency per squared wave vector, so the fundamental physical unit is $(JS)^{-1}a^2$.

We can calculate $S^{xx}(\mathbf{q}, \omega)$ from data on a discrete space–time grid from simulations. The most efficient way to achieve this is via the convolution theorem and using the relation (5.205). For convenience we change to a dimensionless quantity, taking out the units,

$$S^{xx}(\mathbf{q}, \omega) = \left(\frac{JS}{a^2}\right) \times \left(\frac{2\pi}{t_{\text{end}}} \frac{2\pi}{N_x a} \frac{2\pi}{N_y a}\right) \langle |A^x(\mathbf{q}, \omega)|^2 \rangle. \tag{8.67}$$

This is determined from the space–time FT of the unit spin field,

$$A^x(\mathbf{q}, \omega) = \left(\frac{t_{\text{end}}}{2\pi N_t}\right) \sum_{t_l} \left[\left(\frac{a}{2\pi}\right)^2 \sum_{\mathbf{r}} s_{\mathbf{r}}^x(t_l) e^{-i\mathbf{q}\cdot\mathbf{r}}\right] e^{i\omega t_l}. \tag{8.68}$$

Recall from chapter 5 that these are applied to a $N_x \times N_y$ system that is integrated over time out to final time t_{end} using a sequence of N_t time steps, and **q** and ω are discrete variables in a simulation. Obviously the correlations of other spin components, $S^{yy}(\mathbf{q}, \omega)$ and $S^{zz}(\mathbf{q}, \omega)$ can be defined as well. For XY symmetry models, one should have $S^{xx}(\mathbf{q}, \omega) = S^{yy}(\mathbf{q}, \omega)$. In simulations, we calculate both S^{xx} and S^{yy} separately and then average them together into one in-plane structure function.

8.7.1 Hybrid Monte Carlo–spin dynamics simulations

The hybrid MC–SD approach for the thermal time dynamics of spin systems was introduced in section 5.3. Here we describe a few details of that approach and discuss the main results, especially concerning the dynamic correlations.

The data shown here have been produced for a 128×128 system. The initial part of the calculation is the hybrid cluster MC method that combines Wolff cluster moves with over-relaxation and Metropolis single-spin steps. For a given temperature, the system was first equilibrated for $N_{\text{skip}} = 20\,000$ MCS. Then, a total of $500\,000$ MCS were made, saving a configuration of the system every 1000 MCS.

That produces $N_s = 500$ different states that were used to initiate the time integration. N_s is also the number of states from the ensemble that are used for averages.

For the time integration, the system was integrated using RK4 with an algorithm time step $\Delta t_a = 0.04(JS)^{-1}$, which is adequate for good energy conservation. Note that the Landau–Lifshitz equations of motion are integrated *without damping*, such that the motion takes place in the microcanonical ensemble. However, the results from the N_s different initial conditions will be averaged over in the canonical ensemble for the chosen temperature of the MC. The algorithmic time step Δt_a is rather short. Therefore, data for finding time FTs were saved at the data time interval $\Delta t = 6\Delta t_a$, which allows for the investigation of lower frequency responses. In order to take advantage of a fast FT (FFT) algorithm, the total number of data points in time should be a power of 2, for the most commonly used algorithms. Therefore we used total time samples $N_t = 2^p$, with FFT index $p = 12$ producing 4096 time samples. Thus, the interval of time integration for one initial condition is

$$t_{\text{end}} = N_t \Delta t = 2^{12} \times 6 \times 0.04(JS)^{-1} = 983.04(JS)^{-1}. \tag{8.69}$$

This implies that the lowest non-zero frequency to be analyzed in the FFT is

$$\Delta \omega = \omega_1 = \frac{2\pi}{t_{\text{end}}} = 6.3916 \times 10^{-3}(JS). \tag{8.70}$$

The resulting FTs have 4096 frequency points, however, because periodicity in time is imposed, the second half of the FT is symmetric with the first half, and contains no extra information (aliasing). For this reason, only about the first quarter of the final FFT data are useful. We avoid the part affected by aliasing and we focus on the low frequency part of the spectrum.

For the FT in space, a lattice of 128×128 has a very large number of possible wave vectors. There is no practical need to have data for every possible **q**, instead, it is already very useful to produce only the space FT data at selected **q**. We used only $\mathbf{q} = (q_x, 0)$, $(0, q_x)$ and (q_x, q_x), for $q_x = 2\pi m/L$, $m = 0, 1, 2 \ldots L/2$. The limited values of **q** mean that an FFT is not needed, rather, the space FT is performed with the help of a table of the appropriate needed phase factors $\exp(i\mathbf{q} \cdot \mathbf{r})$, which are constant during the calculations.

8.7.2 Low temperature dynamic structure function

At low temperatures $T < T_c$ the primary excitations present are spin waves. This is seen easily in the dynamic structure functions, some examples of which are shown in figure 8.13 for $k_B T = 0.1JS^2$. Note that S^{xx} is has been calculated as the average of xx and yy spin component correlations, due to the XY symmetry. Wave vectors of the form $\mathbf{q} = (q_x, 0)$ with $q_x = 2\pi m/L$ and m incremented by 4 are displayed. Essentially the only feature present in S^{xx} and also in S^{zz} is a very narrow peak, at the same location in both S^{xx} and S^{zz}, whose position shifts with **q**. At low wave

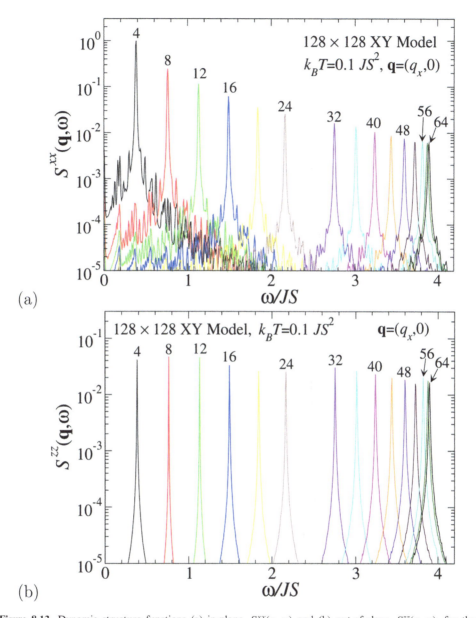

Figure 8.13. Dynamic structure functions (a) in-plane, $S^{xx}(\mathbf{q}, \omega)$ and (b) out-of-plane, $S^{zz}(\mathbf{q}, \omega)$, for the square lattice FM XY model ($L = 128$, $\lambda = 0$) at $k_B T = 0.1 JS^2$, obtained from hybrid cluster MC combined with SD, from averages over 500 initial states, as explained in the text. The wave vectors are $q_x = (2\pi/L)m$ with index m in increments of 4, as indicated.

vectors the in-plane intensity is stronger; at the highest wave vectors, the out-of-plane intensity is more dominant. The peak location is then used to map out the dispersion relation for spin waves, $\omega(\mathbf{q})$, and the simulations give its temperature-dependence. One should also mention that there are many smaller subpeaks present in $S^{xx}(\mathbf{q}, \omega)$; especially for smaller $\mathbf{q}$. These can be attributed to finite size effects and spin wave interference effects [24], that depend on the system size L. For the most part, compared to the spin wave peak the subpeaks are weak and become smeared out with increasing temperature.

Another example of the behavior of the structure functions is given in figure 8.14, at a slightly higher temperature, $k_B T = 0.4JS^2$, still below the transition temperature $k_B T_c \approx 0.70JS^2$. Now, the peaks for any chosen $\mathbf{q}$ have shifted to slightly lower frequency and their widths have increased due to the greater thermal fluctuations. Even so, the spin wave peak is the only significant feature in the structure functions.

A third example is given in figure 8.15 for a temperature approximately the same as the critical temperature, $k_B T = 0.7JS^2$. Now, the behavior of the curves is significantly different, with the growth of much stronger intensity appearing in the low frequency range, below the spin wave peak, primarily for $S^{xx}(\mathbf{q}, \omega)$. This is referred to as a *central peak* (CP). At this temperature, it is relatively weak, compared to the height of the spin wave peak. The in-plane response has the appearance of a spin wave peak riding on top of a wider background. The out-of-plane response has more the appearance of significantly broadened spin waves.

The spin wave dispersion relation $\omega(\mathbf{q})$ for the 2D XY ferromagnet on a square lattice, obtained by MC–SD, can be obtained from the peaks in the structure function $S^{xx}(\mathbf{q}, \omega)$, for a range of $\mathbf{q}$. The results in figure 8.16 have been displayed for the three temperatures below or near T_c, at wave vectors along both the (10)-direction and the diagonal (11)-direction. The solid curves are the $T = 0$ linearized spin wave theory, equation (6.36). The obvious feature of the simulation data is that the spin wave frequency for finite temperature becomes less than the linearized theory predicts. This spin wave *softening* is a typical effect due to thermal fluctuations, which weaken the effective or thermally averaged nearest neighbor exchange interactions, and hence the frequencies. One way to describe this effect theoretically is an approach known as the self-consistent harmonic approximation, which involves a temperature-dependent rescaling of J. Furthermore, the spin wave peak cannot be easily identified for $T > T_c$, especially at small wave vectors, because it becomes lost in the CP of $S^{xx}(\mathbf{q}, \omega)$. This is a rather dramatic effect and again it is closely associated with the BKT transition.

8.7.3 Higher temperature dynamic structure function and central peak

A fourth example of the dynamic structure functions is given in figure 8.17, for a temperature above the critical temperature, $k_B T = 0.9JS^2$. In this case, the response in $S^{xx}(\mathbf{q}, \omega)$ at the lower wave vectors has lost the spin wave peak. The only significant feature is a fairly wide CP. At the larger wave vectors closer to the edge of the Brillouin zone, there is a weak signature of the spin wave. In the out-of-plane correlations of $S^{zz}(\mathbf{q}, \omega)$, the spin wave peak can be identified for all the wave vectors. The curves there have the appearance of a spin wave signature riding on top of a broader but weak CP. Also, at these higher temperatures, the tails are fairly smooth and do not exhibit any features due to finite size effects.

The appearance of the CP as the temperature passes T_c can be illustrated further by replotting some results at particular $\mathbf{q}$'s on a linear scale, for the sequence of

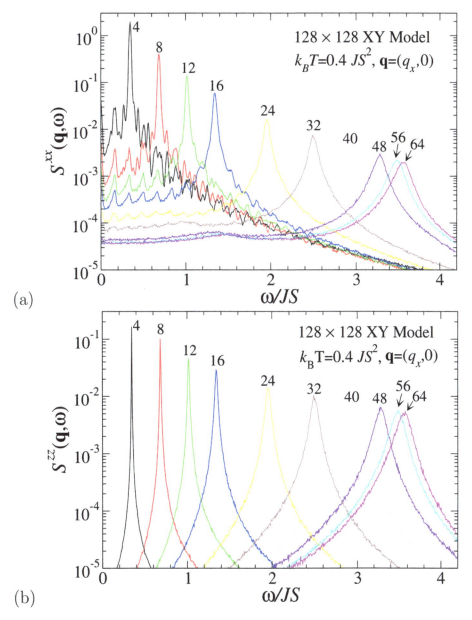

Figure 8.14. Dynamic structure functions (a) in-plane, $S^{xx}(\mathbf{q}, \omega)$ and (b) out-of-plane, $S^{zz}(\mathbf{q}, \omega)$, for the square lattice FM XY model ($L = 128$, $\lambda = 0$) at $k_BT = 0.4JS^2$, from averages over 500 initial conditions, with each $q_x = 2\pi m/L$ for indicated m values. Note the shifts of the peaks to lower frequencies and their increased widths compared to $k_BT = 0.1JS^2$ in figure 8.13.

temperatures. The first example is seen in figure 8.18 for the wave vector index $m = 8$ which gives $\mathbf{q} = (\pi/8, 0)$. The curves are labeled by dimensionless temperature variable $\tilde{T}$, defined from

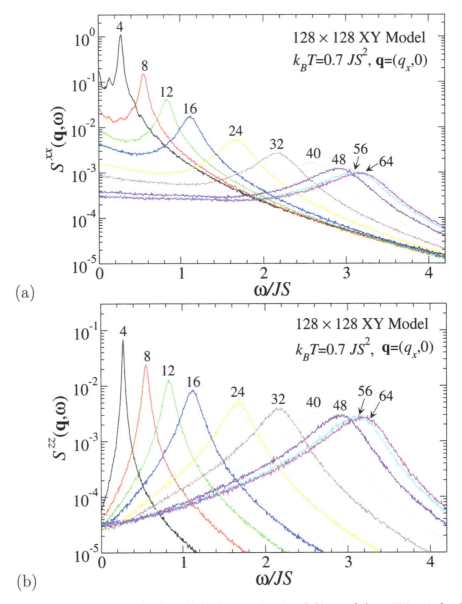

Figure 8.15. Dynamic structure functions (a) in-plane, $S^{xx}(\mathbf{q}, \omega)$ and (b) out-of-plane, $S^{zz}(\mathbf{q}, \omega)$, for the square lattice FM XY model ($L = 128$, $\lambda = 0$) at $k_B T = 0.7 JS^2$, from averages over 500 initial conditions, with each $q_x = 2\pi m/L$ for indicated m values. Note the presence of CP intensity primarily in S^{xx}.

$$\tilde{T} \equiv k_B T/JS^2. \tag{8.71}$$

For $S^{xx}(\mathbf{q}, \omega)$, the relatively sharp spin wave peaks at $\tilde{T} < 0.7$ have minimal intensity in the low frequency region. For $\tilde{T} = 0.7 \approx \tilde{T}_c$, there is a weak low frequency tail of the spin wave that extends to $\omega = 0$. At the higher temperature $\tilde{T} = 0.9$, $S^{xx}(\mathbf{q}, \omega)$ does not exhibit any spin wave peak, but only CP intensity.

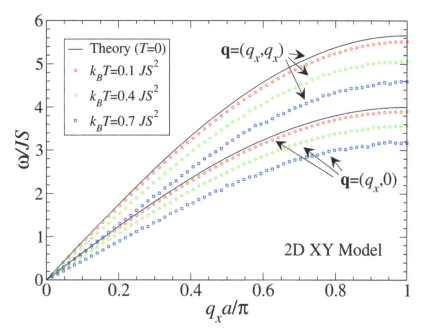

Figure 8.16. The spin wave dispersion relations $\omega(\mathbf{q})$ for the square lattice FM XY model (8.2) with $\lambda = 0$ obtained from hybrid cluster MC combined with SD, using the peak found in $S^{xx}(\mathbf{q}, \omega)$. Solid curves are the $T = 0$ spin wave theory, equation (6.36). The spin wave frequencies at the indicated $T > 0$ were obtained from the peak in $S^{xx}(\mathbf{q}, \omega)$ for wave vectors along both (10)- and (11)-directions. At $k_B T > 0.7JS^2$ the spin wave peak is not distinct from the CP, so $\omega(\mathbf{q})$ was not be estimated.

The spin wave does appear as a reasonably strong but wide peak in $S^{zz}(\mathbf{q}, \omega)$ near $\omega/JS \approx 0.39$. Another example of this behavior is given in figure 8.19, for index $m = 16$ which gives $\mathbf{q} = (\pi/4, 0)$. In this case the CP that appears in $S^{xx}(\mathbf{q}, \omega)$ for $\tilde{T} > \tilde{T}_c$ is wider than that at lower wave vector. Again, the spin wave peak is easily identified in $S^{zz}(\mathbf{q}, \omega)$, although it has a long tail down to zero frequency for $\tilde{T} > \tilde{T}_c$.

8.7.4 Ideal gas model for vortex thermodynamics

Intensity in the dynamic correlations at low frequency and low wave vector is due to larger or extended objects with slow motions. Although spin waves are extended objects, they possess a well-defined $\mathbf{q}$ and we know that their dynamic response at finite temperature is mainly a softened and broadened peak at a reduced spin wave frequency. We have seen that the CP in $S^{xx}(\mathbf{q}, \omega)$ appears only as the temperature surpasses T_c, where we know that a considerable density of vortices and anti-vortices has been generated. The spin waves are still visible in $S^{zz}(\mathbf{q}, \omega)$ for $T > T_c$. Vortices generally have extended in-plane spin profiles, and can be excited with slow velocities, which suggests that they should produce in-plane CP intensity. They may also produce a weaker out-of-plane CP, due to the out-of-plane spin structure in moving vortices. Of course, if only slowly moving vortices are present, then they can produce only small S^z spin components, and small contributions to $S^{zz}(\mathbf{q}, \omega)$. These facts suggest that thermally generated vortices in the BKT transition are primarily

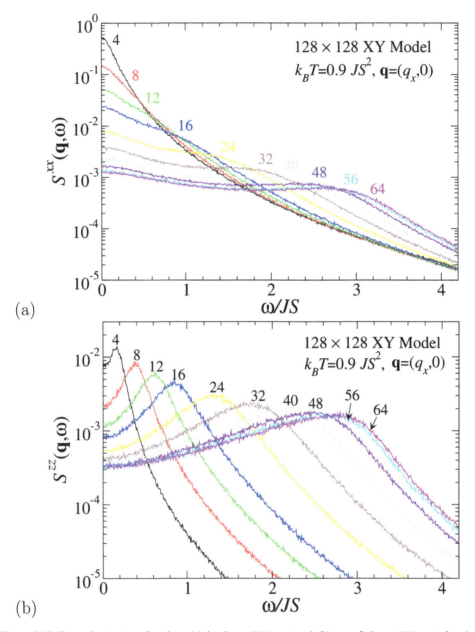

Figure 8.17. Dynamic structure functions (a) in-plane, $S^{xx}(\mathbf{q}, \omega)$ and (b) out-of-plane, $S^{zz}(\mathbf{q}, \omega)$, for the square lattice FM XY model ($L = 128$, $\lambda = 0$) at $k_B T = 0.9JS^2$, from averages over 500 initial conditions, with each $q_x = 2\pi m/L$ for indicated m values. In (a) the colors on m values are used to indicate the corresponding curves. The in-plane correlations now are dominated by the CP signature of vortices. The spin wave peaks are weakened, greatly broadened and mainly visible only in the out-of-plane correlations.

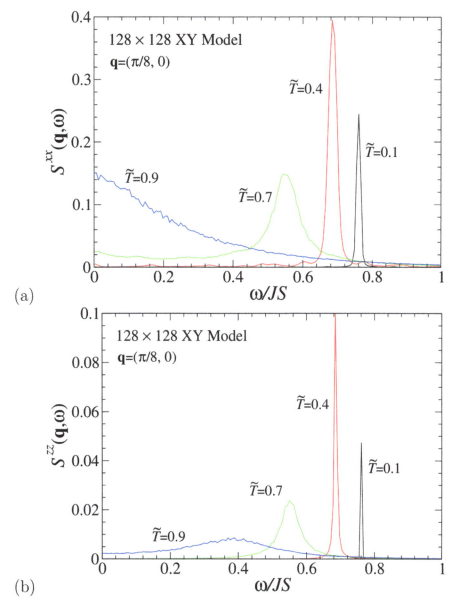

Figure 8.18. Dynamic structure functions for the square lattice FM XY model ($L = 128$, $\lambda = 0$) at $\mathbf{q} = (\pi/8, 0)$, corresponding to wave vector index $m = 8$ in figure 8.17, (a) in-plane, $S^{xx}(\mathbf{q}, \omega)$, and (b) out-of-plane, $S^{zz}(\mathbf{q}, \omega)$, for indicated temperatures $\tilde{T} = k_B T/JS^2$. Note the growth of the CP for $\tilde{T} > \tilde{T}_c \approx 0.7$, and how the spin wave peak is more visible in the out-of-plane correlations at $\tilde{T} = 0.9$.

responsible for the CP in $S^{xx}(\mathbf{q}, \omega)$, and they may make some minor contributions to $S^{zz}(\mathbf{q}, \omega)$ at low frequencies.

These ideas inspired an analysis by Mertens *et al* [25] in terms of an 'ideal gas' of vortices moving in a background of spin waves, which is similar to a model for an

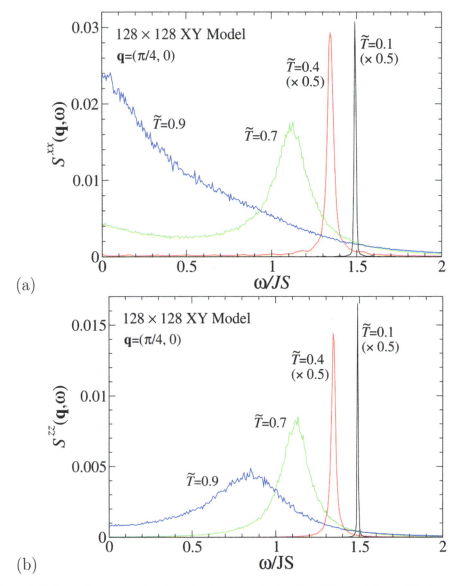

Figure 8.19. Dynamic structure functions for the square lattice FM XY model ($L = 128$, $\lambda = 0$) at $\mathbf{q} = (\pi/4, 0)$, corresponding to wave vector index $m = 16$ in figure 8.17, (a) in-plane, $S^{xx}(\mathbf{q}, \omega)$, and (b) out-of-plane, $S^{zz}(\mathbf{q}, \omega)$, for indicated temperatures $\tilde{T} = k_B T/JS^2$.

ideal gas of solitons in quasi-1D magnetic chains. Here we describe some aspects of this ideal gas model. It can be kept in mind that it makes some assumptions which may not be exactly correct. Further, it is possible to consider other processes that could lead to CP intensity, such as multi-spin wave interference. The main result of the ideal vortex gas model is a prediction of the vortex contribution to the width and height (or total integrated intensity) of the CPs in $S^{xx}(\mathbf{q}, \omega)$ and $S^{zz}(\mathbf{q}, \omega)$.

The basic assumptions of the ideal gas are as follows. We consider primarily $\lambda < \lambda_c$, so the vortices are of the in-plane type. For the in-plane correlations, a vortex at some point causes a reversal of the sign of in-plane spin components on either side of itself. This can be assumed to be the main way in which the correlation functions are affected and, as a result, the density of free vortices (including anti-vortices) ρ has been taken to be related to the correlation length,

$$\rho \approx (2\xi)^{-2}. \tag{8.72}$$

This can be viewed to say that the correlation length ξ is half the mean vortex separation, $\rho^{-1/2}$. Note that ξ enters into the decay of in-plane correlations via (8.56). Concerning the time-dependent effects, the *free vortices* are assumed to move in a ballistic manner at constant velocity, and their population is assumed to be taken from a Maxwellian velocity distribution. These two assumptions may be the most easily called into question, because at the temperatures where many vortices are present, there are likely to be strong vortex–vortex and vortex–spin wave interactions. Any ideal gas approximation usually assumes a low density of particles, so that interactions only help with the accomplishment of ergodicity. This is probably violated for vortices in easy-plane magnets. However, we assume that the temperature is just slightly above T_c, where the lowest number density is present. In addition to this, vortices are not particles of indefinite lifetime, which is another reason that the model can be called into question. Even so, it is interesting to evaluate its predictions.

Here is a synopsis of the calculation of the dynamic correlation function as caused by ballistically moving vortices. See [25] for further details. Consider a simplified definition of the correlations of one in-plane spin component, in continuum notation for unit spins,

$$S^{xx}(\mathbf{r}, t) = \langle s^x(\mathbf{r}, t) s^x(\mathbf{0}, 0) \rangle. \tag{8.73}$$

For an individual vortex with in-plane angle given by (8.7), the spin component is $s^x = \cos \phi(\mathbf{r}, t)$. The vortex field for the in-plane spin components extends over the whole system. For two widely separated points in space with initially correlated or aligned spins, a vortex placed between them will cause a relative sign of -1 in the spin components at those points. Thus, the generation of vortices between two points leads to a *reduction* of their correlations. Then, the calculation of this correlation function requires a type of counting of the number of vortices passing between the two space–time locations. Therefore it is reduced to the form,

$$S^{xx}(\mathbf{r}, t) = \langle \cos^2 \phi(\mathbf{0}, 0) \rangle \langle (-1)^{N(\mathbf{r},t)} \rangle, \tag{8.74}$$

where $N(\mathbf{r}, t)$ counts the number of vortices passing a contour between $(\mathbf{0}, 0)$ and $(\mathbf{r}, t)$. The angle at the initial point, $\phi(\mathbf{0}, 0)$, is uniformly distributed on $[0, 2\pi]$, so that $\langle \cos^2 \phi(\mathbf{0}, 0) \rangle = 1/2$.

The other part involving the average of $(-1)^{N(\mathbf{r},t)}$ has appeared in thermal averages for 1D soliton models. Mertens *et al* [25] found a way to evaluate it analytically in two space dimensions. They considered first a 1D calculation, with

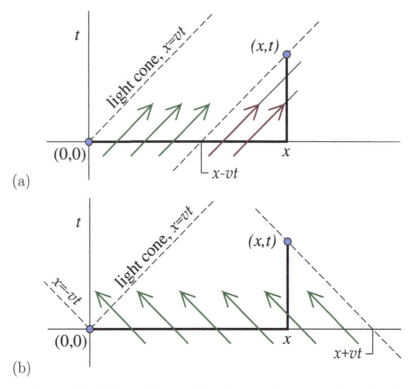

Figure 8.20. Diagram of the 'light cone' for correlations between points (0, 0) and (x, t) by calculation of the number of 1D kinks crossing the connecting contour. Point (x, t) is considered outside the light cone here. (a) For right-moving kinks of speed v, green (red) arrows show those crossing once (twice), which are counted (or not). The width that contributes is $x - vt$, which determines $\bar{n}_r$ in (8.75). (b) For left-moving kinks of speed v, all those over the width $x + vt$ cross the contour once and are counted, which determines $\bar{n}_l$ in (8.78).

kinks (instead of a vortices) of equal speeds v observed to pass the contour $(0, 0) \to (x, 0) \to (x, t)$. The counting of N requires doing two cases, $x > vt$ and $x < vt$, separately. In the first case with $x > vt$, the point (x, t) is said to be *outside the light cone* (where the light cone has $x = vt$). This case is sketched in figure 8.20. In fact, we only need to consider the right-moving kinks (velocity $+v$) separately from the left-moving kinks (velocity $-v$). Later, we sum over the effects of all distinct velocities, and here $+v$ and $-v$ are themselves distinct.

In figure 8.20(a), the kinks in a region of width $x - vt$ cross the contour once and are counted (green arrows); there are also other kinks that cross twice (red arrows) so effectively they did not cross and need not be counted (or, they contribute $(-1)^2$, which makes no difference). Assuming a linear density of kinks as $\rho = (2\xi)^{-1}$, with half moving to the right and half moving to the left, the mean number of kinks to cross to the right is

$$\bar{n}_r = \frac{1}{2}\rho(x - vt) = \frac{x - vt}{4\xi}. \tag{8.75}$$

Because a low density is assumed in ideal gas theory, the fluctuating number n_r follows a Poisson distribution,

$$p(n_r) = \frac{\bar{n}_r}{n_r!} e^{-\bar{n}_r}, \quad n_r = 0, 1, 2, \ldots. \tag{8.76}$$

Then the average needed for the right-moving kinks is calculated from

$$\langle (-1)^{n_r} \rangle = \sum_{n_r} (-1)^{n_r} p(n_r) = e^{-2\bar{n}_r} = \exp\left[-\frac{(x - vt)}{2\xi}\right]. \tag{8.77}$$

A similar calculation is applied for the left-moving kinks, see figure 8.20(b), where kinks in the width $x + vt$ all cross the contour and are counted. With n_l being the number crossing to the left, and mean value,

$$\bar{n}_l = \frac{1}{2}\rho(x + vt) = \frac{x + vt}{4\xi} \tag{8.78}$$

their contribution to the average is a similar result,

$$\langle (-1)^{n_l} \rangle = \sum_{n_l} (-1)^{n_l} p(n_l) = e^{-2\bar{n}_l} = \exp\left[-\frac{(x + vt)}{2\xi}\right]. \tag{8.79}$$

These factors combine into one factor for all kinks. When the calculations are repeated for points (x, t) also inside the light cone, the net result for an arbitrary point (x, t) inside or outside the light cone can be written as

$$\langle (-1)^{N(x,t)} \rangle = \exp\left[-\frac{|x - vt|}{2\xi} - \frac{|x + vt|}{2\xi}\right]. \tag{8.80}$$

The above calculations apply to a 1D system, but we have a 2D system and vortices, not kinks, with many different velocities. The 1D results were generalized to the 2D situation, which produces a sum over the velocity distribution:

$$\langle (-1)^{N(\mathbf{r},t)} \rangle = \exp\left[-\int_0^\infty dv\, P(v)\left[\frac{|\mathbf{r} - vt|}{2\xi} + \frac{|\mathbf{r} + vt|}{2\xi}\right]\right]. \tag{8.81}$$

The Maxwellian speed distribution $P(v)$, giving the same weight to inward-moving and outward-moving vortices, is

$$P(v) = 2\frac{v}{\bar{v}^2} e^{-v^2/\bar{v}^2}. \tag{8.82}$$

Of course, this is where one assumes freely moving vortices in thermal equilibrium, which implies that the root-mean-square speed $\bar{v}$ is dependent on the temperature. Integration over speeds can be shown to lead to the space–time correlation function in the form,

$$S^{xx}(\mathbf{r}, t) = \frac{1}{2}\exp\left[-\frac{r}{\xi} - \frac{\sqrt{\pi}\bar{v}|t|}{2\xi}\mathrm{erfc}\left(\frac{r}{\bar{v}|t|}\right)\right]. \tag{8.83}$$

It turns out that this can be very well approximated by another analytic expression that preserves both its integrated intensity and asymptotic behaviors as $r \to \infty$ and $t \to \infty$,

$$S^{xx}(\mathbf{r}, t) \approx \frac{1}{2} \exp\left(-\sqrt{(r/\xi)^2 + (\gamma t)^2}\right) \tag{8.84}$$

where γ is a frequency-like quantity,

$$\gamma = \frac{\sqrt{\pi}\bar{v}}{2\xi}. \tag{8.85}$$

One can see that γ sets a time scale for vortex motion in the same way that the correlation length ξ sets a typical length scale. The space and time structures of this approximated correlation function are essentially very similar. This allows one to find the space–time FT. The result is a *squared Lorentzian form*,

$$S^{xx}(\mathbf{q}, \omega) = \frac{1}{2\pi^2} \frac{\gamma^3 \xi^2}{\left\{\omega^2 + \gamma^2\left[1 + (q\xi)^2\right]\right\}^2}. \tag{8.86}$$

In fact, this is indeed a CP structure; the maximum intensity occurs at $\omega = 0$. The height of the CP as a function of $\mathbf{q}$ is one quantity that can be analyzed. The ideal gas model then predicts it to be

$$S^{xx}(\mathbf{q}, \omega = 0) = \frac{1}{2\pi^2} \frac{\gamma^{-1}\xi^2}{\left[1 + (q\xi)^2\right]^2} = \frac{A}{\left[1 + (q\xi)^2\right]^p}, \quad p = 2. \tag{8.87}$$

The amplitude $A = S^{xx}(\mathbf{q} = 0, \omega = 0)$ will be taken from the MC–SD data. The power $p = 2$ comes from the ballistic ideal gas theory, but we allow for a different power in the expression for later analysis. The CP also has a $\mathbf{q}$-dependent half-width, defined as the frequency where $S^{xx}(\mathbf{q}, \omega)$ reaches half its peak value, which is found to be

$$\Gamma_x(\mathbf{q}) = \left(\sqrt{2} - 1\right)^{1/2} \gamma \left[1 + (q\xi)^2\right]^{1/2} = B\left[1 + (q\xi)^2\right]^{1/2} \tag{8.88}$$

Again, its scale or amplitude is determined by the constant $B = \Gamma_x(\mathbf{q} = 0)$, which can be taken from MC–SD data. The total intensity integrated over all frequency (assuming symmetry in positive and negative ω) is

$$I_x(\mathbf{q}) = \int_{-\infty}^{+\infty} d\omega\, S^{xx}(\mathbf{q}, \omega) = \frac{1}{4\pi} \frac{\xi^2}{\left[1 + (q\xi)^2\right]^{3/2}} = \frac{C}{\left[1 + (q\xi)^2\right]^{p-1/2}}, \quad p = 2. \tag{8.89}$$

Its scale is the constant $C = I_x(\mathbf{q} = 0)$. These last three equations can be tested by comparison with the MC–SD simulations. The values at $\mathbf{q} = 0$ are factored out, which facilitates comparison with simulation or experiment, and allows one to concentrate more on the dependencies with $\mathbf{q}$. This is helpful because it is difficult for theory to correctly obtain the amplitude of the CP effects, because indeed other processes besides vortex motion can be responsible for some of the intensity. Note that the vortex ideal gas CP intensity is predicted to vary as $\xi^2 \sim \rho^{-1}$ for small wave vectors, which is physically reasonable. The vortices generated thermally as T increases above T_c reduce the long-distance correlations across the system. The predicted CP half-width is only weakly dependent on q for small q (long wavelengths), and eventually approaches an approximate linear relation with q at larger values.

8.7.5 Comparison of vortex ideal gas model to simulations

Mertens *et al* [25] originally made various comparisons of these ideal gas predictions with early simulation data by averaging over only three initial states, and found reasonably good agreement for the **q**-dependencies, but not so much so for the amplitude of the effects. For comparison here we show some additional results on 128×128 systems for just one temperature, $k_B T = 0.9 J S^2$, slightly above T_c, averaging over 500 initial states. At this temperature, $\rho \approx 0.094$, which would imply by (8.72) a correlation length on the order of $\xi \approx 1.63a$, however, that is probably an underestimate. The alternative relation,

$$\rho \approx \xi^{-2} \tag{8.90}$$

gives instead $\xi \approx 3.26a$, which is more consistent with results from static correlations. Indeed, Dimitrov and Wysin [26] found $\xi = 3.95a$ at this temperature from the static correlation function. In addition, they show in figure 1 of [26] that the relation between vortex number density and ξ^{-2} is not even linear.

For the full range of wave vectors along the (10)-direction, the CP height for $S^{xx}(\mathbf{q}, \omega)$ was measured (at $\omega = 0$); results are shown in figure 8.21. The frequency at which $S^{xx}(\mathbf{q}, \omega)$ becomes half the height was used to estimate the half-width Γ_x, which is plotted in figure 8.22. Then, based on the theory expressions for the squared Lorentzian CP, one finds that the height and half-width can be combined directly to estimate the integrated intensity,

$$I_x(\mathbf{q}) = \frac{\pi}{2}\left(\sqrt{2} - 1\right)^{-1/2} S^{xx}(\mathbf{q}, \omega = 0) \Gamma_x(\mathbf{q}) \approx 2.44\, S^{xx}(\mathbf{q}, \omega = 0) \Gamma_x(\mathbf{q}). \tag{8.91}$$

The intensity from the simulations is then a derived quantity calculated this way from the estimates of CP height and half-width, see figure 8.23. This also means that the amplitudes (values at $\mathbf{q} = 0$) are related by

$$C = \approx 2.44 A B. \tag{8.92}$$

In order to compare simulation data with the theory, the values of CP height, half-width and intensity at $\mathbf{q} = 0$ were taken directly from the simulations, and not allowed to be fitted. These values from the simulations give the constants in dimensionless units, $A = 1.26$, $B = 0.087$ and $C = 2.44 AB = 0.268$. One can note that the implied definition of $C = \xi^2/4\pi$ results in the estimate of correlation length, $\xi = 1.83a$. On the other hand, if the alternative definition (8.90) were used in the ideal gas model, the result would instead be $\xi \approx 3.66a$, which is more reasonable and close to that from static correlations [26].

Figure 8.21 shows the MC–SD data for CP height compared with the ideal gas theory expression (8.87), at the power $p = 2$ and also for an alternative value

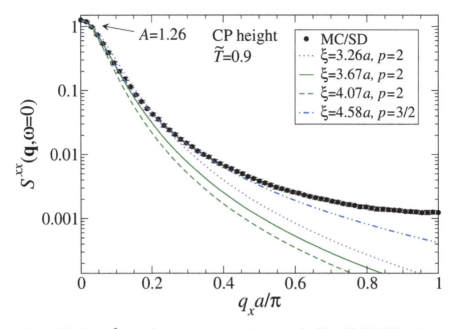

Figure 8.21. CP height at $\tilde{T} = 0.9$ for wave vectors $\mathbf{q} = (q_x, 0)$ on the 128×128 FM XY model system. Symbols are from the hybrid MC–SD simulations. Curves are the ideal gas expression (8.87), using the different values of ξ indicated. The power $p = 2$ is the ballistic ideal gas model; for power $p = 3/2$, however, the data can be fit over a much wider range of q_x.

$p = 3/2$. The theory expression with $p = 2$, $\xi = 3.67a$ fits well to the simulations at low q, but deviates significantly with increasing q. Other values of ξ are shown for comparison. It was noticed that putting $p = 3/2$ allows for a much better fit to large q, using $\xi = 4.58a$. However, the simulation data go to a constant height at the zone boundary, whereas the theory expression continues decaying.

Figure 8.22 shows the MC–SD data for CP half-width $\Gamma_x(\mathbf{q})$ compared with the ideal gas theory, equation (8.88). A value of $\xi = 6.11a$ gives a good match at low q; it seems that there is a change in the slope of $\Gamma_x(\mathbf{q})$ around $q_x \approx 0.32\pi/a$. A value of $\xi = 5.50a$ gives a better fit over the full range of q. Trial curves with smaller values of ξ such as $\xi = 3.67a$ fail to follow the MC–SD data.

Figure 8.23 shows the MC–SD data for CP intensity $I_x(\mathbf{q})$, from using (8.91), compared with the ideal gas theory, (8.89). Various values of ξ for the ballistic theory $p = 2$ can only fit to the data at lower q; the best is around $\xi = 3.67a$. Similar to the peak height, the integrated intensity seems to be better fit over a wider range of q by switching to the power $p = 3/2$, for which $\xi = 3.97a$ gives a good fit. Eventually at large enough q, the MC–SD tends to a constant, then the theoretical expression (8.89) cannot follow the data.

These results indicate that the vortex ideal gas theory does predict a CP, and while the gross features are present, probably the details do not work out completely correctly. It is difficult to obtain all three of the main quantities, peak height, peak width and integrated intensity, to match to the theory with the same correlation

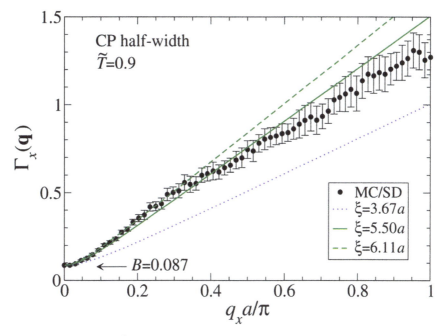

Figure 8.22. CP half-width at $\tilde{T} = 0.9$ for wave vectors $\mathbf{q} = (q_x, 0)$ on the 128×128 FM XY model system. Symbols are from the hybrid MC–SD simulations. Curves are the ideal gas expression (8.88), using the different values of ξ indicated. Note that the value $\xi = 6.11a$ fits well at low q_x, where the spin wave peak in $S^{xx}(\mathbf{q}, \omega)$ is very weak.

length parameter ξ. It is also possible that the structure of $S^{xx}(\mathbf{q}, \omega)$ is not so close to a squared Lorentzian, as indicated that the peak height might better be fit with the power $p = 3/2$ instead of $p = 2$ as in the ideal gas theory. The amplitudes of these quantities were fixed to the simulation results, because they could not be consistently matched, to give one good set of values for ξ and γ. Even so, the idea of slow-moving vortices and the sign-change in a vortex, that causes a contribution to the dynamic correlations, is completely reasonable as an explanation of the vortex part of CP intensity.

We have noted that the theory presumes relation (8.72) between vortex density and correlation length, however, an alternative relation (8.90) without the factor of 2 could be one improvement. In any case, the vortex density is not really linearly proportional to ξ^{-2} [26], as assumed. As mentioned earlier, another probable correction to the results is to take into account the spin wave and multi-spin wave interference contributions, because these also contribute to CP intensity. At higher T and higher vortex density, the ideal gas assumption will become invalid. The process itself of VA pair creation has been proposed as making a contribution to CP intensity [27], especially below T_c [28]. In addition to these effects, other simulations indicate that vortices have significantly finite lifetimes [26]. This is certainly an effect that is mixed up with the scattering of spin waves off vortices and spin waves off other spin waves. This means that the ballistic approximation is probably not as

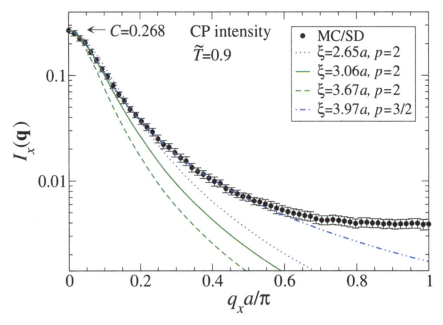

Figure 8.23. CP integrated intensity at $\tilde{T} = 0.9$ for wave vectors $\mathbf{q} = (q_x, 0)$ on the 128×128 FM XY model system. Symbols are from the hybrid MC–SD simulations. Curves are the ideal gas expression (8.89), using the different values of ξ indicated. The power $p = 2$ is the ballistic ideal gas model; for power $p = 3/2$, however, the data can be fit over a much wider range of q_x.

good as it should be, as finite lifetime will lead to a lower frequency limit below which the vortex ideal gas theory needs modification.

The theory was extended also to apply to out-of-plane correlations $S^{zz}(\mathbf{q}, \omega)$, however, this depends strongly on the true structure of moving vortices. Mertens *et al* [25] again found a CP, but of Gaussian form. For interested readers further discussion of that can be found in the literature.

8.7.6 Dynamic correlations with vacancies

Earlier we saw how spin vacancies lead to a reduction of T_c in the XY model, as shown through calculations of the helicity modulus and static correlations. Vacancies tend to attract and pin vortices, even for temperatures below T_c. It has been seen [22] that most of the vortices produced below T_c are in fact, pinned on vacancies. Then, it is interesting to ask if there is a dynamic signature due to pinned vortices, that appears in the dynamic structure function. A vortex pinned on a vacancy has lowered its energy by doing so. It is in a local potential energy minimum, then, it possesses certain dynamic modes of oscillation about its minimum, which should present themselves in $S^{xx}(\mathbf{q}, \omega)$.

In a paper [29] by Paula *et al* this question was addressed by making MC–SD simulations for the XY model with a percentage of randomly placed vacancies. They found that for temperatures below the critical temperature (for a given density of vacancies), there appears in $S^{xx}(\mathbf{q}, \omega)$ not only the spin wave peak, but as well an extra small peak at a low frequency inversely proportional to the lattice size L. This could be typical of finite size effects. However, the frequency found for the mode seems to match that expected for the normal mode oscillations for a pinned vortex, as calculated analytically and by simulations [20]. At $T = 0$ the prediction is for the frequency of a single vortex on a single vacancy in a $L \times L$ system is predicted to be

$$\omega_0 = \frac{13.57}{L} JS. \tag{8.93}$$

When considering higher temperature, the frequency of this type of mode is expected to soften due to thermal effects. Softening is also involved with increasing vacancy density, which weakens the average exchange couplings and therefore lowers the frequencies, see figure 3 of [29].

As an example of these effects, here we show some results from MC–SD simulations with a vacancy density $\rho_{\text{vac}} = 0.16$, for the 128×128 XY model, see figure 8.24. For reference, simulation results for the same pure system, without vacancies, are shown in figure 8.25. The averages have been made over 500 states of the ensemble, including averaging over the random vacancy locations. This was

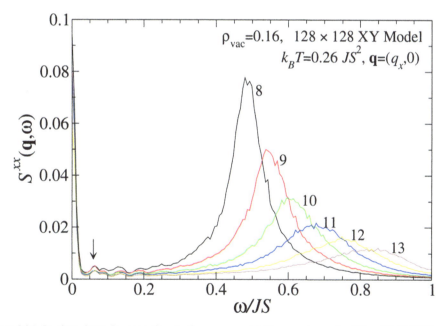

Figure 8.24. In-plane dynamic correlations in the presence of 16% vacancies on the 128×128 FM XY model system, from MC–SD simulations, for wave vectors $\mathbf{q} = (m, 0)(2\pi/L)$ with m values indicated next to the curves. The temperature $k_B T = 0.26 J S^2$ is about 4/7 of T_c. The arrow at $\omega/JS \approx 0.064$ indicates a mode likely due to vortices pinned on vacancies, whose frequency does not change with $\mathbf{q}$. Note that the peak also is present no matter whether m is odd or even. There is also a very strong peak at $\omega = 0$ for all the $\mathbf{q}$ shown, caused by the vacancy disorder; compare figure 8.25.

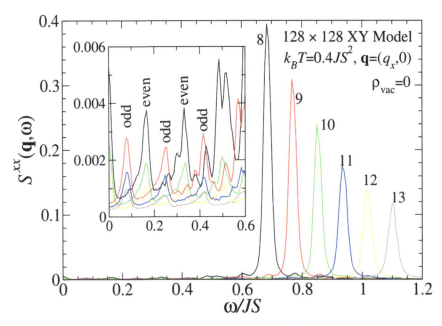

Figure 8.25. In-plane dynamic correlations in the pure 128×128 FM XY model system (no vacancies), from MC–SD simulations, for wave vectors $\mathbf{q} = (m, 0)(2\pi/L)$ with m values indicated next to the curves. The temperature is about 4/7 of T_c. The inset shows a zoom of the weak typical finite size responses below the spin wave peaks, that depend on whether m is odd or even.

achieved by making 25 realizations of the vacancy positions, from which the MC simulations were run to 20 000 MCS for equilibration, and then run another 20 000 MCS, taking a state to be an initial condition for the SD every 1000 MCS. The SD part of the simulations was run the same way as described earlier for the XY model, using $2^{12} = 4096$ time points separated by $\Delta t = 6 \times 0.04(JS)^{-1}$. At a 16% vacancy density, the transition temperature is $k_B T_c \approx 0.453 JS^2$, see figure 8.12. Due to the lower T_c compared to $k_B T_c \approx 0.70 JS^2$ in the pure model, we consider the simulations at relatively lower temperatures. So we show results at $k_B T \approx (4/7) k_B T_c = 0.26 JS^2$ and at $T \approx T_c$.

Referring to figure 8.24 for $T \approx (4/7) T_c$ with 16% vacancies, the spin wave peak is the dominant feature, moving in frequency as the wave vector is changed. However, also a very strong peak appears at $\omega = 0$ for any $\mathbf{q}$. Such a peak at zero frequency of course belies the presence of a frozen-in spin structure. This feature has not been carefully analyzed, but it is certainly caused by the vacancies and the fact that the thermal averages of total S^x and S^y are now not zero when vacancies are present. The vacancies break the full symmetry of the system. The exchange interactions where sites do not all have four nearest neighbors must produce a frozen-in organized spin structure. Note that in figure 8.25 for the corresponding wave vectors in the pure system, naturally there is no zero-frequency peak.

The second obvious feature in figure 8.24 that is directly caused by the presence of vacancies is the weak peak at a low frequency of about $\omega \approx 0.064 JS$, indicated by an arrow. That feature and some other similar small peaks remain present for all the **q** shown. Paula *et al* [29] have indicated that this small feature is ultimately caused by the oscillations of vortices pinned on vacancies. The frequency is quite low compared to the $T = 0$ prediction from (8.93), which gives $\omega_0 \approx 0.106JS$, however, the vacancies distributed over the system weaken the average exchange interactions, which causes a considerable lowering of this mode's frequency. The values of m used to produce the wave vectors, according to $q_x = 2\pi m/L$, are both odd and even, and yet the low frequency features remain nearly the same. Compare the inset in figure 8.25 for the pure system, which shows the very weak response below the spin wave frequency. In the pure system these low frequency peaks alternate by odd/even values of m, which is typical for finite size spin wave effects. This gives a strong indication that the small low frequency structures in the system with vacancies really are due not to spin wave effects, but to pinned vortices.

It is interesting to make another comparison for the same system, with (figure 8.26) and without (figure 8.27) vacancies, but at a temperature more or less equal to T_c. In the system with vacancies, the spin waves are broadened, and there starts to appear a general wide background similar to a CP. The narrow peak at $\omega = 0$ is still present. There is also a very faint but broadened possible peak around $\omega \approx 0.055JS$, that is probably the pinned vortex oscillation. It has been both broadened and

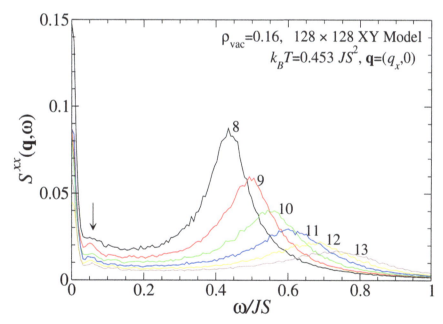

Figure 8.26. In-plane dynamic correlations in the presence of 16% vacancies on the 128 × 128 FM XY model system, from MC–SD simulations, for wave vectors $\mathbf{q} = (m, 0)(2\pi/L)$ with m values indicated next to the curves. The temperature is the critical temperature. The arrow at $\omega/JS \approx 0.055$ indicates the mode probably due to vortices pinned on vacancies, that has shifted downward with increasing temperature. Note that the peak also is present no matter whether m is odd or even. The strong narrow peak at $\omega = 0$ for all the **q** is still present, caused by the vacancy disorder; compare figure 8.27.

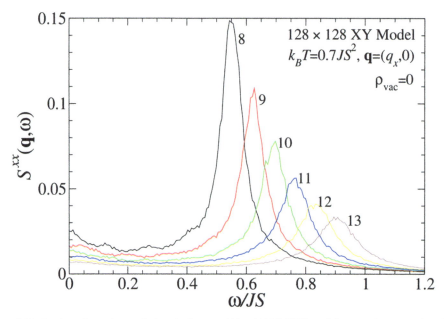

Figure 8.27. In-plane dynamic correlations in the pure 128 × 128 FM XY model system (no vacancies), from MC–SD simulations, for wave vectors $\mathbf{q} = (m, 0)(2\pi/L)$ with m values indicated next to the curves. The temperature is the critical temperature. The weak finite size responses below the spin wave peaks that depend on whether m is odd or even are still present but rather broadened, compare figure 8.25.

moved to a lower frequency compared to that in figure 8.24 for $T \approx (4/7)T_c$, due to greater thermal effects. Again this faint peak is present regardless of whether the wave vector index m is odd or even. In the system without vacancies at $T = T_c$, figure 8.27, the spin wave peaks are narrower, and there is also a general wide background, with weak oscillations. Looking more closely, the curves still exhibit a weak set of finite size effects peaks that depend on whether m is odd or even. These finite size effect peaks are rather broadened due to the temperature. But there is no single weak peak present independent of $\mathbf{q}$ such as that in the system with vacancies.

All these results taken together point to the idea that the fraction of vortices pinned on vacancies make small-amplitude oscillations. The oscillation has a particular signature, which is exhibited as a weak peak at a frequency lower than that given in the zero-temperature estimate of equation (8.93). The softening of its frequency occurs both due to increasing temperature and increasing vacancy density. Each of those effects lowers the effective exchange interaction strength, thereby reducing the mode frequency. Once the temperature is above T_c, this weak peak is gone because of the growth of the much larger CP associated with the much greater vortex density of the BKT transition.

Bibliography

[1] Hikami S and Tsuneto T 1980 Phase transition of quasi-two-dimensional planar system *Prog. Theor. Phys.* **63** 387

[2] Hirakawa K, Yoshizawa H and Ubukoshi K 1982 Neutron scattering study of the phase transition in two-dimensional planar ferromagnet K_2CuF_4 *J. Phys. Soc. Japan* **51** 2151
[3] Gouvêa M E, Wysin G M, Bishop A R and Mertens F G 1989 Vortices in the classical two-dimensional anisotropic Heisenberg model *Phys. Rev.* B **39** 11840
[4] Wysin G M, Gouvêa M E, Bishop A R and Mertens F G 1988 Classical spin dynamics in the two-dimensional anisotropic Heisenberg model In ed D P Landau, K K Mon and H B Schuttler *Computer Simulation Studies in Condensed Matter Physics: Recent Developments* vol 33 (Berlin: Springer) pp 40–8
[5] Takeno S and Homma S 1980 Classical spin systems, nonlinear evolution equations, and nonlinear excitations *Prog. Theor. Phys.* **64** 1193
[6] Nikiforov A V and Sonin É B 1983 Dynamics of magnetic vortices in a planar ferromagnet *Zh. Eksp. Teor. Fiz.* **85** 642
[7] Wysin G M 1996 Magnetic vortex mass in two-dimensional easy-plane magnets *Phys. Rev.* B **54** 15156
[8] Wysin G M 1994 Instability of in-plane vortices in two-dimensional easy-plane ferromagnets *Phys. Rev.* B **49** 8780
[9] Zaspel C E and Godinez D 1996 Vortex energies in the two-dimensional easy-plane ferromagnet *J. Magn. Magn. Mater.* **162** 91
[10] Wysin G M 1998 Critical anisotropies of two-dimensional magnetic vortices *Phys. Lett.* A **240** 95
[11] Berezinskii V L 1971 Destruction of long-range order in one-dimensional and two-dimensional systems having a continuous symmetry group I. Classical systems *Sov. Phys. JETP* **32** 493
[12] Berezinskii V L 1972 Destruction of long-range order in one-dimensional and two-dimensional systems having a continuous symmetry group II. Quantum systems *Sov. Phys. JETP* **34** 610
[13] Kosterlitz J M and Thouless D J 1973 Ordering, metastability and phase transitions in two-dimensional systems *J. Phys. C: Solid State. Phys.* **6** 1181
[14] Cuccoli A, Tognetti V and Vaia R 1995 Two-dimensional XXZ model on a square lattice: a Monte Carlo simulation *Phys. Rev.* B **52** 10221
[15] Harada K and Kawashima N 1997 Universal jump in the helicity modulus of the two-dimensional quantum XY model *Phys. Rev.* B **55** R11949
[16] Minnhagen P 1987 The two-dimensional Coulomb gas, vortex unbinding, and superfluid-superconducting films *Rev. Mod. Phys.* **59** 1001
[17] Pereira A R, Mól L A S, Leonel S A, Coura P Z and Costa B V 2003 Vortex behavior near a spin vacancy in two-dimensional XY magnets *Phys. Rev.* B **68** 132409
[18] Wysin G M 2003 Vortex-vacancy interactions in two-dimensional easy-plane magnets *Phys. Rev.* B **68** 184411
[19] Zaspel C E, McKennan C M and Snaric S R 1996 Instability of planar vortices in layered ferromagnets with nonmagnetic impurities *Phys. Rev.* B **53** 11317
[20] Pereira A R, Leonel S A, Coura P Z and Costa B V 2005 Vortex motion induced by lattice defects in two-dimensional easy-plane magnets *Phys. Rev.* B **71** 014403
[21] Paula F M, Pereira A R and Mól L A S 2004 Diluted planar ferromagnets: nonlinear excitations on a non-simply connected manifold *Phys. Lett.* A **329** 155
[22] Wysin G M 2005 Vacancy effects in an easy-plane Heisenberg model: Reduction of T_c and doubly charged vortices *Phys. Rev.* B **71** 094423
[23] Wysin G M, Pereira A R, Marques I A, Leonel S A and Coura P Z 2005 Extinction of the Berezinskii-Kosterlitz-Thouless phase transition by nonmagnetic disorder in planar symmetry spin models *Phys. Rev.* B **72** 094418

[24] Evertz H G and Landau D P 1996 Critical dynamics in the 2D classical XY-model: A spin dynamics study *Phys. Rev.* B **54** 12302
[25] Mertens F G, Bishop A R, Wysin G M and Kawabata C 1989 Dynamical correlations from mobile vortices in two-dimensional easy-plane ferromagnets *Phys. Rev.* B **39** 591
[26] Dimtrov D A and Wysin G M 1998 Free vortex and vortex-pair lifetimes in classical two-dimensional easy-plane magnets *J. Phys.: Condens. Matter* **10** 7453
[27] Costa J E R and Costa B V 1996 Static and dynamic simulation in the classical two-dimensional anisotropic Heisenberg model *Phys. Rev.* B **54** 994
[28] Pereira A R and Costa J E R 1996 Vortex pairs and central peak in the dynamic structure factor in the 2D XY model below T_{KT} *J. Magn. Magn. Mater.* **162** 219
[29] Paula F M, Pereira A R and Wysin G M 2005 Contribution of impurity-pinned vortices to the response function in a randomly diluted easy-plane ferromagnet on a square lattice *Phys. Rev.* B **72** 094425

IOP Publishing

Magnetic Excitations and Geometric Confinement
Theory and simulations
Gary Matthew Wysin

Chapter 9

Magnetic vortex core motion and internal dynamics

In this chapter some dynamic properties of individual vortices in 2D ferromagnets are presented. Analogous to 1D magnetic solitons, vortices have charges, momentum and mass. In addition, they possess a topological charge or gyrovector **G** that strongly governs the motion of the vortex core in response to applied forces. The Thiele equation, which describes vortex core motion via the gyrovector, is presented and its extension to include mass is discussed. The connection to Lagrangian mechanics for the core motion is also considered. Vortices also interact with spin waves and modify the spin wave spectrum. Oscillation of spin waves on top of a vortex background is considered its *internal dynamics*, which includes effects that can be studied theoretically and with simulations.

9.1 Thiele equations and vortex motion

In a complex system, a vortex is affected by an externally applied magnetic field and the far spin field of other vortices, as well as damping torques. Thiele [1, 2], and also Huber [3], developed a kind of collective coordinate equation that describes the motion of a domain wall or a vortex, treating it as a particle. The Thiele equation of motion gives the dynamics for the motion of the vortex core, which could have some time-dependent position $\mathbf{X}(t)$ and a corresponding time-varying velocity $\mathbf{V}(t) = \dot{\mathbf{X}}(t)$.

In the development of the dynamic equation by Thiele, it was assumed that the vortex structure for a moving vortex is the same as that for a static vortex, $\mathbf{S}(\mathbf{r})$, except for adding a uniform translation, $\mathbf{S} = \mathbf{S}(\mathbf{r} - \mathbf{X}(t))$. Here $\mathbf{r}$ is the location of a spin being measured (field point), whereas $\mathbf{X}$ is the location of the vortex center. If the position $\mathbf{X}(t)$ of the core translates uniformly, it is described by the equation,

$$\mathbf{X}(t) = \mathbf{V}t + \mathbf{X}_0. \tag{9.1}$$

But we have seen earlier that the vortex structure itself is changed in shape by the motion. In particular, the out-of-plane component develops an asymmetric structure linearly proportional to **V**. Thus, we consider here the modification to the Thiele equation, when vortex shape variations that are dependent on **V** are included in the theory [4]. Then, the spin structure is assumed to have a form that has an explicit velocity-dependence, for unit spins,

$$\mathbf{s} = \mathbf{s}(\mathbf{r} - \mathbf{X}(t), \mathbf{V}(t)). \tag{9.2}$$

By translating, the spin at a given location is changed. Also, if the vortex changes its velocity, the vortex changes its structure and a spin at a given location is changed. This latter effect is represented by a dependence on **V**, the second parameter in this collective coordinate spin function. This means that the time derivative of the spin field is obtained from both a convective derivative term and an acceleration term:

$$\frac{d\mathbf{s}}{dt} = -\frac{d\mathbf{X}}{dt} \cdot \frac{\partial \mathbf{s}}{\partial \mathbf{r}} + \frac{d\mathbf{V}}{dt} \cdot \frac{\partial \mathbf{s}}{\partial \mathbf{V}} = -\mathbf{V} \cdot \frac{\partial \mathbf{s}}{\partial \mathbf{r}} + \mathbf{A} \cdot \frac{\partial \mathbf{s}}{\partial \mathbf{V}}. \tag{9.3}$$

The last term, proportional to the vortex acceleration $\mathbf{A} = \dot{\mathbf{V}}$, will lead to a vortex effective mass. The dot products indicate the sum over x- and y-components of both the coordinate and velocity gradients.

Equation (9.2) represents only the contributions to the spin field due to a particular vortex under consideration. In a real application, the spin field will be a nonlinear superposition of parts due to the vortex of interest, and all other contributions from say, other vortices at far away locations. It can be kept in mind, then, that the total spin field will not be translating as in (9.2). Indeed, the vortex of interest is influenced by the spin fields of all other vortices and other independent modes in the system.

9.1.1 Derivation of Thiele's equation of motion

The Thiele equation is derived from the Landau–Lifshitz equations, including Gilbert damping, by considering it as an equation with force-like terms (acting on the vortex). Wysin and Mertens [4] extended it to include acceleration effects, based on analysis of vortex momentum. First, we consider a derivation based on the Thiele approach, in which the damping is easily accounted for. The Landau–Lifshitz–Gilbert (LLG) equation (2.105) for **s** is

$$\dot{\mathbf{s}} = \mathbf{s} \times \gamma \mathbf{B} - \alpha \mathbf{s} \times \dot{\mathbf{s}}, \tag{9.4}$$

where

$$\gamma \mathbf{B} = -\frac{\delta H}{\delta \mathbf{S}} \tag{9.5}$$

is the effective field providing the torque (it could depend on **s**). The LLG equation can be expressed equivalently as

$$\mathbf{s} \times \mathbf{H}_{\text{net}} = 0 \tag{9.6a}$$

$$\mathbf{H}_{\text{net}} \equiv \gamma \mathbf{B} + (\mathbf{s} \times \dot{\mathbf{s}}) - \alpha \dot{\mathbf{s}}. \tag{9.6b}$$

Note that this works because spin length is conserved, which implies $\mathbf{s} \cdot \dot{\mathbf{s}} = 0$, used in the vector identity,

$$\mathbf{s} \times (\mathbf{s} \times \dot{\mathbf{s}}) = (\mathbf{s} \cdot \dot{\mathbf{s}})\mathbf{s} - (\mathbf{s} \cdot \mathbf{s})\dot{\mathbf{s}} = -\dot{\mathbf{s}}. \tag{9.7}$$

Equation (9.6a) proves that the fields $\mathbf{H}_{\text{net}}$ and $\mathbf{s}$ are parallel. Any change in $\mathbf{s}$ must be perpendicular to $\mathbf{s}$, i.e. $(\nabla \mathbf{s}) \cdot \mathbf{s} = 0$, where $\nabla = \hat{\mathbf{e}}_j \partial_j = \hat{\mathbf{e}}_j \frac{\partial}{\partial r_j}$, and the spin components are contracted by dot ($\cdot$) here, resulting in a vector. In this section repeated indices $i, j, k = 1, 2$, are summed over. Because $\mathbf{H}_{\text{net}}$ is parallel to $\mathbf{s}$, one has an expression that contains local force-like quantities as a conservation law:

$$-(\nabla \mathbf{s}) \cdot \mathbf{H}_{\text{net}} = -(\hat{\mathbf{e}}_j \partial_j \mathbf{s}) \cdot \left[\gamma \mathbf{B} + (\mathbf{s} \times \dot{\mathbf{s}}) - \alpha \dot{\mathbf{s}}\right] = 0. \tag{9.8}$$

A minus sign is included so that the resulting terms go directly as written into the Thiele equation being produced. The time derivatives contained here are taken in the convective form (9.3). It is more convenient to express this time derivative using index notation,

$$\dot{\mathbf{s}} = -V_k \partial_k \mathbf{s} + A_k \tilde{\partial}_k \mathbf{s}, \tag{9.9}$$

where $\tilde{\partial}_j \equiv \frac{\partial}{\partial V_j}$ is the velocity gradient and $\partial_k = \frac{\partial}{\partial r_k}$ is a component of the gradient with respect to the field point $\mathbf{r}$. This now gives

$$-\gamma \mathbf{B} \cdot (\hat{\mathbf{e}}_j \partial_j \mathbf{s}) - \mathbf{s} \cdot (\partial_j \mathbf{s} \times \partial_k \mathbf{s}) \hat{\mathbf{e}}_j V_k - \alpha (\partial_j \mathbf{s}) \cdot (\partial_k \mathbf{s}) \hat{\mathbf{e}}_j V_k$$
$$+ \mathbf{s} \cdot (\partial_j \mathbf{s} \times \tilde{\partial}_k \mathbf{s}) \hat{\mathbf{e}}_j A_k + \alpha (\partial_j \mathbf{s}) \cdot (\tilde{\partial}_k \mathbf{s}) \hat{\mathbf{e}}_j A_k = 0. \tag{9.10}$$

The terms in the first line give Thiele's original equation; the second line contains the modifications caused by velocity-dependent structure changes. This then motivates the definitions of integrals over a vortex structure, which are collective properties of a vortex that act to describe the dynamics.

The first term in the first line of (9.10) is related to an effective force $\mathbf{F}$ acting on the vortex. Consider using the definition (9.5) of the effective field $\gamma \mathbf{B}$, and integrating over 2D space,

$$-\int d^2 r \, \gamma \mathbf{B} \cdot (\hat{\mathbf{e}}_j \partial_j \mathbf{s}) = \frac{1}{S} \hat{\mathbf{e}}_j \int d^2 r \, \frac{\delta H}{\delta \mathbf{S}} \cdot \frac{\partial \mathbf{S}}{\partial r_j}$$
$$= \frac{v_{\text{cell}}}{S} \hat{\mathbf{e}}_j \int \frac{d^2 r}{v_{\text{cell}}} \frac{\delta H}{\delta \mathbf{S}} \cdot \left(-\frac{\partial \mathbf{S}}{\partial X_j}\right)$$
$$= -\frac{v_{\text{cell}}}{S} \hat{\mathbf{e}}_j \frac{\partial}{\partial X_j} \int \frac{d^2 r}{v_{\text{cell}}} \mathcal{H} = -\frac{v_{\text{cell}}}{S} \frac{\partial H}{\partial \mathbf{X}}. \tag{9.11}$$

In the second line of (9.11), the gradient $\partial \mathbf{S}/\partial \mathbf{r}$ has been replaced by $-\partial \mathbf{S}/\partial \mathbf{X}$, and the area per spin, v_{cell}, has been included into the integrand. The last line of (9.11) is written as a gradient of the total Hamiltonian, *with respect to the vortex core location* $\mathbf{X}$. That defines the force acting on the vortex,

$$\mathbf{F} = -\frac{\partial H}{\partial \mathbf{X}}. \tag{9.12}$$

Therefore this first term in (9.10) is summarized as

$$-\int d^2 r \, \gamma \mathbf{B} \cdot (\hat{\mathbf{e}}_j \partial_j \mathbf{s}) = \frac{1}{\sigma} \mathbf{F}, \qquad \sigma \equiv \frac{S}{v_{\text{cell}}}. \tag{9.13}$$

The factor σ is spin density per area and v_{cell} is the area occupied by one spin in the system. In many papers the factor $1/\sigma$ on $\mathbf{F}$ is instead expressed equivalently as γ/m_0, where $m_0 = \gamma S/v_{\text{cell}}$ is the magnetic dipole moment per unit area.

The second term in the first line of (9.10) leads to an anti-symmetric gyrotensor G_{jk} and gyrovector $\mathbf{G} = G_i \hat{\mathbf{e}}_i$ derived from it (both taken as dimensionless here):

$$\mathsf{G}_{jk} \equiv \int d^2r \, \mathbf{s} \cdot (\partial_j \mathbf{s} \times \partial_k \mathbf{s}) \tag{9.14a}$$

$$G_i = \frac{1}{2} \epsilon_{ijk} \mathsf{G}_{jk}. \tag{9.14b}$$

Because j and k can only take the values 1 or 2, in application for a 2D system the only non-zero elements of the gyrotensor are $\mathsf{G}_{12} = -\mathsf{G}_{21}$. Then, the gyrovector only has an out-of-plane component, $\mathbf{G} = G_3 \hat{\mathbf{e}}_3$, where $G_3 = \mathsf{G}_{12}$, and it can also be expressed in a form using the canonical variables (see the derivation of (9.28) below),

$$\mathbf{G} = G_3 \hat{\mathbf{e}}_3 = \int d^2r \, \nabla \phi \times \nabla s^z. \tag{9.15}$$

Some algebra is aided by the result of the following exercise.

Exercise 9.1. From (9.14b) and the identity for the Levi–Civita symbol,

$$\epsilon_{ijk} \epsilon_{ilm} = \delta_{jl}\delta_{km} - \delta_{jm}\delta_{kl} \tag{9.16}$$

show the additional relation,

$$\mathsf{G}_{jk} = \epsilon_{ijk} G_i. \tag{9.17}$$

With the result (9.17) for G_{jk}, the corresponding term in (9.10) becomes

$$-\mathsf{G}_{jk}\hat{\mathbf{e}}_j V_k = -\epsilon_{ijk} G_i \hat{\mathbf{e}}_j V_k = +\epsilon_{jik}\hat{\mathbf{e}}_j G_i V_k = \mathbf{G} \times \mathbf{V}. \tag{9.18}$$

As the gyrovector $\mathbf{G}$ points perpendicular to the plane, this *gyrotropic force* $\mathbf{G} \times \mathbf{V}$ then produces forces on the vortex acting within the 2D plane. It is the most important physical effect that results from the gyrovector.

The next collective property from the first line of (9.10) is a symmetric dissipation tensor with components D_{jk} (also dimensionless),

$$\mathsf{D}_{jk} \equiv -\alpha \int d^2r \, (\partial_j \mathbf{s}) \cdot (\partial_k \mathbf{s}). \tag{9.19}$$

This can define a dyadic object, $\mathsf{D} = \mathsf{D}_{jk} \hat{\mathbf{e}}_j \hat{\mathbf{e}}_k$, that operates on $\mathbf{V}$. Then using only these vortex properties, the original form of the Thiele equation is

$$\frac{1}{\sigma} \mathbf{F} + \mathbf{G} \times \mathbf{V} + \mathsf{D} \cdot \mathbf{V} = 0. \tag{9.20}$$

This equation describes the dynamics of the vortex core motion, assuming the internal spin structure of a vortex is unaffected by velocity. All terms in the equation have dimensions of velocity. Note that the gyrovector will vanish for an in-plane vortex. Then, in the absence of dissipation, this equation is inadequate for the dynamics of an in-plane vortex, because it becomes $\mathbf{F} = 0$, which is not necessarily true if the vortex is exposed to some external field. This is an important reason why the acceleration terms must be included.

9.1.2 Including vortex mass effects

Now consider the velocity-dependent vortex structure changes. The first term in the second line of (9.10) motivates definition of an effective mass tensor $\mathbf{M} = \mathsf{M}_{jk}\hat{\mathbf{e}}_j\hat{\mathbf{e}}_k$, with elements

$$\mathsf{M}_{jk} \equiv -\int d^2r \, \mathbf{s} \cdot (\partial_j \mathbf{s} \times \tilde{\partial}_k \mathbf{s}). \tag{9.21}$$

This definition gives $\mathbf{M}$ dimensions of time. A vortex mass tensor $\mathbf{m}$ in units of kilograms would be obtained by including a factor of the spin density,

$$\mathbf{m} \equiv \sigma \mathbf{M}. \tag{9.22}$$

Also one has another term, a new type of dissipation tensor $\tilde{\mathbf{D}} = \tilde{\mathsf{D}}_{jk}\hat{\mathbf{e}}_j\hat{\mathbf{e}}_k$, with elements

$$\tilde{\mathsf{D}}_{jk} \equiv -\alpha \int d^2r \, (\partial_j \mathbf{s}) \cdot (\tilde{\partial}_k \mathbf{s}). \tag{9.23}$$

The addition of these leads to a modified Thiele equation, including acceleration effects,

$$\frac{1}{\sigma}\mathbf{F} + \mathbf{G} \times \mathbf{V} + \mathbf{D} \cdot \mathbf{V} = \mathbf{M} \cdot \mathbf{A} + \tilde{\mathbf{D}} \cdot \mathbf{A}. \tag{9.24}$$

While both $\mathbf{M}$ and $\tilde{\mathbf{D}}$ appear in the same way in the equation, one expects that the 2×2 matrices associated with them have different structures, leading to quite different physical effects. The effect of $\tilde{\mathbf{D}}$ is expected to be small, compared to the damping caused by D.

The factor of σ dividing the force can be moved to the other terms, defining ones with more natural units. Thus we could write this modified Thiele equation with all terms having force dimensions as

$$\mathbf{F} + \mathbf{g} \times \mathbf{V} + \mathbf{d} \cdot \mathbf{V} = \mathbf{m} \cdot \mathbf{A} + \tilde{\mathbf{d}} \cdot \mathbf{A}. \tag{9.25}$$

σ has been used to make the gyrovector, dissipation tensors and mass tensor with physically reasonable units:

$$\mathbf{g} = \sigma \mathbf{G} = S \int \frac{d^2r}{v_{\text{cell}}} \mathbf{s} \cdot (\partial_1 \mathbf{s} \times \partial_2 \mathbf{s})\hat{\mathbf{e}}_3 \tag{9.26a}$$

$$\mathbf{d} = \sigma \mathbf{D} = -\alpha S \int \frac{d^2r}{v_{\text{cell}}} (\partial_j \mathbf{s}) \cdot (\partial_k \mathbf{s})\hat{\mathbf{e}}_j\hat{\mathbf{e}}_k \tag{9.26b}$$

$$\mathbf{m} = \sigma \mathbf{M} = -S \int \frac{d^2r}{v_{\text{cell}}} \mathbf{s} \cdot (\partial_j \mathbf{s} \times \tilde{\partial}_k \mathbf{s})\hat{\mathbf{e}}_j\hat{\mathbf{e}}_k \tag{9.26c}$$

$$\tilde{\mathbf{d}} = \sigma \tilde{\mathbf{D}} = -\alpha S \int \frac{d^2r}{v_{\text{cell}}} (\partial_j \mathbf{s}) \cdot (\tilde{\partial}_k \mathbf{s})\hat{\mathbf{e}}_j\hat{\mathbf{e}}_k. \tag{9.26d}$$

Next, it is necessary to discuss the values of these collective properties of a vortex, contrasting the results for in-plane versus out-of-plane vortices.

9.1.3 Calculation of vortex gyrovector and other properties

The expression in (9.14) is not so convenient for calculation of the gyrovector components, $G_3 = G_{12} = -G_{21}$, just as (9.21) is not particularly good for calculation of M_{jk}. It is better to find their expressions in terms of the in-plane and out-of-plane angles (ϕ, θ). Letting $\mathbf{s} = (\cos\theta \cos\phi, \cos\theta \sin\phi, \sin\theta)$, one has the components of a space or velocity gradient, where ∂ represents any of $\partial/\partial r_j$ or $\partial/\partial V_k$,

$$\partial\mathbf{s} = (-\sin\theta \cos\phi\, \partial\theta - \cos\theta \sin\phi\, \partial\phi, -\sin\theta \sin\phi\, \partial\theta + \cos\theta \cos\phi\, \partial\phi, \cos\theta\, \partial\theta)$$
$$= (-\sin\theta \cos\phi, -\sin\theta \sin\phi, \cos\theta)\partial\theta + (-\sin\phi, \cos\phi, 0)\cos\theta\, \partial\phi$$
$$= \hat{\boldsymbol{\theta}}\, \partial\theta + \hat{\boldsymbol{\phi}}\cos\theta\, \partial\phi. \tag{9.27}$$

The unit vectors $\hat{\boldsymbol{\theta}}$ and $\hat{\boldsymbol{\phi}}$ implicitly defined in the second line are oriented in the directions of increasing θ and ϕ, respectively. Their cross product is $\hat{\boldsymbol{\phi}} \times \hat{\boldsymbol{\theta}} = \hat{\mathbf{r}} = \mathbf{s}$. The relation (9.27) represents how the unit vector $\mathbf{s}$ can only change as its tip moves on the unit sphere. This gradient can be combined with another general gradient component, $\partial'\mathbf{s}$, leading to

$$\mathbf{s} \cdot (\partial\mathbf{s} \times \partial'\mathbf{s}) = \mathbf{s} \cdot \left[\left(\hat{\boldsymbol{\theta}}\, \partial\theta + \hat{\boldsymbol{\phi}}\cos\theta\, \partial\phi \right) \times \left(\hat{\boldsymbol{\theta}}\, \partial'\theta + \hat{\boldsymbol{\phi}}\cos\theta\, \partial'\phi \right) \right]$$
$$= \cos\theta \left(\partial\phi\, \partial'\theta - \partial\theta\, \partial'\phi \right)$$
$$= \partial\phi\, \partial' s^z - \partial s^z\, \partial'\phi. \tag{9.28}$$

Choosing $\partial \to \partial_j$ and $\partial' \to \partial_k$ (two space gradients), expression (9.14) becomes

$$G_{jk} = \int d^2r \left(\partial_j\phi\, \partial_k s^z - \partial_j s^z\, \partial_k\phi \right). \tag{9.29}$$

Finally, using this to find $G_3 = G_{12}$ gives the angular formula for the gyrovector (9.15).

For static in-plane vortices with $S^z = 0$, and even for moving ones, with S^z antisymmetric about the direction of motion, (9.15) gives $\mathbf{G} = 0$. For out-of-plane vortices, consider the static profile, and assume that the change in $\mathbf{G}$ will be essentially zero for moving out-of-plane vortices. The in-plane angle $\phi(\varphi)$ changes by $2\pi q$ around the core, and the out-of-plane spin component $s^z(r)$ takes the value $s^z(0) = p = \pm 1$ at the core and $s^z = 0$ as $r \to \infty$. Then (9.15) becomes

$$\mathbf{G} = \int_0^{2\pi} d\varphi \int_0^\infty r\, dr\, \frac{1}{r}\frac{\partial\phi}{\partial\varphi}\hat{\boldsymbol{\varphi}} \times \frac{\partial s^z}{\partial r}\hat{\mathbf{r}} = \int_0^{2\pi} d\varphi \int_0^\infty r\, dr\, \frac{q}{r}\frac{\partial s^z}{\partial r}(-\hat{\mathbf{z}}) = 2\pi p q \hat{\mathbf{z}}. \tag{9.30}$$

This gives $G = \pm 2\pi$, which is the solid angle on the unit sphere covered by the vortex spin distribution. This defines the topological charge $Q = G = 2\pi p q$ of a vortex.

For components of the mass tensor, choose $\partial \to \partial_j$ (space gradient) and $\partial' \to \tilde{\partial}_k$ (velocity gradient), which then gives from (9.21) the angular formulation,

$$\mathsf{M}_{jk} = -\int d^2r (\partial_j\phi\, \tilde{\partial}_k s^z - \partial_j s^z\, \tilde{\partial}_k\phi). \tag{9.31}$$

One can see that $\mathbf{M}_{jk}$ is not anti-symmetric. As a result it is not possible to write its contraction with the acceleration $\mathbf{A}$ as a cross product. To estimate $\mathbf{M}_{jk}$, we can use the result $s^z \approx \sin\theta_1$ in (8.39) for the leading corrections to the static vortex structure, due to motion. Also, the in-plane angle does not depend on $\mathbf{V}$, to leading order, so $\tilde{\partial}_k \phi \approx 0$. The non-zero velocity gradient is that of s^z, which is

$$\tilde{\partial}_k s^z = \frac{\partial}{\partial V_k}\left[-\frac{\mathbf{V}\cdot\nabla\phi}{JS(4\delta - q^2/r^2)}\right] = \frac{-\partial_k\phi}{JS(4\delta - q^2/r^2)} \approx \frac{-1}{4\delta\, JS}\partial_k\phi. \qquad (9.32)$$

The last approximation uses the field far from the core. This reduces the mass tensor to

$$\mathsf{M}_{jk} \approx \frac{1}{4\delta\, JS}\int d^2 r\, \partial_j\phi\, \partial_k\phi. \qquad (9.33)$$

From (8.8), with $\nabla\phi = q\hat{\varphi}/r$, we have

$$(\partial_1\phi, \partial_2\phi) = \frac{q}{r}(-\sin\varphi, \cos\varphi). \qquad (9.34)$$

Upon integration over the plane, $\sin\varphi$ and $\cos\varphi$ are orthogonal functions, and only the diagonal terms of M_{jk} survive. With lower and upper radial cutoffs r_0 and R, they are

$$\mathsf{M}_{jk} = \frac{\delta_{jk}}{4\delta JS}\int_0^{2\pi} d\varphi\, \sin^2\varphi \int_{r_0}^R r\, dr\, \frac{q^2}{r^2} = \frac{\pi q^2}{4\delta JS}\ln\left(\frac{R}{r_0}\right)\delta_{jk}. \qquad (9.35)$$

This has the same logarithmic dependence on system size as the static in-plane vortex energy. Although calculated from the in-plane vortex structure, the mass tensor for out-of-plane vortices should be nearly identical.

The dissipation tensor D_{jk} can be written with the help of (9.27) as

$$\mathsf{D}_{jk} = -\alpha\int dr^2\left(\hat{\boldsymbol{\theta}}\,\partial_j\theta + \hat{\boldsymbol{\phi}}\cos\theta\,\partial_j\phi\right)\cdot\left(\hat{\boldsymbol{\theta}}\,\partial_k\theta + \hat{\boldsymbol{\phi}}\cos\theta\,\partial_k\phi\right)$$
$$= -\alpha\int dr^2\left(\partial_j\theta\,\partial_k\theta + \cos^2\theta\,\partial_j\phi\,\partial_k\phi\right). \qquad (9.36)$$

For an in-plane vortex, the contribution from gradients of θ will be proportional to V^2, so we consider it small. Then the main contribution comes from the gradients of ϕ, and using $\cos\theta \approx 1$,

$$\mathsf{D}_{jk} \approx -\alpha\int dr^2\, \partial_j\phi\, \partial_k\phi = -\alpha\pi q^2\delta_{jk}\ln\frac{R}{r_0} \qquad \text{(in-plane vortex)}. \qquad (9.37)$$

For an in-plane vortex, the damping tensor is directly proportional to the mass tensor. One can see that the negative sign here is physically reasonable. In the presence of very strong damping, the vortex motion (with $\mathbf{G} = 0$) will follow $\mathbf{V} = \mathbf{F}/|\mathsf{D}|$, which gives a motion of the vortex in the same direction as the force.

For an out-of-plane vortex, both terms in the integrand of (9.36) can contribute. The part with gradients of θ is larger near the core ($r < r_v$) where θ approaches $\pm\pi/2$. Far from the core, at radius $r > r_v$, the part with gradients of ϕ is larger, where $\cos\theta \approx 1$. We can use the asymptotic forms found for the static out-of-plane vortex to obtain an estimate to these two contributions in the separate regions $r < r_v$ and $r > r_v$. For $q^2 = 1$ and using (8.19a), the core region's contribution (integral only over $r < r_v$) is found to be

$$\mathsf{D}_{jk,\text{core}} = -\alpha\pi A_0^2 \delta_{jk}. \tag{9.38}$$

Note that there is no lower cutoff r_0 present, because the out-of-plane vortex does not have a singularity at $r \to 0$. For the region exterior to the core (integral only over $r > r_v$), one obtains

$$\mathsf{D}_{jk,\text{ext}} = -\alpha\pi\delta_{jk}\left\{\ln\frac{R}{r_v} + A_\infty^2\left[\frac{1}{4}e^{-2} + \frac{3}{2}E_1(2)\right]\right\} \tag{9.39}$$

where the last part involves the exponential integral with value $E_1(2) \approx 0.0489$, defined from

$$E_1(x) \equiv \int_x^\infty dt\, \frac{e^{-t}}{t}. \tag{9.40}$$

Note that the logarithmic term does have a dependence on a large radius cutoff R, but it is scaled by r_v rather than r_0 as in the in-plane vortex. Then the total damping constant is the sum of these parts,

$$\mathsf{D}_{jk} = -\alpha\pi\delta_{jk}\left\{A_0^2 + \ln\frac{R}{r_v} + A_\infty^2\left[\frac{1}{4}e^{-2} + \frac{3}{2}E_1(2)\right]\right\}$$

$$\approx -\alpha\pi\delta_{jk}\left(\ln\frac{R}{r_v} + 1.20\right) \quad \text{(out-of-plane vortex).} \tag{9.41}$$

The dominant term here can be the logarithm, unless r_v is fairly large. As a result, one sees that especially as λ approaches closer to 1, causing a larger value of r_v, the tendency will be for out-of-plane vortices to have weaker damping than in-plane vortices.

Finally we can calculate the additional damping factor $\tilde{\mathsf{D}}$, based on transforming the definition to angular variables,

$$\tilde{\mathsf{D}}_{jk} = -\alpha\int dr^2\left(\tilde{\partial}_j\theta\,\tilde{\partial}_k\theta + \cos^2\theta\,\tilde{\partial}_j\phi\,\tilde{\partial}_k\phi\right) \approx -\alpha\int dr^2 \tilde{\partial}_j\theta\,\tilde{\partial}_k\theta. \tag{9.42}$$

This is reduced to the last form because to leading order in $\mathbf{V}$ the in-plane angle does not change with velocity. Assuming the velocity-dependent changes in θ are small and given from (9.32), then

$$\cos\theta\,\tilde{\partial}_k\theta \approx \tilde{\partial}_k s^z \approx \frac{-1}{4\delta JS}\tilde{\partial}_k\phi. \tag{9.43}$$

Then this damping parameter is determined by the integral

$$\tilde{\mathsf{D}}_{jk} = \frac{\alpha}{4\delta JS}\int dr^2\,\sec^2\theta(\partial_j\theta)(\partial_k\phi). \tag{9.44}$$

However, this integral is identically zero, because $\theta = \theta(r)$ while $\phi = \phi(\varphi)$. One has specifically,

$$(\partial_1\theta, \partial_2\theta) = \frac{\partial\theta}{\partial r}(\cos\varphi, \sin\varphi) \tag{9.45}$$

and together with relation (9.34), the orthogonality of the sine and cosine functions integrated over the plane leads to $\tilde{\mathbf{D}} = 0$. Then, for most situations the main modification of the original Thiele equation is only the mass term. Vortex dynamics for both in-plane and out-of-plane vortices should be fairly well described by a *modified* Thiele equation with mass,

$$\frac{1}{\sigma} \mathbf{F} + \mathbf{G} \times \mathbf{V} + \mathbf{D} \cdot \mathbf{V} = \mathbf{M} \cdot \mathbf{A}. \tag{9.46}$$

Exercise 9.2. From the definition (9.36) of the dissipation tensor D_{jk}, (a) verify the results (9.38) and (9.39) for the contributions from interior and exterior regions of an out-of-plane vortex. (b) Check the total result for the damping tensor, equation (9.41). Compare the result to that in (9.37) for in-plane vortices. You may need to assume a cutoff length.

9.2 Relation of vortex momentum to the Thiele equations

The vortex mass and gyrovector can also be related to considerations of vortex linear momentum. There are various uses of the word 'momentum' for slightly different concepts, including canonical momentum, kinetic momentum and momentum as a generator of translations. For instance, the kinetic momentum $\mathbf{p} = m\mathbf{v}$ of an electron in a magnetic field is not the same as the canonical momentum $\mathbf{P}$; these are different by a term dependent on the vector potential. A similar situation applies to vortices, which can cause a lot of confusion. Here we consider how to define both the kinetic and canonical momenta of a vortex, in such a way as to be consistent with the undamped Thiele equation [5].

A vortex kinetic momentum $\mathbf{p}$ (dependent only on vortex velocity and not on position) can be defined tentatively like that for solitons [6], based on the kinetic integral in the Lagrangian (see (2.126b) and its application (7.51) for a 1D chain), for a translating structure,

$$K = \int \frac{\mathrm{d}^2 r}{v_{\text{cell}}} S^z \, \dot{\phi} = \mathbf{V} \cdot \mathbf{p}, \tag{9.47}$$

where the kinetic momentum is then defined as

$$\mathbf{p} = -\int \frac{\mathrm{d}^2 r}{v_{\text{cell}}} S^z \, \nabla \phi = -\sigma \int \mathrm{d}^2 r \, s^z \, \nabla \phi. \tag{9.48}$$

This is the 2D generalization of kink momentum, (7.55) for a magnetic chain, where v_{cell} is the area per spin. One can see that $\mathbf{p}$ is zero for a stationary vortex, where S^z is either zero or circularly symmetric, while $\nabla \phi$ is anti-symmetric around the vortex center. Motion of the vortex leads to an asymmetric part of S^z, that combined with $\nabla \phi$ leads to non-zero momentum. The evaluation is left as an exercise.

> **Exercise 9.3.** Use the small out-of-plane perturbation (8.39) for moving in-plane vortices to obtain the estimate of their kinetic momentum,
> $$\mathbf{p} \approx \frac{\pi q^2 \sigma}{4\delta JS} \ln\left(\frac{R}{r_0}\right) \mathbf{V} = \frac{\pi q^2}{4\delta J v_{\text{cell}}} \ln\left(\frac{R}{r_0}\right) \mathbf{V}. \qquad (9.49)$$

This result for $\mathbf{p}$ is the same as $\mathbf{p} \approx m\mathbf{V}$ using a scalar mass m (units of kg) derived from (9.35),

$$\mathsf{m}_{jk} = \mathsf{m}\delta_{jk} = \sigma \mathsf{M}_{jk}. \qquad (9.50)$$

At this basic level the definitions of mass and kinetic momentum, for in-plane vortices, are consistent. One can also show that the primary contribution to vortex energy due to non-zero velocity is of the expected form, $\frac{1}{2}m\mathbf{V}^2$. While these results point to this momentum being the usual kinetic or mechanical momentum, it is important to analyze its dynamic properties, with the aim of seeing its connection to the canonical momentum, $\mathbf{P}$. At the same time, one needs to see how the momentum for out-of-plane vortices, with non-zero gyrovector, compares with momentum of in-plane vortices.

9.2.1 Poisson bracket of kinetic momentum components

One interesting property of $\mathbf{p}$ is the Poisson bracket (PB, recall the definition (2.63)) between its two components, as considered in [5]. For a canonical momentum $\mathbf{P}$, the two components are expected to be independent and have a zero PB, $\{P_x, P_y\} = 0$. For a continuum system, the fundamental PB that replaces $\{\phi_i, S_j^z\} = \delta_{ij}$ for spins on discrete sites is

$$\{\phi(\mathbf{r}), S^z(\mathbf{r}')\} = v_{\text{cell}}\delta(\mathbf{r} - \mathbf{r}'). \qquad (9.51)$$

Then using the definition (2.131), we want to check the PB of the xy-components of the above kinetic momentum by calculating

$$\{p_x, p_y\} = \int \frac{d^2 r}{v_{\text{cell}}} \left(\frac{\delta p_x}{\delta \phi} \frac{\delta p_y}{\delta S^z} - \frac{\delta p_x}{\delta S^z} \frac{\delta p_y}{\delta \phi} \right). \qquad (9.52)$$

The required functional derivatives of p_x are found by applying (2.129) for two space dimensions to the x-component of (9.48), using subscripts on the fields to indicate partial derivatives,

$$\frac{\delta p_x}{\delta \phi} = \left(\partial_\phi - \partial_x \partial_{\phi_x} - \partial_y \partial_{\phi_y} \right)\left(-S^z \partial_x \phi \right) = \partial_x S^z \qquad (9.53a)$$

$$\frac{\delta p_x}{\delta S^z} = \left(\partial_{S^z} - \partial_x \partial_{S^z_x} - \partial_y \partial_{S^z_y} \right)\left(-S^z \partial_x \phi \right) = -\partial_x \phi. \qquad (9.53b)$$

With similar results for the derivatives of p_y, we have the PB,

$$\{p_x, p_y\} = \int \frac{d^2 r}{v_{\text{cell}}} \left(-\partial_x S^z \partial_y \phi + \partial_x \phi \partial_y S^z \right) = \sigma \mathsf{G}_{12} = \sigma G_3. \qquad (9.54)$$

The RHS is identified with σ times the gyrovector by comparing with relation (9.29). Note that a similar results holds for an electric charge in a magnetic field; the PB of two kinetic momentum components is proportional to the third component of the magnetic induction. This behavior is not typical of a canonical momentum.

One could also obtain this same result by changing to a definition of **p** based on discrete differences of spin components on lattice sites, avoiding the functional derivatives. For instance, for a square lattice with spins at sites labeled by indices (n, m) for (x, y)-directions respectively, the discrete lattice definitions, using central differences, are

$$p_x = -\sum_{nm} S^z_{n,m} \partial_x \phi_{n,m} = -\frac{1}{2} \sum_{n,m} S^z_{n,m} (\phi_{n+1,m} - \phi_{n-1,m}) \qquad (9.55a)$$

$$p_y = -\sum_{nm} S^z_{n,m} \partial_y \phi_{n,m} = -\frac{1}{2} \sum_{n,m} S^z_{n,m} (\phi_{n,m+1} - \phi_{n,m-1}). \qquad (9.55b)$$

Then the needed PB is a sum of terms, using primed indices in the p_y sum,

$$\{p_x, p_y\} = \frac{1}{4} \sum_{n,m,n',m'} \left\{ S^z_{n,m}(\phi_{n+1,m} - \phi_{n-1,m}), S^z_{n',m'}(\phi_{n',m'+1} - \phi_{n',m'-1}) \right\}. \qquad (9.56)$$

Using the canonical PBs (2.101), only limited terms are non-zero here, giving

$$\{p_x, p_y\} = \frac{1}{4} \sum_{n,m,n',m'} \left[S^z_{n,m}(\phi_{n',m'+1} - \phi_{n',m'-1})(\delta_{n+1,n'} - \delta_{n-1,n'})\delta_{m,m'} \right.$$
$$\left. + (\phi_{n+1,m} - \phi_{n-1,m}) S^z_{n',m'} \delta_{n,n'}(\delta_{m,m'+1} - \delta_{m,m'-1}) \right]. \qquad (9.57)$$

Employing the deltas to eliminate the primed indices leads to

$$\{p_x, p_y\} = \frac{1}{4} \sum_{n,m} \left[S^z_{n,m}(\phi_{n+1,m+1} - \phi_{n+1,m-1} - \phi_{n-1,m+1} + \phi_{n-1,m-1}) \right.$$
$$\left. + (\phi_{n+1,m} - \phi_{n-1,m})(S^z_{n,m+1} - S^z_{n,m-1}) \right]. \qquad (9.58)$$

Finally, shifting n by ± 1 on the first line changes nothing, but leads to a result that is equivalent to the discretized gyrovector density, whose sum gives back the gyrovector:

$$\{p_x, p_y\} = \frac{1}{4} \sum_{n,m} \left[-(S^z_{n+1,m} - S^z_{n-1,m})(\phi_{n,m+1} - \phi_{n,m-1}) \right.$$
$$\left. + (\phi_{n+1,m} - \phi_{n-1,m})(S^z_{n,m+1} - S^z_{n,m-1}) \right]$$
$$= \sum_{n,m} \left(-\partial_x S^z_{n,m} \partial_y \phi_{n,m} + \partial_x \phi_{n,m} \partial_y S^z_{n,m} \right) = \sigma G_3. \qquad (9.59)$$

This is a good check that the PB obtained via functional derivatives is correct.

9.2.2 The momentum conjugate to X

Suppose there is another momentum **P** that is conjugate to the vortex position **X**. The canonical PBs between these should then be

$$\{X_i, P_j\} = \delta_{ij}, \qquad \{X_i, X_j\} = 0, \qquad \{P_i, P_j\} = 0. \tag{9.60}$$

It is interesting to suppose a transformation from **p** to **P**, that will ensure these canonical PBs and reproduce (9.54):

$$\mathbf{p} = \mathbf{P} - \frac{1}{2}\sigma \mathbf{G} \times \mathbf{X} = \left(P_x + \frac{1}{2}\sigma G_3 Y, P_y - \frac{1}{2}\sigma G_3 X\right). \tag{9.61}$$

Then one sees that this produces the PB already obtained above for the kinetic momentum components,

$$\{p_x, p_y\} = \left\{P_x + \frac{1}{2}\sigma G_3 Y, P_y - \frac{1}{2}\sigma G_3 X\right\}$$

$$= \frac{1}{2}\sigma G_3 \{Y, P_y\} - \frac{1}{2}\sigma G_3 \{P_x, X\} = \sigma G_3. \tag{9.62}$$

The transformation produces the correct PB, while at the same time giving a new momentum **P** that is conjugate to vortex core position,

$$\mathbf{P} = \mathbf{p} + \frac{1}{2}\sigma \mathbf{G} \times \mathbf{X}. \tag{9.63}$$

9.2.3 Momentum as a generator of translations

Next, consider the translational property of **p**, by evaluating its PB with an arbitrary function of the spin field, $f(\mathbf{x}) \equiv f(\mathbf{S}(\mathbf{x})) = f(\phi(\mathbf{x}), S^z(\mathbf{x}))$. Then following the same rules for PBs, we have the algebra for one component,

$$\{p_j, f(\mathbf{x})\} = \int \frac{d^2 r}{v_{\text{cell}}} \left(\frac{\delta p_j}{\delta \phi} \frac{\delta f}{\delta S^z} - \frac{\delta p_j}{\delta S^z} \frac{\delta f}{\delta \phi}\right). \tag{9.64}$$

The function f can be written as a functional by including a delta function:

$$f(\mathbf{x}) = \int \frac{d^2 r}{v_{\text{cell}}} f(\phi(\mathbf{r}), S^z(\mathbf{r})) v_{\text{cell}} \delta(\mathbf{r} - \mathbf{x}), \tag{9.65}$$

which allows calculation of its derivatives (at the location **r** needed in (9.64)) as

$$\frac{\delta f}{\delta \phi} = \frac{\partial f}{\partial \phi} v_{\text{cell}} \delta(\mathbf{r} - \mathbf{x}), \qquad \frac{\delta f}{\delta S^z} = \frac{\partial f}{\partial S^z} v_{\text{cell}} \delta(\mathbf{r} - \mathbf{x}). \tag{9.66}$$

Then we obtain

$$\{p_j, f(\mathbf{x})\} = \int \frac{d^2 r}{v_{\text{cell}}} \left(\frac{\partial S^z}{\partial r_j} \frac{\partial f}{\partial S^z} + \frac{\partial \phi}{\partial r_j} \frac{\partial f}{\partial \phi}\right) v_{\text{cell}} \delta(\mathbf{r} - \mathbf{x}) = \frac{\partial f}{\partial x_j}. \tag{9.67}$$

If the function depends on a vortex spin field in the form $\mathbf{S}(\mathbf{x} - \mathbf{X})$, where $\mathbf{X}$ is the core location, then with $\partial/\partial x_j = -\partial/\partial X_j$, the result is

$$\{\mathbf{p}, f(\mathbf{x} - \mathbf{X})\} = \frac{\partial f}{\partial \mathbf{x}} = -\frac{\partial f}{\partial \mathbf{X}}. \tag{9.68}$$

Compare the expansion of $f(\mathbf{x} - \mathbf{X})$ for small displacement $\mathbf{X}$,

$$f(\mathbf{x} - \mathbf{X}) \approx f(\mathbf{x}) - \mathbf{X} \cdot \frac{\partial f}{\partial \mathbf{X}}. \tag{9.69}$$

The shift in position $\mathbf{X}$ then acts on $\{\mathbf{p}, f\}$ to produce the change in f due to a small translation. Therefore, the kinetic momentum $\mathbf{p}$ is a generator of space translations on arbitrary functions of the spin field. Note that it is analogous to the translations in time that are generated by the Hamiltonian, according to the PB relation,

$$\{H, f(\mathbf{S})\} = -\frac{df}{dt}. \tag{9.70}$$

The *canonical momentum* $\mathbf{P}$ found earlier in (9.61), which is conjugate to vortex location $\mathbf{X}$, has the same translational property, as found for one component:

$$\{P_j, f(\mathbf{x} - \mathbf{X})\} = \left\{ p_j + \frac{1}{2}\sigma\mathbf{G} \times \mathbf{X}, f(\mathbf{x} - \mathbf{X}) \right\} = \{p_j, f(\mathbf{x} - \mathbf{X})\} = -\frac{\partial f}{\partial X_j}. \tag{9.71}$$

This is the same as expected from a general chain rule such as (2.66a) for PBs,

$$\{P_j, f(\mathbf{x} - \mathbf{X})\} = \{P_j, X_k\}\frac{\partial f}{\partial X_k} = -\delta_{jk}\frac{\partial f}{\partial X_k} = -\frac{\partial f}{\partial X_j}. \tag{9.72}$$

9.2.4 Time derivative of kinetic momentum p

The dynamics of $\mathbf{p}$ is important for comparison to the Thiele equation. Thus, we ask what is $\dot{\mathbf{p}}$? This can be answered by appealing to PB algebra again[1]. The time derivative of a component of $\mathbf{p}$ is determined from a PB with the Hamiltonian, which can be expressed as

$$\dot{p}_j = \{p_j, H\} = \int \frac{d^2 r}{v_{cell}} \left(\frac{\delta p_j}{\delta \phi} \frac{\delta H}{\delta S^z} - \frac{\delta p_j}{\delta S^z} \frac{\delta H}{\delta \phi} \right)$$

$$= \int \frac{d^2 r}{v_{cell}} \left(\frac{\partial S^z}{\partial r_j} \frac{\delta H}{\delta S^z} + \frac{\partial \phi}{\partial r_j} \frac{\delta H}{\delta \phi} \right) = -\frac{\partial}{\partial X_j} \int \frac{d^2 r}{v_{cell}} \mathcal{H} = -\frac{\partial H}{\partial X_j} = F_j. \tag{9.73}$$

Our earlier results (9.53a) for the functional derivatives of p_j were used, as well as the ansatz that the spin field is assumed of the form $\mathbf{S}(\mathbf{r} - \mathbf{X})$, or $\frac{\partial}{\partial r_j} = -\frac{\partial}{\partial X_j}$. This result gives a simple principle,

---

[1] One can also try to differentiate through the integral of the original definition (9.48). This has technical difficulties, however, because of the singularity at the vortex core, which can lead to nonsensical or poorly defined results. This seems to be the difficulty in some earlier works [4, 5].

$$\dot{\mathbf{p}} = \mathbf{F}. \tag{9.74}$$

The rate of change of the defined kinetic momentum is the force on the vortex.

The expression for $\dot{p}_j$ can be expressed differently, using instead the canonical equations (2.130) to replace the functional derivatives of H in the integrand:

$$\dot{p}_j = \int \frac{d^2 r}{v_{\text{cell}}} \left(\frac{\delta p_j}{\delta \phi} \frac{\delta H}{\delta S^z} - \frac{\delta p_j}{\delta S^z} \frac{\delta H}{\delta \phi} \right) = \int \frac{d^2 r}{v_{\text{cell}}} \left((\partial_j S^z)(\dot{\phi}) - (-\partial_j \phi)(-\dot{S}^z) \right). \tag{9.75}$$

Now replacing the spin time derivatives with the convective form (9.9) gives

$$\dot{p}_j = V_k \int \frac{d^2 r}{v_{\text{cell}}} \left(\partial_j \phi \partial_k S^z - \partial_j S^z \partial_k \phi \right) - A_k \int \frac{d^2 r}{v_{\text{cell}}} \left(\partial_j \phi \tilde{\partial}_k S^z - \partial_j S^z \tilde{\partial}_k \phi \right). \tag{9.76}$$

This contains the gyrotensor (9.29) and the mass tensor (9.31), and comparing with (9.18), it is summarized as

$$\dot{\mathbf{p}} = -\sigma \mathbf{G} \times \mathbf{V} + \sigma \mathbf{M} \cdot \mathbf{A}. \tag{9.77}$$

Although no damping is included, these results reproduce the modified Thiele equation (9.46), once $\dot{\mathbf{p}}$ is replaced by the force $\mathbf{F}$ on the vortex. This derivation based on the canonical equations of motion and classical PBs is much more direct than our earlier derivation starting from the LLG equations, although the former allowed the inclusion of damping directly. A slightly different approach was used in [5], but required differentiation of the integral definition for $\mathbf{p}$, which is not a well-defined mathematical operation, because of the singularity at the core. The derivations here have avoided that procedure. Using the scalar value $m = \sigma M$ from a diagonal mass tensor, the modified Thiele equation is

$$\mathbf{F} + \sigma \mathbf{G} \times \mathbf{V} = m\mathbf{A}. \tag{9.78}$$

These last results show that Thiele equation dynamics can be considered from the point of view of momentum conservation. However, the system contains an additional momentum term $\sigma \mathbf{G} \times \mathbf{X}$, whose time derivative appears here in the force equation. This extra momentum depends on the choice of origin, however, a shift in the origin simply gives an irrelevant constant, whose time derivative is zero.

9.2.5 Comparison of vortex dynamics to electric charge dynamics

The modified Thiele equation in the form (9.78), without damping, can be seen to be equivalent to the non-relativistic equation of motion for an electric charge e with velocity $\mathbf{v}$ and acceleration $\mathbf{a}$, affected by combined electric and magnetic forces (Lorentz force):

$$e\mathbf{E} + e\mathbf{v} \times \mathbf{B} = m\mathbf{a}. \tag{9.79}$$

The equivalence requires $\mathbf{F}$ in the Thiele equation to be identified with only the electric part of the Lorentz force, and the gyrovector must be identified with the negative magnetic induction, while spin density σ needs to be equivalent to the charge:

$$\mathbf{F} \to e\mathbf{E}, \qquad \sigma \mathbf{G} \to -e\mathbf{B}. \tag{9.80}$$

This maps vortex dynamics directly onto charge dynamics. We can see that both **F** and $\sigma\mathbf{G} \times \mathbf{V}$ in the vortex problem should be considered forces in some general sense, whose net sum gives the vortex acceleration. The contribution to **F** is only due to space variations of H, while the *gyroforce* $\sigma\mathbf{G} \times \mathbf{V}$ is analogous to the magnetic Lorentz force. We can also see how this equivalence relates to canonical momentum.

For vortices, refer again to the connection between kinetic and canonical momentum found in (9.61). Combining with (9.74) and the modified Thiele equation as (9.77), leads to three equivalent ways to write the time derivative of kinetic momentum:

$$\dot{\mathbf{p}} = \mathbf{F} = \dot{\mathbf{P}} - \frac{1}{2}\sigma\mathbf{G} \times \mathbf{V} = -\sigma\mathbf{G} \times \mathbf{V} + m\mathbf{A}. \tag{9.81}$$

This is then solved to give two equivalent expressions for $\dot{\mathbf{P}}$:

$$\dot{\mathbf{P}} = \mathbf{F} + \frac{1}{2}\sigma\mathbf{G} \times \mathbf{V} \tag{9.82a}$$

$$\dot{\mathbf{P}} = m\mathbf{A} - \frac{1}{2}\sigma\mathbf{G} \times \mathbf{V}. \tag{9.82b}$$

While $\dot{\mathbf{P}}$ does not appear in the equations for vortex or charge dynamics, but rather acts as more of a calculational aid, some insight is obtained by comparing the relations in the two systems.

It is well known that a Lagrangian for a point charge of mass m in non-relativistic motion can be written as

$$L = \frac{1}{2}m\mathbf{v}^2 - e\Phi + e\mathbf{v} \cdot \mathcal{A}, \tag{9.83}$$

where Φ is the scalar potential and $\mathcal{A}$ is the vector potential. The resulting canonical momentum is defined by

$$\mathbf{P} = \frac{\partial L}{\partial \mathbf{v}} = m\mathbf{v} + e\mathcal{A} = m\mathbf{v} + \frac{1}{2}e\mathbf{B} \times \mathbf{x}, \tag{9.84}$$

where the vector potential $\mathcal{A}(\mathbf{x})$ associated with a uniform magnetic induction is

$$\mathcal{A} = \frac{1}{2}\mathbf{B} \times \mathbf{x} = \frac{1}{2}\epsilon_{jkl}\hat{\mathbf{e}}_j B_k x_l. \tag{9.85}$$

The Euler–Lagrange equation of motion for one component is

$$\dot{P}_i = \frac{d}{dt}\left(\frac{\partial L}{\partial \dot{x}_i}\right) = \frac{\partial L}{\partial x_i} = -e\frac{\partial \Phi}{\partial x_i} + ev_j\frac{\partial \mathcal{A}_j}{\partial x_i}. \tag{9.86}$$

The needed gradients of the vector potential are

$$\frac{\partial A_j}{\partial x_i} = \frac{\partial}{\partial x_i}\left(\frac{1}{2}\epsilon_{jkl}B_k x_l\right) = \frac{1}{2}\epsilon_{jki}B_k. \tag{9.87}$$

Then with $v_j \epsilon_{jki} B_k = \epsilon_{ijk} v_j B_k = (\mathbf{v} \times \mathbf{B})_i$, this leads to the first way to write the momentum's time derivative,

$$\dot{\mathbf{P}} = \frac{\partial L}{\partial \mathbf{x}} = -e\frac{\partial \Phi}{\partial \mathbf{x}} - \frac{1}{2}e\mathbf{B} \times \mathbf{v}. \tag{9.88}$$

Note that this equivalent to (9.82a), under the mappings in (9.80). The other way to obtain a result for $\dot{\mathbf{P}}$ is to time differentiate its expression (9.84) directly[2]:

$$\dot{\mathbf{P}} = \frac{\mathrm{d}}{\mathrm{d}t}\frac{\partial L}{\partial \mathbf{v}} = m\mathbf{a} + e\frac{1}{2}\mathbf{B} \times \mathbf{v} + e\frac{\partial \mathcal{A}}{\partial t}. \tag{9.89}$$

Changing $e\mathbf{B} \rightarrow -\sigma\mathbf{G}$, and assuming static fields, this is seen to be equivalent to (9.82b) for the vortex problem. Then equating the two expressions for $\dot{\mathbf{P}}$ eliminates that variable, and leads to the Lorentz force equation (9.79).

These results then indicate that the force $\mathbf{F}$ in the vortex problem is analogous to *only* the electrostatic force for the charge problem. With the mappings in (9.80) the two systems have essentially equivalent time derivatives of canonical momenta. The connection between their kinetic momenta, however, is less obvious. For a charge one can identify a kinetic momentum $\mathbf{p} = m\mathbf{v}$, with time derivative $\dot{\mathbf{p}} = m\mathbf{a}$, however, that is not an element directly used in the derivations of the Lorentz force law dynamics. For *in-plane* vortices that have $\mathbf{G} = 0$, we were able to show that the kinetic momentum is $\mathbf{p} \approx \mathsf{m}\mathbf{V}$, however, that is not necessarily true for out-of-plane vortices. We were able to use PBs to obtain $\dot{\mathbf{p}}$ as in (9.77), containing $\mathsf{m}\mathbf{A}$ as well as a gyrotropic contribution. This strongly contrasts $\dot{\mathbf{p}} = m\mathbf{a}$ in the charge dynamics problem and is a significant difference in the dynamics.

9.2.6 Lagrangian mechanics for vortex core motion

The ideas in the last section have mapped vortex motion approximately onto the motion of a charge in combined electric and magnetic fields. The results can be collected together by determination of a Lagrangian that will reproduce the undamped modified Thiele equation (9.78). To achieve this, make the usual transformation from Hamiltonian to Lagrangian, using the canonical momentum,

$$L = \mathbf{P} \cdot \dot{\mathbf{X}} - H. \tag{9.90}$$

Based on time integration of (9.82b), we take the canonical momentum to be defined up to an arbitrary constant as

$$\mathbf{P} = \mathsf{m}\mathbf{V} - \frac{1}{2}\sigma\mathbf{G} \times \mathbf{X}. \tag{9.91}$$

This is consistent with canonical momentum expression (9.84) for a charge in a field, changing $e\mathbf{B}$ to $-\sigma\mathbf{G}$. This gives the vortex Lagrangian, with $\dot{\mathbf{X}} = \mathbf{V}$,

---

[2] For the electrodynamic problem the term $\frac{\partial \mathcal{A}}{\partial t}$ contributes to $\mathbf{E}(\mathbf{x}, t) = -\nabla\Phi - \frac{\partial \mathcal{A}}{\partial t}$. The contributions to both $\mathbf{E}$ and $\mathbf{B}$ in the Lorentz force law are split between the two ways to write $\dot{\mathbf{P}}$. The mapping from the vortex problem to a charge system only requires a situation with static electric and magnetic fields.

$$L = mV^2 - \frac{1}{2}\sigma(\mathbf{G} \times \mathbf{X}) \cdot \mathbf{V} - H. \tag{9.92}$$

Note that the second term has the form $+\sigma\mathcal{A} \cdot \mathbf{V}$, with an effective vector potential,

$$\mathcal{A} = -\frac{1}{2}\mathbf{G} \times \mathbf{X}. \tag{9.93}$$

The interaction term can also be written as $\mathcal{A} \cdot \mathbf{V} = +\frac{1}{2}(\mathbf{G} \times \mathbf{V}) \cdot \mathbf{X}$, which is useful later for differentiation. To apply L and check that the correct equation of motion results, it is necessary to assume that the Hamiltonian has a kinetic energy contribution of $\frac{1}{2}mV^2$. This will ensure as well that the canonical momentum is correctly derived from L by $\mathbf{P} = \frac{\partial L}{\partial \mathbf{V}}$. In addition, H can have other parts that depend on vortex position, whose negative gradient leads to the force $\mathbf{F}$. The Euler–Lagrange equation for one component is

$$\frac{\partial L}{\partial X_j} = \frac{d}{dt}\left(\frac{\partial L}{\partial V_j}\right). \tag{9.94}$$

This requires the derivatives,

$$\frac{\partial L}{\partial X_j} = -\frac{\partial H}{\partial X_j} + \frac{1}{2}\sigma(\mathbf{G} \times \mathbf{V})_j \tag{9.95a}$$

$$\frac{\partial L}{\partial V_j} = 2mV_j - \frac{\partial H}{\partial V_j} - \frac{1}{2}\sigma(\mathbf{G} \times \mathbf{X})_j = mV_j - \frac{1}{2}\sigma(\mathbf{G} \times \mathbf{X})_j. \tag{9.95b}$$

Then it is clear that the Euler–Lagrange equations (9.94) for all components give the correct equation,

$$-\frac{\partial H}{\partial \mathbf{X}} + \sigma\mathbf{G} \times \mathbf{V} = m\mathbf{A} \tag{9.96}$$

once the negative gradient of H is identified as the force $\mathbf{F}$. Thus the stated Lagrangian (9.92) is appropriate for analyzing dynamics of vortices, including vortex mass effects. In the case of in-plane vortices, the gyrovector is not present, and the mass effects should dominate (for very weak damping). In the case of out-of-plane vortices, the mass effects could be relatively weak; setting m to zero can give a leading approximation for out-of-plane vortex dynamics in some potential contained in H.

9.3 Vortex forces and motions

The basic dynamics of vortex core motion has been interesting to study in terms of comparing numerical simulations to predictions of the modified Thiele equation (9.46). For quasi-2D ferromagnets, a number of simple situations have been studied, including individual vortices near a boundary, and pairs of vortices in a larger system with a far away boundary. Here we give a short summary of some of the possible motions, and indicate when a non-zero vortex mass reveals its presence.

The force **F** acting on a vortex is one of the most relevant factors that determines the motion. There are three direct ways that forces are produced, including externally applied fields, the interaction of a vortex with other vortices or anti-vortices, and the interaction of a vortex with the boundary of the system. Let us estimate **F** for the most obvious situations.

9.3.1 Force due to applied field

The first way to generate force is to apply a magnetic induction $\mathbf{B}_{\text{ext}}$, especially, within the easy place. The interaction of $\mathbf{B}_{\text{ext}}$ with the spins will cause an induced magnetization and also an energy gradient. Although the vortex will become slightly deformed by an applied field, we suppose that deformation is small, which will be true as long as $\mathbf{B}_{\text{ext}}$ is much weaker than the nearest neighbor exchange fields. This means our the calculation will slightly over-estimate the force. The force comes only from the negative space gradient of the Zeeman energy,

$$U_B = -\sum_n \boldsymbol{\mu}_n \cdot \mathbf{B}_{\text{ext}} = -\sum_n \gamma \mathbf{S}_n \cdot \mathbf{B}_{\text{ext}}, \qquad (9.97)$$

which is easy to estimate for the vortex as a whole. For stationary in-plane and out-of-plane vortices, the component S^z is circularly symmetric around the vortex core position **X**, while (8.7) describes the in-plane angle.

Uniform applied field

If the whole system is exposed to a *spatially uniform field*, then only the in-plane components of $\mathbf{B}_{\text{ext}}$ can contribute to the force. Suppose the applied field of strength B_{ext} makes an angle ϕ_B to the x-axis. For the vortex, expression (8.7) for $\phi(\mathbf{r})$ is extended to a vortex with core position (X, Y),

$$\phi(\mathbf{r}) = q \tan^{-1}\left(\frac{x - X}{y - Y}\right) + \phi_0. \qquad (9.98)$$

For an in-plane vortex, the Zeeman energy in continuum theory then becomes

$$\begin{aligned}
U_B &= -\gamma S B_{\text{ext}} \int \frac{d^2r}{v_{\text{cell}}} \cos(q\varphi + \phi_0 - \phi_B) \\
&= -\gamma S B_{\text{ext}} \int \frac{d^2r}{v_{\text{cell}}} \left\{ \cos(q\varphi) \cos(\phi_0 - \phi_B) - \sin(q\varphi) \sin(\phi_0 - \phi_B) \right\} \\
&= -q\gamma S B_{\text{ext}} \int \frac{d^2r}{v_{\text{cell}}} \left\{ \frac{x - X}{\rho} \cos(\phi_0 - \phi_B) - \frac{y - Y}{\rho} \sin(\phi_0 - \phi_B) \right\}, \qquad (9.99)
\end{aligned}$$

where the last line applies only to $q = \pm 1$ and ρ is the radius measured from the vortex center,

$$\rho = \sqrt{(x - X)^2 + (y - Y)^2}. \qquad (9.100)$$

Integration over an infinite system leads to constant U_B, and no force. However, we must consider a finite system, wherein an infinitesimal vortex displacement causes one set of spins to become more aligned with $\mathbf{B}_{\text{ext}}$ and another set of spins to become less aligned with $\mathbf{B}_{\text{ext}}$. The spins in the region of the core are the ones most affected.

For the force F_x, the gradient of U_B with respect to X requires the following derivatives,

$$\frac{d}{dX}\left(\frac{x-X}{\rho}\right) = -\frac{(y-Y)^2}{\rho^3}, \qquad \frac{d}{dX}\left(\frac{y-Y}{\rho}\right) = -3\frac{(x-X)(y-Y)}{\rho^3}. \quad (9.101)$$

Then this gives

$$F_x = -q\gamma S B_{\text{ext}} \int \frac{d^2 r}{v_{\text{cell}}} \left\{ \frac{(y-Y)^2}{\rho^3} \cos(\phi_0 - \phi_B) - 3\frac{(x-X)(y-Y)}{\rho^3} \sin(\phi_0 - \phi_B) \right\}. \quad (9.102)$$

This is still indeterminant, because the boundary of the system must be specified. Let us take the simplest case: a circular system of radius R and the vortex at the center of the system. One can later consider the vortex slightly displaced from the center as a perturbation. Then, putting $X = Y = 0$, then $\rho = r$, one has some simple integrations over the circular system:

$$\int d^2 r \frac{(y-Y)^2}{\rho^3} = \int_0^{2\pi} d\varphi \int_0^R r\, dr\, \frac{1}{r} \sin^2 \varphi = \pi R. \quad (9.103)$$

The second term of (9.102) involves a product $\sin \varphi \cos \varphi$ which integrates to zero. A similar calculation applies to finding F_y, then, the resulting force on an in-plane vortex in a uniform magnetic field is

$$\mathbf{F} = -q\pi R \sigma \gamma B_{\text{ext}} \big(\cos(\phi_0 - \phi_B), -\sin(\phi_0 - \phi_B)\big). \quad (9.104)$$

To interpret this result, recall that ϕ_0 is the angle the spins make with the positive x-axis, along that axis. When $\phi_0 = 0$, the spin field is of a vortex ($q = +1$) is radially outward from the core: it has the appearance of the electric field of a positive electric charge. Then in that case, the force is opposite to the direction of $\mathbf{B}_{\text{ext}}$:

$$\mathbf{F} = -q\pi R \sigma \gamma B_{\text{ext}} (\cos \phi_B, \sin \phi_B) = -q\pi R \sigma \gamma\, \mathbf{B}_{\text{ext}}, \qquad \phi_0 = 0. \quad (9.105)$$

Conversely, the force on an anti-vortex ($q = -1$) will be parallel to $\mathbf{B}_{\text{ext}}$ if $\phi_0 = 0$. It is easy to check that a small displacement along the force direction for these spin configurations will lead to greater alignment of spins with the field.

A more interesting case is when the spin field has $\phi_0 = 90°$, giving the appearance of a magnetic field around a current-carrying wire. Now, the force becomes

$$\mathbf{F} = -q\pi R \sigma \gamma B_{\text{ext}} (\sin \phi_B, -\cos \phi_B) = q\pi R \sigma \gamma\, \hat{\mathbf{z}} \times \mathbf{B}_{\text{ext}}, \qquad \phi_0 = 90°. \quad (9.106)$$

If the vortex were to slide sideways to the field, in the direction of $q\hat{\mathbf{z}} \times \mathbf{B}_{\text{ext}}$, one can see that it will cause more spins to align with the field, lowering the Zeeman energy. Again, the force is oppositely directed on anti-vortices when compared to vortices. In small magnetic particles this is the more probable vortex structure, because of demagnetization energy causing the spins to follow close to the direction of a boundary.

The result (9.104) has been calculated for in-plane vortices only. If this calculation is repeated for out-of-plane vortices, then near the vortex core, the non-zero S^z causes the lengths of the in-plane components to be smaller. It is clear that the contribution to Zeeman energy from the core region is reduced, and hence so is the force. Without an exact expression for $S^z(\rho)$ (measuring from the vortex core location), only an approximate value for $\mathbf{F}$ can be found. The Zeeman energy in the case of out-of-plane vortices interacting with an *in-plane* magnetic induction is now

$$U_B = -\gamma S B_{\text{ext}} \int \frac{d^2 r}{v_{\text{cell}}} [1 - (s^z)^2]^{1/2} \cos(q\varphi + \phi_0 - \phi_B). \quad (9.107)$$

The result of this extra factor is that the core region has a much weaker contribution to the interaction; the spins are nearly perpendicular to $\mathbf{B}_{\text{ext}}$ there. This effect can be incorporated approximately by using the vortex core radius r_v in (8.18) as the lower cutoff of the radial integration. This means that (9.104) can be extended to the case of out-of-plane vortices, provided that R is replaced by $R - r_v$. For a system large compared to r_v, the force on out-of-plane vortices will be nearly the same as on in-plane vortices.

Non-uniform applied field
Another way to exert a force on a vortex is via the out-of-plane spin component, in a *non-uniform* out-of-plane applied field $B_{\text{ext}}^z(\mathbf{r})$. This applies only to out-of-plane vortices. Start from only the z-component's contribution to Zeeman energy,

$$U_B = -q\gamma S \int \frac{d^2 r}{v_{\text{cell}}} s^z(\mathbf{r}) \cdot B_{\text{ext}}^z(\mathbf{r}). \quad (9.108)$$

The s^z spin field is circularly symmetric around the vortex core, so a uniform B_{ext}^z would not produce any force. Suppose B_{ext}^z is a constant plus a uniform gradient along the x-direction[3],

$$B_{\text{ext}}^z(x, y) = B_0 + \left(\frac{\partial B^z}{\partial x}\right)_0 x. \quad (9.109)$$

The zero subscripts refer to the values at the origin. Consider again a circular system, with its center at the origin. Then the estimate of the force is expressed as

$$F_x = -\frac{\partial U_B}{\partial X} = \gamma\sigma \int d^2 r \, \frac{\partial s^z}{\partial \rho} \frac{\partial \rho}{\partial X} \left[B_0 + \left(\frac{\partial B^z}{\partial x}\right)_0 x\right]$$

$$= \gamma\sigma \int d^2 r \, \frac{\partial s^z}{\partial \rho} \left[\frac{-(x - X)}{\rho}\right] \left[B_0 + \left(\frac{\partial B^z}{\partial x}\right)_0 x\right]. \quad (9.110)$$

---
[3] Maxwell's equation $\nabla \cdot \mathbf{B} = 0$ would imply also non-zero in-plane components of $\mathbf{B}_{\text{ext}}$. We suppose that these extra components can be ignored to leading order.

When the vortex is at the center of the system with $X = 0$ and $\rho = r$, only the gradient part is seen to contribute, being dependent on $x^2 = r^2 \cos^2 \varphi$,

$$F_x = -\gamma\sigma \int_0^{2\pi} d\varphi \int_0^R dr\, r \frac{\partial s^z}{\partial r} \cos\varphi \left(\frac{\partial B^z}{\partial x}\right)_0 r\cos\varphi$$

$$= -\pi\gamma\sigma \left(\frac{\partial B^z}{\partial x}\right)_0 \int_0^R dr\, r^2 \frac{\partial s^z}{\partial r} = 2\pi\gamma\sigma \left(\frac{\partial B^z}{\partial x}\right)_0 \int_0^R dr\, r\, s^z(r). \quad (9.111)$$

The last step involved integration by parts, assuming that $s^z(R) \approx 0$ on the system boundary. To evaluate further, it is necessary to have $s^z(r)$. An estimate can be made based on the approximate solutions (8.19) found earlier, for the two regions $r < r_v$ and $r > r_v$, which is an exercise.

Exercise 9.4. From the approximate out-of-plane component (8.19), check the evaluation of the following integrals for force contributions from the core region and the far field region of an out-of-plane vortex,

$$\int_0^{r_v} dr\, r\, s^z(r) \approx p\left(\frac{r_v}{A_0}\right)^2 (A_0 \sin A_0 + \cos A_0 - 1) \approx 0.3943 p r_v^2 \quad (9.112a)$$

$$\int_{r_v}^R dr\, r\, s^z(r) \approx p A_\infty r_v^2 \left(e^{-1} + \frac{1}{2}\sqrt{\pi}\, \mathrm{erfc}(1)\right) \approx 0.8664 p r_v^2. \quad (9.112b)$$

These depend on the core polarization $p = \pm 1$ and the constants A_0 and A_∞ that define the approximate structure of $s^z(r)$; 'erfc' is the complementary error function.

The core contributes about 1/3 and the far region about 2/3 of the total force. Then combining these contributions, and generalizing to a gradient of B^z in any direction, the force is found to be

$$\mathbf{F} \approx 1.26 \times 2\pi\gamma\sigma p r_v^2 (\nabla B^z)_0. \quad (9.113)$$

The 0 subscript means the gradient is measured at the location of the vortex center. Although this has been calculated for a vortex centered in a circular system, one could also find the small corrections if the vortex is offset away from the center. Note the dependence on core polarization p, which shows that the force is in the direction appropriate to increase the alignment of spins with $\mathbf{B}_{\mathrm{ext}}$ for $p = +1$, or reduce the amount of anti-alignment for $p = -1$. Thus, a small gradient in $\mathbf{B}_{\mathrm{ext}}$ might be used to separate vortices of opposite polarizations.

In the case of strong damping, a displacement of the vortex will take place in the same direction as the calculated force. If the damping is weak, then at least initially, an out-of-plane vortex will make some gyrotropic motion, typically perpendicular to the instantaneous force.

9.3.2 Vortex pair forces

The second way forces are generated is through the interactions between pairs of vortices or anti-vortices. First, consider a pair of in-plane vortices, one of vorticity q_1, and phase angle ϕ_{01} at the origin, and a second of vorticity q_2 and phase angle ϕ_{02} at a position $\mathbf{X} = (X, 0)$ along the x-axis (chosen as the axis from one vortex to the other). With only in-plane spin components, the total in-plane angle is a superposition of the two vortices,

$$\phi = \phi_1 + \phi_2, \quad \text{where} \quad \begin{cases} \phi_1 = q_1 \tan^{-1}\left(\dfrac{y}{x}\right) + \phi_{01} \\ \phi_2 = q_2 \tan^{-1}\left(\dfrac{y}{x - X}\right) + \phi_{02}. \end{cases} \quad (9.114)$$

In the Hamiltonian (8.4) the energy density depends only on the square of $\nabla \phi = \nabla \phi_1 + \nabla \phi_2$, which is independent of the phase angles, and determined from

$$\nabla \phi_1 = q_1 \frac{\hat{\mathbf{z}} \times \mathbf{r}}{r^2} = \frac{q_1}{x^2 + y^2}(-y, x) \quad (9.115a)$$

$$\nabla \phi_2 = q_2 \frac{\hat{\mathbf{z}} \times (\mathbf{r} - \mathbf{X})}{|\mathbf{r} - \mathbf{X}|^2} = \frac{q_2}{(x - X)^2 + y^2}(-y, x - X). \quad (9.115b)$$

Its absolute square has two direct terms (single vortex) and cross terms (the interaction),

$$|\nabla \phi|^2 = \frac{q_1^2}{x^2 + y^2} + \frac{q_2^2}{(x - X)^2 + y^2} + 2 q_1 q_2 \frac{x(x - X) + y^2}{\left(x^2 + y^2\right)\left[(x - X)^2 + y^2\right]}. \quad (9.116)$$

When integrated over the system, the first two terms are the single vortex energies; only the last term will be the pair interaction energy that depends on the separation X. This pair interaction potential U_{pair} is obtained, writing the integration in circular coordinates,

$$\begin{aligned} U_{\text{pair}} &= \frac{1}{2} J S^2 \int d^2 r \ |\nabla \phi|^2_{\text{pair}} \\ &= J S^2 q_1 q_2 \int_0^{2\pi} d\varphi \int_0^R r \, dr \frac{r^2 - rX \cos \varphi}{r^2 \left[r^2 - 2rX \cos \varphi + X^2\right]}. \end{aligned} \quad (9.117)$$

Doing the radial integration first, a change of variable to $u = r^2 - 2rX \cos \varphi + X^2$, with $du = 2(r - X \cos \varphi) dr$, transforms the radial integral to logarithmic form,

$$\begin{aligned} U_{\text{pair}} &= J S^2 q_1 q_2 \int_0^{2\pi} d\varphi \int \frac{1}{2} \frac{du}{u} \\ &= J S^2 q_1 q_2 \int_0^{2\pi} d\varphi \, \frac{1}{2} \ln\left[r^2 - 2rX \cos \varphi + X^2\right]_0^R = 2\pi J S^2 q_1 q_2 \ln\left(\frac{R}{X}\right). \end{aligned} \quad (9.118)$$

Note the similarity to the in-plane single vortex energy (8.15), but with an extra factor of 2 and the separation X replacing the short radius cutoff r_0.

Then the negative gradient gives the force, which is along the line connecting the vortices:

$$F_x = -\frac{\partial U_\text{pair}}{\partial X} = \frac{2\pi J S^2 q_1 q_2}{X}. \tag{9.119}$$

Not surprisingly, the dependence on $1/X$ is analogous to the force (per unit length) between a pair of current-carrying wires, which produce a magnetic field of the same geometry as the vortex in-plane spin field. Opposite signed vortices produce attractive forces while like-signed vortices produce repulsive forces.

For out-of-plane vortices, the result must be corrected due to the non-zero value of $S^z(r)$. These corrections will come mostly from the core region, $r < r_\text{v}$. If the core regions of the pair overlap, then this correction could be significant, yet there is no simple calculation based on the energy (8.4). Generally speaking, as long as the core regions do not overlap, expression (9.119) should be a good approximation for the force between out-of-plane vortices.

Exercise 9.5. Two out-of-plane vortices with opposite polarizations $p_2 = -p_1$ are separated by distance X_{12}. Suppose their gyrovectors are anti-parallel; they have equal vorticities. That arrangement causes them to make a parallel translational motion, as found from the Thiele equation, see figure 9.8. Discuss whether it would be possible to use some applied magnetic field arrangement to stop that motion and freeze the vortices in place. If there is a possible solution, also analyze whether it would be stable.

9.3.3 Vortex image forces

A third effect that produces a force on a vortex is the interaction with the boundary of the system. A boundary has a tendency to polarize the spins in such a way that is equivalent to the field of an image vortex across the boundary, outside the system. Depending on boundary conditions, it is possible to determine the location and charge of the image and then determine the force as that for a pair interaction.

Free boundary conditions

The first type of boundary condition is *free boundary conditions*. This means that the system is sharply cut off at the boundary, and the spins along the edge are lacking a nearest neighbor in the direction $\hat{n}$ pointing outward from the system. In a discrete lattice system, the equations of motion are of a form,

$$\dot{\mathbf{S}}_i = \mathbf{S}_i \times \mathcal{F}_i, \tag{9.120}$$

where $\mathcal{F}_i$ is the effective field due to the sites that are neighbors of site i. Across the boundary, it is conceptually useful to imagine fictitious spins on the other side, from which the boundary condition and the image vortex can be analyzed. A missing neighbor due to a boundary can be set to the value of the spin $\mathbf{S}_i$ itself, without changing the dynamics, since $\mathbf{S}_i \times \mathbf{S}_i = 0$. This is equivalent to saying that the spin field within the system will adjust itself quasi-statically so that its gradient perpendicular to the boundary becomes zero. This is the concept of minimal changes in $\mathbf{S}(\mathbf{r})$ as one moves from the real spins within the system to the fictitious ones outside. So for free boundary conditions, we assume the quasi-static boundary conditions, where subscript 'b' indicates evaluation at the boundary,

$$(\hat{\mathbf{n}} \cdot \nabla)\mathbf{S}|_b = \left.\frac{\partial \mathbf{S}}{\partial n}\right|_b = 0 \tag{9.121}$$

In fact, $\mathbf{S}(n)$ is continuous and has a zero slope across a boundary, where n is a coordinate measured through the boundary. This is equivalent to two equations for the angular coordinates,

$$\left.\frac{\partial \theta}{\partial n}\right|_b = 0, \quad \left.\frac{\partial \phi}{\partial n}\right|_b = 0. \tag{9.122}$$

Applied to the component $S^z = S\sin\theta$, it means that an out-of-plane vortex with polarization p has an image vortex across the boundary with the *same polarization*.

For either an out-of-plane or in-plane vortex, the condition on ϕ determines the vorticity of the image vortex, as follows. A vortex produces a field $\nabla\phi$ given in (8.8) that can be derived from a vector potential as $\nabla\phi = \nabla \times \mathcal{A}$, which determines $\mathcal{A}$ according to

$$\nabla\phi = \frac{q}{r}\hat{\varphi} = \nabla \times \mathcal{A} = \left(\frac{\partial \mathcal{A}_r}{\partial z} - \frac{\partial \mathcal{A}_z}{\partial r}\right)\hat{\varphi}. \tag{9.123}$$

Only a z-component of $\mathcal{A}$ is required, which is solved as

$$\mathcal{A}_z = -q \ln r. \tag{9.124}$$

That is the potential for a vortex at the origin; it is easy to shift this to another center. Then, assume a vortex q at position $\mathbf{X}$ inside the system requires an image q' outside the system at position $\mathbf{X}'$. Their combined vector potential is

$$\mathcal{A}_z = -q \ln|\mathbf{r} - \mathbf{X}| - q' \ln|\mathbf{r} - \mathbf{X}'|. \tag{9.125}$$

It is helpful to apply this to the case of a *straight boundary*, and then generalize to other cases. Suppose there is an infinite straight boundary along the line $x = x_b$ parallel to the y-axis, as in figure 9.1, while the vortex is placed at $x = X, y = 0$. The boundary condition at $x = x_b$ is obtained by the curl component,

$$\left.\frac{\partial \phi}{\partial x}\right|_{x_b} = (\nabla \times \mathcal{A})_x = \frac{\partial \mathcal{A}_z}{\partial y} = 0. \tag{9.126}$$

This says that the value of $\mathcal{A}_z$ *along the boundary* is a constant. Then expressing it in Cartesian coordinates, we have

$$\mathcal{A}_z(x_b, y) = -\frac{1}{2}\left\{q \ln\left[(x_b - X)^2 + y^2\right] + q' \ln\left[(x_b - X')^2 + y^2\right]\right\} = \text{constant}. \tag{9.127}$$

One can verify that (9.126) and this last expression are satisfied, with $\mathcal{A}_z(x_b, y) = 0$, only if

$$q' = -q \quad \text{and} \quad X' - x_b = x_b - X. \tag{9.128}$$

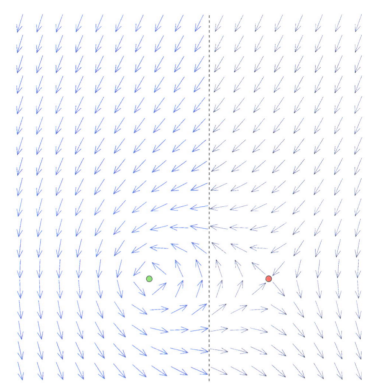

Figure 9.1. The effect of free boundary conditions on a vortex (green dot, for $q = +1$) within a system, which is the region left of the dashed line, is as if there are fictitious spins outside the system (gray arrows) whose field is that of an anti-vortex (red dot, for $q' = -1$) equidistant from the boundary. This causes the gradient of the spin field perpendicular to the boundary to be zero there. The vortex is attracted to the boundary.

Then, the image vorticity is *opposite* to that of the original vortex, and the image is located equidistant from the boundary on the outside of the system. In figure 9.1 the $q = +1$ vortex is depicted with a green dot, while its $q' = -1$ image is displayed as a red dot. This result is identical to an electrostatics image for an electric charge near a grounded infinite plane. If we let the distance from vortex to the boundary be denoted $d_b \equiv |x_b - X|$, then our previous result (9.119) for the interaction of a pair of vortices separated by $2d_b$ gives an estimate of the force due to the boundary,

$$F_x = \frac{\pi J S^2 q^2}{d_b}. \tag{9.129}$$

A positive value here indicates a force of attraction towards the boundary.

In the case of an arbitrary *curved boundary*, the condition on the curl of $\mathcal{A}$ contains both components relative to the normal to the boundary, $\hat{\mathbf{n}}$, according to

$$\left.\frac{\partial \phi}{\partial n}\right|_b = \hat{\mathbf{n}} \cdot \nabla \phi = \hat{\mathbf{n}} \cdot (\nabla \times \mathcal{A}) = n_x \partial_y \mathcal{A}_z - n_y \partial_x \mathcal{A}_z = (\hat{\mathbf{n}} \times \nabla)_z \mathcal{A}_z = 0. \tag{9.130}$$

If $\hat{\mathbf{n}}$ is normal to the boundary, then $\hat{\mathbf{n}} \times \nabla$ is the component of gradient directed *along the boundary*. Then for any boundary, this demonstrates that a constant value

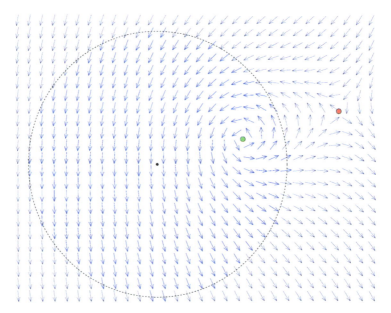

Figure 9.2. The effect of free boundary conditions on a vortex ($q = +1$, green dot) within a circular system. One can imagine fictitious spins outside the system whose field is that of an anti-vortex ($q' = -1$, red dot) at radius given by (9.132) outside the boundary. The gradient of the spin field perpendicular to the boundary is zero at the boundary and the vortex is attracted to the boundary.

of $\mathcal{A}_z$ along the boundary will be required. The image problem then requires making $\mathcal{A}_z$ constant on the boundary, regardless of its exact shape.

An example of a curved boundary is a vortex q at position $\mathbf{X} = (X, 0)$ within a circular system of radius R, similar to that in figure 9.2. The image charge q' is assumed by symmetry to be at $\mathbf{X}' = (X', 0)$. In circular coordinates the total vector potential can be expressed as

$$\mathcal{A}_z = -q \ln[r^2 - 2rX \cos \varphi + X^2] - q' \ln[r^2 - 2rX' \cos \varphi + X'^2]. \quad (9.131)$$

This needs to be independent of angular location φ at the system edge $r = R$. The derivative with respect to φ being zero leads to an expected result, $q' = -q$. Thus the image is of opposite vorticity. Imposing the boundary condition at radius R then gives the result for the position of the image,

$$X' = \frac{R^2}{X}, \quad (9.132)$$

which is similar to that for an electrostatic image outside a grounded cylinder or even a grounded sphere. Thus we can then find that the image force on a vortex in a circular system is radially outward, for free boundary conditions, determined by the vortex–image separation,

$$F_r = F_x = \frac{2\pi J S^2 q^2}{X' - X} = \frac{2\pi J S^2 q^2}{\dfrac{R^2}{X} - X}. \quad (9.133)$$

Again, the vortex is attracted towards the boundary, with a spin field such as that in figure 9.2. Of course, the motion itself need not be towards the boundary, due to gyrotropic effects. The result is only valid for out-of-plane vortices if the vortex is not within a vortex core radius (r_v) of the system edge. If the vortex is close to the center of the system ($X \ll R$), one sees that the force is linearly proportional to X, away from the center.

> **Exercise 9.6.** Consider the vector potential due to a vortex and its image given in (9.131). (a) Show that $q' = -q$ is needed to make the potential constant on the circular boundary at $r = R$. (b) Show that the image location is related to the vortex location by $XX' = R^2$.

Strong demagnetization or fixed boundary conditions

Another type of boundary condition is that where demagnetization effects are strong along the boundary. This may not be important for an individual layer of spins, but could be important as more layers are added to build up a mesoscopic thin film (i.e. a micromagnetic system). If one supposes that the demagnetization energy should be reduced to its minimum, this implies that no surface magnetic charges should be present on the system boundaries. Based on the definition (3.23), for our purpose here we take an edge magnetic charge density as $\sigma_S = \mathbf{S} \cdot \hat{\mathbf{n}}$. The requirement of $\sigma_S = 0$ means that the spins need to follow the path of the boundary at an edge, which can be summarized as

$$\hat{\mathbf{n}} \cdot \mathbf{S}|_b = 0. \tag{9.134}$$

It should be kept in mind that this is a rather strong assumption. It may require a certain minimum system size in order to be true.

As an example, consider first a *linear edge*, again at $x = x_b$ along a line parallel to the y-axis, with a charge q vortex at position $\mathbf{X} = (X, 0)$ near this boundary. This is depicted in figure 9.3. In contrast to free boundary conditions where A_z was found to be constant (and zero) on the boundary, now the x-component of $\mathbf{S}(\mathbf{r})$ is assumed to be zero. Note that for free boundary conditions only $\nabla\phi$ was determined; an arbitrary constant angle could be added to that solution to fully determine the spin field. Here, consider the in-plane spin angle (not $\nabla\phi$) as a superposition of the original vortex field and an image, including a constant:

$$\phi = \phi_1 + \phi_2, \quad \text{where} \begin{cases} \phi_1 = q \tan^{-1} \dfrac{y}{x - X} + \phi_0 \\ \phi_2 = q' \tan^{-1} \dfrac{y}{x - X'}. \end{cases} \tag{9.135}$$

The image q' is assumed to be at $\mathbf{X}' = (X', 0)$. With $\hat{\mathbf{n}} = \hat{\mathbf{x}}$, we want here only the x-component of $\mathbf{S}$ to be forced to be zero at $x = x_b$. It is possible to imagine that this requires either $q' = q$ or $q' = -q$. Further, it simplifies the calculation by already assuming $\phi_0 = \pm \frac{\pi}{2}$, which makes the original vortex field point in a circular right- or

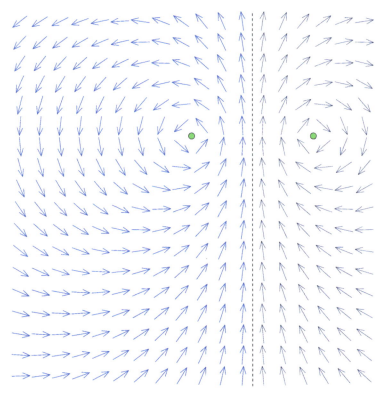

Figure 9.3. The effect of strong demagnetization boundary conditions (or fixed boundary conditions) on a vortex (green dot for $q = +1$) within a system, which is the region left of the dashed line. One can consider fictitious spins outside the system (gray arrows) whose field is that of an image vortex (also $q' = +1$) equidistant from the boundary. This causes the spin field to follow the boundary, eliminating the pole density there. The vortex is repelled from the boundary.

left-hand sense around the core point **X**. Rotating the original vortex field by $\pm\frac{\pi}{2}$, it becomes

$$\phi_1 = q \tan^{-1} \frac{x - X}{-y}. \tag{9.136}$$

Then the total in-plane angle is obtained from the tangent sum formula,

$$\tan \phi = \frac{\tan \phi_1 + \tan \phi_2}{1 - \tan \phi_1 \tan \phi_2} = \frac{\dfrac{q(x - X)}{-y} + \dfrac{q'y}{x - X'}}{1 + qq' \dfrac{x - X}{x - X'}}$$

$$= \frac{q'y^2 - q(x - X)(x - X')}{y[(x - X') + qq'(x - X)]}. \tag{9.137}$$

To force the spins to follow this straight boundary parallel to $\hat{y}$, one needs $\phi = 90°$ along $x = x_b$, meaning the denominator must vanish. That is seen to be possible only if $q' = q$, in which case the image is found to be equidistant from the boundary as the original vortex:

$$d_b = X' - x_b = x_b - X. \quad (9.138)$$

The result is shown in figure 9.3, where both the real vortex and its image are depicted with green dots, to indicate their positive vorticity. Then, as the image has the same sign of vorticity as the vortex, the force between them is *repulsive*, but of the same magnitude as in expression (9.129). Thus we see the details of the boundary conditions will have a strong influence on dynamics.

This example can easily be extended to the case of a *curved boundary*, for a vortex q at position $\mathbf{X} = (X, 0)$ within a circular system of radius R, similar to that in figure 9.4. The image q' is again assumed to be at position $\mathbf{X}' = (X, 0)$. Expression (9.137) still applies in this case, but now the resulting angle ϕ evaluated on the circular boundary at $r = R$ must follow the boundary direction, which is the angular coordinate φ shifted by $\phi_0 = \pm\frac{\pi}{2}$, according to the sense of the spins' rotation:

$$\phi|_b = \varphi \pm \frac{\pi}{2}. \quad (9.139)$$

This means the tangent needed is

$$\tan \phi|_b = \frac{\sin\left(\varphi \pm \frac{\pi}{2}\right)}{\cos\left(\varphi \pm \frac{\pi}{2}\right)} = \frac{\cos \varphi}{-\sin \varphi} = -\cot \varphi = \frac{x}{-y}. \quad (9.140)$$

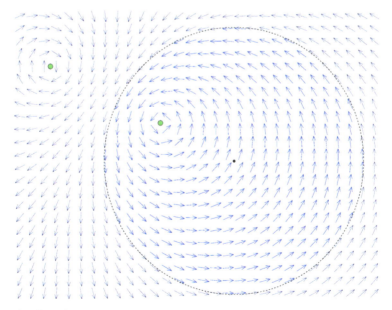

Figure 9.4. The effect of strong demagnetization boundary conditions on a $q = +1$ vortex (green for positive vorticity) in a circular system. The fictitious spins outside the system (gray arrows) make a field of an image vortex (also $q' = +1$) at the radius given by (9.144) outside the boundary. The spin field is forced to follow the boundary and the vortex is repelled from the boundary.

Then the boundary condition (9.134) will be satisfied provided at the boundary we have

$$\tan\phi\,|_b = \frac{q'y^2 - q(x-X)(x-X')}{y[(x-X') + qq'(x-X)]} = \frac{x}{-y}. \tag{9.141}$$

With $x = R\cos\varphi$ and $y = R\sin\varphi$, this can be rearranged into

$$(1+qq'-q)R^2\cos^2\varphi + q'R^2\sin^2\varphi + \left[q(1-q')X + (q-1)X'\right]R\cos\varphi = qXX'. \tag{9.142}$$

The equation needs to be made valid and independent of φ by appropriate choices of q' and X'. The coefficient of $\cos\varphi$ must be zero to accomplish this, which is seen to give

$$q = q' = 1. \tag{9.143}$$

It is an intriguing result, that shows that the boundary condition can be satisfied *only for a vortex excitation*, not an anti-vortex. This makes sense, because the anti-vortex field is not capable of having its spin direction follow a circular boundary. Then with the image of the same sign, the vortex is repelled from the boundary, just as for a straight edge. Once the charges are determined, the remaining terms lead to a familiar result for the image location,

$$X' = \frac{R^2}{X}. \tag{9.144}$$

The situation is depicted in figure 9.4, where the combination of the fields of the original vortex and image vortex do follow the path of the boundary. Note that in the most general case, this requires a particular choice of the phase angle ϕ_0 as well. Then the force on a vortex ($q = +1$ only) due to its image is in the inward radial direction and is obtained from (9.119) as

$$F_r = F_x = -\frac{2\pi JS^2}{X' - X} = -\frac{2\pi JS^2}{\dfrac{R^2}{X} - X}. \tag{9.145}$$

In the limit of $X \ll R$, this is a linear restoring force that confines the vortex close to the center of the circle,

$$F_r \approx -\frac{2\pi JS^2}{R^2}X, \quad \text{for } X \ll R. \tag{9.146}$$

This is a result that could be of great use in the study of vortex motion in magnetic nanodots. It is worth noting that the force constant varies as the inverse square of the radius.

Exercise 9.7. (a) Check that the strong demagnetization boundary condition in the form (9.141) is transformed to (9.142). (b) Show that $q' = q = +1$ is required to satisfy the boundary condition at $r = R$. Then, it is impossible to satisfy the boundary condition with an anti-vortex. (c) Show that the image location is related to the vortex location by $XX' = R^2$, the same as for free boundary conditions.

9.4 Some simple examples of vortex dynamics

Here we look at some simple few-vortex situations and consider the dominant forces and the resulting motions, as described by the modified Thiele equation (9.46). Out-of-plane vortices are considered, because their dynamics is most interesting due to the presence of their non-zero gyrovector $\mathbf{G}$. For single vortices, the simplest situations are when they are interacting with their images caused by a boundary. The boundary may be straight or curved; in the latter case we assume a circular shape. For pairs of vortices in a system (two vortices or a vortex and an anti-vortex), there would be the direct force between the two as the primary driving force, and the image effects could be secondary, if the vortices are far from the system boundary. These kinds of motions have been verified in simulations of vortex pairs in rectangular systems [7] and individual vortices in circular systems.

9.4.1 An individual vortex near a straight boundary

Suppose a vortex of charge q, polarization p and gyrovector $\mathbf{G} = G\hat{\mathbf{z}} = 2\pi pq\hat{\mathbf{z}}$ finds itself a distance d_b from a straight boundary such as that in figure 9.1 or figure 9.3. In the case of free boundary conditions, the force on the vortex due to its image is towards the boundary, regardless of the sign of q. In the case of strong demagnetization (fixed) boundary conditions, the force is away from the boundary. Strong demagnetization requires the spins to point along the direction of the boundary, and that is possible only for vortices with $q = +1$; it is impossible to satisfy the requirement of zero pole density on the boundary if $q = -1$.

Suppose $\hat{\mathbf{n}}$ is a normal vector pointing *outward* from the system (from q towards its image q'). In both the cases of free and fixed boundary conditions, we can write the force on the vortex, due to its image outside the system in the following form, based on (9.129),

$$\mathbf{F} = -\frac{\pi J S^2 q q'}{d_b}\hat{\mathbf{n}}, \qquad q' = \nu_{bc} q, \qquad (9.147)$$

where the index ν_{bc} that determines the image charge depends on boundary conditions,

$$\nu_{bc} = \begin{cases} -1 & \text{free boundaries} \\ +1 & \text{fixed boundaries}. \end{cases} \qquad (9.148)$$

A free (fixed) boundary condition produces an image of the opposite (same) vorticity, attracting (repelling) the vortex to (from) the boundary[4].

---

[4] Note that the unit vector $\hat{\mathbf{n}}$ points towards the image vortex. Then, $-\hat{\mathbf{n}}$ points from the image towards q. The notation $\hat{\mathbf{R}}$ is used later in (9.181), in place of $-\hat{\mathbf{n}}$, for the unit vector pointing towards the charge of interest, when forces between vortex pairs inside a system are considered.

Steady-state uniform motion without damping

If there is no damping, a steady-state solution of the Thiele equation at constant velocity $\bar{\mathbf{V}}$ is then found by applying a cross product with the gyrovector (with acceleration $\mathbf{A} = 0$),

$$\mathbf{G} \times \left[\frac{1}{\sigma} \mathbf{F} + \mathbf{G} \times \bar{\mathbf{V}} \right] = 0, \quad (9.149)$$

which then leads to the steady-state undamped velocity

$$\bar{\mathbf{V}} = \frac{\mathbf{G} \times \mathbf{F}}{\sigma G^2} = -\nu_{bc} \frac{\pi J S^2 q^2}{\sigma 2 \pi p q d_b} \hat{\mathbf{z}} \times \hat{\mathbf{n}}. \quad (9.150)$$

For later analysis, it is useful to put this result in the following form:

$$\bar{\mathbf{V}} = -\frac{C}{G d_b} \hat{\mathbf{z}} \times \hat{\mathbf{n}}, \quad (9.151)$$

where C is a force constant between the vortex q and its image $q' = \nu_{bc} q$,

$$C = q q' \pi \sigma^{-1} J S^2. \quad (9.152)$$

In terms of this constant, the force on the vortex, per unit spin density, is

$$\mathbf{f} = \sigma^{-1} \mathbf{F} = -\frac{C}{d_b} \hat{\mathbf{n}}. \quad (9.153)$$

This stationary solution, depicted in figure 9.5, represents a uniform translational motion parallel to the boundary. Its direction is determined by the combination of q, p and ν_{bc}. The speed will be all the greater for a vortex closer to the boundary. However, the result will apply only as long as the distance d_b is greater than the vortex core radius r_v. For given $G = 2\pi q p$, the two different types of boundary conditions give motions in opposite directions. Some interesting examples can be

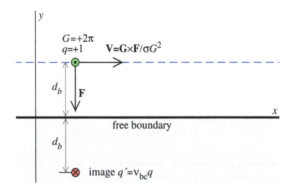

Figure 9.5. Coordinate system for a vortex q moved by the force of its image q' outside the system boundary. The vortex has $q = +1$ (shown as green), $p = +1$ and $G = +2\pi$ (shown with ⊙), being attracted to the boundary with a force according to (9.147). For a free boundary ($\nu_{bc} = -1$), its image is of opposite sign, $q' = -1$ (shown as red), with $p' = +1$ and then $G' = -2\pi$ (shown with ⊗). The steady-state vortex velocity results from the Thiele equation (9.20) without damping; see also (9.150).

seen in figure 3 of [7], where the closer a vortex is to a boundary, the faster it moves as it interacts with its image outside the system.

Steady-state motion with damping
Now, consider the effect of damping, including a damping term $\mathsf{D}_{jk}V_k$ in the Thiele equation. As seen earlier, the damping matrix is diagonal, with negative elements, so it is helpful to write it as $\mathsf{D}_{jk} \equiv -\mathsf{D}\delta_{jk}$, including the minus sign so that the constant D is positive. A little consideration shows that in fact, the presence of damping already forces there to be components of **V** both parallel and perpendicular to the boundary. A mass term also leads to both velocity components being necessary. Let us consider the individual effects of damping and mass first separately, then later it is easy to conclude what their combined effects are.

For a concrete example (such as figure 9.1 rotated 90° clockwise), take the boundary to be the x-axis, with the vortex in the first quadrant at position $\mathbf{X}(t) = (X(t), Y(t))$ and velocity $\mathbf{V}(t) = (V_x(t), V_y(t))$. The outward normal is then $\hat{\mathbf{n}} = -\hat{\mathbf{y}}$, and the only force component is F_y. Define $f_y = \sigma^{-1}F_y$ to simplify the notation. Then the modified Thiele equation (9.46) in Cartesian components is written for this example as

$$-GV_y - \mathsf{D}V_x = \mathsf{M}\dot{V}_x \tag{9.154a}$$

$$f_y + GV_x - \mathsf{D}V_y = \mathsf{M}\dot{V}_y. \tag{9.154b}$$

The scaled force is

$$f_y = \sigma^{-1}F_y = \nu_{\text{bc}}q^2 \frac{\pi JS^2}{\sigma Y} = \frac{C}{Y}. \tag{9.155}$$

Now the distance to the boundary, Y, is not necessarily constant, due to $V_y \neq 0$. However, the velocity component V_x parallel to the boundary can be eliminated, solving (9.154b) as

$$GV_x = -f_y + \mathsf{D}V_y + \mathsf{M}\dot{V}_y. \tag{9.156}$$

This can be substituted into (9.154a) to give a separated equation for the component of the motion perpendicular to the boundary,

$$G^2 V_y = \mathsf{D}\left(f_y - \mathsf{D}V_y - \mathsf{M}\dot{V}_y\right) + \mathsf{M}\left(\dot{f}_y - \mathsf{D}\dot{V}_y - \mathsf{M}\ddot{V}_y\right). \tag{9.157}$$

Due to the force varying with Y^{-1}, an exact solution for arbitrary D and M is difficult. At this point we consider separately the cases of damping without mass and mass without damping.

Damping without mass
Suppose the mass is dropped from (9.157), with the damping being the only effect to modify the gyrovector force. This could be considered the strong damping limit. Then the equation of motion is:

$$\left(G^2 + \mathsf{D}^2\right)V_y = \mathsf{D}f_y \quad \Longrightarrow \quad \frac{dY}{dt} = \frac{\mathsf{D}C}{G^2 + \mathsf{D}^2}\frac{1}{Y} \tag{9.158}$$

or
$$\int_{Y_0}^{Y(t)} dY' \, Y' = \int_0^t dt' \, \frac{1}{2}\Gamma, \qquad \text{where } \Gamma \equiv \frac{2DC}{G^2 + D^2}. \tag{9.159}$$

Starting at an initial distance from the boundary, Y_0, the exact solution is easily found,

$$Y(t) = \left(Y_0^2 + \Gamma t\right)^{1/2}. \tag{9.160}$$

Within the approximations to arrive at this, it shows that the vortex accelerates as it approaches closer to the boundary, due to the increasing force. At the same time, (9.154a) for zero mass indicates that the trajectory has a constant slope:

$$\frac{V_y}{V_x} = -\frac{D}{G} \quad \Longrightarrow \quad \frac{dY}{dX} = -\frac{D}{G}. \tag{9.161}$$

Then this shows that the vortex path is a straight line, with the constant $-D/G$ being the slope relative to the boundary. From this or by integration of (9.154a) the position along the boundary is

$$X(t) = X_0 - \frac{G}{D}\left[\left(Y_0^2 + \Gamma t\right)^{1/2} - Y_0\right]. \tag{9.162}$$

Figure 9.6 shows two examples of these results for free boundary conditions, one for a vortex with positive G (taking $q = p = +1$) and one with negative G (using $q = +1$ but $p = -1$), with different damping constants D. Note that the slope $-D/G$ does not depend on the type of boundary condition; D is positive while $G = 2\pi q p$, positive or negative, determines the slope. The force constant $C \propto \nu_{bc}$, however, is negative for free boundary conditions but positive for fixed boundary conditions, see (9.155). Then with $\Gamma \propto C$, the vortex approaches the boundary for free boundaries ($\Gamma < 0$) but it moves away from the boundary for fixed boundaries ($\Gamma > 0$). The translation

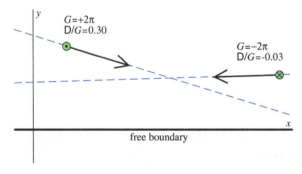

Figure 9.6. Examples of vortex motion near a boundary with free boundary conditions ($\nu_{bc} = -1$), damping parameter D and no mass, according to the results in (9.160) and (9.162). The vortex on the left has $q = +1$ (shown as green), $p = +1$ and $G = +2\pi$ (shown with $\odot$), being attracted to the boundary and moving at the descending slope $-D/G = -0.30$. The vortex on the right has $q = +1$ (green), $p = -1$ and then $G = -2\pi$ (shown with $\otimes$), but it is also attracted towards the boundary, moving in the reverse direction along the slope $-D/G = +0.03$.

parallel to the boundary can be seen to be in the direction of $-qp\nu_{bc}\hat{x}$, as found without damping. For the examples in figure 9.6, both vortices are attracted to the boundary but move in opposite senses along the boundary because they have opposite signs of G.

Mass without damping
At this point suppose the mass is included, but not the damping. Later the correction $D > 0$ causes can be addressed. The force varies with position and hence with time, and we make an approximation,

$$\dot{f}_y = \frac{df_y}{dY}\frac{dY}{dt} = -\frac{C}{Y^2}\dot{Y} \approx -\frac{C}{Y_0^2}V_y, \qquad (9.163)$$

where $Y_0 = Y(0)$ is the initial separation from the boundary. The vortex is assumed to be far enough from the boundary that this approximation of a nearly constant force makes sense. Then from (9.157), the transverse velocity component approximately satisfies a harmonic oscillator equation,

$$\ddot{V}_y \approx -\left(\frac{G^2}{M^2} + \frac{C}{MY_0^2}\right)V_y. \qquad (9.164)$$

Assuming some initial velocity away from the boundary, $V_y(0)$, a solution with a convenient phase is

$$V_y(t) = V_y(0)\cos \omega_c t \qquad (9.165a)$$

$$Y(t) = \int_0^t dt'\, V_y(t') = Y_0 + Y_1 \sin \omega_c t, \qquad Y_1 = V_y(0)/\omega_c, \qquad (9.165b)$$

where the undamped cyclotron frequency is

$$\omega_c = \sqrt{\frac{G^2}{M^2} + \frac{C}{MY_0^2}}. \qquad (9.166)$$

The corresponding motion parallel to the boundary is then found from (9.156),

$$V_x(t) = G^{-1}\left(-f_y + M\dot{V}_y\right) \approx -\frac{C}{GY_0} - V_y(0)\frac{M\omega_c}{G}\sin \omega_c t. \qquad (9.167)$$

This is summarized as

$$V_x(t) = \bar{V}_x - V_y(0)\frac{M\omega_c}{G}\sin \omega_c t, \qquad \bar{V}_x = -C/(GY_0) \qquad (9.168a)$$

$$X(t) = \int_0^t dt'\, V_x(t') = X_0 + \bar{V}_x t + X_1 \cos \omega_c t, \qquad X_1 = V_y(0)M/G. \qquad (9.168b)$$

Note that $\bar{V}_x$ is the velocity $\bar{V}$ of uniform motion found in (9.151). This is a rather interesting solution, because it shows that the vortex core makes a small-amplitude cycloidal motion on top of the uniform motion parallel to the boundary.

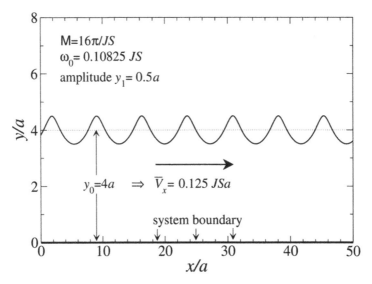

Figure 9.7. The type of undamped cycloidal oscillations expected for a vortex at distance $y_0 = 4a$ from a system boundary while interacting with its image across the boundary. The plot is made by assuming a typical vortex mass $M = 16\pi/JS$, gyrovector $G = 2\pi$, square lattice spin density $\sigma = S/a^2$ and force constant of magnitude $|C| = \pi JSa^2$, which gives the drift speed from $\bar{V}_x = -C/(GY_0)$ and cycloidal frequency from (9.166).

The cyclotron frequency becomes $\omega_c \to G/M$ if the vortex is far from the boundary; otherwise, $\omega_c < G/M$. Then, the amplitude Y_1 of the transverse oscillations is larger than the amplitude X_1 of the longitudinal oscillations, producing an elliptical oscillatory motion as the vortex translates parallel to the boundary. An example of the motion is shown in figure 9.7, for a vortex on a square lattice at an initial distance of $Y_0 = 4a$ from the boundary. The mass has been assumed to be the value $M = 16\pi/JS$, which is typical for small systems [8] with $R < 20a$. For this distance, the drift velocity is found to be $\bar{V}_x = 0.125JSa$, and the cycloidal frequency is $\omega_c = 0.10825JS$, whereas the limiting value of the frequency for a vortex very far from the boundary is $\omega_c \to G/M = 0.125JS$. Figure 9.7 is made for cycloidal oscillations in the counterclockwise sense, as would be the case for a vortex with $G = +2\pi$ and free boundary conditions. An amplitude $Y_1 = 0.5a$ was assumed; the amplitude of the cycloidal motion is not determined by this calculation. In a sense, it is a free parameter that represents something very similar to the amplitude of a classical spin wave mode.

Weak damping for a vortex with mass
Now consider if a weak damping $D \ll G$ is included together with the mass. First, we can see a simple mechanical law results from the modified Thiele equation, multiplying (9.154a) by V_x and (9.154b) by V_y, and adding, to give,

$$\frac{d}{dt}\text{KE} = f_y V_y - \frac{2D}{M}\text{KE}, \qquad \text{KE} = \frac{1}{2}M\left(V_x^2 + V_y^2\right). \tag{9.169}$$

This indicates that a usual vortex kinetic energy associated with the mass changes in time either due to the damping, or due to the work done by the force. Without damping, we had $V_y = 0$ and there was no work, and the kinetic energy was constant. With damping, there could even be positive work, as in the case of a vortex moving towards its image across the boundary with free boundary conditions. The competition between the damping and the work will determine whether $\mathbf{V}^2$ is increasing or decreasing.

Going back to (9.157), and again assuming that the vortex is far from the boundary ($Y \approx Y_0$ plus oscillations), the first term on the RHS can be approximated (for weak damping),

$$\mathrm{D} f_y \approx \frac{\mathrm{D} C}{Y_0}. \tag{9.170}$$

The equation now is arranged as

$$\left(G^2 + \mathrm{D}^2 + \frac{\mathrm{M} C}{Y_0^2} \right) V_y = \frac{\mathrm{D} C}{Y_0} - 2\mathrm{D}\mathrm{M} \dot{V}_y - \mathrm{M}^2 \ddot{V}_y. \tag{9.171}$$

Then one sees a quasi-steady-state solution with $\dot{V}_y \approx 0$ would be present with a slow drift velocity towards (away from) the boundary for free (fixed) boundary conditions,

$$\bar{V}_y = \frac{\mathrm{D} C / Y_0}{G^2 + \mathrm{D}^2 + \mathrm{M} C / Y_0^2}. \tag{9.172}$$

This drift can be taken out of $V_y(t) = \bar{V}_y + \tilde{V}_y(t)$ to obtain an equation for the oscillatory part, $\tilde{V}_y(t)$,

$$\mathrm{M}^2 \ddot{\tilde{V}}_y + 2\mathrm{D}\mathrm{M} \dot{\tilde{V}}_y + \left(G^2 + \mathrm{D}^2 + \frac{\mathrm{M} C}{Y_0^2} \right) \tilde{V}_y = 0. \tag{9.173}$$

Then assuming a usual exponential type of solution, $\tilde{V}_y \propto \exp(-i\omega t)$, leads to an equation for the complex frequency,

$$-\mathrm{M}^2 \omega^2 - 2\mathrm{D}\mathrm{M} i\omega + G^2 + \mathrm{D}^2 + \frac{\mathrm{M} C}{Y_0^2} = 0. \tag{9.174}$$

Using the definition (9.166) of the undamped frequency, this is

$$\omega^2 + \frac{2\mathrm{D}}{\mathrm{M}} i\omega - \omega_c^2 - \frac{\mathrm{D}^2}{\mathrm{M}^2} = 0. \tag{9.175}$$

We obtain two solutions from the quadratic formula,

$$\omega = \frac{1}{2} \left[-\frac{2i\mathrm{D}}{\mathrm{M}} \pm \sqrt{\left(\frac{2i\mathrm{D}}{\mathrm{M}}\right)^2 + 4\left(\omega_c^2 + \frac{\mathrm{D}^2}{\mathrm{M}^2}\right)} \right] = -\frac{i\mathrm{D}}{\mathrm{M}} \pm \omega_c. \tag{9.176}$$

Then this will lead to decaying sine and cosine solutions, whose amplitudes diminish with $\exp(-\gamma_D t)$, where the decay constant is $\gamma_D = D/M$. We can see that the overall core motion will be described by

$$V_y(t) \approx \bar{V}_y + \tilde{V}_y(0)e^{-\gamma_D t}\cos\omega_c t. \tag{9.177}$$

From this, one can go back and also find the motion $V_x(t)$ from (9.156). It exhibits a slight reduction in the uniform velocity parallel to the boundary and a damping of the oscillatory motion. For free boundary conditions, the vortex will indeed tend to move slowly towards its image of opposite vorticity on the other side of the boundary, as the damping takes out energy from the system. This causes the vortex to leave the system, eventually. For fixed or strong demagnetization boundary conditions, the tendency will be instead to move away from the boundary, but again in a process that lowers the energy, this time moving away from its image of the same vorticity. In that case, the damping effect will push the vortex back into the system and tend to stabilize it.

9.4.2 A pair of vortices in a large system

Consider a pair of vortices with charges q_1, p_1 and q_2, p_2 far from any boundaries in a large system, so that the forces due to images can be neglected, to leading order. The vortex locations are denoted as $\mathbf{X}_1$ and $\mathbf{X}_2$. They attract or repel each other with a Coulomb-like force given in (9.119). This situation was considered in [9] without damping. Here we show the main results, which are rather simple.

At first we seek uniform or stationary solutions, where the vortices move at constant speeds. Later we allow for possible small perturbations about these states. Looking at the Thiele equation for just one vortex affected by constant force, we know the steady-state velocity in (9.150) is in the direction of $\mathbf{G} \times \mathbf{F}$. The vortices exert equal and opposite forces on each other. Then the vortex velocities are in opposite directions if their gyrovectors are parallel, and conversely, the velocities are parallel (and equal) if the gyrovectors are anti-parallel. Figure 9.8 shows the result of

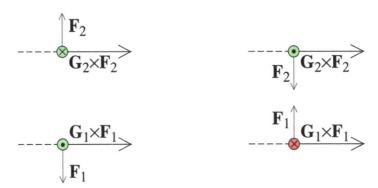

Figure 9.8. Examples of vortex pairs with anti-parallel gyrovectors performing linear translational motion. The gyrovector component out of the page is indicated by $\odot$ for $G = +2\pi$ and by $\otimes$ for $G = -2\pi$. The VV pair on the left has equal vorticities $q_1 = q_2 = +1$ (green) but opposite polarizations. The VA pair on the right has opposite vortices (red for $q = -1$) but equal polarizations. The translational velocity of a vortex is in the direction of $\mathbf{G} \times \mathbf{F}$, according to (9.150).

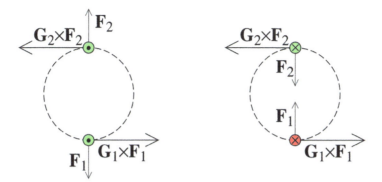

Figure 9.9. Examples of vortex pairs with parallel gyrovectors performing orbital motion. The gyrovector component $G = 2\pi qp$ out of the page is indicated by $\odot$ for $G = +2\pi$ and by $\otimes$ for $G = -2\pi$. The VV pair on the left has equal vorticities $q_1 = q_2 = +1$ (green) and equal polarizations ($p_1 = p_2 = +1$). The VA pair on the right has opposite vorticities (red for $q = -1$) and opposite polarizations. Each vortex moves instantaneously in the direction of $\mathbf{G} \times \mathbf{F}$, according to (9.150).

applying this rule to a vortex–vortex (VV) pair and a vortex–anti-vortex (VA) pair with opposite gyrovectors: the pair will translate together in a straight line. Figure 9.9 shows the result of this rule for a VV pair and a VA pair with parallel gyrovectors: the pair will orbit one about the other in uniform circular motion. This analysis leads to the conclusion that the motion has only these two possibilities:

$$\text{Anti-parallel } \mathbf{G}: q_1 p_1 = -q_2 p_2 \implies \text{parallel translational motion} \quad (9.178a)$$

$$\text{Parallel } \mathbf{G}: q_1 p_1 = q_2 p_2 \implies \text{circular orbital motion.} \quad (9.178b)$$

To make this result quantitative, we can calculate the speed in each case. It is convenient to let the separation of the pair be a parameter $2R_0$, and suppose that a unit vector $\hat{\mathbf{R}}_{12}$ points from vortex 2 towards vortex 1.

For *anti-parallel* gyrovectors, the parallel translation of the two vortices takes place at the steady-state velocity $\bar{\mathbf{V}} = \mathbf{V}_1 = \mathbf{V}_2$ with

$$\bar{\mathbf{V}} = \frac{\mathbf{G}_1 \times \mathbf{F}_1}{\sigma G_1^2} = \frac{2\pi JS^2 q_1 q_2/(2R_0)}{\sigma 2\pi p_1 q_1}(\hat{\mathbf{z}} \times \hat{\mathbf{R}}_{12}) = \frac{C}{G_1 R_0}(\hat{\mathbf{z}} \times \hat{\mathbf{R}}_{12}). \quad (9.179)$$

This uses the generalized definition of the force constant in (9.152),

$$C = q_1 q_2 \pi \sigma^{-1} JS^2 \quad (9.180)$$

and the force is defined by

$$\mathbf{F}_1 = \frac{C}{R_0}\hat{\mathbf{R}}_{12}. \quad (9.181)$$

Compare with (9.147) for the force between a charge and its image outside the system, which points in the opposite direction. Of course, larger separation results in slower motion. Changing from a VV pair to a VA pair makes no difference once the polarizations are reversed to keep the gyrovectors anti-parallel, see figure 9.8.

When the gyrovectors are *parallel*, the resulting circular motion at some assumed steady-state frequency ω makes it necessary to include the acceleration term. The Thiele equation is applied just to vortex 1, while vortex 2 makes the same motion but half a period out-of-phase. We can write the acceleration as $\mathbf{A}_1 = -\omega^2 \mathbf{X}_1$, assuming $\mathbf{X}_1$ is relative to the center of the motion, and the velocity as $\mathbf{V}_1 = \omega \hat{\mathbf{z}} \times \mathbf{X}_1$, in

$$\sigma^{-1} \mathbf{F}_1 + \mathbf{G}_1 \times \mathbf{V}_1 = M \mathbf{A}_1. \tag{9.182}$$

With $\mathbf{G}_1 \times \mathbf{V}_1 = -\omega G_1 \mathbf{X}_1$, this gives for the radial component,

$$\frac{C}{R_0} - G_1 \omega R_0 = -M\omega^2 R_0. \tag{9.183}$$

Note that in the limit of zero mass, the solution (of the original Thiele equation) must be

$$\omega_0 = \omega(M=0) = \frac{C}{G_1 R_0^2}. \tag{9.184}$$

The general quadratic equation with mass (9.183) can be rearranged as

$$\omega^2 - \frac{G_1}{M}\omega + \frac{C}{MR_0^2} = 0, \tag{9.185}$$

whose general solution is

$$\omega = \frac{G_1}{2M} - \sqrt{\left(\frac{G_1}{2M}\right)^2 - \frac{C}{MR_0^2}}. \tag{9.186}$$

The negative root was selected to give the result (9.184) in the limit $M \to 0$. This looks singular as far as the dependence on small M. Using the identity, $(a - \sqrt{a^2 - b})(a + \sqrt{a^2 - b}) = b$, another way to write it is

$$\omega = \frac{JS^2}{\sigma R_0^2}(q_2 p_1) \left(1 + \sqrt{1 - \frac{q_2}{q_1} \frac{JS^2 M}{\pi \sigma R_0^2}}\right)^{-1}. \tag{9.187}$$

Although small mass $MJS \ll 1$ is probably not very likely [8], we can consider finding the small M limit (also because the ratio of M/R_0^2 is what is more relevant). An expansion leads to a result with a linear dependence on mass,

$$\omega \approx \omega_0 \left(1 + \frac{q_2}{q_1} \frac{JS^2 M}{4\pi \sigma R_0^2}\right), \qquad \frac{JS^2 M}{4\pi \sigma R_0^2} \ll 1. \tag{9.188}$$

These are curious results, showing that a pair with equal vorticities (VV or AA) will have a faster steady-state rotation than a pair with opposite signs (VA). In order to have parallel gyrovectors, a VV or AA pair must have equal polarizations $p_1 = p_2$. For a VA pair, the polarizations would have to be in opposite directions. Then the VV or AA pair has a particularly different out-of-plane spin structure than a VA pair, which will be responsible for the change in the rotational frequency. Earlier we assumed a mass

$M = 16\pi/JS$, for a system around $R_0 = 20a$, which comes from studies of single vortex oscillations in circular systems [8]. For these parameters, the expansion parameter is

$$\frac{JS^2 M}{4\pi\sigma R_0^2} \approx \frac{16\pi S}{4\pi S \cdot 20^2} = \frac{1}{100}. \tag{9.189}$$

Therefore the mass effect on rotation rate can be expected to be rather small.

Exercise 9.8. (a) Verify that (9.183) results by application of the modified Thiele equation to a pair of vortices with parallel gyrovectors. (b) Show that the orbital frequency is given by general solution (9.186). (c) Make an appropriate expansion for small mass $MJS \ll 1$ and confirm the result (9.188), which shows that the relative vorticity of the pair q_2/q_1 affects the steady-state rate of rotation.

Cycloidal oscillations of a pair

Now consider fluctuations on the stationary solutions just found. Suppose there is an additional small-amplitude displacement superimposed. This extra displacement is taken to be in opposite directions for each vortex of the pair. To make this clear, define the half-separation vector $\mathbf{R}_{12}$ and the center vector $\mathbf{X}_c$ of the vortices as

$$\mathbf{R}_{12} = \frac{1}{2}(\mathbf{X}_1 - \mathbf{X}_2) \tag{9.190a}$$

$$\mathbf{X}_c = \frac{1}{2}(\mathbf{X}_1 + \mathbf{X}_2). \tag{9.190b}$$

Then $\mathbf{R}_{12}(t)$ measures the vortex positions relative to the center of the pair, i.e.

$$\mathbf{X}_1(t) = \mathbf{X}_c(t) + \mathbf{R}_{12}(t) \tag{9.191a}$$

$$\mathbf{X}_2(t) = \mathbf{X}_c(t) - \mathbf{R}_{12}(t). \tag{9.191b}$$

The center position translates uniformly at velocity $\bar{\mathbf{V}}$ for anti-parallel gyrovectors, see (9.179), or it is fixed in location for parallel gyrovectors. Either way, it does not contribute an acceleration. The half-separation can be assumed to be that from the stationary solution, $\mathbf{R}_0(t)$, combined with small oscillations $\mathbf{r}(t)$, writing,

$$\mathbf{R}_{12}(t) = \mathbf{R}_0(t) + \mathbf{r}(t). \tag{9.192}$$

For *anti-parallel* gyrovectors, the stationary solution of parallel translation has $\mathbf{X}_c(t) = \mathbf{c}_0 + \bar{\mathbf{V}}t$, and constant separation $2\mathbf{R}_0$. Supposing that the translation is along the x-axis, $\bar{\mathbf{V}} = (\bar{V}_x, 0)$, one could take the vortex locations described by $\mathbf{R}_0 = (0, Y_0)$ and fluctuations $\mathbf{r}(t) = (X(t), Y(t))$. The velocity of vortex 1 is then $\mathbf{V}_1 = (\bar{V}_x + \dot{X}, \dot{Y})$, and for vortex 2 the velocity is $\mathbf{V}_2 = (\bar{V}_x - \dot{X}, -\dot{Y})$. To leading order in the small fluctuations, this physical situation is the same as that already considered in section 9.4.1, for a vortex interacting with its image, moving along a boundary parallel to the x-axis. When the vortex moves to slightly larger Y, its image moves farther away in the opposite direction. The corresponding small displacements in X do not affect the force, and play a limited role. Therefore we already have solved for the frequency of the oscillations, in (9.166). Using the force constant in (9.180), the small-amplitude cycloidal oscillations have the undamped frequency,

$$\omega_c = \sqrt{\frac{G^2}{M^2} + \frac{C}{My_0^2}} = \frac{G}{M}\sqrt{1 + \frac{q_1 q_2 JS^2 M}{\pi\sigma(2R_0)^2}}. \tag{9.193}$$

Note that this depends on $2R_0$, the average separation of the pair, and the relative signs of the vorticities $q_1 q_2$. A VV or AA pair will have a higher frequency than a VA pair. The resulting trajectories for this parallel translation in the x-direction are still given by (9.165) and (9.168).

For **parallel** gyrovectors, the stationary solution of rotational motion is described with a fixed center $\mathbf{X}_c = \mathbf{c}$, and a constant separation distance $2R_0$. Of course, the half-separation vector itself is rotating at the steady-state frequency ω in (9.187), with $\dot{\mathbf{R}}_0 = \omega\hat{\mathbf{z}} \times \mathbf{R}_0$. Therefore suppose this part of the motion has components,

$$\mathbf{R}_0 = R_0(\cos\omega t, \sin\omega t) \tag{9.194a}$$

$$\dot{\mathbf{R}}_0 = \omega R_0(-\sin\omega t, \cos\omega t). \tag{9.194b}$$

The components of the modified Thiele equation (for vortex 1) are

$$f_x - G_1 V_y = M\dot{V}_x \tag{9.195a}$$

$$f_y + G_1 V_x = M\dot{V}_y, \tag{9.195b}$$

where the scaled force, using C of (9.180), is

$$\mathbf{f} = \sigma^{-1}\mathbf{F} = \frac{C}{R_{12}^2}\mathbf{R}_{12}. \tag{9.196}$$

It is assumed that $\mathbf{R}_0(t)$ keeps a fixed length, whereas $\mathbf{r}(t)$ must make an elliptical cycloidal motion to solve the equations. It is somewhat tedious but with appropriate expansions and keeping the leading linear terms in $\mathbf{r}(t)$ and its time derivatives, one can find the small-amplitude oscillations at a frequency ω_c about the uniform rotational motion at frequency ω. The resulting cyclotron frequency for those oscillations is similar to that for the case of parallel translation, becoming

$$\omega_c = \frac{G}{M}\sqrt{1 - \frac{q_1 q_2 JS^2 M}{2\pi\sigma R_0^2}}. \tag{9.197}$$

Compare (9.193) for the case of anti-parallel gyrovectors, which has the opposite sign in the radical and an extra factor of 2 dividing the mass. Of course, these mass effects can be expected to be rather small; the dominant contribution to this cyclotron oscillation frequency is G/M. These types of oscillations can be reproduced in simulations of vortex pairs, especially for parallel gyrovectors, such as seen in the work of Mertens *et al* [9]. Further discussion of a more elaborate *collective coordinate* theory for describing the motion of vortex pairs can be found in Völkel *et al* [7]. In this collective coordinate theory, the ideas on vortex mass presented here have been extended to a more general theory where a mass tensor is developed that is a multi-vortex object, rather than the mass discussed here, which has been a single vortex property.

9.4.3 An individual vortex in a circular system

A situation of great interest is a single vortex within a circular system of radius R, especially because this can be the ground state in the case where demagnetization effects are included. At this point, we do not explicitly include demagnetization, but instead, consider how a vortex moves in a circular quasi-2D magnet with either free ($\nu_{bc} = -1$) or fixed ($\nu_{bc} = +1$) boundary conditions. Recall that the latter type, which forces the spins to point along the edge of the circle on the boundary, is the limit of strong demagnetization. It also can be satisfied only if the vortex has a positive charge $q = +1$.

Suppose a vortex q is inside the circle, at time-varying position $\mathbf{X}(t)$ measured from the center of the circle. For either type of boundary condition, there is an image vortex $q' = \nu_{bc} q$ outside the system at a radius $X' = R^2/X$. The vortex will move in a circle of radius X, under the force caused by its image. From the earlier result (9.133), the radial component of force scaled by σ is

$$f_r = -qq' \frac{2\pi J S^2/\sigma}{R^2/X - X} = \frac{-2CX}{R^2 - X^2}. \tag{9.198}$$

This is equivalent to using the force definition (9.147) with $\hat{\mathbf{n}} = \hat{\mathbf{r}}$, with the definition of C in (9.152). If the vortex is very close to the center of the circle, with $X \ll R$, the force will be linear in X. For free (fixed) boundary conditions the force points outward (inward), with $C < 0$ ($C > 0$). The direction of the force is *opposite* to that when we considered a pair of vortices rotating about each other, because the image charge is outside the system, on the same radius as the charge q itself. Either boundary condition will have a stationary solution that is an orbital motion around the center of the circle, although the outward force might be expected to lead to an unstable orbit.

Let the stationary solutions have a constant speed with velocity given by

$$\mathbf{V} = \omega \hat{\mathbf{z}} \times \mathbf{X} \tag{9.199}$$

similar to the pair of vortices with parallel gyrovectors orbiting about each other. The modified Thiele equation (9.46) requires $\mathbf{G} \times \mathbf{V} = -\omega G \mathbf{X}$. Then the modified Thiele equation without damping is

$$\frac{-2C}{R^2 - X^2} \mathbf{X} - \omega G \mathbf{X} = -M\omega^2 \mathbf{X}. \tag{9.200}$$

The sign on C is opposite to that in (9.183) for the rotating vortex pair, again because now q interacts with its image which is outside the system. For a chosen orbital radius X the frequency is determined, similar to the result (9.187) for a rotating vortex pair,

$$\omega = 2\omega_0 \left(1 + \sqrt{1 - 4M\omega_0/G}\right)^{-1} \tag{9.201a}$$

$$\omega_0 = \frac{-2C/G}{R^2 - X^2}. \tag{9.201b}$$

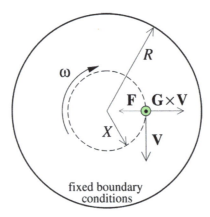

Figure 9.10. Circular path of a vortex with $q = +1$, $G = +2\pi$ (indicated by ⊙) moving in a circular orbit of radius X in a circular system of radius R. Assuming fixed boundary conditions ($\nu_{bc} = +1$), the image vortex produces a force towards the center. The sense of the velocity is in the direction of $\mathbf{G} \times \mathbf{F}$. The gyrotropic orbital frequency $\omega = V/X$ is given in (9.201), using a positive force constant C, leading to clockwise rotation ($\omega < 0$). The rotation would be reversed for free boundary conditions or by reversing the sign of G.

The frequency ω_0 is the result at zero mass. The root from the quadratic formula was chosen to give the correct result at zero mass. The force constant $C \propto qq' = \nu_{bc}q^2$ is negative for free boundary conditions ($\nu_{bc} = -1$). This then leads to a positive value of ω_0 when $G = +2\pi$, corresponding to a counterclockwise vortex motion. The outward force on the vortex is balanced by $\mathbf{G} \times \mathbf{V}$ and $-M\omega^2\hat{\mathbf{r}}$ both pointing inward. For the other case of fixed boundary conditions ($\nu_{bc} = +1$), the constant C is positive and $\omega_0 < 0$ for $G = +2\pi$, and the vortex rotates in the clockwise sense. This is sketched in figure 9.10. Now the force and $-M\omega^2\hat{\mathbf{r}}$ are both inward but $\mathbf{G} \times \mathbf{V}$ is outward. The velocity is in the same sense as the cross product $\mathbf{G} \times \mathbf{F}$, see figure 9.10.

If the mass is small relative to G/ω_0, an expansion then gives the approximate result,

$$\omega \approx \omega_0\left(1 + \frac{M\omega_0}{G}\right) = \omega_0\left[1 - \frac{q'}{q}\frac{JS^2M}{2\pi\sigma(R^2 - X^2)}\right], \qquad \frac{JS^2M}{2\pi\sigma(R^2 - X^2)} \ll 1. \quad (9.202)$$

The last form is written to compare with (9.188) obtained for a rotating pair of vortices. The main difference is the opposite sign on the mass term, due to the location of the image. The other important difference is the dependence on radial location of the vortex, which must subtract from the size of the circle. However, in the limit $X \to 0$, this gyrotropic frequency reaches a finite limit inversely proportional to the squared radius of the circle. Thus these results are important for the development of magnetic oscillators based on this gyrotropic mode of oscillation.

A real system will have some small damping. As seen earlier in the case of a vortex near a straight boundary, the damping will tend to cause the vortex to drift towards the boundary for free boundary conditions. In that case, the vortex is unstable in the system, and after some number of revolutions, it will leave the system, as it moves to lower energy (being attracted to its image). Therefore, the case of an outward force on the vortex is certainly unstable. On the other hand, for fixed boundary conditions (strong demagnetization), damping will cause the vortex to move away from the boundary, towards the center of the system. The system will be stable. This is the situation needed for the design of a stable oscillator.

9.5 Vortex–spin wave interactions and normal modes

Up to this point a magnetic vortex has been treated almost as if it has a fixed structure, for a given magnetic anisotropy described by parameter λ. The only variations considered in the structure were those due to a change in velocity, which led to the idea of a vortex mass. In this section we consider more general natural oscillatory changes in vortex structure associated with the fact that it has a certain flexibility, just as a guitar string or the membrane of a drum. If the spin field of a vortex is slightly perturbed, say, by thermal fluctuations, it has natural modes of oscillation about the original structure. These oscillations make up what might be called the vortex *internal dynamics*. These normal modes of a vortex also characterize the interaction of the vortex with the spin waves of the system.

The mode spectrum was first calculated numerically for small ferromagnetic (FM) systems in [10] and extended to the case of antiferromagnetic systems in [11] and [12]. Some of the modes of oscillation appear as a result of imparting motion to the vortex; these are called *translational modes*. They tend to have a fairly low frequency, perhaps inversely proportional to the system size, basically because the energy associated with a shift of vortex position is close to zero. Then, other *instability modes* are possible that become strong when the anisotropy parameter λ is close to the critical value λ_c for crossover from stable in-plane vortices to stable out-of-plane vortices. This type of mode has a wave function concentrated near the vortex core region, where the out-of-plane fluctuations become strong. Finally there are many other modes that correspond in general to a continuum of spin waves *scattered* from the vortex. Far from the vortex, they may be nearly the same as found without a vortex; the core region of the vortex changes the spectrum. As in quantum scattering problems, one can analyze the modes in terms of radial and azimuthal dependencies and quantum numbers. The FM system has also been analyzed with the presence of an applied field perpendicular to the easy plane, in the so-called cone state [13].

Here the calculational procedure will be described for FM systems, and then some results described for the FM easy-plane model without applied fields. The easy-plane model is studied on a square lattice, for the discrete Hamiltonian,

$$H = -J \sum_{\mathbf{n}} \sum_{j=1}^{z} \left(S_{\mathbf{n}}^x S_{\mathbf{n}+\mathbf{a}_j}^x + S_{\mathbf{n}}^y S_{\mathbf{n}+\mathbf{a}_j}^y + \lambda S_{\mathbf{n}}^z S_{\mathbf{n}+\mathbf{a}_j}^z \right). \tag{9.203}$$

The spectrum of modes will be computed for small systems by numerical diagonalization of a linearized perturbation problem. Results will be analyzed for the system with and without a vortex, for a full range of the anisotropy parameter λ. The calculations are performed on a circular system of radius R, with the sites on a square lattice grid, for a static vortex. Fixed or vortex-Dirichlet boundary conditions (as in strong demagnetization) are assumed, which makes the vortex have its minimum energy when centered in the circular system.

9.5.1 The initial discrete lattice vortex solution

To initiate the perturbation calculations, first a static in-plane or out-of-plane vortex solution on the lattice is required. It may be close to the approximate analytic solutions already described, such as (8.7) for the in-plane angle, and (8.19) for the out-of-plane component if $\lambda > \lambda_c$. As mentioned earlier, these become modified slightly on a grid, and we need to use the grid solution, which will be a local minimum energy configuration. The discrete structure can be calculated by the spin alignment technique, see for examples the curves found in figure 8.1. Even for an in-plane vortex, the in-plane angles in the discrete vortex change in the core region, when compared to the continuum arctangent solution of (8.7). The discrete structure can be found just as well for triangular and hexagonal lattices just as for a square lattice, without extra complications.

The static state achieved is described by the set of planar spherical in-plane and out-of-plane angles (ϕ_n^0, θ_n^0), or the corresponding spins $\mathbf{S}_n^0 = (S_n^{0x}, S_n^{0y}, S_n^{0z})$. To simplify some notation, it is helpful in the perturbation analysis to define some variables that appear repeatedly,

$$m_n^0 = \sin \theta_n^0, \qquad p_n^0 = \cos \theta_n^0. \tag{9.204}$$

Profiles of m_n^0 as obtained by spin alignment relaxation were given in figure 8.1. The factors p_n^0 represent the projection of the spins onto the xy-plane. In the core region where m_n^0 is large, the deviation of p_n^0 below unity becomes important.

9.5.2 Perturbations of a static vortex

The oscillations about the static vortex solution are found from perturbation analysis, just as spin waves were analyzed in chapter 6. There are two main differences here. First, the unperturbed state is not uniform, so we need to define different local axes at each site. Second, we consider a finite system, so that translational symmetry is not present. The modes will not be simply distinguished by wave vectors.

To deal with the non-uniform unperturbed state, we use the static directions $\mathbf{S}_n^0$ as the local quantization axes $\tilde{\mathbf{z}}$ for each site. The spins make small fluctuations about those directions. Therefore only the spin components perpendicular to each $\mathbf{S}_n^0$ will be involved in the oscillations. It is those spin components for which the equations of motion will be linearized. As in [10], one can make a transformation to new site-dependent coordinate axes $\tilde{\mathbf{x}}, \tilde{\mathbf{y}}, \tilde{\mathbf{z}}$, where $\tilde{\mathbf{z}}$ is along $\mathbf{S}_n^0$, $\tilde{\mathbf{x}}$ is in the original xy-plane and perpendicular to $\tilde{\mathbf{z}}$, and then $\tilde{\mathbf{y}} = \tilde{\mathbf{z}} \times \tilde{\mathbf{x}}$ is the third member of the mutually orthogonal triplet. These local axes are given in terms of the global $\hat{\mathbf{x}}, \hat{\mathbf{y}}, \hat{\mathbf{z}}$ basis as (see figure 9.11)

$$\begin{pmatrix} \tilde{\mathbf{x}} \\ \tilde{\mathbf{y}} \\ \tilde{\mathbf{z}} \end{pmatrix} = \begin{pmatrix} -\sin \phi_n^0 & \cos \phi_n^0 & 0 \\ -m_n^0 \cos \phi_n^0 & -m_n^0 \sin \phi_n^0 & p_n^0 \\ p_n^0 \cos \phi_n^0 & p_n^0 \sin \phi_n^0 & m_n^0 \end{pmatrix} \begin{pmatrix} \hat{\mathbf{x}} \\ \hat{\mathbf{y}} \\ \hat{\mathbf{z}} \end{pmatrix}. \tag{9.205}$$

Magnetic Excitations and Geometric Confinement

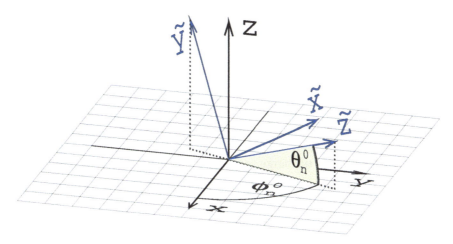

Figure 9.11. Sketch of the relation between the original global xyz-coordinate axes and the local $\tilde{x}\tilde{y}\tilde{z}$-coordinate axes for a particular spin site, as defined in (9.205). The $\tilde{z}$-axis is taken along the direction (ϕ_n^0, θ_n^0) of the unperturbed spin at the site. Then $\tilde{x}$ is perpendicular to $\tilde{z}$ and in the xy-plane, while $\tilde{y} = \tilde{z} \times \tilde{x}$ to make the third member of the right-handed triplet.

With $(p_n^0)^2 + (m_n^0)^2 = 1$, the matrix here is of orthogonal form and the transpose is its inverse. The local spin components are obtained by $S_n^{\tilde{x}} = \mathbf{S}_n \cdot \tilde{\mathbf{x}}$, and so on, which shows that the spins are transformed the same way,

$$\begin{pmatrix} S_n^{\tilde{x}} \\ S_n^{\tilde{y}} \\ S_n^{\tilde{z}} \end{pmatrix} = \begin{pmatrix} -\sin\phi_n^0 & \cos\phi_n^0 & 0 \\ -m_n^0 \cos\phi_n^0 & -m_n^0 \sin\phi_n^0 & p_n^0 \\ p_n^0 \cos\phi_n^0 & p_n^0 \sin\phi_n^0 & m_n^0 \end{pmatrix} \begin{pmatrix} S_n^x \\ S_n^y \\ S_n^z \end{pmatrix}. \quad (9.206)$$

The dynamic equations of motion without damping are $\dot{\mathbf{S}}_n = \mathbf{S}_n \times \mathcal{F}_n$. These can be written out in the local coordinates, which are transformed into the global coordinates as needed, by the previous equation. The time derivatives should be linearized in $S_n^{\tilde{x}}, S_n^{\tilde{y}}, S_n^{\tilde{z}}$. The derivation of the equations of motion is a simple but tedious exercise.

Exercise 9.9. Starting from the equations of motion from Hamiltonian (9.203), transform them to the local rotated coordinates (9.206). Show that after linearization in all spin components, the components along the quantization axes are nearly conserved quantities, $S_n^{\tilde{z}}$, while the transverse components form a set of linear differential equations,

$$\dot{S}_n^{\tilde{x}} = JS \sum_{m=n+a} \left\{ m_n^0 \sin(\phi_n^0 - \phi_m^0) S_m^{\tilde{x}} + \left[p_n^0 p_m^0 \cos(\phi_n^0 - \phi_m^0) + \lambda m_n^0 m_m^0 \right] S_n^{\tilde{y}} \right.$$
$$\left. - \left[m_n^0 m_m^0 \cos(\phi_n^0 - \phi_m^0) + \lambda p_n^0 p_m^0 \right] S_m^{\tilde{y}} \right\} \quad (9.207a)$$

$$\dot{S}_n^{\tilde{y}} = JS \sum_{m=n+a} \left\{ m_m^0 \sin(\phi_n^0 - \phi_m^0) S_m^{\tilde{y}} - \left[p_n^0 p_m^0 \cos(\phi_n^0 - \phi_m^0) + \lambda m_n^0 m_m^0 \right] S_n^{\tilde{x}} \right.$$
$$\left. + \cos(\phi_n^0 - \phi_m^0) S_m^{\tilde{x}} \right\}. \qquad (9.207b)$$

These equations (9.207) can be solved by assuming oscillatory time-dependence, after which they become an eigenvalue problem. Wysin and Völkel [10] analyzed and solved the problem in the language of quantum eigenmodes, by supposing there are linear combinations of the local spin coordinates that form creation and annihilation operators. While that is not essential it makes a closer connection to the problem as studied in quantum terminology. The creation operator for a mode labeled by some index k is

$$B_k^\dagger = \sum_n \left(w_{k,n}^{(1)} S_n^{\tilde{x}} + w_{k,n}^{(2)} S_n^{\tilde{y}} \right), \qquad (9.208)$$

where pairs $w_{k,n}^{(1)}$, $w_{k,n}^{(2)}$ are complex numbers that give amplitudes of the oscillations at each site, and now $S_n^{\tilde{x}}$ and $S_n^{\tilde{y}}$ are treated as operators. There is a corresponding annihilation operator, which is just the conjugate,

$$B_k = \sum_n \left(w_{k,n}^{(1)*} S_n^{\tilde{x}} + w_{k,n}^{(2)*} S_n^{\tilde{y}} \right). \qquad (9.209)$$

For the creation operator, a time-dependence of the form $\exp(i\omega_k t)$ is assumed, so that its time derivative is

$$\dot{B}_k^\dagger = i\omega_k B_k^\dagger. \qquad (9.210)$$

The annihilation operator has the negative of this frequency, $-\omega_k$. With this time-dependence, the linearized equations (9.207) become an eigenvalue problem of similar structure involving the amplitudes,

$$i\omega_k w_{k,n}^{(1)} = JS \sum_{m=n+a} \left\{ -m_m^0 \sin(\phi_n^0 - \phi_m^0) w_{k,m}^{(1)} + \cos(\phi_n^0 - \phi_m^0) w_{k,m}^{(2)} \right.$$
$$\left. - \left[p_n^0 p_m^0 \cos(\phi_n^0 - \phi_m^0) + \lambda m_n^0 m_m^0 \right] w_{k,n}^{(2)} \right\} \qquad (9.211a)$$

$$i\omega_k w_{k,n}^{(2)} = JS \sum_{m=n+a} \left\{ -m_n^0 \sin(\phi_n^0 - \phi_m^0) w_{k,m}^{(2)} - \left[m_n^0 m_m^0 \cos(\phi_n^0 - \phi_m^0) + \lambda p_n^0 p_m^0 \right] w_{k,m}^{(1)} \right.$$
$$\left. + \left[p_n^0 p_m^0 \cos(\phi_n^0 - \phi_m^0) + \lambda m_n^0 m_m^0 \right] w_{k,n}^{(1)} \right\}. \qquad (9.211b)$$

This sets up the eigenvalue problem. The matrix on the RHS has all real elements, but it is not Hermitian. The problem, however, has real eigenfrequencies ω_k, hence the matrix on the RHS is expected to have pure imaginary eigenvalues. The system can be solved by a routine such as RG (real general matrix) from the EISPACK eigensystem subroutine package, although the system size that can be solved will be rather limited (say, a maximum $R \sim 20a$ on a laptop, due to memory restrictions). In other work by Ivanov *et al* [14] one can find a better way to describe the matrix and use its symmetries to solve it on larger systems up to $R \sim 100a$. Also, that work showed that the eigenvalues are indeed pure imaginary and hence give real eigenfrequencies. However, such large systems may lead to numerical instabilities and even if the memory issue is resolved the required CPU time may be prohibitive. Whatever numerical diagonalization is used, double precision is necessary for stability. The full set of $w_{k,n}^{(1)}$, $w_{k,n}^{(2)}$ over the lattice then is the eigenvector associated with a mode of frequency ω_k. While diagonalization could give a large number of modes, in usual practice we are interested more so in those of the lowest frequencies.

Once a mode has been determined, normalization for quantum creation/annihilation operators requires a unit commutator, $[B_k, B_{k'}^\dagger] = \delta_{k,k'}$. Based on a canonical commutator,

$$\left[S_\mathbf{n}^{\tilde{x}}, S_\mathbf{m}^{\tilde{y}}\right] = i\hbar S_\mathbf{n}^{\tilde{z}} \delta_{\mathbf{n},\mathbf{m}} \to i\hbar S \delta_{\mathbf{n},\mathbf{m}}, \qquad (9.212)$$

this results in the normalization condition,

$$\left[B_k, B_{k'}^\dagger\right] = \hbar S \sum_\mathbf{n} \left[\left(iw_{k,\mathbf{n}}^{(1)*} w_{k',\mathbf{n}}^{(2)}\right) + \left(iw_{k',\mathbf{n}}^{(1)*} w_{k,\mathbf{n}}^{(2)}\right)^*\right] = \delta_{k,k'}. \qquad (9.213)$$

Although for classical modes the normalization is somewhat arbitrary, for the quantum system a correct normalization is needed if one wants to use these results to give the spin fluctuations in thermal equilibrium. Once the modes are known and correctly normalized, the original spin operators in the local coordinates can be recovered. Inverting (9.208) and (9.209), they are

$$S_\mathbf{n}^{\tilde{x}} = iS \sum_k \left(w_{k,\mathbf{n}}^{(2)} B_k - w_{k,\mathbf{n}}^{(2)*} B_k^\dagger\right) \qquad (9.214a)$$

$$S_\mathbf{n}^{\tilde{y}} = -iS \sum_k \left(w_{k,\mathbf{n}}^{(1)} B_k - w_{k,\mathbf{n}}^{(1)*} B_k^\dagger\right). \qquad (9.214b)$$

These can be used to analyze the spin fluctuations caused by a given pure mode or in a combination of the modes.

The original unperturbed vortex state can be denoted $|0\rangle$, being like the ground state of the spin wave modes being found here. A state with a single mode excited is created by its creation operator, acting on the ground state,

$$|k\rangle = B_k^\dagger |0\rangle. \qquad (9.215)$$

With these being normalized operators with unit commutator (9.212), the excited states are also unit normalized:

$$\langle k|k\rangle = \langle 0|B_k B_k^\dagger|0\rangle = 1. \qquad (9.216)$$

Using as reference the vortex ground state, there is also the relation for the expectation value of the number operator, $N_k \equiv B_k^\dagger B_k$, $\langle 0|B_k^\dagger B_k|0\rangle = 0$, because $|0\rangle$ has no excitations present. However, if the state of the system is a *mixed state*, corresponding to thermal equilibrium, then it is important to realize that some population of the modes will be present. In thermal equilibrium, the expectation value of the number operator leads to the well-known Bose–Einstein occupation number, and we write

$$\langle N_k \rangle = \langle B_k^\dagger B_k \rangle = \frac{1}{\exp(\beta \hbar \omega_k) - 1}. \tag{9.217}$$

Generally, we can discuss the spin fluctuations in a state of thermal equilibrium, but then isolate each contribution of each particular mode k.

The spin fluctuations take place transverse to the $\tilde{z}$-directions at each site, but average out to zero so that $\langle S_n^{\tilde{x}} \rangle = \langle S_n^{\tilde{y}} \rangle = 0$ in any state. The expectation value of the longitudinal component, however, can be estimated using the Holstein–Primakoff transformation [15], which is used to ensure a conserved total spin length when fluctuations are present,

$$S_n^{\tilde{z}} = S - \frac{1}{2S}\left(S_n^{\tilde{x}} - iS_n^{\tilde{y}}\right)\left(S_n^{\tilde{x}} + iS_n^{\tilde{y}}\right). \tag{9.218}$$

The last term involves the product of the spin lowering and raising operators, S_n^- and S_n^+. After expressing this in terms of B_k and $B_k^\dagger$ by (9.209) and (9.208), a short exercise leads to the result

$$\langle S_n^{\tilde{z}} \rangle = S - \frac{1}{2}S\sum_k \left[\left|w_{k,n}^{(1)} + iw_{k,n}^{(2)}\right|^2 \langle B_k^\dagger B_k \rangle + \left|w_{k,n}^{(1)} - iw_{k,n}^{(2)}\right|^2 \langle B_k B_k^\dagger \rangle \right]. \tag{9.219}$$

To arrive at this result, one uses the fact that $\langle B_k \rangle = \langle B_k^\dagger \rangle = 0$ as well as $\langle B_k B_{k'} \rangle = \langle B_k^\dagger B_{k'}^\dagger \rangle = 0$. This shows that the spin length along $\tilde{z}$ is slightly reduced by the fluctuations.

Eventually the squared fluctuations away from the mean, in the original global coordinates xyz, are desired. They can be partitioned into in-plane and out-of-plane types, from the definitions,

$$\left\langle \left(\delta S_n^{\text{in}}\right)^2 \right\rangle = \left\langle \left(S_n^x - \langle S_n^x \rangle\right)^2 + \left(S_n^y - \langle S_n^y \rangle\right)^2 \right\rangle \tag{9.220a}$$

$$\left\langle \left(\delta S_n^{\text{out}}\right)^2 \right\rangle = \left\langle \left(S_n^z - \langle S_n^z \rangle\right)^2 \right\rangle. \tag{9.220b}$$

By using the inverse of the coordinate transformation in (9.206), these can be expressed using the local coordinates,

$$\left\langle \left(\delta S_n^{\text{in}}\right)^2 \right\rangle = \left\langle S_n^{\tilde{x}2} \right\rangle + \left\langle S_n^{\tilde{y}2} \right\rangle \sin^2 \theta_n^0 \tag{9.221a}$$

$$\left\langle \left(\delta S_n^{\text{out}}\right)^2 \right\rangle = \left\langle S_n^{\tilde{y}2} \right\rangle \cos^2 \theta_n^0. \tag{9.221b}$$

This is reasonable; the $\tilde{y}$-axis makes an angle θ_n^0 to the $\hat{z}$-axis, and all the out-of-plane motions take place in $S_n^{\tilde{y}}$. The $\tilde{x}$-axis lies in the original xy-plane, and so $S_n^{\tilde{x}}$ contributes only to in-plane fluctuations. There is also a contribution to in-plane fluctuations from the component of $S_\mathbf{n}^{\tilde{y}}$ projected onto the xy-plane. Now, the expectations contained here can be found from using (9.214) to express all parts in terms of the creation/annihilation operators,

$$\langle S_\mathbf{n}^{\tilde{x}2} \rangle = (\hbar S)^2 \sum_k \left| w_{k,\mathbf{n}}^{(2)} \right|^2 \langle 2B_k^\dagger B_k + 1 \rangle \tag{9.222a}$$

$$\langle S_\mathbf{n}^{\tilde{y}2} \rangle = (\hbar S)^2 \sum_k \left| w_{k,\mathbf{n}}^{(1)} \right|^2 \langle 2B_k^\dagger B_k + 1 \rangle. \tag{9.222b}$$

Consistent with (9.214), these show how the $w^{(1)}$ amplitudes give a sense of the $S_\mathbf{n}^{\tilde{y}}$ variations and the $w^{(2)}$ amplitudes relate to the $S_\mathbf{n}^{\tilde{x}}$ variations. Then, the expressions of (9.221) become

$$\left\langle \left(\delta S_\mathbf{n}^{\text{in}}\right)^2 \right\rangle = (\hbar S)^2 \sum_k \left(\left| w_{k,\mathbf{n}}^{(2)} \right|^2 + \left| w_{k,\mathbf{n}}^{(1)} \right|^2 \sin^2 \theta_\mathbf{n}^0 \right) \langle 2B_k^\dagger B_k + 1 \rangle \tag{9.223a}$$

$$\left\langle \left(\delta S_\mathbf{n}^{\text{out}}\right)^2 \right\rangle = (\hbar S)^2 \sum_k \left| w_{k,\mathbf{n}}^{(1)} \right|^2 \cos^2 \theta_\mathbf{n}^0 \langle 2B_k^\dagger B_k + 1 \rangle. \tag{9.223b}$$

Each mode contributes a separate term on a chosen spin site. Because this has been obtained through a semiclassical approach, the last '1' after the number operator $B_k^\dagger B_k$ actually is the zero-point term, present even when the mode is not excited. It might be ignored for the analysis of the classical system.

9.5.3 Examples of the spin wave mode spectrum on a vortex

An example of these exact diagonalization calculations is to use a circular system of radius $R = 15a$, discretized on a square lattice, see figure 9.12 for the frequency spectrum. This results in a system with $N = 716$ sites, which means that with two variables at each site, the exact diagonalization involves a real matrix of size 1432×1432, with 1432 eigenstates. One must also save the 1432 complex eigenvectors, all with 1432 elements. This size is easily solved for a selected value of λ on a laptop computer, and it is not too difficult to go to larger sizes. Of course, we are mostly interested in some subset, usually only the lowest few dozen modes. To obtain a complete understanding of the properties or symmetries of the modes, one should calculate the spectrum at a sequence of anisotropies λ. Then with the repetition of the calculations over a range of λ the total CPU time can grow to become less convenient.

When the calculations are performed for two closely spaced values of λ, it is important to be able to identify the eigenmodes at the first value of λ with the same modes (of the same symmetry) at the next value of λ. This process can be achieved by calculating the overlaps of the two sets of eigenvectors with each other, and looking for the values closest to unity. For fixed λ, modes at distinct eigenfrequencies are orthogonal. The normalization integral or overlap is the sum that one finds in (9.213) for the expression of the commutator, $[B_k, B_{k'}^\dagger]$. When modes at two different

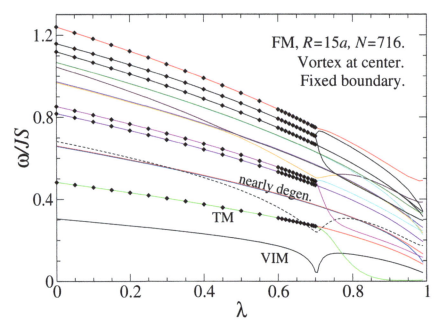

Figure 9.12. Spectrum of normal modes on a static vortex at the center of a circular square lattice system of radius $R = 15a$ (with 716 sites, fixed boundary conditions), obtained by exact diagonalization of the equations (9.211a). Modes marked with diamonds are doubly degenerate. All of those appear for $\lambda < \lambda_c$. A near degeneracy (split slightly by the lattice) is also marked. The lowest mode for $\lambda < \lambda_c$ is associated with the instability of in-plane vortices (VIM), seen by the downward cusp at $\lambda = \lambda_c \approx 0.7$. The lowest doubly degenerate mode contains the translational modes (TM), which split for $\lambda > \lambda_c$.

λ are compared, their overlaps do not need to be very close to unity, so it is important to avoid changing λ by a step that is too large. This technical detail can be difficult to apply because some modes are degenerate and become split in frequency and, as well, some modes cross over other modes.

The changing spectrum of frequencies with λ for the $R = 15a$ system with a vortex is shown in figure 9.12. Mostly the anisotropy constant was incremented by $\Delta\lambda = 0.05$, except for values near the critical value around $\lambda_c \approx 0.7$, where smaller increments were used. Only the lowest 20 modes have been kept here. When more modes are tracked with changing λ, it becomes more difficult to unambiguously identify them as λ changes. There are doubly degenerate modes that have been marked with diamond symbols in figure 9.12; solid curves without symbols are non-degenerate modes.

The eigenvectors or wave functions of some of the modes are presented in figures 9.13 through 9.24, using red line arrows (such as →) to represent the complex amplitudes $w_{k,n}^{(1)}$ and blue hollow head arrows (such as ⟶▷) to represent the complex amplitudes $w_{k,n}^{(2)}$. The horizontal (vertical) components of the arrows are the real (imaginary) parts. For the simplest case of an in-plane unperturbed vortex, recall that $w_{k,n}^{(1)}$ gives out-of-plane oscillations and $w_{k,n}^{(2)}$ gives in-plane oscillations. If the unperturbed vortex is of out-of-plane type, however, then out-of-plane oscillations are

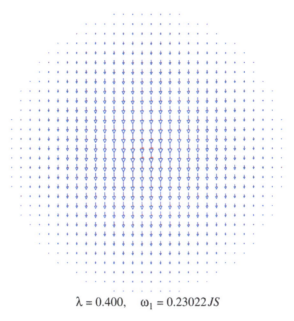

$\lambda = 0.400, \quad \omega_1 = 0.23022\,JS$

Figure 9.13. Eigenvector of the lowest mode for a vortex in the $R = 15a$ system, at $\lambda = 0.40$. The red line arrows (→) represent the amplitude and phase of the complex values $w^{(1)}_{k,\mathbf{n}}$ (mostly out-of-plane oscillations). Blue hollow arrows (—▷) represent $w^{(2)}_{k,\mathbf{n}}$ in the same scale (in-plane oscillations). This is the mode labeled as VIM in figure 9.12. It has rotational symmetry and is predominantly of in-plane fluctuations at this value of λ.

determined only by $w^{(1)}_{k,\mathbf{n}}$, but in-plane oscillations are determined by both $w^{(1)}_{k,\mathbf{n}}$ and $w^{(2)}_{k,\mathbf{n}}$, depending on $\theta^0_{\mathbf{n}}$ at the site, see (9.223).

The lowest mode frequency for $\lambda < \lambda_c$ has a strong downward cusp at $\lambda = \lambda_c$, which is associated with the instability of in-plane vortices found earlier. It is marked as VIM for *vortex instability mode*. The cusp does not go exactly to zero frequency here only because the calculations were not performed at the exact λ_c for the given system size. The presence of this soft mode in the spectrum is a significant confirmation of the vortex instability calculations. Its eigenfunction is shown for $\lambda = 0.40, 0.705, 0.90$ in figures 9.13, 9.17 and 9.21, respectively. Note that the mode is circularly symmetric and very strongly dominated by out-of-plane fluctuations, especially at $\lambda = 0.705 \approx \lambda_c$, as expected from vortex stability analysis. Interestingly, there are a number of higher modes that also have downward cusps at the same λ, but it is not likely that these would go to zero frequency at the exact value of λ_c. An example is 'mode 6' (the sixth lowest mode at $\lambda = 0$), whose eigenvectors are shown at the same values of λ in figures 9.16, 9.20 and 9.24. Mode 6 is also circularly symmetric, but reverses sign at a particular radius: it has a circular nodal line. This is very similar to higher states in any quantum system, such as hydrogen atom wave functions. At $\lambda = 0.705 \approx \lambda_c$, mode 6 reaches its minimum frequency and has a strong central amplitude in $w^{(1)}_{k,\mathbf{n}}$, see figure 9.20.

The lowest of the doubly degenerate modes contains the pair of *translational modes*, marked as TM in the spectrum, figure 9.12. For $\lambda < \lambda_c$, the wave functions of these translational modes have structures comparable to the moving vortex solution (8.40) found earlier, see figures 9.14 and 9.15. Double degeneracy of the TM modes makes sense as there are two independent directions of motion possible. The degeneracy is split once $\lambda > \lambda_c$, with one mode of the doublet moving to very low frequency as $\lambda \to 1$. The eigenfunctions are shown just above the critical anisotropy in figures 9.18 and 9.19. Now they have a dependence on the azimuthal angle φ of the forms $\exp(+i\varphi)$ for the TM mode with higher frequency, and $\exp(-i\varphi)$ for the TM mode of lower frequency. These two functions were mixed in the eigenvectors for $\lambda < \lambda_c$. The splitting becomes much stronger at $\lambda = 0.90$, see the eigenvectors in figures 9.22 and 9.23. The lower of the two modes moves to a very low frequency, and its eigenvector is strongly concentrated in the vortex core.

The other doubly degenerate modes for $\lambda < \lambda_c$ also split once the anisotropy constant surpasses λ_c. This effect is reasonable, because the combination of a vorticity $q = +1$ and a polarization $p = +1$, giving non-zero G, breaks the *rotational symmetry* of the spin structure. To state this in other words, translational modes could also be combined into circular motions of two opposite senses. Once there is a non-zero gyrovector present, the two rotational senses of motions have different energies or frequencies. The Thiele equation shows that non-zero G gives motion in a well-defined direction in response to a force. If $G \neq 0$, functions of the form $\exp(im\varphi)$ with integer quantum numbers $m = 0, \pm 1, \pm 2, \ldots$ become the good eigenstates. They were degenerate and therefore mixed when $G = 0$ for in-plane vortices.

The presence of these features such as the vortex instability mode and mode 6, together with the translational modes, can be considered the internal dynamics of

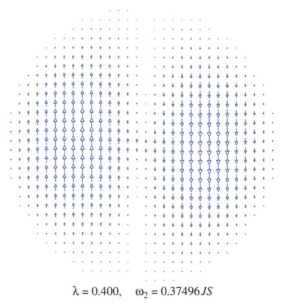

$\lambda = 0.400, \quad \omega_2 = 0.37496 JS$

Figure 9.14. Eigenvector of one of the translational modes (TM in figure 9.12) for a vortex in the $R = 15a$ system, at $\lambda = 0.40$, with notation as in figure 9.13. It would be a linear combination of functions $\exp(i\varphi)$ and $\exp(-i\varphi)$.

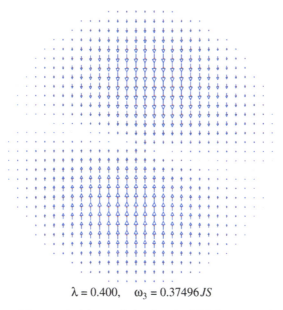

$\lambda = 0.400, \quad \omega_3 = 0.37496\,JS$

Figure 9.15. Eigenvector of the second of the translational modes (TM) for a vortex in the $R = 15a$ system, at $\lambda = 0.40$, with notation as in figure 9.13. It is degenerate with mode 2 shown in figure 9.14, but as a different linear combination of functions $\exp(i\varphi)$ and $\exp(-i\varphi)$.

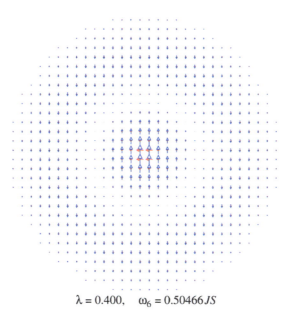

$\lambda = 0.400, \quad \omega_6 = 0.50466\,JS$

Figure 9.16. Eigenvector of mode 6 for a vortex in the $R = 15a$ system, at $\lambda = 0.40$, with notation as in figure 9.13. It is the second of the modes in the spectrum of figure 9.12 that exhibits a downward cusp at λ_c. It has rotational symmetry but a circular node where the sign reverses.

9-55

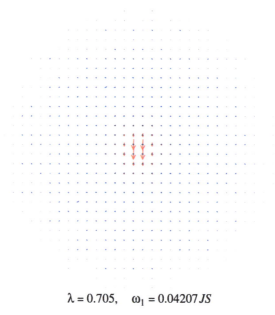

$\lambda = 0.705, \quad \omega_1 = 0.04207\,JS$

Figure 9.17. Eigenvector of the lowest mode for a vortex in the $R = 15a$ system, at $\lambda = 0.705 \approx \lambda_c$ with notation as in figure 9.13. This is the VIM in figure 9.12. Here it is predominantly of out-of-plane fluctuations concentrated in the vortex core.

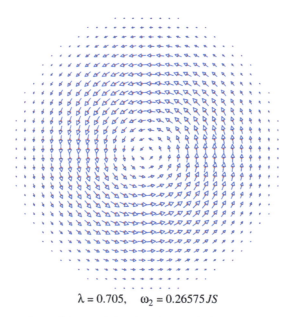

$\lambda = 0.705, \quad \omega_2 = 0.26575\,JS$

Figure 9.18. Eigenvector of one of the translational modes (TM in figure 9.12) for a vortex in the $R = 15a$ system, at $\lambda = 0.705$, with notation as in figure 9.13. The mode's amplitude varies as $\exp(i\varphi)$. It is named mode T_+ later in figure 9.28.

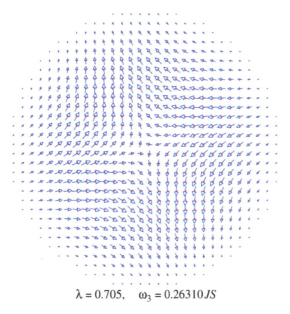

$\lambda = 0.705$, $\omega_3 = 0.26310 JS$

Figure 9.19. Eigenvector of the second translational mode (TM) for a vortex in the $R = 15a$ system, at $\lambda = 0.705$, with notation as in figure 9.13. The degeneracy with mode 2 has been split, see figure 9.18. The mode's amplitude varies as $\exp(-i\varphi)$. It is named mode T_ later in figure 9.28.

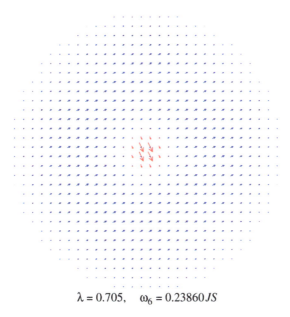

$\lambda = 0.705$, $\omega_6 = 0.23860 JS$

Figure 9.20. Eigenvector of mode 6 for a vortex in the $R = 15a$ system, at $\lambda = 0.705$, with notation as in figure 9.13. In the spectrum of figure 9.12, it is the second downward cusp at λ_c.

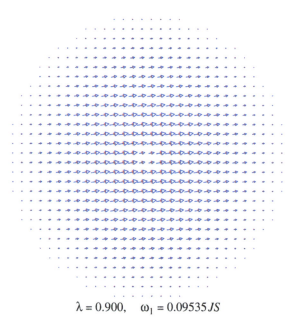

$\lambda = 0.900, \quad \omega_1 = 0.09535\, JS$

Figure 9.21. The VIM eigenvector for a vortex in the $R = 15a$ system, at $\lambda = 0.90$, with notation as in figure 9.13.

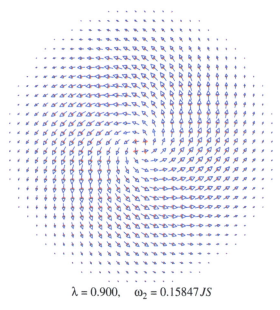

$\lambda = 0.900, \quad \omega_2 = 0.15847\, JS$

Figure 9.22. Eigenvector of the higher translational mode (TM in figure 9.12 or T_+ in figure 9.28) for a vortex in the $R = 15a$ system, at $\lambda = 0.90$, with notation as in figure 9.13. The mode's amplitude varies as $\exp(i\varphi)$.

the vortex. These features show how a vortex should not be considered an object with a fixed structure as it is moving.

For comparison, the spectrum for the same system but without a vortex present is shown in figure 9.25. In this case, the unperturbed state is the state of ferromagnetically

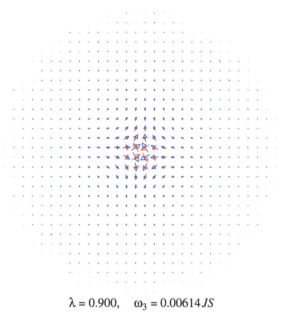

$\lambda = 0.900, \quad \omega_3 = 0.00614\,JS$

Figure 9.23. Eigenvector of the lower translational mode (TM if figure 9.12 or T_- in figure 9.28) for a vortex in the $R = 15a$ system, at $\lambda = 0.90$, with notation as in figure 9.13. The mode's amplitude varies as $\exp(-i\varphi)$. Compare to figure 9.22, which has the opposite azimuthal symmetry and is more spread out.

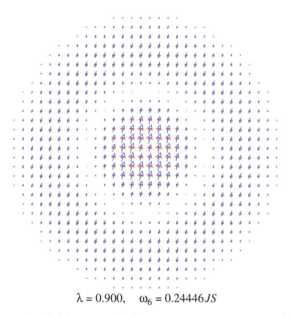

$\lambda = 0.900, \quad \omega_6 = 0.24446\,JS$

Figure 9.24. Eigenvector of mode 6 for a vortex in the $R = 15a$ system, at $\lambda = 0.90$, with notation as in figure 9.13. The sign reversal through the circular node is more obvious.

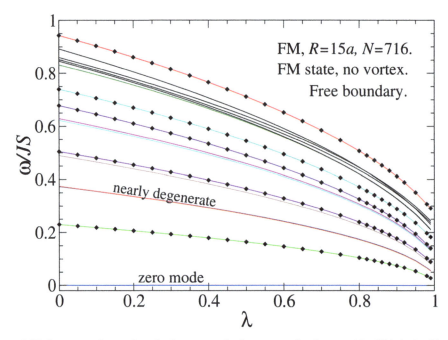

Figure 9.25. Spectrum of normal modes in a square lattice system of radius $R = 15a$ (716 sites) without a vortex, obtained by exact diagonalization of the equations (9.211a). The unperturbed state is that of aligned spins in the xy-plane, free boundary conditions must be used here. Modes marked with diamonds are doubly degenerate. The *zero mode* with $\omega = 0$ appears due to the freedom to rotate all spins together at no energy cost.

aligned spins, with free boundary conditions then being necessary. While there are still certain doubly degenerate modes among the predominantly non-degenerate ones, the spectrum is rather uniform and featureless. Nothing in particular happens around λ_c, because there is no vortex present. There is, however, an extra mode at zero frequency. This is due to the freedom of rotating all spins together around the z-axis, without any energy cost. It is a direct consequence of the free boundary conditions.

A further comparison can be made by looking at the spectrum with a vortex present, but with *free boundary conditions*, see figure 9.26, again for the vortex in a square lattice system of radius $R = 15a$. If damping were present, the vortex would probably be unstable, and its dynamics would cause it to move towards the boundary. The diagonalization calculations, however, do not include damping, and only give a measure of the small-amplitude harmonic oscillations that can appear. Even so, the spectrum changes slightly from that for fixed boundary conditions. First, the zero frequency mode now appears, because the free boundary condition allows all spins to rotate together around the z-axis, with no energy cost, even when a (centered) vortex is present. Second, the translational mode frequencies acquire a small imaginary part, on the order of $10^{-6} JS$, due to round-off errors, that would correspond to the same effect expected if damping were present. This shows that the instability of the vortex remaining at the system center actually does appear in the calculation. Third, the vortex instability mode that would go to $\omega = 0$ at λ_c

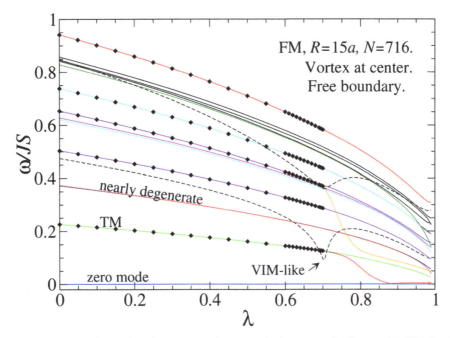

Figure 9.26. Spectrum of normal modes on a vortex in a square lattice system of radius $R = 15a$ (716 sites, free boundary conditions), obtained by exact diagonalization of the equations (9.211a). Modes marked with diamonds are doubly degenerate. Modes shown as dashed are similar to the VIM but do not go to zero frequency at λ_c. The *zero mode* with $\omega = 0$ appears due to the freedom to rotate all spins together at no energy cost. Note how all modes tend to have lower frequencies than for fixed boundary conditions, figure 9.12. The translational modes and others degenerate for $\lambda < \lambda_c$ become split for $\lambda > \lambda_c$, due to non-zero G.

now is not present, however, there are other higher modes with a strong downward cusp at λ_c (shown as dashed curves). The free boundary condition is softer than a fixed boundary, and that is responsible for this change.

It is very clear that the presence or absence of a vortex, together with the type of boundary conditions, has a substantial effect on the spin wave spectrum.

9.6 Vortex mass obtained from vortex normal modes

The normal mode spectrum can be connected to calculations of the vortex mass [8], for an isolated vortex in a small circular system such as that just considered above. The idea is that the translational modes are directly related to the vortex motion and, hence, their eigenfrequencies are connected to the frequencies one finds for gyrotropic motion of the vortex in the circular system. The mass can be estimated by making the results of the modified Thiele equation consistent with the eigenfrequencies found from the perturbation analysis.

We have already considered the motion of a vortex in a circular system, from the point of view of the modified Thiele equation (9.46). The force was assumed to be the force due to an image vortex. For a system discretized on a lattice, the force may be somewhat different from the analytic expressions already found. For the moment, assume that with fixed boundary conditions, the force is a radial restoring force with some force constant k_F,

$$\mathbf{F} = -k_F \mathbf{X}, \tag{9.224}$$

where $\mathbf{X}(t)$ is the instantaneous vortex location, measured from the center of the circular system. With an assumed rotational motion $\mathbf{V} = \omega \hat{\mathbf{z}} \times \mathbf{X}$ and then $\ddot{\mathbf{X}} = -\omega^2 \mathbf{X}$, the equation of motion is just like (9.200) found earlier,

$$-k_F \mathbf{X} - \omega G \mathbf{X} = -M\omega^2 \mathbf{X}. \tag{9.225}$$

The two frequencies that result for orbital motion are written here as

$$\omega_\pm = \frac{1}{2M}\left(G \pm \sqrt{G^2 + 4k_F M}\right), \qquad \text{for } M > 0, k_F > 0. \tag{9.226}$$

At this point we allow both possible solutions, which are assumed to be mapped to the two possible translational modes. For an out-of-plane vortex, these connect to different rates and senses of rotation around the system center. One can see that with $M > 0$ and $k_F > 0$, one obtains $\omega_+ > 0$ (counterclockwise motion) but $\omega_- < 0$ (clockwise motion), regardless of the sign of G. If instead an in-plane vortex is considered, with $G = 0$, the solutions are those for simple harmonic motion,

$$\omega_\pm = \pm\sqrt{k_F/M}, \qquad \text{for } G = 0. \tag{9.227}$$

This can also be interpreted as saying that circular motions in opposite senses will have the same frequencies. The results so far have required a mass. If the mass were not present, the Thiele equation then gives only one frequency,

$$\omega_0 = -k_F/G, \qquad \text{for } M = 0. \tag{9.228}$$

If both $M = 0$ and $G = 0$, as would be the case for massless in-plane vortices, there is no solution. Thus, we suppose that a mass is always present, in order to have consistency between the Thiele dynamics and the spin wave spectrum.

From the quadratic equation that leads to $\omega_\pm$, one has a relation for the product of the two frequencies, that can be solved for an estimate of the mass,

$$\omega_+ \omega_- = -\frac{k_F}{M} \quad \Longrightarrow \quad M = -\frac{k_F}{\omega_+ \omega_-}. \tag{9.229}$$

Therefore, this can be used to estimate M if the force constant is already known, assuming that the frequencies are taken to be the translational mode frequencies from the perturbation analysis.

We can estimate the force constant essentially by a variation of the spin alignment iteration approach, but enforcing a constraint on the vortex position. The total quasi-static energy of the vortex might be expected to be of a quadratic potential form,

$$U(\mathbf{X}) \approx \frac{1}{2} k_\text{F} X^2. \tag{9.230}$$

To impose a desired position, the in-plane angles ϕ_i for four sites in the core of the vortex can be fixed to have the values given by (8.7), for a desired location, not exactly at the system center. The spin configuration can be relaxed by the spin alignment technique, resulting in a calculation of $U(\mathbf{X})$ for that position. This can be repeated for different small displacements (fractions of a lattice constant) away from the system center. As a result, one finds an approximate quadratic dependence on the displacement from the center, and the force constant k_F is easily determined. This is performed for the same boundary conditions (fixed) as those applied to obtain the normal modes. Results for k_F are shown in figure 9.27, which are rather surprising. For $\lambda < \lambda_\text{c}$, the force constant is very close to constant, independent of the system size. This is in contrast to what we could expect from the continuum expression (9.198) based on the forces due to vorticity between the vortex and its image. Expression (9.198) would predict that $k_\text{F} \propto R^{-2}$ as $X \to 0$. But that is not what is found for the system on a square lattice. This may be because there are discrete pinning forces, especially in the core region, that dominate the force. As $\lambda \to 1$ for out-of-plane vortices, the force constant shown in figure 9.27 decreases considerably, and is all the smaller for larger R, but not as an inverse square law.

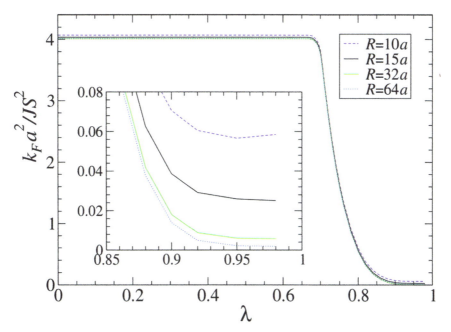

Figure 9.27. The effective vortex force constant k_F as estimated from (9.230) and energy calculated from the spin alignment relaxation scheme, for indicated system radii. The inset gives an indication of the dependence on system radius for out-of-plane vortices.

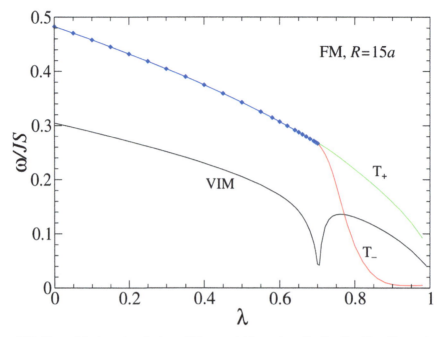

Figure 9.28. Three of the lowest modes in an FM square lattice system of radius $R = 15a$ with a vortex with fixed boundary conditions, obtained by exact diagonalization of equations (9.211a). The VIM is responsible for the transformation of an in-plane vortex into out-of-plane form at $\lambda_c \approx 0.70$. The translational modes T_+ and T_- are degenerate for in-plane vortices but become distinctly split in out-of-plane vortices, due to the non-zero gyrovector that plays such an important role in the dynamics.

The translational mode frequencies for the $R = 15a$ system are shown again in figure 9.28, together with the vortex instability mode as a reference. The lower of the translational modes, labeled as T_-, goes to very low frequency as λ approaches unity. That is the mode whose frequency becomes $\omega_0 = -k_F/G$ in the limit of zero mass; it is the more 'natural' motion for a vortex in a confined circular system. Examples of its wave function were given in figures 9.19 and 9.23. The higher mode T_+ has wave functions such as those in figures 9.22 and 9.18. Based on the result (9.229) above, the translational mode frequencies are identified as ω_+ and ω_-, and used to estimate the mass, whose result is shown in figure 9.29 for various system sizes, all with fixed boundary conditions. Note that M has been calculated in units of $(JS)^{-1}$; the physical mass would be $m = \sigma M$, which has units of $(Ja^2)^{-1}$, which can be expressed in kilograms. The mass has a considerable variation with λ, especially a large peak around λ_c. Also, M increases with system radius, but not according to the logarithmic form in (9.35). It is difficult to reconcile this calculation of M on the discrete lattice system with that continuum theory expression. One can suspect that there are intense discreteness effects that could be responsible for the considerable differences. The continuum expression employed an artificial cutoff in the core region, but that is probably the region with the most important dynamics, which has somehow been over-approximated in the continuum limit.

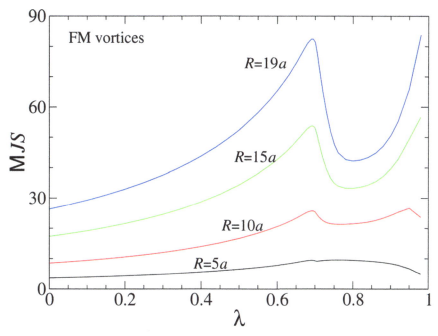

Figure 9.29. FM vortex masses as estimated from expression (9.229), which uses the translational mode eigenfrequencies, calculated on circular square lattice systems of the indicated radii. These values tend to be considerably larger than the predictions of the continuum theory, expression (9.35). The physical mass m is obtained by scaling with the spin density, m = σM.

Bibliography

[1] Thiele A A 1973 Steady-state motion of magnetic domains *Phys. Rev. Lett.* **30** 230
[2] Thiele A A 1974 Applications of the gyrocoupling vector and dissipation dyadic in the dynamics of magnetic domains *J. Appl. Phys.* **45** 377
[3] Huber D L 1982 Dynamics of spin vortices in two-dimensional planar magnets *Phys. Rev.* B **26** 3758
[4] Wysin G M and Mertens F G 1991 Equations of motion for vortices in 2D easy-plane magnets *Nonlinear Coherent Structures in Physics and Biology* vol 393. In ed M Remoissenet and M Peyrard (Berlin: Springer) pp 1–4
[5] Wysin G M, Mertens F G, Völkel A R and Bishop A R 1994 Mass and momentum for vortices in two-dimensional easy-plane magnets *Nonlinear Coherent Structures in Physics and Biology* vol 329. In ed K H Spatschek and F G Mertens (New York: Plenum) pp 177–86
[6] Tjon J and Wright J 1977 Solitons in the continuous Heisenberg spin chain *Phys. Rev.* B **15** 3470
[7] Völkel A R, Wysin G M, Mertens F G, Bishop A R and Schnitzer H J 1994 Collective variable approach to the dynamics of nonlinear magnetic excitations with application to vortices *Phys. Rev.* B **50** 12711
[8] Wysin G M 1996 Magnetic vortex mass in two-dimensional easy-plane magnets *Phys. Rev.* B **54** 15156

[9] Mertens F G, Wysin G M, Völkel A R, Bishop A R and Schnitzer H J 1994 Cyclotron-like oscillations and boundary effects in the 2-vortex dynamics of easy-plane magnets *Nonlinear Coherent Structures in Physics and Biology* vol 329. In ed K H Spatschek and F G Mertens (New York: Plenum) pp 191–8
[10] Wysin G M and Völkel A R 1995 Normal modes of vortices in easy-plane ferromagnets *Phys. Rev.* B **52** 7412
[11] Wysin G M and Völkel A R 1996 Comparison of vortex normal modes in easy-plane ferromagnets and antiferromagnets *Phys. Rev.* B **54** 12921
[12] Ivanov A K, Kolezhuk B A and Wysin G M 1996 Normal modes and soliton resonance for vortices in 2D classical antiferromagnets *Phys. Rev. Lett.* **76** 511
[13] Ivanov B A and Wysin G M 2002 Magnon modes for a circular two-dimensional easy-plane ferromagnet in the cone state *Phys. Rev.* B **65** 134434
[14] Ivanov B A, Schnitzer H J, Mertens F G and Wysin G M 1998 Magnon modes and magnon-vortex scattering in two-dimensional easy-plane ferromagnets *Phys. Rev.* B **58** 8464
[15] Holstein T and Primakoff H 1940 Field dependence of the intrinsic domain magnetization of a ferromagnet *Phys. Rev.* **58** 1098

IOP Publishing

Magnetic Excitations and Geometric Confinement
Theory and simulations
Gary Matthew Wysin

Chapter 10

Vortices in thin ferromagnetic nanodisks

Nanodisks of ferromagnetic (FM) media such as Fe, Co and permalloy-79 (Py, or $Ni_{79}Fe_{21}$) primarily have isotropic exchange interactions competing with the geometric anisotropy due to demagnetization effects (that is, the dipolar interactions). Depending on the aspect ratio of such a cylinder, as well as the material parameters, the ground state can be either of quasi-single-domain (QSD) type, or a topological vortex. The case of vortices is considered in this chapter. Such a vortex moves in an effective potential, and has dynamics with a gyrotropic force similar to the magnetic Lorentz force on a moving electric charge. A vortex in a nanodisk will exhibit a natural rotational motion (gyrotropic precession) as a result. This motion can even be excited by thermal fluctuations. The controlled reversal of vorticity charge and polarization charge of a vortex is an important goal for technological applications.

10.1 Vortex states in magnetic nanodisks

In the previous two chapters we discussed the static and dynamic properties of magnetic vortices in a quasi-2D crystal lattice, with negligible coupling to neighboring planes. Any assumed easy-plane anisotropy was intrinsic to the exchange interactions. In this chapter, we consider a slightly different type of system: a FM continuum medium where the dominant interaction is isotropic FM exchange, modified by a weaker but very important demagnetization field that produces an anisotropic interaction. The demagnetization effect is directly caused by the shape of the system and its boundaries, which produce geometrical confinement. For technological reasons it is most interesting to consider *islands* or *nanodisks* of magnetically permeable materials that are usually grown on some type of non-magnetic substrate. If the shapes of these islands can be controlled when grown, perhaps even with interior holes, one can imagine different ways to control their magnetic responses. As islands, the tendency is for the height to be much smaller than the transverse dimensions.

For the most part in this chapter, we consider magnetic nanodisks of cylindrical shape, with some vertical height or length L in the z-direction, and a circular cross sectional shape in the *xy*-plane, of radius R. Typically for island-like structures, one has a thin-film geometry, with a small aspect ratio, or $L \ll 2R$. The main reason for consideration of this shape is that it favors the formation of a ground state [1] in the form of a vortex. However, later the circular assumption will be generalized to consideration of an elliptical cross sectional shape. The reason for this is to obtain some measure of how the strong changes in demagnetization will affect the vortex structure and dynamics.

A FM material is assumed, with saturation magnetization M_s (magnetic moment per volume) and isotropic FM exchange stiffness A (dimensions of energy per length). For the continuum description of the magnetics, we make the micromagnetics assumption that the local magnetization keeps a constant magnitude M_s but varies in direction, with $\mathbf{M}(\mathbf{r}) = M_s \hat{\mathbf{m}}(\mathbf{r})$, see the discussion with (2.20). The analysis then concentrates on the variation of the unit vector $\hat{\mathbf{m}}(\mathbf{r})$ with position $\mathbf{r}$. This was introduced in chapter 3, and we review the essential tools needed now. The definition of exchange stiffness appeared implicitly in (2.26), which gives the exchange part of the system energy according to

$$H_{\text{ex}} = A \int dV \, |\nabla \hat{\mathbf{m}}|^2. \tag{10.1}$$

Recall also that if the continuum system is discretized for micromagnetics calculations onto a 3D cubic grid of cells with size $a \times a \times a$ and locations $\mathbf{r}_i$, then the cells have an effective exchange interaction energy given by (2.27), or $J_{\text{cell}} = 2aA$. In (2.28), the exchange energy instead is written for a 2D grid of $a \times a \times L$ cells, which can be good for thin-film 2D micromagnetics problems such as those considered here. In that case, the effective exchange constant for a 2D micromagnetics approach [2] becomes

$$J_{\text{cell}} = 2LA. \tag{10.2}$$

Regardless of the way some discretization is set up, the competition between exchange and demagnetization leads to the exchange length presented in (3.149), defined by

$$\lambda_{\text{ex}} \equiv \sqrt{\frac{2A}{\mu_0 M_s^2}}. \tag{10.3}$$

This produces the micromagnetics effective Hamiltonian presented earlier and repeated here,

$$H = -J_{\text{cell}} \left[\sum_{(i,j)} \hat{\mathbf{m}}_i \cdot \hat{\mathbf{m}}_j + \sum_i \left\{ \frac{a^2}{\lambda_K^2} (\hat{\mathbf{m}}_i \cdot \hat{\mathbf{x}})^2 + \frac{a^2}{\lambda_{\text{ex}}^2} \left(\tilde{\mathbf{H}}_{\text{ext}} + \frac{1}{2} \tilde{\mathbf{H}}_M \right) \cdot \hat{\mathbf{m}}_i \right\} \right]. \tag{10.4}$$

All quantities within the square brackets are in dimensionless forms. The anisotropy length scale λ_K is determined from any weak uniaxial anisotropy (see (3.150)). The dimensionless demagnetization field $\tilde{\mathbf{H}}_M \equiv \mathbf{H}_M / M_s$ is obtained from the demagnetization field $\mathbf{H}_M$, which can be found by the techniques presented in chapter 3. If there is an externally applied magnetic field $\mathbf{H}_{\text{ext}}$ it is also converted to dimensionless form as $\tilde{\mathbf{H}}_{\text{ext}} = \mathbf{H}_{\text{ext}} / M_s$. Both $\tilde{\mathbf{H}}_M$ and $\tilde{\mathbf{H}}_{\text{ext}}$ can depend on the cell location $\mathbf{r}_i$, but for clarity the index i is suppressed. The demagnetization field calculation consumes most of the CPU time; it can be accelerated somewhat especially for larger lattices by the use of the fast Fourier transform techniques described in chapter 3. For the thin-film assumption here, recall that the problem becomes 2D, and is solved by using the appropriate 2D Green's functions, refer back to chapter 3 for a review of the techniques for the numerical calculation of the demagnetization field $\mathbf{H}_M$.

Physically, the exchange interaction energy is reduced when neighboring magnetic moments are aligned. The demagnetization interaction energy tends to become greater when large groups of moments are aligned, because the field $\mathbf{H}_M$ may extend out of the system and contains an energy density. Or stated differently, aligned magnetic moments generate magnetic pole density on the system boundaries, increasing the energy. The two effects of exchange and demagnetization compete, as demagnetization tends to break up the aligned regions into smaller domains. Thus, the exchange length λ_{ex} gives the distance over which the exchange forces are capable of maintaining magnetic moments strongly aligned. The typical value of λ_{ex} is of the order of a few nanometers. For Py, the parameters are $A \approx 13$ pJ m^{-1} and $M_s \approx 860$ kA m^{-1}, which gives $\lambda_{ex} \approx 5.3$ nm. It is important to be aware of this length scale when setting up any of the micromagnetics calculations, because the size of the cells needs to be smaller than the exchange length. Otherwise, all important spatial variations in $\hat{\mathbf{m}}(\mathbf{r})$ will not be represented correctly.

Quasi-single-domain state versus vortex state
Primarily, we are interested in the lowest energy state and excitations above that state. Depending on the material and aspect ratio $L/2R$ of the nanodisk, the lowest state may either be of QSD (all moments nearly aligned) or of vortex type (magnetic moments following the direction of the circular edge). For instance, Cowburn *et al* [3] studied the hysteresis loops of *supermalloy* ($Ni_{80}Fe_{14}Mo_5$) disks with diameters of 55–500 nm and thicknesses of 6–15 nm, and found that both types of states or *phases* can appear in samples. Their results were supported by micromagnetics simulations.

Metlov and Guslienko [4] made an interesting stability analysis of the vortex state, assuming that it should produce no poles on the circular edge of the disk. By an analysis of the total system energetics, they determined that three low energy states are possible: (i) a vortex state, (ii) a QSD with $\hat{\mathbf{m}}$ in the xy-plane and (iii) a QSD with $\hat{\mathbf{m}}$ along the $\hat{\mathbf{z}}$-axis. The latter will appear when the length of the cylinder is longer than its diameter.

The results can be explained with simple physical arguments. For very small nanodisks, a QSD state will be preferred, to avoid the extra exchange energy caused by spatial variations in $\hat{\mathbf{m}}(\mathbf{r})$ in a vortex configuration. Once the disk radius is made larger, and for adequate thickness, there is a planar anisotropy due to demagnetization (equivalent to dipolar interactions). That is because magnetic poles should be avoided on the large flat (parallel to xy) surfaces. Thus, at adequately large disk sizes, the vortex state will be preferred, because it has little pole density on any of the surfaces. It may only generate a weak pole density in the vortex core region, on the large flat surfaces. Of course, if the disk is made even larger, even the vortex state will be unstable, breaking into smaller domains, an effect not included in [4]. There are other various complex states possible such as the C-state, S-state, leaf-states, and so on. We are mainly interested here in the vortex state, for its high symmetry and the possibility to control its topological charges.

As examples of the types of states that can be obtained, we obtain some results of the spin alignment relaxation [5] described in chapter 3, applied to the micromagnetics Hamiltonian (10.4) for circular cylindrical nanodisks. The exchange length is set to $\lambda_{ex} = 5.3$ nm as for Py and we ignore any small uniaxial anisotropy K_v, taking $\lambda_K = \infty$. The lattice parameter is taken to be $a = 2.0$ nm, somewhat smaller[1] than λ_{ex}. There is no applied field. Starting from an initial set of moments $\hat{m}_i$ for the cells on a 2D square grid, the sites' magnetization unit vectors are iteratively pointed to the directions of their effective magnetic fields, including the exchange and demagnetization contributions. Depending on the scale of discretization, the obtained vortex or QSD states can be quite (meta)stable, even when not in the lowest energy configuration. The vortex minimizes its energy when its core is at the center of the disk.

Thus, the simulation can be carried out by starting from either a vortex state or an aligned state, modified by small random deviations to break the symmetry and allow for relaxation into the opposite state. Then, we can find the energy of either the vortex or QSD states as functions of the film thickness L, for selected nanodisk radii. These calculations are performed under the thin-film assumption, and the use of the 2D Green's functions for converting the original 3D problem into an effective thin 2D problem, see chapter 3. Examples of the energies that result are displayed in figure 10.1 for R from 20 nm to 80 nm. The results show that typically, the energy of the QSD state (in units of J_{cell}) increases at a rate less than linearly with L, while the vortex state energy (units of J_{cell}) is close to constant or slightly decreasing with L. There is a well-defined crossover point where the two energies are equal and the most stable state switches. The QSD state is the minimum for very thin disks, because of the lack of surface area and magnetic pole formation on the curved edge. For thicker disks the demagnetization energy takes over and the centered vortex state becomes preferred, avoiding the formation of magnetic poles on the curved edge. Thus it becomes clear that the vortex state can be present as the ground state of this system, offering the potential for studies of its interesting dynamics.

10.2 Vortex-in-disk effective potentials

Let us suppose that a nanodisk is of appropriate size to be in the vortex state. Now consider the effective potential $U(\mathbf{X})$ that the vortex experiences, according to its instantaneous 2D location $\mathbf{X} = (X, Y)$. At this point we want to calculate $U(\mathbf{X})$ in a quasi-static approximation, by supposing that some constraint holds the vortex in the desired location, and computing the system energy under that constraint. A simple example of this quasi-static approach was the calculation of the force constant k_F in the previous chapter, via (9.230). Here we expand on that idea and describe an approach [2] that uses a more general constraint invoked using Lagrange's method of undetermined multipliers. This procedure can also be used in the presence of vacancies that can attract and pin vortices with a strong effective potential. By using

---
[1] A value of lattice parameter a smaller than 2.0 nm would be much better, but then the number of 2D grid sites is increased, slowing considerably the calculation of the demagnetization field $\mathbf{H}_M$ for larger disk radii.

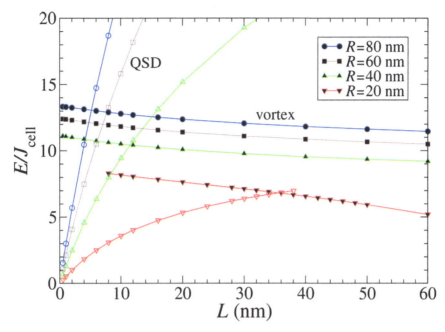

Figure 10.1. From the spin alignment relaxation scheme applied to the 2D approximate micromagnetics, a comparison of the energies of the centered vortex state (filled symbols) with the QSD state ($\hat{\mathbf{m}}$ in the xy-plane, open symbols), as functions of the film thickness L. The energy unit is $J_{\text{cell}} = 2AL$, which also depends on the thickness. One sees that the vortex state is lower in energy above a limiting thickness that decreases with increasing disk radius R. Even when the QSD state is lower in energy, a metastable vortex higher in energy is possible. The calculations should be considered reliable only for $L < R$.

a Lagrange multiplier approach to impose a position constraint on the vortex, information about the constraining force and its effective field is obtained.

The method works as follows. Having discretized the system on the micromagnetics lattice, a certain number N_c of cells closest to the vortex core position are considered to be *core sites*, on which a constraint will be applied. The number N_c could be as small as 4, although in some work we have applied as many as 96. It is best if the diameter of the core region is similar in size to the exchange length. Outside the core region the cell dipoles are free to move according to the Landau–Lifshitz equation torques, without any constraints. We still use a spin alignment scheme as in [5], but it now includes an extra constraint field in the core region, which is the undetermined Lagrange multiplier. One actually needs to impose two constraints: 1) a constraint on spin length and 2) a constraint on vortex core location. Each of these has its own Lagrange multiplier. We consider the spin length constraint first and then later incorporate that into the general problem. The details of the length constraint makes it easier to understand how the position constraint can be implemented.

10.2.1 Spin-length constraint, as an example

At this point the micromagnetic unit vector cell dipoles $\hat{\mathbf{m}}_i$ will be generalized to have an adjustable length, writing them as $\mathbf{m}_i$. Each corresponds to a physical magnetic dipole of magnitude $\mu_{\text{cell}} = M_s v_{\text{cell}} = M_s L a^2$. We consider how the method of Lagrange multipliers would be applied to constrain these dipoles to a desired length $|\mathbf{m}_i| \to 1$. We start from the given micromagnetics Hamiltonian $H[\mathbf{m}_i]$ as in (10.4), which is a functional of all the dipoles $[\mathbf{m}_i]$. To impose a constraint on spin length, extra terms must be added to this, one for each site, with an undetermined multiplier on each site, denoted as α_i (not a damping parameter). This defines a Lagrange function that is to be minimized,

$$\Lambda[\mathbf{m}_i] = H[\mathbf{m}_i] + J_{\text{cell}} \sum_i \alpha_i \left(\mathbf{m}_i^2 - m^2 \right). \tag{10.5}$$

The factor of J_{cell} is included to make α_i dimensionless. When the dipoles have been constrained to the length m, it is clear that Λ becomes equivalent to the original Hamiltonian. Now consider variations in the dipole components, and also in the undetermined multipliers, to implement the constraint. One has

$$\frac{\partial \Lambda}{\partial \alpha_i} = J_{\text{cell}} \left(\mathbf{m}_i^2 - m^2 \right) = 0 \tag{10.6}$$

which obviously gives the correct lengths, using $m = 1$. For the derivatives with respect to dipole components indexed by $\beta = x, y, z$,

$$\frac{\partial \Lambda}{\partial m_i^\beta} = \frac{\partial H}{\partial m_i^\beta} + 2 J_{\text{cell}} \alpha_i m_i^\beta = 0. \tag{10.7}$$

The partial derivative of H is a general effective field. Comparing to (2.84), we have here a dimensionless effective field,

$$f_i^\beta \equiv -\frac{1}{J_{\text{cell}}} \frac{\partial H}{\partial m_i^\beta} = -\frac{\mu_{\text{cell}}}{J_{\text{cell}}} \frac{\partial H}{\partial \mu_i^\beta} = \frac{a^2}{\lambda_{\text{ex}}^2} \frac{\mathcal{F}_i^\beta}{\mu_0 M_s}. \tag{10.8}$$

$\mathcal{F}_i$ is identical to the effective field appearing in (2.84), changing the spin $\mathbf{S}$ into a cell magnetic dipole $\boldsymbol{\mu}_i$. With this one has

$$-f_i^\beta + 2\alpha_i m_i^\beta = 0. \tag{10.9}$$

This gives the solution for the required dipole component,

$$m_i^\beta = \frac{f_i^\beta}{2\alpha_i}. \tag{10.10}$$

This can be combined with the length constraint, summing over components for one dipole,

$$\mathbf{m}_i^2 = \frac{1}{4\alpha_i^2} \left[\left(f_i^x \right)^2 + \left(f_i^y \right)^2 + \left(f_i^z \right)^2 \right] = m^2, \tag{10.11}$$

which then gives the values needed for the undetermined multipliers:

$$\alpha_i = \frac{1}{2m} |\mathbf{f}_i|. \tag{10.12}$$

With this, one now finds by inserting (10.12) into (10.10) that the procedure to minimize the Hamiltonian, under the length constraint, is

$$m_i^\beta = m \frac{f_i^\beta}{|\mathbf{f}_i|}. \tag{10.13}$$

Regardless of the length m, this is seen to be the usual spin alignment relaxation scheme, pointing each dipole along the direction of its effective field $\mathbf{f}_i$. In practice we set $m = 1$. This gives the condition for a local minimizing energy state. For computation, it needs to be iterated until a desired precision in energy or some other quantity is obtained. Usually the energy may converge to some final value before the actual dipole components have relaxed to a corresponding precision.

10.2.2 Constraint on vortex position

Now the constraint of core position can also be included into the scheme. In the simplest approximation, with $N_c = 4$ for the number of constrained core sites, suppose the vortex core is held at the center of a plaquette, as in figure 8.2. Then the N_c core sites have a net zero magnetic moment,

$$\sum_{i=1}^{N_c} m_i^x = \sum_{i=1}^{N_c} m_i^y = 0. \tag{10.14}$$

This is only true for a symmetric location of the core; later we consider how to relax that assumption and use an arbitrary core position. Now a corresponding term is included into the functional $\Lambda[\mathbf{m}_i]$ to incorporate this constraint:

$$\Lambda[\mathbf{m}_i] = H[\mathbf{m}_i] + J_{\text{cell}}\left[\sum_i \alpha_i(\mathbf{m}_i^2 - m^2) - \lambda \cdot \sum_{n=1}^{N_c} \mathbf{m}_n\right]. \tag{10.15}$$

This includes an in-plane vector undetermined Lagrange multiplier $\lambda = (\lambda^x, \lambda^y)$ which is the same for all core sites; there is no such term on other sites. If the constraint is satisfied, then $\Lambda = H$. Application of the Lagrange relations in the core region gives the following minimization conditions,

$$\frac{\partial \Lambda}{\partial m_n^x} = \frac{\partial H}{\partial m_n^x} + J_{\text{cell}}(2\alpha_n m_n^x - \lambda^x) = 0 \tag{10.16a}$$

$$\frac{\partial \Lambda}{\partial m_n^y} = \frac{\partial H}{\partial m_n^y} + J_{\text{cell}}(2\alpha_n m_n^y - \lambda^y) = 0 \tag{10.16b}$$

$$\frac{\partial \Lambda}{\partial m_n^z} = \frac{\partial H}{\partial m_n^y} + J_{\text{cell}}(2\alpha_n m_n^y) = 0. \tag{10.16c}$$

With the partials of the Hamiltonian being effective force components, one obtains the results

$$m_n^x = \frac{1}{2\alpha_n}\left(f_n^x + \lambda^x\right) \tag{10.17a}$$

$$m_n^y = \frac{1}{2\alpha_n}\left(f_n^y + \lambda^y\right) \tag{10.17b}$$

$$m_n^z = \frac{1}{2\alpha_n}f_n^z. \tag{10.17c}$$

This result is applicable anywhere, taking $\lambda = 0$ outside the core region. Then one sees that λ is an extra constraining effective field that acts only on the core sites. To determine its strength, use the spin length constraint as found earlier,

$$\mathbf{m}_n^2 = \frac{1}{4\alpha_n^2}\left[\left(f_n^x + \lambda^x\right)^2 + \left(f_n^y + \lambda^y\right)^2 + \left(f_n^z\right)^2\right] = m^2. \tag{10.18}$$

Then the length constraint parameter is given from

$$\frac{1}{\alpha_n} = \frac{2m}{\sqrt{\left(f_n^x + \lambda^x\right)^2 + \left(f_n^y + \lambda^y\right)^2 + \left(f_n^z\right)^2}}. \tag{10.19}$$

To completely determine the solution, the position constraint (10.14) must be applied to obtain λ. For components $\beta = x, y$, that gives a result using sums over the core region,

$$\sum_{n=1}^{N_c} \frac{1}{2\alpha_n}\left(f_n^\beta + \lambda^\beta\right) = 0 \implies \lambda^\beta = -\frac{\sum_{i=1}^{N_c} f_n^\beta/\alpha_n}{\sum_{i=1}^{N_c} 1/\alpha_n}. \tag{10.20}$$

This looks somewhat circular, as the α_n are used to determine λ and vice versa. This turns out to cause no difficulty, as the parameters must be solved iteratively and self-consistently. The procedure can start from an initial approximation, $\lambda = 0$. Then in an iteration step, first α_n are determined from (10.19), followed by finding λ from (10.20). Then the new direction for a dipole is obtained from the normalization condition (10.17), giving

$$\mathbf{m}_n = m\frac{\left(f_n^x + \lambda^x\right)\hat{\mathbf{x}} + \left(f_n^y + \lambda^y\right)\hat{\mathbf{y}} + f_n^z\hat{\mathbf{z}}}{\sqrt{\left(f_n^x + \lambda^x\right)^2 + \left(f_n^y + \lambda^y\right)^2 + \left(f_n^z\right)^2}}. \tag{10.21}$$

The result is clearly of desired length m, and because λ was determined from the constraint (10.20), the position constraint is obviously satisfied. This is the method to obtain a vortex positioned in the center of a plaquette.

To obtain the vortex at a more arbitrary off-center position in a plaquette, the constraint needs to be modified. Now the core sums (10.14) do not need to be zero if the core location is offset slightly from the grid of cells. Therefore it makes sense to change the position constraints to be

$$\sum_{i=1}^{N_c} m_i^x = T^x, \qquad \sum_{i=1}^{N_c} m_i^y = T^y. \tag{10.22}$$

The 2D vector $\mathbf{T} = (T^x, T^y)$ is a pair of numbers set by the core sums as determined by the initial assumed vortex configuration for position $\mathbf{X}$ at a desired location. Thus, the numbers (T^x, T^y) are initially set but allowed to vary during the calculation. The functional being minimized is now changed into

$$\Lambda[\mathbf{m}_i] = H[\mathbf{m}_i] + J_{\text{cell}} \left[\sum_i \alpha_i \left(\mathbf{m}_i^2 - m^2 \right) - \lambda \cdot \left(\sum_{n=1}^{N_c} \mathbf{m}_n - \mathbf{T} \right) \right]. \tag{10.23}$$

Now the core sums must satisfy the conditions in the core region,

$$\sum_{n=1}^{N_c} \frac{1}{2\alpha_n} \left(f_n^\beta + \lambda^\beta \right) = T^\beta, \tag{10.24}$$

which then leads to the slightly modified constraining fields,

$$\lambda^\beta = \frac{T^\beta - \sum_{n=1}^{N_c} f_n^\beta / \alpha_n}{\sum_{n=1}^{N_c} 1/\alpha_n}. \tag{10.25}$$

An iteration is performed, continuously changing the values of λ^β at each step, and evaluating the dipoles according to expression (10.21). During the iteration, the dipoles, especially in the core region, evolve out-of-plane components, causing their in-plane components to be diminished. This means that the core sums which give $\mathbf{T}$ will diminish as the iterations proceed. To accommodate this, it is found necessary to allow $\mathbf{T}$ to be readjusted in magnitude, according to an algorithm,

$$\mathbf{T} = \left[\sum_{n=1}^{N_c} \mathbf{m}_n(0) \right] \left\langle \sqrt{1 - \left(\frac{m_n^z}{m} \right)^2} \right\rangle. \tag{10.26}$$

Here $\mathbf{m}_n(0)$ are the dipole values at the start of the iteration, which were used to define the core position. The second factor contains the average projection onto the xy-plane within the core region. As the dipoles move out-of-plane in the core, that projection diminishes, thus rescaling $\mathbf{T}$ appropriately. Normally the values of $\mathbf{T}$ produced in this algorithm are quite small. They become larger only when the core position $\mathbf{X}$ falls between cell centers. The presence of $\mathbf{T}$ is needed to accommodate the vortex into a position that does not line up perfectly within a plaquette of the grid. The number of core cells can must be at least four, but other numbers such as $N_c = 12, 16, 24, 48$ and 96 give a symmetrical set of sites around a desired core position. Usually $N_c = 24$ works out fairly well. For this method to work well, the size of the constrained core region should be larger than the exchange length.

The method can be applied even when the vortex is exposed to an external magnetic field $\mathbf{H}_{\text{ext}}$, which gives the dimensionless field $\tilde{\mathbf{H}}_{\text{ext}} = \mathbf{H}_{\text{ext}}/M_s$. It is interesting to compare the (dimensionless) constraining field λ to the applied field. This is accomplished by expressing the applied field in an equivalent dimensionless form. The applied field term for one cell is

$$-J_{\text{cell}} \frac{a^2}{\lambda_{\text{ex}}^2} \tilde{\mathbf{H}}_{\text{ext}} \cdot \mathbf{m}_n = -\mu_0 \mathbf{H}_{\text{ext}} \cdot \boldsymbol{\mu}_n = -\mu_0 \mathbf{H}_{\text{ext}} \cdot \mu_{\text{cell}} \mathbf{m}_n. \tag{10.27}$$

The corresponding constraining term for that cell has been written as

$$-J_{\text{cell}} \, \lambda \cdot \mathbf{m}_n. \tag{10.28}$$

Comparing there, this suggests the definition of the corresponding dimensionless applied field as

$$\mathbf{h}_{\text{ext}} \equiv \frac{a^2}{\lambda_{\text{ex}}^2} \tilde{\mathbf{H}}_{\text{ext}} = \frac{a^2}{\lambda_{\text{ex}}^2} \frac{\mathbf{H}_{\text{ext}}}{M_s}. \tag{10.29}$$

This is then the field with which to compare the strength of the constraining field λ. Alternatively, we can say that a particular value of λ must correspond to a physical magnetic field strength $\mathbf{H}_\lambda$, defined by a similar relation,

$$\lambda \equiv \frac{a^2}{\lambda_{\text{ex}}^2} \frac{\mathbf{H}_\lambda}{M_s}. \tag{10.30}$$

These relations depend on the micromagnetics cell size a, in relation to the exchange length. The fields should be perturbative, which again reminds us that the ratio a/λ_{ex} must be kept small enough so that the spatial variations in all fields can be correctly calculated. In some typical simulations for Py, with $\lambda_{\text{ex}} = 5.3$ nm, we have used $a = 2.0$ nm, which results in the factor $a^2/\lambda_{\text{ex}}^2 \approx 1/7$, a value that is acceptable.

10.2.3 Some calculated vortex potentials

The typical potentials $U(\mathbf{X})$ for a vortex in a nanodisk of thickness L and radius R can be found by the procedure just spelled out, by applying the energy minimization procedure with position constraint for a sequence of vortex locations $\mathbf{X}$. For example, the vortex core can be held at a position $\mathbf{X} = (X, 0)$ by the Lagrange constraint technique, measured from the disk center. The location X can be allowed to range across the diameter of the disk. For the examples shown here, the core region was taken to be $N_c = 24$ cells. The constraining field acts only on those core cells.

The initial state of the iteration can be a planar vortex (all $m_i^z \approx 0$); we used a circulation of the dipoles in the counterclockwise sense ($\phi_0 = +90°$) around the vortex core. This means that a displacement at positive (negative) X requires a negative (positive) value of the constraining field λ^y. The signs would be reversed for a vortex with a clockwise circulation ($\phi_0 = -90°$) of the magnetization.

A single iteration step corresponds to redirecting all cells in the system once. Usually it is necessary to repeat the iterations thousands of times to obtain a desired energy precision; we used approximately 1 part in 10^7. However, the iterations were actually stopped when the average changes in the components of the $\mathbf{m}_i$ satisfied a requirement,

$$\langle |\Delta m_i^x| + |\Delta m_i^y| + |\Delta m_i^z| \rangle < 10^{-7}. \tag{10.31}$$

As for any spin alignment scheme, this is necessary because the energy tends to converge to a final value faster than the magnetization distribution. As the system relaxes to its energy minimizing state, the constraining field grows, and it may cause a slight deformation of the core.

The potentials found are close to parabolic form, as long as the vortex is close to the disk center. An example of these results and the change in energy due to the relaxation scheme is shown in figure 10.2 for a Py disk, as calculated using $\lambda_{\text{ex}} = 5.3$ nm and cell size $a = 2.0$ nm. The initial state is a planar vortex state, hence it has a higher energy. During the relaxation an out-of-plane magnetization component develops in the vortex core region, which is greatly responsible for the lowering of the energy.

In figure 10.3 different relaxed potentials for circular disks with thickness $L = 12$ nm and various radii are shown. Here we have scaled the position X by the radius for

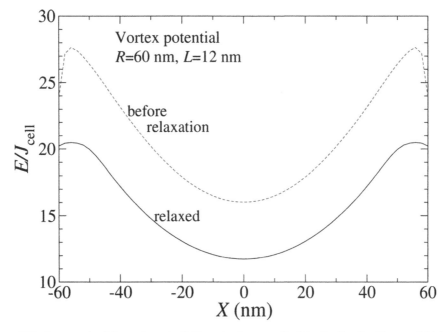

Figure 10.2. An example of the vortex potential in a circular magnetic disk of radius $R = 60$ nm and thickness $L = 12$ nm, using Py parameters and cell size $a = 2.0$ nm. The energy unit is $J_{\text{cell}} = 2AL$. The vortex position $\mathbf{X} = (X, 0)$ (relative to the disk center) was moved along a diameter to map out the potential. The state before relaxation was planar with $m^z = 0$. The relaxed state, which develops a non-zero out-of-plane component, was found by the Lagrange constrained spin alignment technique described in the text.

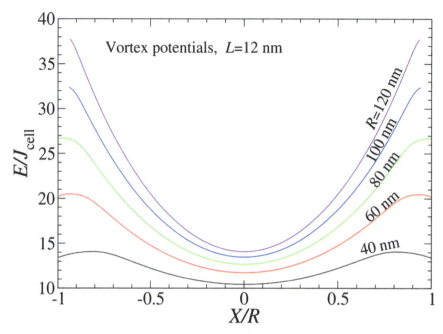

Figure 10.3. Relaxed vortex potentials in circular magnetic disks of indicated radii R, at thickness $L = 12$ nm, using Py parameters and cell size $a = 2.0$ nm. Note that the vortex position has been scaled relative to the disk radius.

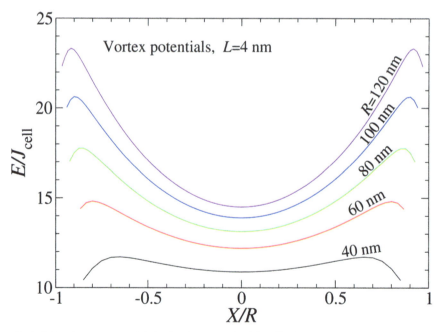

Figure 10.4. Relaxed vortex potentials in circular magnetic disks of indicated radii R, at thickness $L = 4$ nm, using Py parameters and cell size $a = 2.0$ nm.

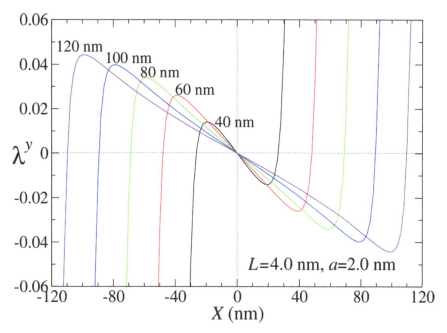

Figure 10.5. Behavior of the vortex position constraining field λ^y as a function of the vortex location, for indicated disk radii R, at thickness $L = 4$ nm, using Py parameters and cell size $a = 2.0$ nm. The rotation of the in-plane magnetization is counterclockwise around the core, which leads to the particular sign on λ^y. The reversal of λ^y near the disk edge signifies the lack of stability there.

convenience of plotting all $U(\mathbf{X})$ together. A larger range of energies is possible in the disks of greater radius, however, that change takes place over a greater distance. If the vortex is constrained very close to the edge of a disk, it can be unstable towards leaving the system, as the potential bends downward in that region. For the larger systems the algorithm could not produce a stable relaxed vortex all the way to the disk edge. In figure 10.4, similar results are presented, but for disk thickness $L = 4.0$ nm. The primary modification of the potentials is that they are shallower for the thinner disks.

Constraining field
For this last case with $L = 4.0$ nm, the constraining field λ^y is also shown as a function of vortex location in figure 10.5. It is interesting to see that indeed the negative slopes of the curves (for vortices well within the disk) makes sense. A displacement along positive $\hat{\mathbf{x}}$ results in an increase in the number of cell dipoles with negative y-components. This is depicted in figure 10.6, for the case of a vortex with a positive (counterclockwise) sense of circulation, corresponding to the phase angle $\phi_0 = +90°$. Thus, a constraining field along $-\hat{\mathbf{y}}$ over the core region will indeed tend to displace the positive circulation vortex towards the $+\hat{\mathbf{x}}$-direction, as expected. It is clear that the field λ and the displacement it causes will always be perpendicular. This further implies that an externally applied field, also along $-\hat{\mathbf{y}}$, will displace the vortex along $+\hat{\mathbf{x}}$.

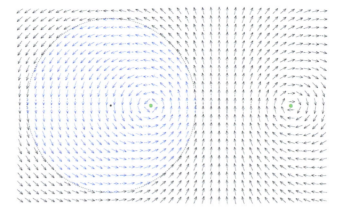

Figure 10.6. Sketch of the in-plane magnetization components expected for a vortex with positive (counter-clockwise) circulation when displaced along $+\hat{x}$. The system dipoles are the blue arrows within the circle; gray arrows represent fictitious dipoles outside the system used for boundary conditions, including the image vortex. As in chapter 9, the image repels the vortex in the system, producing the central restoring force. This displacement can be aided by an applied field or a core constraining field along $-\hat{y}$.

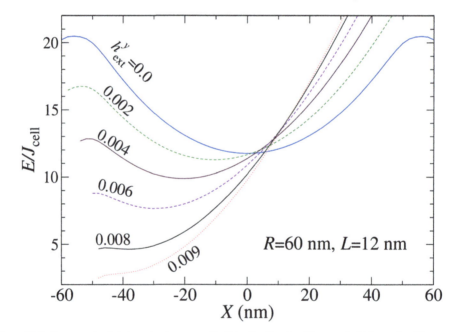

Figure 10.7. Vortex potentials in a circular magnetic disk of radius $R = 60$ nm and thickness $L = 12$ nm, using Py parameters and cell size $a = 2.0$ nm, for the indicated applied field strengths h_{ext}^y, in dimensionless units, see (10.29). A vortex of positive circulation (or $\phi_0 = +90°$) was used, so that field applied along $\hat{y}$ causes the minimum to displace along $-\hat{x}$.

Applied field

The effect of an applied field on the nanodisk can also be tested by this method. It is well known that an external field $\mathbf{H}_{\text{ext}}$ will tend to shift the equilibrium position of the vortex. To see that, consider the results in figure 10.7, for a Py nanodisk with

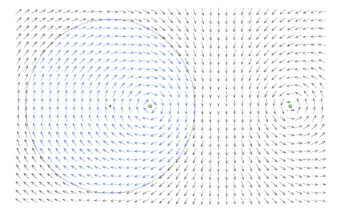

Figure 10.8. Sketch of the in-plane magnetization components expected for a vortex with negative (clockwise) circulation when displaced along $+\hat{\mathbf{x}}$. The system dipoles are the blue arrows; gray arrows are fictitious dipoles outside the system used for boundary conditions, including the image vortex which repels the vortex in the system. This displacement can be aided by an applied field or a core constraining field along $+\hat{\mathbf{y}}$, in contrast to the case of positive circulation, figure 10.6.

$R = 60$ nm and $L = 12$ nm, for a set of different dimensionless field strengths h_{ext}^y, as defined in (10.29). The simulations were performed for a vortex of positive (counterclockwise) circulation sense, with $\phi_0 = +90°$, figure 10.6. Hence, a field along $\hat{\mathbf{y}}$ will tend to displace the vortex along $-\hat{\mathbf{x}}$, which is seen in the overall shifts of the potential curves with increasing h_{ext}^y. That position shift causes a larger fraction of the dipoles to become aligned with $\mathbf{H}_{\text{ext}}$. Conversely, a vortex with a negative (clockwise) sense of circulation, figure 10.8, will be displaced in the $+\hat{\mathbf{x}}$-direction by a field along $+\hat{\mathbf{y}}$. For these parameters, one finds that beyond a field strength around $h_{\text{ext}}^y \approx 0.008$, the vortex will be displaced until it is pushed out of the system, regardless of its direction. Thus, there is a limit to its stability in the presence of the applied field. Using (10.29) and the simulation parameters, this upper limit on the field is converted to units of M_s as

$$H_{\text{ext}}^y = \left(\frac{\lambda_{\text{ex}}}{a}\right)^2 M_s h_{\text{ext}}^y \approx 0.056 M_s. \quad (10.32)$$

This is physically a rather weak field. It shows that the vortex responds easily to an applied field, or it has a high susceptibility for this effect.

The vortex minimum energy position responds to applied field slightly differently as the disk size is varied. To see this, it is interesting to plot the minimizing position X_0 as a function of applied field h_{ext}^y, for different disk sizes, see figure 10.9. For thinner disks ($L = 4.0$ nm), the vortex is more easily displaced than for thicker disks ($L = 12$ nm). Of course, this implies that rather weak fields already move the vortex out of the system. The curves in figure 10.9 also clearly show that in disks of larger radii the susceptibility for vortex displacement by a given field is higher. Some degree of nonlinearity also is apparent for the larger radius disks.

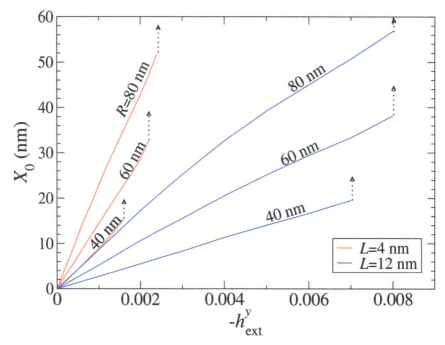

Figure 10.9. The minimum energy vortex location X_0 as a function of an applied external field component h_{ext}^y, for various disk sizes, using Py parameters and $a = 2.0$ nm. Numbers next to curves are the disk radii. Upward arrows at the ends of the curves show the limits of vortex stability, beyond which the vortex is expelled from the disk.

Exercise 10.1. The vortex displacement response in figure 10.9 results from a competition between the force of the applied field and a restoring demagnetization force coming mostly from the curved boundary. For the case of a single layer of atomic spins, both were evaluated approximately in chapter 9, see (9.106) for the force due to applied field, and (9.145) for the force due to the boundary. Check whether the theory in chapter 9 can be used to explain the results of figure 10.9 in some approximate way or limit.

Effective vortex force constant k_F

One can see from the different system sizes that the potential is stronger for greater disk thickness, but weaker for larger disk radius. A consequence of this is that an effective force constant k_F can be found [6], describing the pull of the vortex towards the disk center. That is based on a simple parabolic assumption for fitting the potentials, assumed also to be isotropic,

$$U(\mathbf{X}) = U(0) + \frac{1}{2} k_F \mathbf{X}^2. \tag{10.33}$$

The potential can be fitted using a sequence of locations $\mathbf{X} = (X, 0)$ to estimate k_F. Or a simple or faster method is to assume the parabolic form holds, and then evaluate the

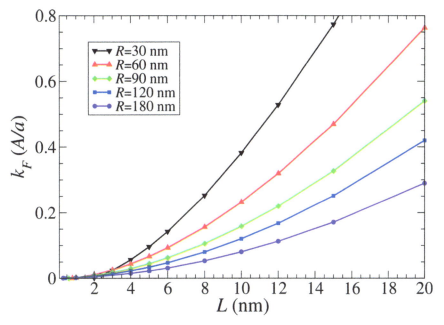

Figure 10.10. The vortex force constant k_F versus disk thickness L for Py parameters, using cell size $a = 2.0$ nm. The force constant is given in terms of the exchange stiffness A of the medium.

energy for the vortex at the center and for a small displacement on the order of one cell size from the center, using those to obtain k_F. In this latter method it is possible to obtain reasonable information about the changes in k_F with disk size, using a limited set of calculations. Some results of those types of calculations are shown in figures 10.10 and 10.11, where the force constants were calculated by using the relaxed system energy for a displacement of $\Delta X = 2a$. One sees in figure 10.10 that the force constant increases with disk thickness, close to a quadratic function of L, and k_F is smaller for larger disk radii. In figure 10.11 the dependence on R is better indicated, with different curves corresponding to choices of L. For the smallest thickness there ($L = 2$ nm), the vortex is barely stable, especially at the smallest R, leading to a very weak force constant. Aside from the smallest radii, k_F decreases with increasing disk radius, possibly close to a $1/R$ relationship. These dependencies of k_F on disk dimensions are important for the determination of the vortex gyrotropic dynamics in this type of system.

A force constant was already found in chapter 9 for the case of a vortex in an easy-plane atomic model ferromagnet, in a system of circular shape. Using the strong demagnetization boundary conditions, expression (9.145) was found for the force between the vortex at radial position X and its image outside the system. For the limiting case of small displacement from the disk center, it reduces to a linear force law (9.146), or

$$F \approx -k_F X, \qquad k_F \approx \frac{2\pi J S^2}{R^2}. \qquad (10.34)$$

To make a comparison to the micromagnetics notation, it is necessary to let the spin–spin exchange energy JS^2 go over into the inter-cell exchange energy

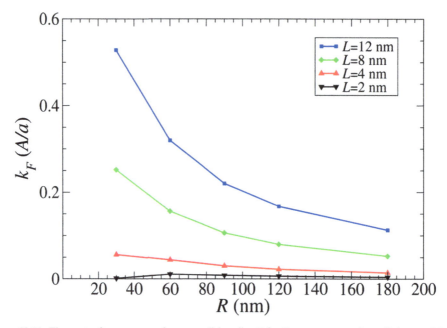

Figure 10.11. The vortex force constant k_F versus disk radius R for Py parameters, using cell size $a = 2.0$ nm. At the smallest thickness shown the vortex is rather unstable, especially for the smallest radii, leading to a very small force constant.

$J_{\text{cell}} = 2AL$. This means that the strong demagnetization boundary conditions gives an estimate for the force constant as

$$k_F \approx \frac{4\pi AL}{R^2}. \tag{10.35}$$

This linear increase with L is not seen in the numerical results such as figure 10.10 and 10.11. Instead, it appears that k_F probably increases at a power somewhat higher than linear in L, and it diminishes more like R^{-1} rather than an inverse square form. In previous work [6], k_F has been assumed to be proportional to L^2/R, which is a reasonable approximation, but even that form does not give a completely accurate representation.

10.3 $T = 0$ Gyrotropic vortex motion in thin nanodisks

The topic of zero-temperature gyrotropic vortex motion has already been discussed in chapter 9 for the case of vortices in systems where the demagnetization field was not explicitly included. In particular, there it was seen that generally a circular vortex motion can result, due to the presence of a non-zero gyrovector $\mathbf{G}$ for out-of-plane vortices. The combination of the gyrovector force $\sigma \mathbf{G} \times \mathbf{V}$ and a restoring force towards the center of a circular system leads to gyrotropic circular motion. It was shown that the restoring force points towards the center of a circular system if the strong demagnetization boundary conditions are applied, forcing spins to point along the circular edge along the boundary.

Essentially a very similar situation exists in a nanometer sized disk of soft FM material, treating the boundary not by a somewhat arbitrary boundary condition, but instead including the demagnetization field at its natural strength. The tendency of the demagnetization effects to force $\hat{\mathbf{m}}(\mathbf{r})$ to follow the boundary at the disk edge will depend on the disk size, being stronger for larger disks. There is still the possibility of generation of weak magnetic pole density on the boundaries. Thus, the simulation of the realistic demagnetization effects could give slightly different results than the simple assumption of strong demagnetization boundary conditions. That is the situation considered here, for the case of a thin-film soft permeable magnetic material. In this analysis, we consider the dynamics expected from a Thiele equation for the vortex motion, and compare it with the motion obtained by numerical simulation of a zero-temperature Landau–Lifshitz–Gilbert (LLG) equation for the equivalent system.

10.3.1 Thiele equation dynamics for a thin-film system

The Thiele equation for a nanodisk system is considered without damping and also without a vortex mass due to vortex structural deformations with velocity. We have seen earlier in (9.20) that the Thiele equation can be written for this simplified case as

$$\mathbf{F} + \sigma \mathbf{G} \times \mathbf{V} = 0. \tag{10.36}$$

In this notation the gyrovector is a dimensionless quantity defined in expression (9.15), and it was evaluated in (9.30), leading to

$$\mathbf{G} = G\hat{\mathbf{z}} = 2\pi p q \hat{\mathbf{z}} \tag{10.37}$$

where $p = \pm 1$ is the core polarization and $q = +1$ is the only stable choice for a vortex affected by demagnetization forces. An anti-vortex, with $q = -1$, would be unstable, because its magnetization field does not follow the circular edge of the disk. The Thiele equation also contains the parameter σ, which is the spin density per unit area. It is commonly expressed equivalently in terms of the electron's gyromagnetic ratio γ and the magnetic dipole moment per unit area, denoted as

$$m_0 = \mu_{\text{cell}}/a^2 = LM_s. \tag{10.38}$$

From the basic definition for the magnetic dipole of a spin, $\mu = \gamma S$, the spin density per unit area is then given by

$$\sigma = m_0/\gamma = LM_s/\gamma. \tag{10.39}$$

In some works, this factor might be included into the definition of gyrovector with dimensions of spin density.

For the simplest analysis, the vortex restoring force is assumed to be linear, with some force constant k_F,

$$\mathbf{F} = -\frac{\partial U}{\partial \mathbf{X}} = -k_F \mathbf{X}. \tag{10.40}$$

Then the steady-state gyrotropic solution is an orbital motion at a constant angular velocity $\boldsymbol{\omega}_G = \omega_G \hat{\mathbf{z}}$, such that the vortex velocity is

$$\mathbf{V} = \boldsymbol{\omega}_G \times \mathbf{X}. \qquad (10.41)$$

Using that in the Thiele equation gives

$$-k_F \mathbf{X} + \sigma \mathbf{G} \times (\boldsymbol{\omega}_G \times \mathbf{X}) = 0. \qquad (10.42)$$

With both $\mathbf{G}$ and $\boldsymbol{\omega}_G$ oriented along the $\hat{\mathbf{z}}$-axis, one has a simple result for the angular velocity

$$\omega_G = -\frac{k_F}{\sigma G}. \qquad (10.43)$$

Written also in terms of the more fundamental parameters, this is

$$\omega_G = -\frac{\gamma k_F}{2\pi pqLM_s}. \qquad (10.44)$$

Although the vorticity is included here for completeness, in this system it can only take the value $q = +1$. Then the sign of ω_G is determined only by the sign of the vortex core polarization, i.e. the sign of m^z in the vortex core region. A positive value of p gives also a positive G, resulting in a negative value of ω_G, or clockwise vortex motion in the xy-plane. For negative p, the motion instead will be counterclockwise. While L and M_s are fairly well-known material parameters, the force constant must be estimated theoretically. It has been predicted by some different analytic approximations such as the *rigid vortex approximation* [7] and an approach called the *two-vortex model* [8]. In the work presented here we use k_F found numerically from the spin alignment scheme together with the Lagrange constraint for vortex position, described earlier in section 10.2.3 and displayed in figures 10.10 and 10.11. This approach allows for extra deformations of the vortex structure, and should be expected to give a different force constant than the analytic approaches.

Exercise 10.2. The demagnetization at the curved boundary of a nanodisk makes $q = +1$ the preferred vorticity. Even so, consider the possibility to obtain a state with two $q = +1$ vortices, symmetrically located on a diameter, equidistant from the disk center. Assume each vortex is acted on by a restoring force from the boundary (some force constant k_F) and a repulsive force from the other vortex such as that in (9.119). Analyze the Thiele equations for the pair. Compare the dynamics expected in the limits of strong damping and weak damping. Discuss whether the assumption of a restoring force due to the boundary is reasonable.

10.3.2 Numerical simulation of vortex dynamics at zero temperature

The fairly simple prediction (10.44) of the Thiele equation can be tested using numerical simulations for the zero temperature LLG dynamics, starting from an appropriate vortex initial condition. The initial condition can be obtained from the spin alignment procedure with a Lagrange position constraint, finding a good description for an initial vortex at some small displacement $\mathbf{X}$ from the disk center. Then, this is used in LLG dynamics solved in the sense of micromagnetics. The system is partitioned into a 2D array of cells each with a magnetic dipole $\boldsymbol{\mu}_i = \mu_{\text{cell}}\hat{\mathbf{m}}_i = La^2 M_s \hat{\mathbf{m}}_i$. Each cell obeys LLG dynamics, according to an equation that can include damping if desired,

$$\frac{d\boldsymbol{\mu}_i}{dt} = \boldsymbol{\mu}_i \times \mathcal{F}_i - \frac{\alpha}{\mu_{\text{cell}}}\boldsymbol{\mu}_i \times (\boldsymbol{\mu}_i \times \mathcal{F}_i). \tag{10.45}$$

Taking out a factor of the dipole magnitude μ_{cell}, this is equivalent to the same form of equation for the unit vector local magnetization dipoles,

$$\frac{d\hat{\mathbf{m}}_i}{dt} = \hat{\mathbf{m}}_i \times \mathcal{F}_i - \alpha \hat{\mathbf{m}}_i \times (\hat{\mathbf{m}}_i \times \mathcal{F}_i). \tag{10.46}$$

The effective field $\mathcal{F}_i$ acting on each site is equivalent to the local magnetic induction by $\mathcal{F}_i = \gamma \mathbf{B}_i$, which is obtained from the gradient of the micromagnetics Hamiltonian. It has been evaluated earlier in (3.155), and we rewrite it again here,

$$\mathbf{B}_i = B_0 \left[\sum_{(i,j)} \hat{\mathbf{m}}_j + 2\frac{a^2}{\lambda_K^2}(\hat{\mathbf{m}}_i \cdot \hat{\mathbf{x}})\hat{\mathbf{x}} + \frac{a^2}{\lambda_{\text{ex}}^2}\left(\tilde{\mathbf{H}}_{\text{ext}} + \tilde{\mathbf{H}}_M\right) \cdot \hat{\mathbf{m}}_i \right]. \tag{10.47}$$

This employs a basic unit for magnetic induction,

$$B_0 = \frac{J_{\text{cell}}}{\mu_{\text{cell}}} = \mu_0 M_s \left(\frac{\lambda_{\text{ex}}}{a}\right)^2. \tag{10.48}$$

Then a dimensionless magnetic induction $\mathbf{b}_i$ is defined with this,

$$\mathbf{b}_i \equiv \mathbf{B}_i / B_0. \tag{10.49}$$

Although any system of units for time, magnetic induction, *etc*, is permissible, it is somewhat convenient for numerical work to combine the gyromagnetic ratio γ with B_0, both appearing on the RHS of the dynamical equations, to define a natural unit of frequency, $f_0 = \gamma B_0$. The reciprocal defines a natural unit of time for simulations, $t_0 = 1/f_0 = (\gamma B_0)^{-1}$. This suggests the definition of a dimensionless time variable for the simulations,

$$\tau \equiv \gamma B_0 t = t/t_0. \tag{10.50}$$

With these definitions, the numerical simulations are performed in dimensionless units for the equations,

$$\frac{d\hat{\mathbf{m}}_i}{d\tau} = \hat{\mathbf{m}}_i \times \mathbf{b}_i - \alpha \hat{\mathbf{m}}_i \times (\hat{\mathbf{m}}_i \times \mathbf{b}_i). \tag{10.51}$$

In this system of units one then performs the integration forward in time by some appropriate scheme, such as fourth-order Runge–Kutta (RK4). The time step can be of the order of $\Delta\tau = 0.04$ for reliable results; by using dimensionless variables there is a better sense of numerical error control. Based on a number of simulations, this size of time step used with the RK4 scheme (see chapter 5) without damping can conserve the total system energy to better than 12 digit precision, over as many as 5.0×10^5 time steps. This is typical even for simulations with around 4000 cells. To see precise energy conservation, however, requires the evaluation of the demagnetization field $\mathbf{H}_M$ at all four sub-steps of the RK4 scheme.

For Py, the material parameters are $A = 13$ pJ m^{-1}, $M_s = 860$ kA m^{-1}, which we have seen give $\lambda_{ex} = 5.3$ nm. With a cell parameter of $a = 2.0$ nm and the gyromagnetic ratio $\gamma = e/m_e = 1.76 \times 10^{11}$ T^{-1}s^{-1}, the computational units are based on $\mu_0 M_s = 1.08$ T and $B_0 = 7.59$ T. Note that B_0 gives the order of magnitude of the exchange field between neighboring cells. Then one also has $f_0 = 1.336$ THz and $t_0 = 0.75$ ps. Some published articles quote the parameters in CGS units. In that case, it is common to quote frequencies in units of the CGS expression, γM_s, which is equivalent to the SI expression $\frac{\mu_0}{4\pi}\gamma M_s = 15.1$ GHz.

The numerical integration results in the time-dependent cell dipoles, $\hat{\mathbf{m}}_i(\tau)$. From this information it is necessary to detect the motion of the vortex core. This can be achieved in various ways. One simple way to detect the effect of vortex core motion is the calculation of the spatially averaged magnetization as

$$\langle \hat{\mathbf{m}}_i \rangle = \frac{1}{N} \sum_{i=1}^{N} \hat{\mathbf{m}}_i. \tag{10.52}$$

Clearly this will oscillate in time if the vortex core $\mathbf{X}(t)$ moves in circular motion. For numerical calculation, this does not require a direct measurement of $\mathbf{X}(t)$. The downside is that spin wave oscillations could also lead to an oscillatory response in $\langle \hat{\mathbf{m}}_i \rangle$, however, most likely at much higher frequencies. The gyrotropic frequency will be much lower than most of the spin wave spectrum.

For more exact information about the vortex motion, it is necessary to directly measure $\mathbf{X}(t)$. The core location is the point around which the in-plane magnetization components have a divergent curl, in the sense of (8.13). It is also the point at which the out-of-plane component is a maximum. Either of these might be used to locate $\mathbf{X}(t)$. In [6], a hybrid of these two ideas has been used to define the fluctuating core location. Initially, one locates the set of four cells where there is a net 2π rotation of the in-plane component, based on

$$\oint \nabla \phi \cdot d\ell = 2\pi q. \tag{10.53}$$

On a square grid of cells, that corresponds to searching each plaquette of four neighboring cells, and summing the changes in in-plane angle around the plaquette, see figure 10.12. Each change $\Delta \phi_i = \phi_{i+1} - \phi_i$ must be placed on the same branch, such as $-\pi \leqslant \Delta \phi_i \leqslant \pi$, by using a 2π shift if it is outside of this range. The discrete vorticity in a plaquette is then

$$q = \frac{1}{2\pi} \sum_{i=1}^{4} \left[\Delta \phi_i \right]. \tag{10.54}$$

The changes are taken in the counterclockwise sense and interpreted with periodic boundary conditions around the plaquette. The square bracket indicates shifting into the main branch from $-\pi$ to $+\pi$. Most of the time, the sum is zero, such as the

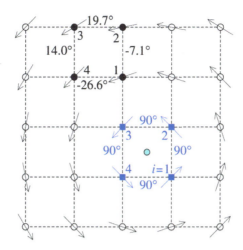

Figure 10.12. Notation and example of locating the plaquette containing the vortex core, according to (10.54). The vortex core (turquoise dot) is centered in the blue plaquette, leading to the angular changes $[\Delta\phi_i]$ as shown (in degrees here) for each bond, which sum to 360° or 2π rad. In other plaquettes, such as the black plaquette, the angular changes sum to zero.

black plaquette in figure 10.12. Assuming a simulation of an individual vortex, there will be only one plaquette in the system, in which the sum on the RHS is found to be 2π (blue plaquette of figure 10.12), signifying that the vorticity center has been located to a precision of one cell length a. The center of that plaquette defines an estimate of the vorticity center, $\mathbf{r}_v$. Unfortunately, this position estimate then jumps in increments of $\pm a$, which is not acceptable for careful tracking of the vortex core.

To obtain a better estimate of $\mathbf{X}(t)$, one can use an average position weighted by the squared m_i^z-components of only the cells nearest to the vorticity center found from (10.54). In [6] only the cells within four exchange lengths were used to make an average,

$$\mathbf{X} = \frac{\sum_{|\mathbf{r}_i-\mathbf{r}_v|<4\lambda_{ex}} (m_i^z)^2 \mathbf{r}_i}{\sum_{|\mathbf{r}_i-\mathbf{r}_v|<4\lambda_{ex}} (m_i^z)^2}. \tag{10.55}$$

This definition gives a position $\mathbf{X}(t)$ that changes smoothly with time and is within one lattice constant of the initial estimate, $\mathbf{r}_v$. By using a weight dependent on m_i^z, it incorporates primarily the sites with the largest out-of-plane magnetization, which should be in the region of the vortex core. Furthermore, the restriction to a distance of four exchange lengths helps to avoid other regions not associated with the vortex, where spin waves may produce larger values of m_i^z. This definition for the core location also tends to correlate well with a point of local maximum magnetization energy. In the simulations, plots of the components of $\hat{\mathbf{m}}(t)$ on the grid can be used to verify the correctness of (10.55).

10.3.3 Typical gyrotropic motions from simulations

One can verify the predictions of the Thiele equation for the vortex gyrotropic frequencies by making simulations for circular disks with various thicknesses and

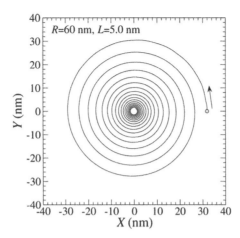

Figure 10.13. Example of vortex core motion with damping parameter $\alpha = 0.02$ obtained from zero-temperature time evolution using RK4 integration of the LLG equations. The vortex with negative core polarization ($p = -1$) was initially relaxed at the initial position $\mathbf{X}_0 = (32, 0)$ nm marked by the dot. The counterclockwise motion is consistent with the prediction of the Thiele equation. Damping causes the vortex to move downhill in the effective potential, towards the minimum at the disk center. The final time is $\tau = 50\,000$. Another view of the motion appears in figure 10.14.

radii. In [6] thickness values $L = 5.0$ nm, 10 nm and 20 nm were used, together with radii values $R = 30$ nm, 60 nm, 90 nm and 120 nm, all using lattice cell size $a = 2.0$ nm and Py material parameters. An initial set of calculations can be performed simply to check that the vortex is stable and moves in the correct direction, starting from a Lagrange-relaxed initial state at some position displaced from the disk center. By including a small damping $\alpha = 0.02$, a short simulation is sufficient to observe the sense of the rotational motion. A stable vortex solution will move towards the disk center as it loses energy; and an unstable solution typically will move out of the disk at its outer edge.

An example of stable vortex motion with damping is shown in figure 10.13, for $R = 60$ nm, $L = 5.0$ nm, and the vortex was initiated at position $\mathbf{X}_0 = (32, 0)$ nm, measured from the disk center. The damping parameter $\alpha = 0.02$ was continuously applied and the dynamics was calculated using the RK4 scheme. The vortex was given a negative core polarization $p = -1$, and thus rotates in the counterclockwise (positive) sense, consistent with the Thiele prediction (10.44). The damping causes a decreasing orbital radius as the vortex moves downward in its effective potential like those in figure 10.4. The time-dependence of the diminishing radius is apparent in figure 10.14, where X/R and the perpendicular component of in-plane magnetization, $\langle m^y \rangle$ are shown as functions of time. These quantities have similar amplitudes together with a 180° phase difference. The period of the motion for the first three orbits is $\tau_G \approx 2850, 2860, 2890$ and tends closer to $\tau_G \approx 2930$ for later times, as the vortex further relaxes and moves into the interior part of the potential. With the time unit being $t_0 = 0.75$ ps, the final period is near $T_G = 2930 t_0 \approx 2.20$ ns, corresponding to a frequency of $f_G = T_G^{-1} = 0.455$ GHz for Py.

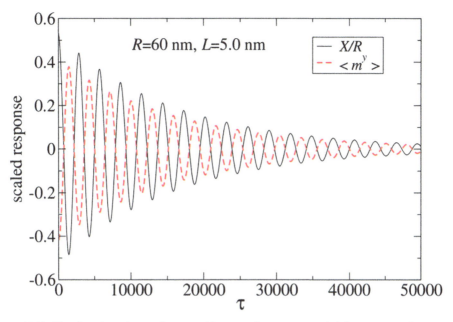

Figure 10.14. The time-dependence of one position coordinate $X(\tau)$ scaled by system radius and the perpendicular average magnetization component $\langle m^y \rangle(\tau)$ for the vortex gyrotropic motion shown in figure 10.13, with damping strength $\alpha = 0.02$. Note the 180° phase relationship.

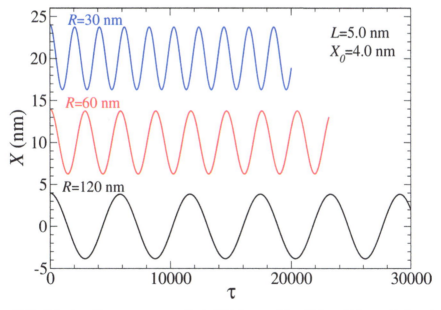

Figure 10.15. Behavior of the vortex core coordinate $X(\tau)$ for nanodisks of 5.0 nm thickness at indicated radii, shifted vertically for clarity. A damping of $\alpha = 0.02$ was turned on only for $\tau < 1000$. The periods can be calculated precisely during the energy-conserving motion after that.

To check whether the periods and frequencies obtained corroborate with the Thiele equation, the simulations can be performed by using the damping only for a limited time interval ($\tau = 1000$) to smooth out the initial vortex state. After that, with the damping turned off, the energy-conserving motion can be tracked and the frequency calculated. For consistency with the estimation of the force constants, the motion is initiated at a smaller radius, $\mathbf{X}_0 = (4, 0)$ nm, so that the interior part of the potential is acting on the vortex. These simulations result in very smooth motion of the vortex center, such as that shown in figure 10.15, for different disk radii with thickness $L = 5.0$ nm. By averaging over five to ten rotations for a range of disk sizes, precise estimates of the gyrotropic frequency $f_G = \omega_G/2\pi$ can be calculated. Some results for f_G in units of $\frac{\mu_0}{4\pi}\gamma M_s$, versus disk thickness for several R's, are plotted in figure 10.16. The frequencies have also been scaled to apply for Py material parameters on the right-hand axis. For the larger disk radii used, the frequency diminishes approximately with $1/R$.

While there is no simple functional dependence of f_G for all sizes, the tendency is for f_G to increase with aspect ratio L/R, but not exactly in a linear relationship. The frequency results are replotted in figure 10.17 versus the aspect ratio L/R. This gives close to a linear relationship with a slope of 0.25 when the frequency is plotted in units of $\frac{\mu_0}{4\pi}\gamma M_s$. If this is compared to the Thiele equation prediction (10.44), we have an approximate relationship,

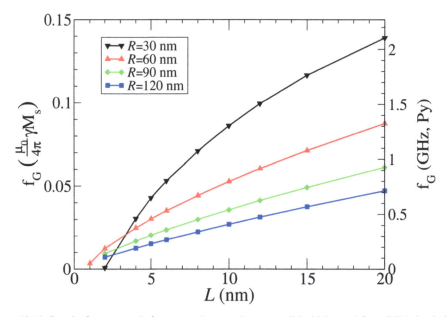

Figure 10.16. Results for gyrotropic frequency $f_G = \omega_G/2\pi$ versus disk thickness L from RK4 simulations at zero temperature, using cell size $a = 2.0$ nm. On the right-hand axis the unit conversion $\frac{\mu_0}{4\pi}\gamma M_s \approx 15.1$ GHz for Py has been applied. The vortex state is unstable below a minimum thickness, where no data are present.

$$f_G \approx \frac{1}{4}\left(\frac{\mu_0}{4\pi}\gamma M_s\right)\frac{L}{R} = \frac{\gamma k_F}{4\pi^2 pqLM_s}. \tag{10.56}$$

This then requires that the force constant follows the functional form,

$$k_F \approx \frac{\pi}{4}\mu_0 M_s^2 \frac{L^2}{R}. \tag{10.57}$$

Although we have not plotted the force constant in this way, it does appear in figure 5 of [6] which shows that this dependence is very roughly correct. Probably k_F should depend on a slightly different power of L. Disregarding that slight correction, the L^2/R-dependence has also been predicted by Guslienko *et al* [8] via the analytic two-vortices model, but with a numerical factor larger than $\pi/4$, *viz.*

$$k_F \approx 1.11\mu_0 M_s^2 \frac{L^2}{R}. \tag{10.58}$$

The results here indicate a somewhat weaker potential, as can be expected because the numerical simulations allow for a greater flexibility of the magnetization configuration, resulting in a smaller effective force constant.

Regardless of the precise size-dependence of the force constant, it is possible to see that the Thiele equation is very adequate in describing the gyrotropic frequencies, when those are compared to the force constants. If (10.44) has the

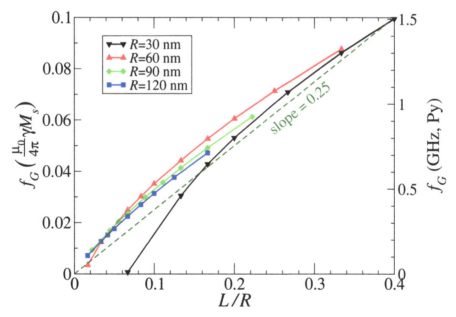

Figure 10.17. Results for gyrotropic frequency $f_G = \omega_G/2\pi$ versus disk aspect ratio L/R from RK4 simulations at zero temperature, using cell size $a = 2.0$ nm. On the right-hand axis the unit conversion $\frac{\mu_0}{4\pi}\gamma M_s \approx 15.1$ GHz for Py is used. The green dashed line suggests a rough slope of 0.25 in the relationship, except for the smaller disk radius.

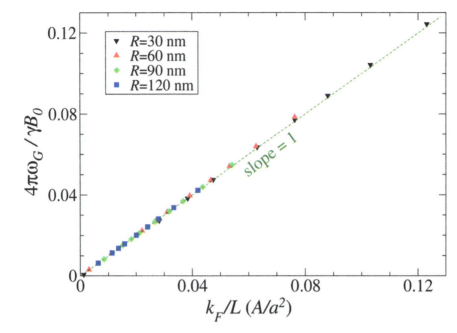

Figure 10.18. Results for gyrotropic frequency scaled by the time unit $t_0 = (\gamma B_0)^{-1}$ versus force constant divided by disk thickness, for different disk radii indicated (cell size is $a = 2.0$ nm). The green dashed line represents (10.59) from the Thiele equation. The results indicate complete consistency between the force constants obtained from static vortex calculations and the frequencies obtained via RK4 dynamics.

frequency combined with the time unit $t_0 = (\gamma B_0)^{-1}$, there results the equivalent relationship,

$$\frac{\omega_G}{\gamma B_0} = -\frac{1}{4\pi} \frac{k_F a^2}{qpLA}. \tag{10.59}$$

The frequency should be linearly proportional to k_F/L. This is indeed the case, as seen in a plot of $\omega_G/\gamma B_0$ versus k_F/L in figure 10.18, where the data for different disk sizes are plotted together. All the different system sizes collapse onto the same relationship, very close to linear with a slope consistent with the relation (10.59). Note that this result combines the *static* calculations of k_F, under the assumption of parabolic potentials, together with the *dynamic* calculations of ω_G. Any deviations from a linear relationship might be due to the variation of the potentials away from parabolic form, for vortex motion too far from the disk center, especially on smaller disks.

Exercise 10.3. The gyrotropic frequencies f_G in figures 10.16 and 10.17 become zero at small but non-zero disk thickness L, below which the vortex state is unstable. These minimum thicknesses depend on the disk radius. Give physical reasons why this should be so, and explain why the minimum thickness is greater for smaller R.

10.4 Thermalized vortex motion

The simulations just described at zero temperature required the vortex to be placed at an initial location away from the disk center. That is a somewhat artificial set up, although it is useful for obtaining the gyrotropic frequencies. It has been seen in micromagnetics studies [9] for finite temperature that a vortex will undergo *spontaneous gyrotropic motion*, without the application of an applied magnetic field. The thermal fluctuations cause a vortex to move away from the disk center (even if initiated there) by several nanometers, and once it is off center, the natural gyrotropic motion takes place, as modified by random thermal forces. It is somewhat surprising that the orbital motion can easily dominate over the stochastic fluctuations. Here we describe further simulations of this type and discuss the statistics of the vortex core location. The theoretical discussion of the statistics is based on an effective Lagrangian for the vortex core in the effective vortex potential.

10.4.1 The simulation method

For simulations that include temperature and time dynamics, the simulation method that most closely resembles the real physical situation is the magnetic Langevin equation approach. The basic ideas for solving general Langevin equations were discussed in chapter 5, and the particular case of spin systems was summarized in section 5.6. The effect of temperature is to produce stochastically fluctuating random magnetic induction fields that are allowed to affect the magnetization dynamics. The dynamics of (10.51) is now modified by including these extra fields, in their dimensionless form $\mathbf{b}_s(\tau)$, so that we are solving the following system of Langevin equations,

$$\frac{d\hat{\mathbf{m}}_i}{d\tau} = \hat{\mathbf{m}}_i \times \left(\mathbf{b}_i + \mathbf{b}_{s,i}(\tau)\right) - \alpha \hat{\mathbf{m}}_i \times \left[\hat{\mathbf{m}}_i \times \left(\mathbf{b}_i + \mathbf{b}_{s,i}(\tau)\right)\right]. \quad (10.60)$$

The stochastic fields $\mathbf{b}_{s,i}(\tau)$ vary from site to site and are added directly to the deterministic fields $\mathbf{b}_i$ for each site, obtained from the system Hamiltonian. Thermal equilibrium requires that the stochastic field components (labeled by $\lambda = x, y, z$) are correlated according to the fluctuation-dissipation (FD) theorem, which gives for the dimensionless fields

$$\left\langle b_{s,i}^\lambda(\tau) b_{s,i}^{\lambda'}(\tau') \right\rangle = 2\alpha \, \mathcal{T} \delta_{\lambda\lambda'} \delta(\tau - \tau'). \quad (10.61)$$

This is equivalent to expression (5.138), using a dimensionless temperature variable,

$$\mathcal{T} = \frac{k_B T}{J_{\text{cell}}} = \frac{k_B T}{2AL}. \quad (10.62)$$

At each cell of the micromagnetics grid, the individual magnetic dipole is of strength $\mu_{\text{cell}} = La^2 M_s$, equivalent to the expression $\mu = \gamma S$ in an atomic model. Thus we could also write the FD theorem in fully dimensional form,

$$\gamma\mu_{\text{cell}}\langle B_{s,i}^\lambda(t) B_{s,i}^{\lambda'}(t')\rangle = 2\alpha\, k_B T\, \delta_{\lambda\lambda'}\, \delta(t-t'). \tag{10.63}$$

These relations are then used to find the statistics of the random magnetic induction fields.

The Langevin equations can be solved, as explained in section 5.6, by the second-order Heun method. This requires that the stochastic fields integrated over a time step ($\Delta\tau$) are replaced by random numbers of an appropriate distribution. Numerically, we carry out the replacement,

$$\int_{\tau_n}^{\tau_n+\Delta\tau} d\tau\, b_{s,i}^\lambda(\tau) \longrightarrow \sigma_s w_{n,i}^\lambda. \tag{10.64}$$

Here σ_s is the variance, which is determined from the FD theorem, and each number $w_{n,i}^\lambda$ is a random deviate of zero mean and unit variance, as might be provided from a random number generator. The actual distribution of random numbers is not strongly confined by this requirement, they only must have zero mean and the correct variance. The variance is determined by integration and squaring, then averaging over the ensemble,

$$\sigma_s^2 = \left\langle \left(\int_{\tau_n}^{\tau_n+\Delta\tau} d\tau\, b_{s,i}^\lambda(\tau)\right)^2\right\rangle = \int_{\tau_n}^{\tau_n+\Delta\tau} d\tau \int_{\tau_n}^{\tau_n+\Delta\tau} d\tau'\, \langle b_{s,i}^\lambda(\tau) b_{s,i}^{\lambda'}(\tau')\rangle. \tag{10.65}$$

Use of the FD theorem in the form (10.61) here leads easily to the result,

$$\sigma_s = \sqrt{2\alpha\mathcal{T}\Delta\tau}. \tag{10.66}$$

Then, longer time steps or higher temperature increases the variance, and tends to amplify the typical sizes of the stochastic impulses. The products $\sigma_s w_{n,i}^\lambda$ for $\lambda = x, y, z$, are used in the Heun solution, where each $w_{n,i}^\lambda$ is most simply taken as a uniform random number from -0.5 to $+0.5$. Because that range of random numbers already has a variance of $1/\sqrt{12}$, see (5.94), one can correct it by multiplying $w_{n,i}^\lambda$ by a factor of $\sqrt{12}$ to change its variance to unity. The rest of the Heun scheme, viewed as a predictor–corrector method combining an Euler predictor stage with a trapezoid rule corrector stage, is described in section 5.6.1.

For some example simulations here, we again use Py parameters, including exchange stiffness $A = 13$ pJ m^{-1}. The unit by which thermal energy is scaled (equal to J_{cell}) depends on the disk thickness. For a thickness $L = 10$ nm, one has $J_{\text{cell}} = 2AL = 260$ zJ, whereas a room temperature of 300 K corresponds to a thermal energy of $k_B T = 4.14$ zJ. Therefore even room temperature is very low, in the sense that $\mathcal{T} = k_B T / J_{\text{cell}} = 0.016$ for the $L = 10$ nm thickness. A thinner sample has a higher scaled temperature $\mathcal{T}$; it would be more greatly affected by thermal fluctuations for a given fixed temperature.

10.4.2 Spontaneous gyrotropic vortex motion in stochastic dynamics

One can perform simulations of the stochastic dynamics, starting with a vortex at the center of the disk, as found from its energy minimized state. Here we show some

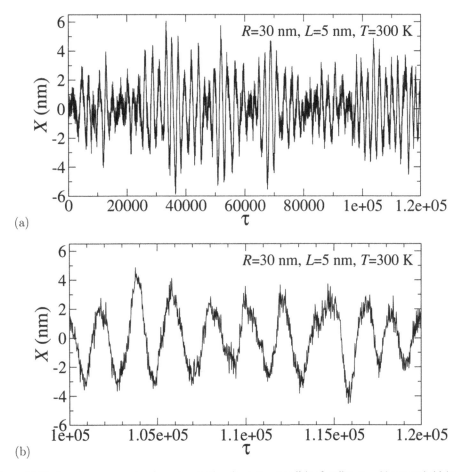

Figure 10.19. Spontaneous gyrotropic vortex motion for a Py nanodisk of radius $R = 30$ nm and thickness $L = 5.0$ nm at temperature 300 K (dimensionless temperature $\mathcal{T} = 0.032$). In (a) the motion on the X-axis only is shown over a long time interval, starting from a centered vortex initial condition, $\mathbf{X}_0 = (0, 0)$. In (b) the motion at the end of the simulation is displayed; the period is approximately $\tau_G = 2250$.

typical evolutions of the magnetization dynamics, including the thermal fluctuations.

The first case is a disk of thickness $L = 5.0$ nm and a relatively small radius, $R = 30$ nm. This corresponds to the scaled temperature variable $\mathcal{T} = 0.032$. If the temperature were zero, a vortex at the disk center would remain there, being the energy minimizing state. Once the temperature effects are included, the vortex can be affected by some fluctuations and be pushed from the center. This causes it to spontaneously begin a gyrotropic motion. This natural tendency for the vortex to oscillate is rather impressive, in that it requires no outside applied field or external energy input.

Some results can be seen in figure 10.19, where one component of its location, $X(\tau)$ is plotted. $Y(\tau)$ is similar, but shifted in-phase by a quarter of a period. It is possible to follow the motion for dozens of periods and, if desired, extract frequency

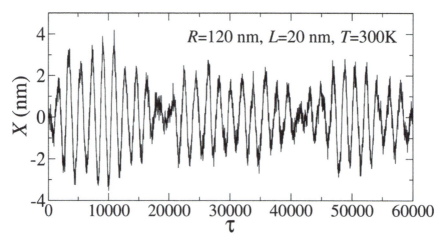

Figure 10.20. Spontaneous gyrotropic vortex motion for a Py nanodisk of radius $R = 120$ nm and thickness $L = 20$ nm at temperature 300 K (dimensionless temperature $\mathcal{T} = 0.008$), starting from the disk center. The average period is approximately $\tau_G = 1870$ for the 32 revolutions in this simulation.

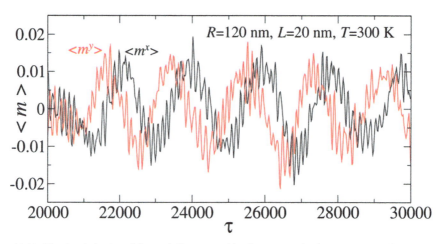

Figure 10.21. The time behavior of the spatially averaged in-plane magnetization components for spontaneous gyrotropic vortex motion for a Py nanodisk of radius $R = 120$ nm and thickness $L = 20$ nm at temperature 300 K. Note the high frequency spin wave oscillations that are excited on top of the gyrotropic response.

information via Fourier transforms. An average of 24 of the smoother rotations gives an estimate of the period as $\tau_G = 2250$, resulting in a real period of $T_G = 1.68$ ns and a frequency of $f_G = 0.594$ GHz. This is consistent with what is found in a power spectrum for the same data. The amplitude of these motions is on the order of 2–6 nm, while at the same time, there are ±15% variations caused in the magnetization components.

Another example is shown in figure 10.20, for a system with four times the thickness and also four times the radius ($R = 120$ nm, $L = 20$ nm). In this case the motion has a somewhat shorter period, estimated as $\tau_G = 1870$, leading to $f_G = 0.713$ GHz. Another interesting feature is present, in that there are points in

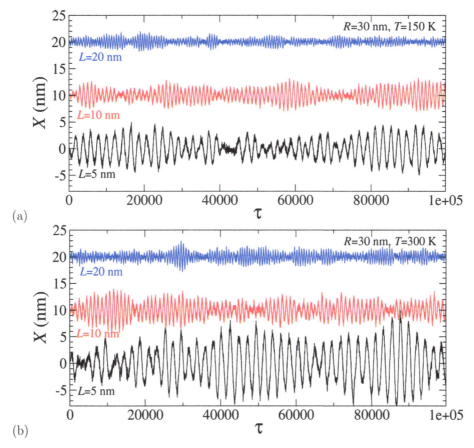

Figure 10.22. Spontaneous gyrotropic vortex motions (of one coordinate $X(\tau)$) for Py nanodisks of radius $R = 30$ nm and indicated thicknesses at temperatures (a) 150 K and (b) 300 K. The vortex was initially relaxed at the disk center. Curves are shifted vertically for clarity. The thicker disks have a stronger restoring force and therefore smaller vortex deviations from the center. The dependence of gyrotropic period on disk thickness is also apparent.

time where the amplitude is greatly diminished. That appears to be caused by an interference of two close frequencies forming beats in the motion. It is probably caused by a superposition of spin wave modes on top of the vortex structure, see [10] where a doublet of spin waves is discussed, and [6], where a power spectrum is used to analyze the data. Some simple evidence of spin wave effects, however, can be seen by looking closely at the average system magnetization components for this simulation, as in figure 10.21. It is noted that both $\langle m^x \rangle$ and $\langle m^y \rangle$ contain high frequency oscillations on top of the low frequency gyrotropic response.

The scale of the thermal fluctuations depends on the temperature, and on the depth of the effective potential experienced by the vortex. With the potential depending on the nanodisk size, it is interesting to compare the scale of the fluctuations for different disk sizes and temperatures. Examples of this for $R = 30$ nm at several disk thicknesses are shown in figure 10.22, at the two temperatures of 150 K and 300 K, for

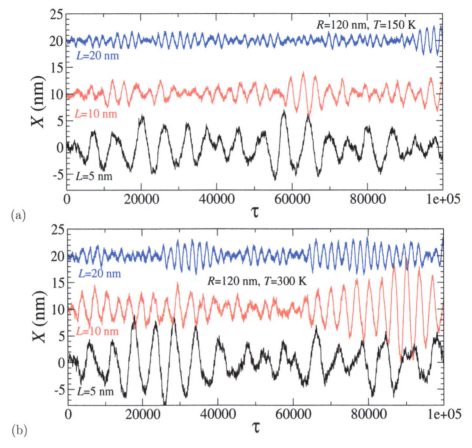

Figure 10.23. Spontaneous gyrotropic vortex motions (of one coordinate $X(\tau)$) for Py nanodisks of radius $R = 120$ nm and indicated thicknesses at temperatures (a) 150 K and (b) 300 K, with the vortex initiated at the disk center. Curves are shifted vertically for clarity. The gyrotropic periods can be seen to be larger here than for the $R = 30$ nm disks in figure 10.22.

Py material parameters. The vortex was initially relaxed at the disk center, $X(0) = Y(0) = 0$. Although only $X(\tau)$ is displayed, $Y(\tau)$ is very similar but with a quarter period phase shift. The gyrotropic motion is spontaneous as seen earlier. The period is larger for thinner disks; the frequency increases with disk thickness. This can be expected based on the stronger restoring potential for larger L. The other obvious feature is that the *amplitude* of the motion is larger for thinner disks, which can again be attributed to their weaker force constant k_F. Of course, with a more shallow restoring potential, there will be greater freedom of the vortex to move away from the disk center as it is affected by thermal fluctuations. It is also clear in figure 10.22 that the higher temperature leads to greater amplitude of the spontaneous motion, as more thermal energy is shared into that degree of freedom.

The corresponding spontaneous gyrotropic motions for $R = 120$ nm Py disks are shown in figure 10.23, again for the two temperatures 150 K and 300 K. The primary difference that can be seen is the lower gyrotropic frequencies compared to the $R = 30$ nm disks. However, another feature apparent in all these cases is that the amplitude drifts up and down and periodically becomes rather small. This again appears to be due to an interference of some nearby spin wave modes that are excited together with the vortex motion. Aside from that, it can be interesting to make a statistical analysis of the distribution of vortex positions.

10.4.3 Hamiltonian dynamics and statistics of vortex core position

It is possible to analyze the statistical mechanics of the vortex core position $\mathbf{X}$ and velocity $\mathbf{V} = \dot{\mathbf{X}}$, starting from an effective Lagrangian for the Thiele equation, (9.92). In chapter 9 it was shown that the Lagrangian is equivalent to that for a charged particle in a magnetic induction $\mathbf{B} = -\mathbf{G}$, with electric charge e replaced by σ, so that the vector potential is

$$\mathcal{A} = \frac{1}{2}\mathbf{B} \times \mathbf{X} \longrightarrow -\frac{1}{2}\mathbf{G} \times \mathbf{X}. \tag{10.67}$$

The primary features of the statistics can be obtained by assuming the intrinsic vortex mass $\mathsf{m} = 0$, then the Hamiltonian contains only a parabolic potential,

$$H = U(\mathbf{X}) = \frac{1}{2}k_F \mathbf{X}^2. \tag{10.68}$$

By (9.91), the vortex canonical momentum is given by

$$\mathbf{P} = \mathsf{m}\mathbf{V} - \frac{1}{2}\sigma \mathbf{G} \times \mathbf{X} \implies \mathbf{P} = -\frac{1}{2}\sigma \mathbf{G} \times \mathbf{X}. \tag{10.69}$$

This is an interesting limiting case, because by letting the intrinsic mass go to zero, the usual Hamiltonian dynamics to be derived from H, which does not depend on $\mathbf{P}$, does not lead directly to the Thiele equation. One can verify, however, that the Lagrangian dynamics works out correctly, from the Lagrangian

$$L = \mathbf{P} \cdot \dot{\mathbf{X}} - H = -\frac{1}{2}\sigma(\mathbf{G} \times \mathbf{X}) \cdot \dot{\mathbf{X}} - \frac{1}{2}k_F \mathbf{X}^2. \tag{10.70}$$

This can be written in 2D components as

$$L = -\frac{1}{2}\sigma G(X\dot{Y} - Y\dot{X}) - \frac{1}{2}k_F(X^2 + Y^2) \tag{10.71}$$

where $G = G_z = 2\pi pq$. The Euler–Lagrange equations derived from this Lagrangian are

$$\frac{\partial L}{\partial X} - \frac{\mathrm{d}}{\mathrm{d}t}\frac{\partial L}{\partial \dot{X}} = -k_F X - \sigma G \dot{Y} = 0 \tag{10.72a}$$

$$\frac{\partial L}{\partial Y} - \frac{d}{dt}\frac{\partial L}{\partial \dot{Y}} = -k_F Y + \sigma G \dot{X} = 0. \tag{10.72b}$$

With the force being $\mathbf{F} = -k_F \mathbf{X} = (-k_F X, -k_F Y)$ and gyrovector term $\mathbf{G} \times \mathbf{V} = G\hat{z} \times \dot{\mathbf{X}} = (-G\dot{Y}, G\dot{X})$, these are equivalent to the Thiele equation,

$$\mathbf{F} + \sigma \mathbf{G} \times \mathbf{V} = 0. \tag{10.73}$$

This verifies that the Lagrangian is a valid one.

Statistical mechanics is usually based on knowledge of the Hamiltonian. The Hamiltonian here, depending only on the potential $U(\mathbf{X})$, gives the Hamiltonian equations of motion,

$$\dot{\mathbf{X}} = \frac{\partial H}{\partial \mathbf{P}} = 0 \tag{10.74a}$$

$$\dot{\mathbf{P}} = -\frac{\partial H}{\partial \mathbf{X}} = -k_F \mathbf{X}. \tag{10.74b}$$

This is clearly inconsistent with the Thiele equation and the actual gyrotropic dynamics. The difficulty can be traced to the fact that the position coordinates $\mathbf{X} = (X, Y)$ and canonical momentum coordinates $\mathbf{P} = \frac{1}{2}\sigma G(Y, -X)$ are proportional to each other and are therefore redundant. However, all the components of coordinate and momentum should be considered as linearly independent in the Hamiltonian formalism, and all should appear in the definition of H. This is required, because the gyrotropic motion conserves no component of $\mathbf{X}$ or $\mathbf{P}$. A way out of this dilemma is to reformulate H as equal parts that act as potential (depending on $\mathbf{X}^2$) and kinetic (depending on $\mathbf{P}^2$) terms. This is true even if one works in another choice of gauge, see [11] where the Landau gauge is applied to quantum effects in this type of problem. As a result, with $\mathbf{X}^2 = (2\mathbf{P}/\sigma G)^2$, we divide the original Hamiltonian into equivalent kinetic and potential parts, as

$$H = \frac{1}{2}k_F \mathbf{X}^2 = \frac{1}{4}k_F \mathbf{X}^2 + \frac{1}{4}k_F \left(\frac{2\mathbf{P}}{\sigma G}\right)^2. \tag{10.75}$$

Now consider the dynamics that results, say, for one Cartesian axis, which is

$$\dot{X} = \frac{\partial H}{\partial P_x} = \frac{2k_F}{\sigma^2 G^2} P_x \tag{10.76a}$$

$$\dot{P}_x = -\frac{\partial H}{\partial P_x} = -\frac{1}{2}k_F X. \tag{10.76b}$$

These do become the Thiele equations in the form (10.72), once the momentum is replaced by its equivalent in terms of position, (10.69). If instead the momentum is eliminated, they lead to a second-order equation for a simple harmonic oscillator,

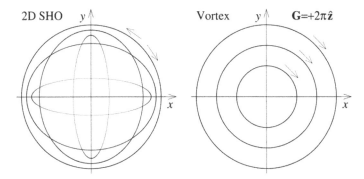

Figure 10.24. Comparison of the possible circular and elliptical orbits in a 2D simple harmonic oscillator (2D SHO) with the purely circular orbits of a vortex in a circularly symmetric restoring potential. For the 2D SHO, any elliptical shape and orbital direction is allowed. For a vortex with positive gyrovector (along $\hat{z}$), only circular orbits in the clockwise sense are possible, see (10.43).

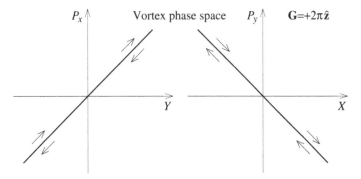

Figure 10.25. Sketch of the possible vortex orbits as viewed in slices of the phase space. The constraint (10.69) forces the position and momentum components to be related by $P_x = \tfrac{1}{2}GY$ and $P_y = -\tfrac{1}{2}GX$, forcing the reduction of a 4D phases space to only two effective dimensions.

$$\ddot{X} = -\frac{k_F^2}{\sigma^2 G^2}X. \qquad (10.77)$$

The variations of H with respect to Y and P_y also reproduce the Thiele equation, and this same 1D simple harmonic oscillator equation for $\ddot{Y}$, with frequency

$$\omega_G = -\frac{k_F}{\sigma G}. \qquad (10.78)$$

The negative sign is chosen to be consistent with the earlier result, (10.43).

The Hamiltonian (10.75) is actually that for a 2D simple harmonic oscillator with an effective spring constant $k_{SHO} = \tfrac{1}{2}k_F$ and an effective mass $m_{SHO} = \frac{1}{2k_F}\sigma^2 G^2$. The natural frequency can be calculated directly from the standard oscillator result, using the negative root,

$$\omega_G = -\sqrt{\frac{k_{SHO}}{m_{SHO}}} = -\sqrt{\frac{\frac{1}{2}k_F}{\frac{1}{2k_F}\sigma^2 G^2}} = -\frac{k_F}{\sigma G}, \qquad (10.79)$$

which is necessarily the same as obtained from the equations for the components.

Equipartition in thermalized vortex motion
A 2D harmonic oscillator normally has the possibility of linear motions along either coordinate axis, as well as a full range of elliptical motions all the way to the limit of circular motions. Typical orbits are compared to those for the vortex problem in figure 10.24. The 4D phase space has two coordinates and their respective canonical momenta. In the vortex problem, however, relation (10.69) represents a constraint between the coordinates and momenta, that reduces the phase space to 2D, see figure 10.25. The nature of the constraint is that it allows for only the circular motions of vortices, as long as the potential is circularly symmetric. Therefore, for thermal equilibrium, for each *independent* coordinate or momentum variable there is an average thermal energy of $\frac{1}{2}k_B T$ available. For the two independent degrees of freedom here, we have

$$\langle H(\mathbf{X}, \mathbf{P}) \rangle = k_B T. \qquad (10.80)$$

The Hamiltonian is expressed in (10.68) as purely potential energy. This makes it possible to calculate the rms vortex displacement from the disk center, on each coordinate axis, by

$$\left\langle \frac{1}{2}k_F X^2 \right\rangle = \left\langle \frac{1}{2}k_F Y^2 \right\rangle = \frac{1}{2}k_B T. \qquad (10.81)$$

From this the resulting rms displacements are obtained,

$$X_{rms}^2 = Y_{rms}^2 = \frac{k_B T}{k_F}. \qquad (10.82)$$

In addition to this, the net rms displacement in an arbitrary direction is found from

$$r_{rms}^2 = \langle X^2 + Y^2 \rangle = \langle r^2 \rangle = \frac{2k_B T}{k_F}. \qquad (10.83)$$

Extended vortex simulations up to times $\tau = 2.5 \times 10^5$ have been used to test this relation on different disk sizes, see figure 10.26. The results correspond to a set of different force constants. Even so, the average squared displacements generally increase linearly with the temperature, although together with considerable statistical fluctuations, especially for systems with the smallest force constants.

The probability distribution for vortex location is obtained directly from a Boltzmann factor $\exp(-\beta H)$, with appropriate normalization factor, giving a Gaussian distribution on each axis. The likelihood to find the vortex in some area element $dX\, dY$ is

$$p(X, Y) dX\, dY = \frac{\beta k_F}{2\pi} e^{-\frac{1}{2}\beta k_F (X^2 + Y^2)} dX\, dY. \qquad (10.84)$$

Distribution (10.84) can be seen as equivalent to the circularly symmetric radial distribution,

$$p(r)\mathrm{d}r = \beta k_F r e^{-\frac{1}{2}\beta k_F r^2} \mathrm{d}r. \qquad (10.85)$$

The distribution has a maximum probability at the radius

$$r_{\max} = \frac{r_{\mathrm{rms}}}{\sqrt{2}} = \sqrt{\frac{k_B T}{k_F}} \qquad (10.86)$$

and that maximum value is

$$p(r_{\max}) = \sqrt{\beta k_F}\, e^{-1/2} = r_{\max}^{-1} e^{-1/2}. \qquad (10.87)$$

Based on long simulations (to $\tau = 2.5 \times 10^5$) of a single vortex in different sized disks, it is possible to calculate a histogram of vortex radial positions, and compare that with this predicted distribution. One must use simulations of sufficient duration in τ so that the vortex makes a large number of revolutions over which averages are made. As seen in figures 10.22 and 10.23, there are fluctuations of considerable magnitude. There are time intervals over which the motion exhibits smooth gyrotropic motion, interrupted by intervals of a somehow

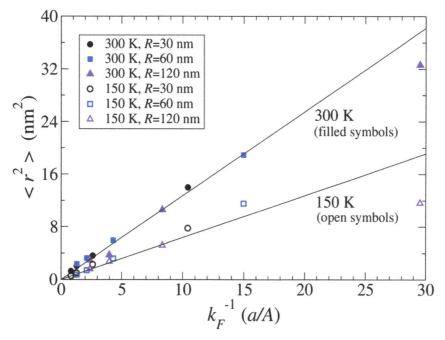

Figure 10.26. Average squared displacement of the vortex core from the disk center, as a function of the inverse force constant. Symbols come from the simulations out to dimensionless time $\tau = 2.5 \times 10^5$ for indicated disk sizes. Solid lines are the theory expression (10.83) derived from equipartition. Parameters for Py have been used, with cell constant $a = 2.0$ nm. Force constants were estimated from the static relaxed vortex structures, as summarized in figures 10.10 and 10.11.

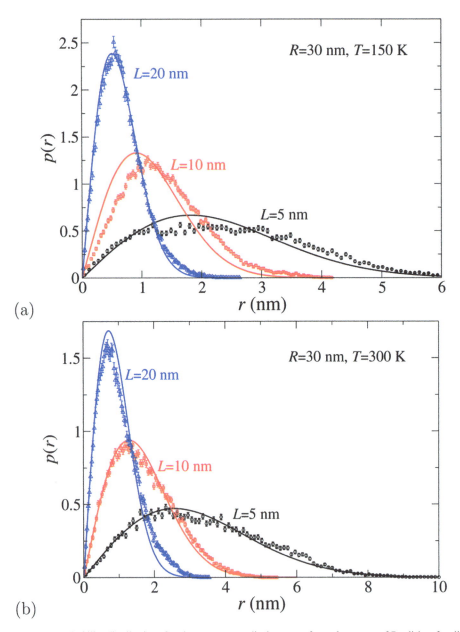

Figure 10.27. Probability distributions for the vortex core displacement from the center of Py disks of radius $R = 30$ nm, for temperatures (a) 150 K and (b) 300 K, from simulations (symbols) to $\tau = 2.5 \times 10^5$ and from the equipartition theory (solid curves), equation (10.85). The static force constants as in figures 10.11 and 10.10 were used to produce the theory curves.

impeded motion of smaller amplitude and possibly a strong phase shift. For disks of radius $R = 30$ nm, some distributions at two temperatures are shown in figure 10.27. Again, the parameters are those for Py material, where a temperature of

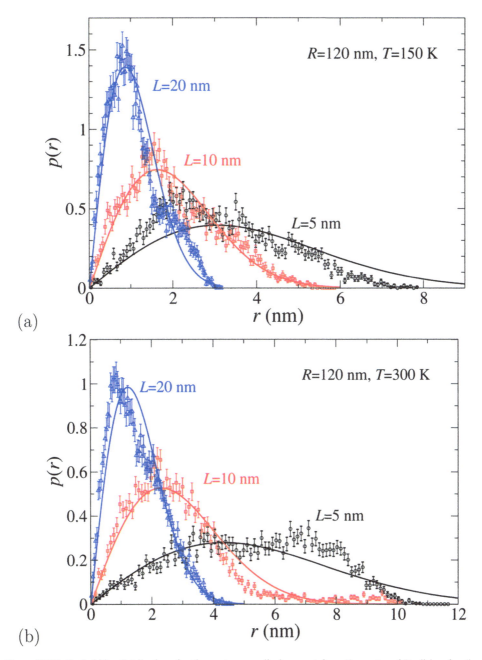

Figure 10.28. Probability distributions for the vortex core displacement from the center of Py disks of radius $R = 120$ nm, for temperatures (a) 150 K and (b) 300 K, from simulations (symbols) to $\tau = 2.5 \times 10^5$ and from the equipartition theory (solid curves), (10.85), using static force constants.

300 K corresponds to $k_B T = 0.1592 Aa$, using exchange stiffness $A = 13$ pJ m^{-1} for Py, and cell size $a = 2.0$ nm. For this relatively small disk radius, the simulation data fall reasonably close to the Boltzmann predictions (10.85), shown as solid curves in figure 10.27.

Data are also shown in figure 10.28 for disks of larger radius, $R = 120$ nm, where the fluctuations tend to be noticeably greater. This can be attributed to the longer orbital periods in larger disks, thereby reducing the total number of revolutions that were averaged. This means that it becomes more difficult to eliminate the effects of the somewhat erratic changes in orbital amplitude that occur randomly. Simulations over much longer time sequences should work to reduce such statistical errors.

In thinner disks, the restoring potential is weaker, giving lower gyrotropic frequencies. Then, the number of revolutions will also be fewer for a chosen fixed time, leading again to greater fluctuations in the numerical measurement of the distribution. Besides this, the weaker (stronger) force constant in thinner (thicker) disks also implies a larger (smaller) average amplitude of spontaneous gyrotropic motion. This is the most obvious feature of the probability distributions found, and of course it is also apparent in the relation (10.86) for the location of the maximum, $r_{\max} = (\beta k_F)^{-1/2}$, which additionally exhibits the greater amplitude with increasing temperature. Some deviations from these expected results might be present for thinner disks if the larger amplitude moves the vortex out of the region where the potential is parabolic. Even so, the general shape of the distribution is reasonably well described.

In the analysis here, we have a circularly symmetric potential, hence the distribution has this same symmetry. The more general case of disks with elliptical shape [12] leads to different force constants along the two principal axes, and as a result, a probability distribution with non-equal rms displacements on those axes. The radial distribution is then only found to be circularly symmetric if an appropriate radial coordinate is used. Still, the probability distribution can be described reasonably well as derived from a Boltzmann factor, see [12] for further details.

Both the analytic and simulation results unequivocally demonstrate how the *disorganized* random thermal fluctuations are able to produce an *organized* and partially coherent gyrotropic periodic motion. The gyrotropic motion is a naturally organized dynamic state of the vortex, that can be directly excited by thermal disorder. This is an impressive result. It is rather similar to the presence of linear modes such as spin waves of a partial coherence that can also be excited by thermal disorder. The persistent vortex motion clearly makes a basis for the design of magnetic oscillators, that may require only a limited energy input per cycle to maintain the oscillations.

Exercise 10.4. The Boltzmann probability distribution $p(X, Y)$ in (10.84) or in circularly symmetric form $p(r)$ in (10.85) were confirmed in the simulations, but such measurements in experiments may be difficult. Instead, consider the distribution of the total in-plane magnetic moment $\mathbf{M} = (M_x, M_y)$ of the nanodisk, $p(M_x, M_y)$. Assuming a perfect vortex structure for the in-plane components, such as (8.7), what are the predictions for probability

> distributions such as $p(M_x)$ and $p(M_{xy})$, where $M_{xy} = \sqrt{M_x^2 + M_y^2}$? One could expect that the width of this distribution will be affected by spin waves. This effect should be ignored in leading order.

An effective vortex mass induced by the potential

It was noted earlier that the 4D space of two coordinates and two momenta is reduced to 2D due to the constraint (9.91). This implies the direct connection (10.41) between the vortex velocity **V** and its position **X** via the vector angular frequency ω_G. Then the magnitudes are connected by the relation, $|\mathbf{V}| = |\omega_G| r$, where r is the orbital radius and $|\omega_G| = k_F/\sigma|G|$. This can be used to transform the rms radius (10.83) into a rotational rms velocity,

$$V_{\text{rms}} = |\omega_G| r_{\text{rms}} = \frac{k_F}{\sigma|G|} \sqrt{\frac{2k_B T}{k_F}} = \frac{\sqrt{2k_F k_B T}}{\sigma|G|}. \tag{10.88}$$

Recalling that $\sigma = LM_s/\gamma$, the rms velocity is proportional to $\sqrt{k_F}/L$, which may increase weakly with the disk thickness. More importantly, this gives the transformation of the radial probability distribution $p(r)dr$ into the equivalent vortex speed distribution $f(V)dV$, by equating the likelihoods,

$$p(r)dr = f(V)dV \tag{10.89}$$

and using the relation, $V = |\omega_G| r$. Then there results the vortex speed distribution,

$$f(V) = \frac{p(r)}{dV/dr} = \frac{p(V/|\omega_G|)}{|\omega_G|} = \frac{\beta k_F}{|\omega_G|} \frac{V}{|\omega_G|} e^{-\frac{1}{2}\beta k_F \frac{V^2}{\omega_G^2}}. \tag{10.90}$$

The factor in the exponent suggests the definition of an effective gyrotropic dynamical mass appearing there, that is in some sense induced by the potential,

$$m_G = \frac{k_F}{\omega_G^2} = \frac{\sigma^2 G^2}{k_F}. \tag{10.91}$$

The potential can be said to be responsible for this mass, because of its dependence on the force constant. This puts the distribution into the form,

$$f(V) = \beta m_G V e^{-\frac{1}{2}\beta m_G V^2}, \tag{10.92}$$

which has the kinetic energy term $\frac{1}{2} m_G V^2$ combined with $-\beta$ in the Boltzmann factor. The mass that is implied here is extremely small in physical units. It can be estimated from the earlier approximate result (10.57) for the force constants. Combining that expression with the definition (10.91) and including the definition of σ, we have some interesting cancellations,

$$m_{\rm G} \approx \frac{(LM_{\rm s}/\gamma)^2(2\pi pq)^2}{\frac{\pi}{4}\mu_0 M_{\rm s}^2 L^2 \big/ R} = \frac{16\pi R}{\mu_0 \gamma^2} \approx \left(1.29 \times 10^{-15} \text{ kg m}^{-1}\right) \times R. \quad (10.93)$$

With $\mu_0 = 4\pi \times 10^{-7}$ N A^{-2} and $\gamma = 1.76 \times 10^{11}$ Am (Ns)$^{-1}$, the constant present was obtained from the value $1/(\mu_0 \gamma^2) \approx 2.57 \times 10^{-17}$ kg m^{-1}. The result is independent of both the disk thickness and saturation magnetization, leaving only a dependence on radius in the approximations used. At a disk radius of $R = 100$ nm for example, the mass is estimated as $m_{\rm G} \approx 1.3 \times 10^{-22}$ kg, an extremely small value independent of the material. Although it is an apparent mass which is a consequence of moving in the disk's potential, it plays the role of describing the vortex speed distribution.

Bibliography

[1] Usov N A and Peschany S E 1993 Magnetization curling in a fine cylindrical particle *J. Magn. Magn. Mater.* **118** L290

[2] Wysin G M 2010 Vortex-in-nanodot potentials in thin circular magnetic dots *J. Phys.: Condens. Matter* **22** 376002

[3] Cowburn R P, Koltsov D K, Adeyeye A O, Welland M E and Tricker D M 1999 Single-domain circular nanomagnets *Phys. Rev. Lett.* **83** 1042

[4] Metlov K L and Guslienko K Y 2002 Stability of magnetic vortex in soft magnetic nano-sized circular cylinder *J. Magn. Magn. Mater.* **242-5** 1015

[5] Wysin G M 1996 Magnetic vortex mass in two-dimensional easy-plane magnets *Phys. Rev. B* **54** 15156

[6] Wysin G M and Figueiredo W 2012 Thermal vortex dynamics in thin circular ferromagnetic nanodisks *Phys. Rev. B* **86** 104421

[7] Yu Guslienko K, Novosad V, Otani Y, Skima H and Fukamichi K 2001 Field evolution of magnetic vortex state in ferromagnetic disks *Appl. Phys. Lett.* **78** 3848

[8] Yu Guslienko K, Ivanov B A, Novosad V, Otani Y, Skima H and Fukamichi K 2002 Eigenfrequencies of vortex state excitations in magnetic submicron-size disks *J. Appl. Phys.* **91** 8037

[9] Machado T S, Rappoport T G and Sampaio L C 2012 Vortex core magnetization dynamics induced by thermal excitation *Appl. Phys. Lett.* **100** 112404

[10] Ivanov B A, Avanesyan G G, Khvalkovskiy A V, Kulagin N E, Zaspel C E and Zvezdin K A 2010 Non-Newtonian dynamics of the fast motion of a magnetic vortex *JETP Lett.* **91** 178

[11] Ivanov B A, Galkina E G and Galkin A Y 2010 Quantum dynamics of vortices in small magnetic particles *Low Temp. Phys.* **36** 747

[12] Wysin G M 2015 Vortex dynamics in thin elliptic ferromagnetic nanodisks *Low Temp. Phys.* **41** 1009

Magnetic Excitations and Geometric Confinement
Theory and simulations
Gary Matthew Wysin

Chapter 11

Spin ices and geometric frustration

Certain types of magnetic crystals have a very distinct geometric structure, combined with a strong uniaxial anisotropy on the individual magnetic ions. An example in 3D is a spin ice on a tetragonal lattice. A uniaxial anisotropy favors two primary directions for the dipole on a site, whereas the dipolar interactions with nearest neighbor sites place a constraint on how dipoles point into or out of a vertex on the lattice. The resulting 'two-in two-out' ice rule controls the low energy states. The low energy states cannot minimize all nearest dipole–dipole interactions simultaneously, so the system is said to exhibit *frustration*. The higher energy states involve deviations from the ice rule, producing magnetic monopoles connected by strings. Similar novel excitations appear in artificial spin ice models in 2D. In this chapter these kinds of ordered or engineered magnetic materials with frustrated dynamics are described.

11.1 Spin ice and frustrated states

In the previous chapter the dynamics of vortices in magnetic nanodisks was considered. With circular symmetry, the ground state of a disk of appropriate size could be a vortex, which is seen to have interesting physical properties and especially dynamics. Typically those nanodisks could be grown as islands on a non-magnetic substrate, in large numbers, but sufficiently separated so that they do not interact magnetically among themselves.

In this chapter we primarily consider a similar type of multi-island system, but in a situation where the islands are not circularly symmetric, but rather shaped liked a race track or some elliptical shape. They are close enough together that the interactions among them are very important. This is an array of interacting mesoscopic magnetic particles. With appropriate shapes and arrangements, the system is called *artificial spin ice*, because such an engineered system mimics the physics of atomic dipoles in 3D spin ice crystals. In 3D atomic spin ice, the lattice structure is such that the magnetic dipoles interact strongly with their neighbors, and

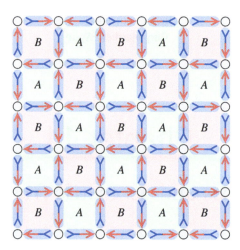

Figure 11.1. A ground state configuration of square lattice artificial spin ice, composed from islands of race-track shape. They are drawn in blue to indicate low energy for the ground state. Each island has a magnetic dipole moment μ_n that strongly tends to align on the long axis of its island, becoming Ising-like. In a ground state, all vertices (junctions of four neighboring islands, shown as gray dots) obey the two-in two-out ice rule, as type I vertices, see figure 11.2. There are two sublattices of cells marked A and B. All dipoles alternate direction from site to site, along any row or column. In neighboring cells the sense of rotation of the dipoles changes between clockwise and counterclockwise. A second ground state degenerate with this one is obtained by reversing all the dipoles.

with fields of other atoms, causing dipolar interactions as well as locally anisotropic interactions. These can be modeled by the usual long-range dipolar interaction together with single ion uniaxial anisotropy terms.

One important 3D lattice in spin ice materials is the pyrochlore lattice, where neighboring dipoles μ_n meet in groups of four at vertices. Another, of course, is the natural 3D lattice of water ice, from which the name for magnetic *spin ice* is derived. The same is true for 2D square lattice *artificial spin ice*, first studied by Wang *et al* [1], whose ground state is depicted in figure 11.1. This ground state was observed in experiment in 2011 by Morgan *et al* [2]. The uniaxial anisotropy tends to make each dipole point towards or away from a selected vertex, with equivalent energies. If the uniaxial anisotropy is very strong compared to thermal energy $k_B T$, only those two in/out states are available, thermodynamically speaking, to each dipole. This would give a total of 2^4 possible configurations relative to the one vertex. Once the dipolar interactions are included, however, it is found that the lowest energy configurations require two of the dipoles to point inward towards the vertex and two should point outward away from the vertex. This is known as the two-in two-out ice rule. Out of the 16 vertex configurations, there will be six that satisfy this rule, as well as eight configurations where three point in and one points out, and two configurations where all the dipoles point in or out. These four different types of vertices are depicted in figure 11.2. The reason for the two-in two-out rule is very simple: the north/south poles of the dipoles prefer to arrange themselves (considered as charges) so there is not any excess local density of north or south poles at a vertex. The net 'pole charge' at a vertex should be as close to zero as possible, for the minimum energy. The usual discussion of the possible states of a spin ice assumes the discretization of the states of the individual dipoles. They primarily can only align or anti-align with the axis of the uniaxial anisotropy term, which has been assumed to be rather strong compared to available microscopic thermal energy per dipole. Thus, the dipoles that in principle are 3D vectors, are thermodynamically reduced to an Ising-like phase space, with a single dimension and two allowed states for each.

In an artificial 2D spin ice, the situation is more at the control of the experimentalists. The uniaxial anisotropy can be built in with a desired strength, to a certain extent, by controlling the shapes of the 2D islands. Typically all that is necessary is to form elongated islands, whose shape anisotropy will exhibit itself as primarily a uniaxial anisotropy. The dipolar interactions among the dipoles then comes in naturally.

An artificial spin ice could be built on other lattices such as triangular and Kagomé lattices in 2D. In triangular lattice ice, there is a three-in three-out rule, because each vertex is the junction of six dipoles. In Kagomé ice, a vertex is the junction of an odd number of dipoles; there is always a monopole charge present and no ice rule. Even so, these different lattices have a well-ordered ground state structure, controlled by some different rules at the vertices. With strong shape anisotropy of the islands, the simplest model is still that each dipole becomes an Ising-like degree of freedom, unless the temperature is high.

11.1.1 About frustration

For either real or artificial spin ice, the system is assumed to be dominated by the uniaxial anisotropy energy. With that being high compared to thermal energy, the dipoles align parallel to the long axes of the islands, with a choice of two energetically equivalent directions.

The dipoles also interact among themselves via the magnetostatic dipole–dipole interaction. The dipoles minimize their interaction energy when the north poles are close to south poles. There are six dipole pairs in the chosen vertex: four diagonal nearest neighbor pairs and two second nearest neighbor pairs across the center. None of the vertex configurations of figure 11.2 *simultaneously* minimize all of these

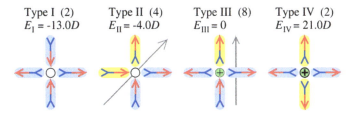

Figure 11.2. Sketches of the possible configurations of a vertex formed at the junction of four neighboring islands in square lattice artificial spin ice. The magnetic islands here are of race-track shape, with dipole moment aligned on the long axis. Multiplicities of the states are given in parenthesis. Blue indicates ground-state-like; yellow shows the changes away from ground state configuration. Types I and II obey the two-in two-out ice rule, but type I is of lower energy. Types II and III have a non-zero magnetic moment, shown as a gray arrow. Types III and IV have single and double monopole charges, indicated by normal and bold plus signs, respectively.

pair interactions. Even in the lowest energy configuration (type I), the diagonal pairs are minimized but the pairs across the center are not. This is the idea of *frustration*. If the system happens to be relaxing from a high energy state, the dipoles first align on the long axes to minimize the anisotropy energy, but then find themselves not able to simultaneously minimize all of their dipolar interactions. This makes it easy for the system to become stuck in a state far from a ground state.

At finite temperature, the system as a whole becomes *dynamically frustrated*, meaning that thermodynamically, it cannot easily access all of its phase space, which would otherwise be available. The dominance of a uniaxial anisotropy means that dipoles cannot reverse direction without going over strong energy barriers, that are typically higher than room temperature thermal energies. This exaggerates the effects of frustration. It means that even if the system is somehow close to its ground state, as viewed in the abstract phase space, it may have great difficulty in relaxing in thermal equilibrium into the ground state. The natural sampling of all accessible states that is usually assumed in statistical mechanics may not hold in any strong sense. As a consequence, a low temperature configuration could be quite far from what is expected for the ground state. As a further consequence, it could have considerable frozen-in entropy, giving a measure of the disorder in the system caused by the frustration.

It should be noted that, in artificial spin ice, the ground state could be degenerate, provided the system has been built with a perfect symmetry (and no external applied magnetic field). The degeneracy should be at least two-fold, because a second ground state can always be obtained from a first one by reversing all the dipoles, which will not change any of the energies. That is apparent in figure 11.1 for square lattice artificial spin ice. This is also related to the fact that the square lattice ground state has a perfect antiferromagnetic (AFM) ordering. Moving along any horizontal or vertical straight path through the lattice, the dipole directions alternate up/down relative to an axis perpendicular to that path. As a result, an imaginary operation of translation of the entire system through one lattice constant a along $\hat{x}$ or $\hat{y}$ also is equivalent to reversing all dipoles. This again confirms the AFM type of ordering; the state would have a static structure function with strong weight at a wave vector $\mathbf{q} = (\pi/a, \pi/a)$. While the two ground states are nearly indistinguishable, differing just by a phase shift, the global reversal of all dipoles is not an easy process. Due to anisotropy coupled with frustration, it may be very challenging for that reversal process to take place in thermal equilibrium at low temperature.

Exercise 11.1. Consider a triplet of three classical equal magnitude spin vectors $\mathbf{S}_1, \mathbf{S}_2, \mathbf{S}_3$, that interact with each other equally with an AFM exchange coupling J, according to a Hamiltonian

$$H = J(\mathbf{S}_1 \cdot \mathbf{S}_2 + \mathbf{S}_2 \cdot \mathbf{S}_3 + \mathbf{S}_3 \cdot \mathbf{S}_1). \tag{11.1}$$

Show that in the minimum energy configuration none of the pairs are anti-aligned, but instead, each makes a 120° angle to the other two in a co-planar state.

11.2 Magnetic monopoles and string excitations in spin ice

The type III and type IV vertices for square artificial spin ice have unequal numbers of inward and outward dipoles, see figure 11.2. Those of type III have a net of either two outward or of two inward; these can be considered to have a single unit of a type of *monopolar* charge, with either a positive (+1 unit) or negative (−1 unit) sign. This charge corresponds to an excess of either south or north pole density at the vertex, respectively. This concentration of pole density contains extra localized energy. The type IV vertices take it a step further, having either four outward or four inward dipoles, which can be considered as +2 or −2 units of monopolar charge. The type IV vertices can be expected to have the highest localized energy. These fundamental monopole excitations out of the ground state have possible topological charges $q = 0, \pm 1, \pm 2$. In this method of counting, each north (south) pole of a dipole contributes an amount of charge Δq equal to $+1/2$ ($-1/2$) to its nearest vertex.

Castelnovo, Moessner and Sondhi [3] suggested that the appearance of magnetic monopoles in spin ice presents a new physical process, that of the emergence of new types of particles due to *fractionalization*. The idea is that although the fundamental entities in the system are dipoles (ultimately due to atomic electronic states), they can be considered to split into fractional parts that recombine at the vertices to form the monopoles of unit magnetic charges. Essentially, the individual dipoles are imagined to split into their north/south poles, which can be recombined into objects that can behave as new particles on the vertices. With this view in mind, the idea of physical systems with such quasi-particles that could have interesting thermodynamics and even real time dynamics attracts a lot of interest. This is analogous to the appearance of fractional electric charges for the quasi-particles that control the quantum Hall effect.

In figure 11.2, islands drawn in blue are in a configuration like that in a ground state, which is the type I vertices. Starting from that configuration, the other vertices shown can obtained by reversing one or two islands' dipoles. Those that were reversed starting from the type I state are drawn in yellow to indicate an excitation. A sense of the excitation is seen in the generation of a net magnetic moment for type II and type III vertices, shown by gray arrows. Note that a single reversal takes type I into type III, already generating a monopole charge $q = \pm 1$. Another reversal next to the first reversed island brings the vertex to type II, which lowers its energy, but gives it a net magnetic moment. If the second reversal is on the opposite side to the first one, there will be no net magnetic moment, but a type IV vertex with a $q = \pm 2$ monopolar charge and higher energy is reached. Note that if all four islands are reversed, each type is mapped back into itself. Negative charge signs (not shown in figure 11.2) can result by reversing different dipoles in the vertex.

Starting from a ground state, one can imagine the types of states which can be generated by reversing dipoles in an extended spin ice system. The reversal of a single dipole already changes the two vertices to which it is connected. An example of these individual changes is given in figure 11.3, where dipoles that have been reversed, starting from the ground state, are drawn in yellow-shaded islands. Near the top of the diagram, two examples of individual flipped dipoles are shown; at each of their connected vertices, a single unit of monopole charge $q = \pm 1$ is generated.

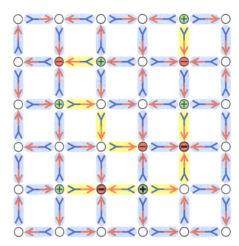

Figure 11.3. Examples of excitations caused by dipole reversals in square lattice artificial spin ice. Starting from a ground state of all type I vertices as in figure 11.1, some islands have had their dipoles reversed, shown here in yellow. This results in a number of type III vertices with single monopole charges (normal plus and minus signs in green or red circles), and also in double monopolar charges (bold plus and minus signs in green or red circles). Note that singly charged monopoles form at the ends of the flipped paths (strings), and doubly charged monopoles form in the straight segments of flipped dipoles. Type II vertices appear at each 90° bend in a yellow path.

Each individual flipped dipole connects to oppositely charged monopoles. Near the center of the diagram in figure 11.3, a *string* of three flipped dipoles along a diagonal path terminates on singly charged monopoles of opposite signs. All the intervening vertices along that path are type II, without charge but with non-zero magnetic moments; the string of flipped dipoles as a whole has a net magnetic moment (relative to the ground state). Finally, in the lower right section of figure 11.3, a path of flipped dipoles makes a single 90° turn (on a type II vertex). Along its straight sections there are doubly charged monopoles of alternating signs. At its ends, the case shown has singly charged monopoles of the same sign. The relative signs of the monopoles at the ends can be seen to depend on whether the number of flipped dipoles in the path is odd or even.

Another example of an excitation out of the ground state is shown in figure 11.4, where closed paths of islands have had their dipoles reversed, shown in yellow. At the top right of the diagram a small square loop is reversed, which changes the vertices at its corners from type I to type II. This will cost an energy change of at least $\Delta E = 4 \times 9.0D$ for the four changed vertices, where the energetics and definition of dipolar energy factor D are given below in section 11.3. In the center of the diagram, a larger loop of 12 dipoles has been reversed. With the path closing on itself, there are no monopoles of single charges $q = \pm 1$, because there are no open ends. In two short straight segments of the path, just as in figure 11.3, the reversal of a pair of dipoles on opposite sides of a vertex has generated monopoles of double charge, $q = \pm 2$. All of the other vertices have become type II, without any charges but with net magnetic moments. Besides the energetics of the monopole charges, this closed loop also has a net magnetic moment, due to the reversals. Although we initially consider no applied

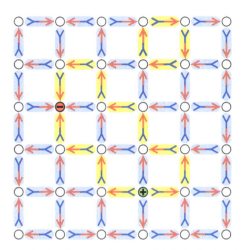

Figure 11.4. Examples of loop excitations in square lattice artificial spin ice. Starting from a ground state of all type I vertices as in figure 11.1, a set of islands have had their dipoles reversed, which are shown here in yellow, forming two closed paths. This results in a number of type II vertices (with magnetic moments) and also in one +2 monopolar charge and one −2 monopolar charge, indicated with plus and minus signs in green and red circles, respectively. These excitations cost considerable energy even when calculating the local dipolar energies only within the changed vertices.

magnetic field, that magnetic moment can interact with the long-range demagnetization field of the system, thereby contributing an additional energy.

These simple examples of excitations in square lattice ice give an idea of the geometrical constraints on the configurations of its vertices. Mainly their energies depend on the local energy in the vertices, however, there will also be a long-range dipolar energy and demagnetization field that determines the global system energy. Mól *et al* [4] have made a deeper analysis of the interactions between pairs of monopoles as a function of their separation, or the strings connecting them. In addition, they have proposed ideas about how the connecting strings contain a tension which further contributes to the thermodynamics.

With this introduction to artificial spin ice systems, next it is necessary to discuss a Hamiltonian to describe the interactions, and from that, discuss the energetics and ways to classify the order and deviation of a state from the ground states. Later, the same Hamiltonian will be used to analyze the possibilities for motion in phase space and real time dynamics from Landau–Lifshitz–Gilbert (LLG) equations. In particular, we will consider an approach to study the dynamics that allows for deviations from the Ising-like behavior of the dipoles. The dipoles themselves are due to the collective magnetic behavior of many atomic dipoles in the islands. As such, a model is developed that considers the possibility of the island dipoles to behave as Heisenberg-like dipoles, that can point in arbitrary directions, albeit strongly constrained by anisotropic magnetic interactions. Therefore, we set up the Hamiltonian as one based on Heisenberg dipoles with 3D freedom of angular movement, that will follow from LLG equations. This is in contrast to an Ising limit, which does not have usual time dynamics.

11.3 Square lattice spin ice energetics and order parameters

Square lattice artificial spin ice is designed by placing the islands as shown in figures 11.1–11.4 in an xy-plane. The vertices that can hold monopole charges are on sites of a square lattice with lattice parameter a, at the points,

$$\mathbf{r}_k = (x_k, y_k) = (m_k a, n_k a). \tag{11.2}$$

The index k labels the kth vertex site, located according to the choice of integers (m_k, n_k). In an unbounded lattice, each unit cell contains one vertex. Within each unit cell, however, there are two islands: one oriented horizontally and one oriented vertically. The two islands per cell act as a two-atom basis. Their locations are

$$\mathbf{r}_{k,a} = \left(m_k + \frac{1}{2}, n_k\right)a, \quad \text{on the } a\text{-sublattice} \tag{11.3}$$

$$\mathbf{r}_{k,b} = \left(m_k, n_k + \frac{1}{2}\right)a, \quad \text{on the } b\text{-sublattice}. \tag{11.4}$$

The lattice has two magnetic sublattices denoted a and b, see figure 11.5. On the a-sublattice the islands are horizontal and on the b-sublattice they point vertically. In a finite system, suppose there are N_c unit cells, or equivalently, N_c charge vertices (labels $k = 1, 2, 3 \ldots N_c$). Then there are $N = 2N_c$ islands, which could be labeled by vertex k and sublattice a or b, or, just by some index $i = 1, 2, 3, \ldots N$. The dipole on island i is a 3D vector, $\boldsymbol{\mu}_i$. Let the dipoles be written as $\boldsymbol{\mu}_i = \mu\hat{\boldsymbol{\mu}}_i$, in terms of a fixed magnitude μ and a unit direction vector $\hat{\boldsymbol{\mu}}_i$. Each island has an easy anisotropy axis pointing along its long axis, $\hat{\mathbf{u}}_i$, which can be $\hat{\mathbf{u}}_i = \pm\hat{\mathbf{x}}$ on the a-sublattice and $\hat{\mathbf{u}}_i = \pm\hat{\mathbf{y}}$

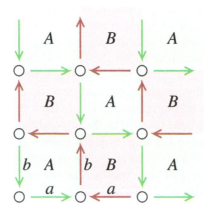

Figure 11.5. Notation for the primitive cells of square lattice artificial spin ice, in a 3 × 3 grid of vertices (gray dots). Each cell can be associated with the vertex at its lower left corner and two dipoles on the magnetic sublattices a and b. The cells are divided into A and B sublattices as indicated with green and red, respectively. The dipoles here are drawn in a ground state configuration; reversal of the dipoles in the B primitive cells brings them to the same state as in the A primitive cells. This fact is useful for the definition of order parameters, see (11.21) and (11.22).

on the b-sublattice. This anisotropy is the shape anisotropy as expected for an elongated particle. It will be characterized by an energy constant denoted as K_1.

In the basic assumption of a 2D system of islands, it is also supposed that the islands are thin perpendicular to the substrate, compared to their dimensions in the xy-plane. This generates an additional shape anisotropy, making the $\hat{z}$-axis a hard anisotropy axis, with some energy constant denoted here as K_3. The constants K_1 and K_3 are assumed to be identical for all islands, although that is probably difficult to achieve in a real sample. The energy associated with these two local anisotropies can be written as

$$H_A = \sum_i \left\{ K_1 \left[1 - (\hat{\mu}_i \cdot \hat{u}_i)^2 \right] + K_3 (\hat{\mu}_i \cdot \hat{z})^2 \right\}. \tag{11.5}$$

This has been written including a constant so that the anisotropy energy is zero in the ground state, where $\hat{\mu}_i$ is parallel or anti-parallel to $\hat{u}_i$. It is important to notice that if the dipoles only rotate within the xy-plane, the energy cost is determined only by K_1. If they move up out of the xy-plane, however, the energy cost involves a combination of K_1 and K_3, depending on the particular direction of that motion.

The next largest term in a Hamiltonian will be the interactions among the dipoles. This is equivalent to a demagnetization effect for the whole system; it is a long-range dipole–dipole interaction. We know that the interaction of a dipole with a magnetic induction is given from (2.59). Summed over the islands, this interaction is

$$H_{\text{int}} = -\sum_{i=1}^{N} \mu_i \cdot \mathbf{B}_i, \tag{11.6}$$

where $\mathbf{B}_i$ should be the net magnetic induction acting on site i. Just to be general here, it could include not only the dipolar fields of all the other islands, but in addition, any external field $\mathbf{B}_{\text{ext}} = \mu_0 \mathbf{H}_{\text{ext}}$ applied to the system. The magnetic induction caused by a point dipole $\mathbf{p}$ at the origin is given in (3.18). Assuming source dipoles $\mathbf{p} = \mu_j$ at locations $\mathbf{r}_j$, and using $\mathbf{B}_i = \mu_0 \mathbf{H}_i$, the field produced at location $\mathbf{r}_i$ is

$$\mathbf{B}_i = \frac{\mu_0}{4\pi} \sum_{j \neq i} \left[\frac{3(\mathbf{r}_i - \mathbf{r}_j)(\mathbf{r}_i - \mathbf{r}_j) \cdot \mu_j - \mu_j}{|\mathbf{r}_i - \mathbf{r}_j|^3} \right] + \mathbf{B}_{\text{ext}}. \tag{11.7}$$

Now we can let the inter-site displacements be written with magnitude and direction,

$$\mathbf{r}_i - \mathbf{r}_j = r_{ij} \hat{\mathbf{r}}_{ij}. \tag{11.8}$$

Then the field interaction energies can be expressed as

$$H_{\text{int}} = -\frac{\mu_0 \mu^2}{4\pi a^3} \sum_{i>j} \left[\frac{3(\hat{\mu}_i \cdot \hat{\mathbf{r}}_{ij})(\hat{\mu}_j \cdot \hat{\mathbf{r}}_{ij}) - \hat{\mu}_i \cdot \hat{\mu}_j}{(r_{ij}/a)^3} \right] - \mu B_{\text{ext}} \sum_i \hat{\mu}_i \cdot \hat{\mathbf{B}}_{\text{ext}}. \tag{11.9}$$

The sum in the dipole–dipole term is over $i > j$ to avoid double counting of those interactions. A unit vector in the direction of $\mathbf{B}_{\text{ext}}$ has also been introduced, so that all energy terms can be compared, according to the sizes of the factors in front of the sums. Then, the total system Hamiltonian is taken to be the sum of the anisotropy and interaction terms:

$$H = H_A + H_{\text{int}}. \tag{11.10}$$

The dipolar terms are scaled overall by a dipolar energy factor,

$$D \equiv \frac{\mu_0}{4\pi} \frac{\mu^2}{a^3}. \tag{11.11}$$

This is controlled, to a certain extent, by the size and separation a of the islands; their magnetic moments are proportional to their volumes, i.e. $\mu = M_s V$. Then of course the external field interaction scale is determined by the product μB_{ext}. Each energy term has its particular scale, which should be compared with the thermal energy scale, $k_B T$. The anisotropy energy scale is by far, the largest of all of these scales. Although the dipolar interaction D can be adjusted somewhat by design of the islands and choice of materials, *etc*, it may remain similar to or larger than $k_B T$. The thermal energies for ordinary room temperature tend to be the weakest energy scale in this problem. Note also that D enters in front of a sum, whose long-range structure can become rather large, thereby enhancing the contribution of dipolar interactions.

It should be kept in mind that this model determines the physics of *point dipoles* with dipole moments of fixed magnitude. The magnetic islands are extended objects, and if placed very close together, they can be expected to have an internal magnetization dynamics that will not preserve the dipole strength μ.

11.3.1 Reduction to an Ising limit

Although a Heisenberg-like model is being considered here so that the system has magnetic dynamics following LLG equations, much of the theoretical work on spin ice models uses the limiting case of an Ising model for the dipoles. As the anisotropy constants K_1 and K_3 are the dominating energies, this makes good sense for some calculations, such as determination of ground states and determination of frozen-in entropy in a state. It is accomplished by introducing Ising variables $\sigma_i = \pm 1$ at the sites, related to the original magnetic dipoles by

$$\boldsymbol{\mu}_i = \sigma_i \mu \hat{\mathbf{u}}_i, \tag{11.12}$$

The mapping to Ising variables is aided by the (arbitrary) choice of unit direction vectors $\hat{\mathbf{u}}_i$ for the islands as displayed in figure 11.6. It is assumed that any vertex is associated to one lattice cell, and each lattice cell contains two dipoles (the a and b basis dipoles at the lower and left edges of a cell). The vertex at the lower left corner of a cell is identified with that cell, see figure 11.5. Then the configuration of figure 11.6 is a diagram of the directions of the $\hat{\mathbf{u}}_i$. At the lower left corner of A cells, all $\hat{\mathbf{u}}_i$ point outward. At the lower left corner of B cells, all $\hat{\mathbf{u}}_i$ point inward. While this state is being used to define the $\hat{\mathbf{u}}_i$, it cannot be lost that it is the state of maximum energy, because all vertices have a doubly-charged monopole. If we

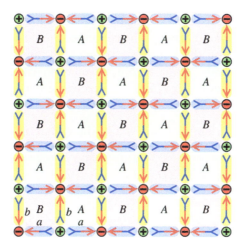

Figure 11.6. A very high energy configuration of square lattice artificial spin ice, obtained from the ground state of figure 11.1 by reversing the islands highlighted in yellow. Each cell can be identified with the vertex at its lower left corner; the islands marked a and b are the primitive basis for the cells. The $A(B)$ cells' vertices have a $+2(-2)$ monopole charge, which is the highest possible density of charges. These dipole directions are taken to define the unit direction vectors $\hat{\mathbf{u}}_i$ for mapping to the Ising limit via $\boldsymbol{\mu}_i = \sigma_i \mu \hat{\mathbf{u}}_i$. This state corresponds to the choice, all $\sigma_i = +1$.

define $\ell = +1 (\ell = -1)$ for the basis sites associated to the $A(B)$ cells, then the island direction vectors are

$$\hat{\mathbf{u}}_i = \begin{cases} \ell \hat{\mathbf{x}} & i \in a\text{-sublattice} \\ \ell \hat{\mathbf{y}} & i \in b\text{-sublattice} \end{cases}. \quad (11.13)$$

The state with maximum monopole density in figure 11.6 has all Ising variables $\sigma_i = +1$, by definition. For comparison, the ground state in figure 11.1 has $\sigma_i = +1$ on the a magnetic sublattice but $\sigma_i = -1$ on the b magnetic sublattice, or vice versa.

Now, in all possible states, the dipoles are aligned or anti-aligned to the island axes, which means that the anisotropy energy H_A is always zero[1], as defined in (11.5). Then, the only energies in the Hamiltonian are the interaction terms, which take the form

$$H_{\text{Ising}} = -D \sum_{i>j} \left[\frac{3(\hat{\mathbf{u}}_i \cdot \hat{\mathbf{r}}_{ij})(\hat{\mathbf{u}}_j \cdot \hat{\mathbf{r}}_{ij}) - \hat{\mathbf{u}}_i \cdot \hat{\mathbf{u}}_j}{(r_{ij}/a)^3} \right] \sigma_i \sigma_j - \mu B_{\text{ext}} \sum_i \sigma_i \hat{\mathbf{u}}_i \cdot \hat{\mathbf{B}}_{\text{ext}}. \quad (11.14)$$

This is a 2D Ising model with couplings beyond nearest neighbors. If there is no applied field, it reduces to a very simple form,

$$H_{\text{Ising}} = \sum_{i>j} J_{ij} \sigma_i \sigma_j, \qquad J_{ij} \equiv -D \left[\frac{3(\hat{\mathbf{u}}_i \cdot \hat{\mathbf{r}}_{ij})(\hat{\mathbf{u}}_j \cdot \hat{\mathbf{r}}_{ij}) - \hat{\mathbf{u}}_i \cdot \hat{\mathbf{u}}_j}{(r_{ij}/a)^3} \right]. \quad (11.15)$$

---

[1] Even though it has been shifted by a large constant to a net zero value, the anisotropy energy H_A dominates the system, forcing dipoles to align with their islands.

The pair couplings J_{ij} are fixed according to the relative locations of the sites, and the fixed definitions of the island axes $\hat{u}_i$. The couplings should fall with $1/r_{ij}^3$, however, if all are taken into account, there can be a strong long-range effect.

It is instructive to evaluate the J_{ij} among the four sites surrounding a charge vertex k, using the notation $i_k = 1, 2, 3, 4$ depicted in figure 11.7.

> **Exercise 11.2.** Referring to figure 11.7, show that the effective exchange interactions of site 1 with its neighbors at the same vertex are
> $$J_{12} = 3\sqrt{2}\,D = J_{14}, \quad J_{13} = 2D. \tag{11.16}$$

Other interactions within a vertex are found by symmetry. These are all positive, corresponding to an AFM coupling within the vertex, which tends to make the $\sigma_i \sigma_j$ products negative, for lowest energy. This can be seen to favor the type I vertices the most, followed by types II, II and IV. Consider the vertex energies in an exercise.

> **Exercise 11.3.** Including only the dipolar energies within a chosen vertex, verify that the energies of the four types are
> $$E_{\mathrm{I}} = -4J_{12} + 2J_{13} \approx -13D \tag{11.17a}$$
> $$E_{\mathrm{II}} = -2J_{13} = -4D \tag{11.17b}$$
> $$E_{\mathrm{III}} = 0 \tag{11.17c}$$
> $$E_{\mathrm{IV}} = 4J_{12} + 2J_{13} \approx 21D. \tag{11.17d}$$

These are the energies that dominate, after the anisotropy terms, and hence this shows the considerable energetic advantage for all vertices to tend towards type I, provided they find a path through phase space. The interactions with dipoles in a neighboring vertex will be of similar magnitude, and ones further away will diminish

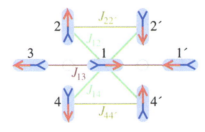

Figure 11.7. Sketch showing the notation for sites within a vertex (at left, an A-site) as labeled by an index $i_k = 1, 2, 3, 4$. Colored lines indicate some effective exchange interaction constants between sites for the Ising limit, see (11.15) and (11.16). There are also interactions with a neighboring vertex, such as $J_{22'} = J_{44'} = -D$. The arrows drawn here show the dipole directions if the Ising variables are all set to $\sigma_i = +1$.

with $1/r^3$. Two inter-vertex interactions are shown in figure 11.7, which have the value $J_{22'} = J_{44'} = -D$. This favors the opposite alignment ($\sigma_2 = \sigma_{2'}$) as in figure 11.7, so that even though this is a weak energy, it is beneficial to lowering the total ground state energy.

In a Monte Carlo (MC) simulation of the Ising model (11.15) with periodic boundary conditions, Silva *et al* [5] were able to find the specific heat as a function of temperature. Their results show a very sharp peak in $C(T)$ near the temperature $T_c \approx 7.2D$. The peak gives strong evidence for a type of phase transition. In the low temperature phase, most vertices are type I, and there are only a few singly charged monopoles that tend to be close together. In the high temperature phase, monopoles are much more numerous (unbound) and connected by strings, which also fluctuate. The results indicate that by performing single spin flips and multi-spin updates such as 'worm moves' the system is able to fall into a ground state as $T \to 0$. In terms of the Ising dynamics, the square lattice spin ice model is not found to exhibit any strong frustration. The motion in its full phase space does not appear to be prohibited in any way, and all expected thermodynamic states are apparently accessible. Of course, Ising spin flips do not need to move any spins over energy barriers, because there is no contribution from the strong anisotropy in this reduced model.

The Ising limit is mentioned because it can bring some simplification. In what follows, we return to the more general idea that the island dipoles are more Heisenberg-like, able to point in any direction, but strongly restricted by anisotropy that causes large energy barriers.

Exercise 11.4. Suppose a square ice system is initially in a ground state, and then one dipole is reversed, such as in the upper left of figure 11.3. Both connected vertices switch from type I to type II. How large is the energy change in the two affected vertices? What if the next nearest dipolar contributions are also included?

Exercise 11.5. Suppose a square ice system is initially in a ground state, and then a small square loop is reversed, such as in the upper right of figure 11.4 but far from any boundary. Four vertices are changed from type I to type I. How large is the energy change in the four affected vertices? What if the next nearest dipolar contributions are also included?

11.3.2 Measurement of order in square spin ice

We can talk about order in the system, by comparing any dipole configuration with a ground state. A configuration that is close to one of the ground states has a high degree of ordering. As there are two degenerate ground states, we need to construct an *order parameter* that can distinguish both when there is ordering, and indicate which ground state is the closest to a given state, in some sense.

In the ground state shown in figure 11.1, all dipoles alternate direction from site to site. One can further see that each unit cell is reversed from its neighboring cells. Therefore, we can consider that the cells also fall on sublattices, as indicated by the green/red colors in figure 11.5, that are arbitrarily named the *A*-and *B*-sublattices (as white/black squares on a checkerboard). Then, using the unit dipoles, a ground state can be expressed as follows:

$$\hat{\mu}_{k,a} = \begin{cases} +s\hat{x} & k \in A \\ -s\hat{x} & k \in B \end{cases}, \quad \hat{\mu}_{k,b} = \begin{cases} -s\hat{y} & k \in A \\ +s\hat{y} & k \in B. \end{cases} \quad (11.18)$$

The parameter $s = \pm 1$ is used to give the two distinct ground states, with dipoles reversed relative to the other state. The ground state of figure 11.1 has $s = +1$. This formula expresses the AFM structure both with respect to the magnetic sublattices (*a* and *b*) and the unit cell sublattice (*A* and *B*). As mentioned earlier, reversal of all dipoles is equivalent to shifting the whole system one lattice constant along either coordinate axis. One can check that equivalently in terms of the Ising variables, a ground state has

$$\sigma_{k,a} = \begin{cases} +s & a\text{-sublattice} \\ -s & b\text{-sublattice}. \end{cases} \quad (11.19)$$

This reflects the fact that a ground state is obtained from the state will all $\sigma_i = +1$ by reversing just the *b*-sublattice ($s = +1$) or just the *a*-sublattice ($s = -1$).

An order parameter can be made from an overlap with a ground state. It can be noticed in figure 11.5 that reversal of the two primitive sites (at the lower and left edges) of all *B* cells brings them to the same configuration as the *A* cells. Let this be a defined operation R, that changes only the *B* cells (by a π rotation). The new configuration of dipoles is

$$\mu'_i = R\mu_i = \begin{cases} +\mu_i & \text{on } A \text{ cells} \\ -\mu_i & \text{on } B \text{ cells}. \end{cases} \quad (11.20)$$

This allows for the definition of an order parameter on each magnetic sublattice, by averages over the cells,

$$Z_a = \frac{1}{N_c} \sum_k \mu'_{k,a} \cdot \hat{x}, \quad Z_b = \frac{1}{N_c} \sum_k \mu'_{k,b} \cdot \hat{y}. \quad (11.21)$$

As defined, the $s = +1$ ground state has $Z_a = +1$, $Z_b = -1$; these values are interchanged for the $s = -1$ ground state. For the high energy state of figure 11.6, the values are $Z_a = Z_b = +1$. Reversing all dipoles in that figure would lead to $Z_a = Z_b = -1$. This pair of order parameters can be quite useful for an overview of the averaged structure on each magnetic sublattice. Both parameters can be combined into a single order parameter that averages the sublattices,

$$Z = \frac{1}{2}(Z_a - Z_b). \quad (11.22)$$

This takes the values $Z = s = \pm 1$ for the two ground states. It also takes the value $Z = 0$ in the high energy state of figure 11.6. Therefore, it gives a sense of the nearness of an arbitrary configuration to either one or the other ground state, whenever Z approaches ± 1. Conversely, Z may be near zero if the configuration is greatly different from a ground state. For instance, a random configuration of the dipoles (each $\hat{\mu}_i$ being free to point anywhere on the unit sphere) will give $Z_a = Z_b = Z = 0$. It is simple to calculate Z during any numerical simulation, either of MC type or magnetic dynamics.

Probability distributions of magnetic sublattice local ordering

The variables Z_a, Z_b give averages over the whole system of some local variables in the individual cells, which are the projections of the dipoles on the local axes $\hat{x}$ on the a-sublattice and $\hat{y}$ on the b-sublattice, see (11.21). This suggests the definitions of two local variables in an individual cell,

$$z_a = \hat{\mu}'_{k,a} \cdot \hat{x}, \quad z_b = \hat{\mu}'_{k,b} \cdot \hat{y}. \tag{11.23}$$

These are distributed over the range from -1 to $+1$. In a simulation, the probability distributions for them, $p_a(z_a)$ and $p_b(z_b)$, can be calculated numerically by making histograms for the variables in the allowed range. The range can be divided into a number of bins B (on the order of 100–200) with widths $w = 2/B$, and by averaging over enough data samples this is a simple calculation of the relative probabilities and errors in each bin.

At high temperature, one might expect great disorder even at the local level, then the probability distributions $p_a(z_a)$ and $p_b(z_b)$ could be close to uniform over the allowed range. For this to be true 'high temperature' would have to mean that the thermal energy $k_B T$ is much greater than all energy scales in the problem. At low temperature, the dipoles could tend to point parallel or anti-parallel to the $\hat{x}$ and $\hat{y}$ axes on the two sublattices. This would concentrate the probability distributions near $z_a \approx \pm 1$ and $z_b \approx \pm 1$. If the system moves into one of its two ground states labeled by $s = \pm 1$, then the distributions should be peaked at $z_a = +s$ and $z_b = -s$. The two distributions taken together will give a full view of the local ordering even when not in a ground state.

It makes sense to take this a step further and also define their combined or net local order parameter,

$$z \equiv \frac{1}{2}(z_a - z_b). \tag{11.24}$$

The probability distribution of z can be obtained indirectly from the distributions $p_a(z_a)$ and $p_b(z_b)$ by summing over both, with an appropriate constraint imposed by a delta function,

$$p(z) = \int_{-1}^{+1} dz_a\, p_a(z_a) \int_{-1}^{+1} dz_b\, p_b(z_b) \delta\left[z - \frac{1}{2}(z_a - z_b)\right]. \tag{11.25}$$

The delta function constrains the sum so that only points satisfying (11.24) for a chosen value z will contribute. In the $z_a z_b$-plane, this corresponds to a line with a unit

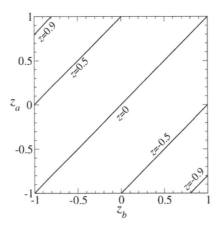

Figure 11.8. Illustration of the $z_a z_b$-plane for the calculation of the derived probability distribution $p(z)$ for square ice using (11.25) from raw probability distributions for z_a and z_b. The solid lines result from (11.24) for the indicated values of z.

slope, as sketched in figure 11.8. For numerical application, the values of z are partitioned into B bins with width $w = 2/B$ just as for z_a and z_b. Values of z within a bin fall into a narrow strip in the $z_a z_b$-plane. Counting the relative frequencies for z to fall into different strips leads to the calculated probabilities for each bin, once normalized. The resulting distribution $p(z)$ will now average over both magnetic sublattices and give a good sense of whether the system has moved into a state approximating one of the ground states. That would be indicated by the probability becoming concentrated near $z = +1$ or near $z = -1$ for the two separate ground states.

11.3.3 Monopole charge densities in square spin ice

The monopole charges that appear in spin ice give another measure of the changes in the order present. The ground states have no monopoles; we have already seen that any dipole reversals out of a ground state will likely create single or doubly charged monopoles. A monopole charge density can be defined as another type of order parameter for the system as a whole. This is accomplished with a simple algorithm for the counting of the charges, similar to that used for calculating vortex density in 2D easy-plane magnets.

At each vertex one can assume a set of outward directed unit vectors $\hat{v}_{i_k}$, $i_k = 1, 2, 3, 4$, pointing towards each of the four surrounding islands. Then each pole of a dipole pointing out (in) contributes a positive (negative) half-unit of monopole charge on that vertex. Then the total charge q_k for vertex k can be expressed as from a discrete definition,

$$q_k = \frac{1}{2} \sum_{i_k=1}^{4} \left[2H(\hat{\mu}_{i_k} \cdot \hat{v}_{i_k}) - 1 \right], \quad (11.26)$$

where $H(x)$ is the Heaviside step function. A Heisenberg-like dipole is counted as outward (inward) by the function in square brackets when the angle between the dipole and $\hat{\mathbf{v}}_{i_k}$ is less than (greater than) 90°. In square ice this results in possible values $q_k = 0, \pm1, \pm2$, corresponding, respectively, to vertex types I and II, type III, and type IV. In a totally random state one can expect these to occur with probabilities of 6/16, 8/16 and 2/16, respectively, based on the multiplicities from figure 11.2. While this definition is very good for the Ising limit, it may make sudden jumps as a dipole rotates from outward to inward or vice versa. Indeed, using it for dipoles free to point in any direction, once a dipole passes the point where it is perpendicular to the island axis, the value of q_k on its two neighboring vertices will change. There will be a jump of monopole charge $\Delta q = \pm 1$ from one vertex to the other, while the total charge is conserved.

Another charge-like definition has been made [6] that is continuously varying and uses just the scalar product,

$$q_k^* = \frac{1}{2} \sum_{i_k=1}^{4} \hat{\boldsymbol{\mu}}_{i_k} \cdot \hat{\mathbf{v}}_{i_k}. \tag{11.27}$$

As a sum of projections of the dipoles on the outward axes, it can take all values continuously from −2 to +2, however, the likelihood of obtaining $q_k^* = \pm 2$ is greatly limited in the presence of any fluctuations. This definition will still give a conserved quantity, because a small rotation of a dipole will cause one vertex to gain an amount Δq that is exactly canceled by an opposite amount for the other vertex connected to that dipole. Now, of course, the flow of charges between the neighboring vertices is continuous.

For the discrete charge definition, one can now define a magnetic charge density per vertex by averaging the absolute valued charges over the N_c vertices of the system, including all types $q_k = 0, \pm1, \pm2$,

$$\rho = \langle |q| \rangle = \frac{1}{N_c} \sum_{k=1}^{N_c} |q_k|. \tag{11.28}$$

A similar definition can be made for a continuous charge density, $\rho^* = \langle |q^*| \rangle$. These quantities give a global sense of the effects of temperature in the generation of monopoles and disorder away from the ground state.

For the discrete charge definition, it is also helpful to distinguish the contributions of single charges from double charges, which can easily be achieved in simulations. Suppose a state has an average of $\langle n_1 \rangle$ singly charged and $\langle n_2 \rangle$ doubly charged monopoles (of either sign) per vertex. Then the charge density due to single charges is $\rho_1 = \langle n_1 \rangle$ and that due to double charges, which contribute two units, is $\rho_2 = 2\langle n_2 \rangle$. This gives the total charge density as the sum,

$$\rho = \rho_1 + \rho_2 = \langle n_1 + 2n_2 \rangle. \tag{11.29}$$

There is not a corresponding formula for the continuous definition, since there is a continuum of possible vertex charges q_k^*.

In the ground states, $\rho = \rho^* = 0$, as there are no monopoles present. As temperature is increased above zero, there can be small angular deviations in the dipoles, creating small values of q_k^*, while all q_k remain zero. This will give a finite value of ρ^*, while the discrete ρ can be expected to remain zero. Only when the temperature is high enough to create discrete monopoles will ρ deviate away from zero.

At the other extreme of very high temperature, the system is at maximum entropy and disorder. All allowed configurations of the vertices will become equally probable. To be specific, this means that any chosen dipole's direction becomes distributed uniformly on the unit sphere (we are still assuming dipoles of fixed magnitudes). This also implies that the 16 different discrete vertex configurations become equally probable. For those configurations, there are six with no charge ($q_k = 0$), eight with a unit charge ($q_k = \pm 1$) and two with a double charge ($q_k = \pm 2$). This gives the average charge density in the high temperature limit,

$$\rho = \langle |q_k| \rangle = \frac{6}{16} \times 0 + \frac{8}{16} \times 1 + \frac{2}{16} \times 2 = \frac{3}{4}. \qquad (11.30)$$

Together with this, one sees the limiting values for the singly charged monopole contribution is $\rho_1 = \langle n_1 \rangle = \frac{8}{16} = \frac{1}{2}$, and for doubly charged monopoles it is $\rho_2 = 2\langle n_1 \rangle = \frac{4}{16} = \frac{1}{4}$, with the probability for double charges being $\langle n_2 \rangle = \frac{1}{8}$. These values serve to check the high temperature limit in simulations.

For the continuous charge definition, the projection of one component of a unit dipole $x_{i_k} = \hat{\boldsymbol{\mu}}_{i_k} \cdot \hat{\mathbf{v}}_{i_k}$ is required, averaged over the four dipoles at a vertex. Any such component of a unit vector on a sphere has a uniform distribution from -1 to $+1$. Then the high temperature limit for the continuous charge density can be found from an integration exercise,

$$\rho^* = \langle |q_k^*| \rangle = \left\langle \frac{1}{2} \left| \sum_{i_k=1}^{4} x_{i_k} \right| \right\rangle$$
$$= \frac{1}{2} \frac{\int dx_1 \int dx_2 \int dx_3 \int dx_4 \, |x_1 + x_2 + x_3 + x_4|}{\int dx_1 \int dx_2 \int dx_3 \int dx_4 \, 1} = \frac{7}{15}. \qquad (11.31)$$

This is less than that for the discrete charge density, even though the variation of ρ^* away from zero at low temperature begins before any change in the discrete ρ takes place.

11.4 Dynamics in square lattice artificial spin ice

Next we consider the application of Langevin-LLG dynamics for the calculation of thermodynamic averages in spin ice. The idea is that in a frustrated system, a MC approach needs to use arbitrary changes in the dipoles. With dynamics, the changes that take place are closer to those naturally occurring in a real system. A dynamic simulation takes into account how the dipoles must climb over energy barriers (under the assumption of fixed dipole magnitudes). In order to perform a dynamic simulation it is necessary to use a model where the dipoles have three components, hence we apply the model with Heisenberg-like dipoles. The temperature is included by using the Langevin equation for LLG dynamics, as was discussed in chapter 5. The approach is similar to that used for the internal vortex dynamics in nanodisks as described in the previous chapter.

At zero temperature, the undamped dynamics follows a torque equation like (2.61), namely,

$$\frac{d\boldsymbol{\mu}_i}{dt} = \boldsymbol{\mu}_i \times \gamma \mathbf{B}_i, \quad \mathbf{B}_i = -\frac{\partial H}{\partial \boldsymbol{\mu}_i}, \tag{11.32}$$

where the effective fields are obtained from the Hamiltonian (11.10) by a gradient such as (2.99). The fields are

$$\mathbf{B}_i = \frac{D}{\mu} \sum_{j \neq i} \left[\frac{3(\hat{\boldsymbol{\mu}}_j \cdot \hat{\mathbf{r}}_{ij})\hat{\mathbf{r}}_{ij} - \hat{\boldsymbol{\mu}}_j}{(r_{ij}/a)^3} \right] + \frac{2K_1}{\mu}(\hat{\boldsymbol{\mu}}_i \cdot \hat{\mathbf{u}}_i)\hat{\mathbf{u}}_i - \frac{2K_3}{\mu}(\hat{\boldsymbol{\mu}}_i \cdot \hat{\mathbf{z}})\hat{\mathbf{z}} + \mathbf{B}_{\text{ext}}. \tag{11.33}$$

For numerical work it is best to change to dimensionless quantities by an appropriate choice of units. Here the lattice parameter is the unit of length. The saturation magnetization M_s gives a convenient unit of magnetic induction, $B_0 = \mu_0 M_s$. For islands made from permalloy (Py), with $M_s = 860$ kA m^{-1}, that gives $B_0 \approx 1.08$ T. Therefore one can define the dimensionless magnetic fields as

$$\mathbf{b}_i \equiv \frac{\mathbf{B}_i}{\mu_0 M_s} = d \sum_{j \neq i} \left[\frac{3(\hat{\boldsymbol{\mu}}_j \cdot \hat{\mathbf{r}}_{ij})\hat{\mathbf{r}}_{ij} - \hat{\boldsymbol{\mu}}_j}{(r_{ij}/a)^3} \right] + 2k_1(\hat{\boldsymbol{\mu}}_i \cdot \hat{\mathbf{u}}_i)\hat{\mathbf{u}}_i - 2k_3(\hat{\boldsymbol{\mu}}_i \cdot \hat{\mathbf{z}})\hat{\mathbf{z}} + \mathbf{h}_{\text{ext}}. \tag{11.34}$$

The dimensionless couplings within this are

$$d = \frac{D}{\mu \mu_0 M_s} = \frac{\mu}{4\pi a^3 M_s} \tag{11.35a}$$

$$k_1 = \frac{K_1}{\mu \mu_0 M_s}, \quad k_3 = \frac{K_3}{\mu \mu_0 M_s}, \quad \mathbf{h}_{\text{ext}} = \frac{\mathbf{B}_{\text{ext}}}{\mu_0 M_s} = \frac{\mathbf{H}_{\text{ext}}}{M_s}. \tag{11.35b}$$

The corresponding natural energy unit is the factor,

$$\varepsilon \equiv \mu \mu_0 M_s. \tag{11.36}$$

The magnetic moment of an island of volume V is $\mu = M_s V$, which affects all of these energy scales. In particular, the dipolar coupling constant takes an interesting form,

$$d = \frac{V}{4\pi a^3}, \tag{11.37}$$

which is related to the relative volume fraction of the grid that is filled by magnetic material. A tighter packing of magnetic islands into the grid will lead to greater dipolar interactions; it is clear that d is well below unity, under the assumption of thin islands. The thermal energy should also be scaled in the same way, leading to the dimensionless temperature variable,

$$\mathcal{T} \equiv \frac{k_{\mathrm{B}} T}{\varepsilon}. \tag{11.38}$$

The value of $\mathcal{T}$ is possibly the most important for consideration of the spin ice properties by this dynamic method. Larger islands, which have larger values of μ and hence ε, could give quite small values of $\mathcal{T}$, leading to a dynamics not unlike that at zero temperature. That would imply that the system will have great difficulty to evolve in the phase space, i.e. the frustration affects will be greatly exhibited, even in the dynamics.

For the Langevin-LLG dynamics, damping and stochastic terms are included into (11.32), and the equation is brought to dimensionless form, including a dimensionless time variable,

$$\tau \equiv t/t_0, \quad t_0 \equiv (\gamma \mu \mu_0 M_{\mathrm{s}})^{-1}. \tag{11.39}$$

The resulting equation is like that in (10.60) for the stochastic dynamics of vortices,

$$\frac{d\hat{\mu}_i}{d\tau} = \hat{\mu}_i \times (\mathbf{b}_i + \mathbf{b}_{\mathrm{s},i}(\tau)) - \alpha \hat{\mu}_i \times [\hat{\mu}_i \times (\mathbf{b}_i + \mathbf{b}_{\mathrm{s},i}(\tau))]. \tag{11.40}$$

This includes the stochastic fields $\mathbf{b}_{\mathrm{s},i}(\tau)$ needed in the Langevin approach, which have the time correlations according to expression (10.61). The natural unit of time is $t_0 \approx 5.26$ ps for Py, using $\gamma = e/m_{\mathrm{e}} \approx 1.76 \times 10^{11}$ T^{-1} s^{-1}. The time step for numerical integration must be a small fraction of t_0 if one is to expect accurate results. For a scheme based on fourth-order Runge-Kutta, a time step around $\Delta \tau \sim 0.04$ usually gives acceptable results. Conversely, this means that a very large number of time steps may be necessary to study the equilibrium situation over long time intervals.

11.4.1 Island geometry and coupling parameters

To set up simulations, the parameters d, k_1, k_3 and ε should first be determined, based on some assumptions about the magnetic islands. In a micromagnetics study of thin elliptically shaped islands Wysin et al [7] estimated the anisotropy energies per unit volume corresponding to K_1 and K_3, based on the shape of the islands. In their model, an island of size denoted $L_x \times L_y \times L_z$ has major axis L_x, minor axis L_y and thickness L_z perpendicular to the xy-plane (semi-major axis $A = L_x/2$, semi-minor axis $B = L_y/2$). For a chosen overall size L_x (the longest axis), the shape is characterized by the lateral aspect ratio $g_1 \equiv L_x/L_y > 1$ and the vertical aspect ratio $g_3 \equiv L_x/L_z > 1$. The higher these ratios are above unity, the stronger is the anisotropy energy associated with that ratio. Figure 11.9 shows the general trends in the energies per volume, K_1/V and K_3/V, versus island thickness. For a reasonable range of island sizes, these results can be scaled with the island volume $V = \pi A B L_z$ to

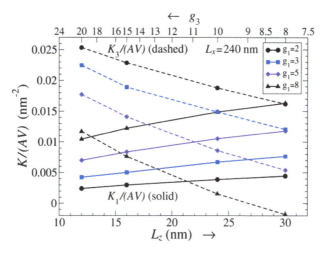

Figure 11.9. The shape anisotropy constants K_1 (solid lines) and K_3 (dashed lines) per island volume and scaled by the exchange stiffness A of the material, from micromagnetics simulations [7] of islands of long axis $L_x = 240$ nm, with elliptic shapes (major axis = L_x, minor axis = L_y, thickness L_z, volume $V = \frac{1}{4}\pi L_x L_y L_z$). The islands are described by lateral aspect ratio $g_1 = L_x/L_y$ and vertical aspect ratio $g_3 = L_x/L_z$. Generally speaking, K_1/V increases with g_1, and K_3/V increases with g_3, provided the other aspect ratio is held fixed.

estimate K_1 and K_3 for a chosen island size. Note that the anisotropy constants are proportional to the exchange stiffness A of the material.

In some of the original work on square spin ice by Wang et al [1], quasi-rectangular islands of Py were used with dimensions 220 nm × 80 nm × 25 nm, where the last number is the thickness. The islands were separated by lattice parameter a ranging from 320 nm to 880 nm. If we assume that these can be described approximately by the model of elliptic islands, then we have $L_x = 220$ nm and aspect ratios $g_1 = 220/80 = 2.75$ and $g_3 = 220/25 = 8.8$, with volume $V = \pi A B L_z = 3.46 \times 10^5$ nm^3. From the saturation magnetization $M_s = 860$ kA m^{-1} for Py, one obtains a dipole moment per island of $\mu = M_s V = 2.97 \times 10^{16}$ A m^2.

Interpolation of the results in figure 11.9 gives the anisotropy values $K_1/(AV) \approx 0.0064$ nm^{-2} and $K_3/(AV) \approx 0.014$ nm^{-2}. For Py, using $A \approx 13$ pJ m^{-1} leads to the estimates of easy-axis anisotropy $K_1 \approx 2.9 \times 10^{-17}$ J and hard axis anisotropy $K_3 \approx 6.4 \times 10^{-17}$ J. Note that these are considerably greater than room temperature (300 K) thermal energy, $k_B T \approx 4.14 \times 10^{-21}$ J. This implies that the Wang et al [1] particles are well described by Ising-like dipoles. The simulation energy unit is $\varepsilon = \mu_0 \mu M_s = 3.21 \times 10^{-16}$ J, also large compared to $k_B T$. This leads to the dimensionless coupling parameters,

$$k_1 = \frac{K_1}{\varepsilon} = 0.0897, \quad k_3 = \frac{K_3}{\varepsilon} = 0.200, \quad \mathcal{T} = \frac{k_B T}{\varepsilon} = 1.29 \times 10^{-5}. \quad (11.41)$$

The fact that k_3 is the largest term here helps to keep the dipoles pointing well within the xy-plane.

At the closest island separation, $a = 320$ nm, used by Wang *et al*, the dipolar coupling constant is found to be $D = 2.68 \times 10^{-19}$ J, which leads to the dimensionless value,

$$d = \frac{D}{\varepsilon} = 8.35 \times 10^{-4}, \quad (11.42)$$

very weak compared to k_1. For a separation of $a = 880$ nm, d will be weaker by the factor $(320/880)^3$. Even so, one needs to remember that the long range of the dipolar interactions involves summing over many terms, which amplifies their effects.

With $k_3 > k_1 > d > \mathcal{T}$, the dipoles naturally have the frustrated behavior expected of a spin ice system. Unfortunately, this makes simulations difficult, primarily because the temperature is so low, and the k_1 anisotropy will maintain the system very Ising-like. Over the number of time steps that can be reasonably calculated in a simulation, the possibility for reversal of dipoles is rather low, except in the presence of an external applied magnetic field. As a result, it will be interesting also to consider some other sets of parameters (according to possibly different models for the islands, new materials, and so on) that could have a more thermalized dynamics at room temperature.

11.4.2 Simulations of equilibrium for Wang *et al* [1] islands

Langevin-LLG simulations can be used to estimate thermodynamic quantities including the average system energy $\langle E \rangle$, total magnetic moment $\langle \boldsymbol{\mu} \rangle = \langle \sum_i \boldsymbol{\mu}_i \rangle$ and monopole densities ρ and ρ^*. From the fluctuations in instantaneous system energy and magnetic moment, the specific heat C and components of magnetic susceptibility χ_{ab} can also be found, using expressions (4.64) and (4.110), respectively. For the susceptibility of the spin ice model, a short exercise shows that we can write specifically,

$$\chi_{ab} = \chi_0[\beta\varepsilon\, N(\langle m_a m_b \rangle - \langle m_a \rangle\langle m_b \rangle)], \quad (11.43)$$

where **m** is the dimensionless average dipole moment per site,

$$\mathbf{m} = \frac{1}{N}\sum_{i=1}^{N} \hat{\mu}_i \quad (11.44)$$

and the natural (and dimensionless!) unit of susceptibility here is,

$$\chi_0 = \frac{2\mu_0 \mu^2}{a^2 L_z \varepsilon}. \quad (11.45)$$

This formula assumes that the system volume is taken as $V_{\text{sys}} = \frac{1}{2}Na^2 L_z$, for N islands. Refer to chapter 4 for further information and an approach to estimate the statistical errors of these quantities. Because fluctuations contribute to C and χ, they are much more susceptible to statistical errors than $\langle E \rangle$ and $\langle \boldsymbol{\mu} \rangle$. This imposes the use of long time sequences for adequate control of the errors, especially if the system is at a point where dipole reversals are more likely.

> **Exercise 11.6.** By using (4.110), verify that expression (11.43) is applicable to spin ice problems and that the unit of susceptibility is given by (11.45).

We show results of Langevin–LLG simulations [7] where the system is initiated in a random state, and the temperature is first set to the highest value of interest. The system is to be relaxed at a sequence of diminishing temperatures, with averages made at each temperature. The last state at the current temperature becomes the initial state at the next lower temperature. This might be considered as a simulated annealing. The second-order Heun algorithm (chapter 5) can be used, with a time step $\Delta\tau = 0.01$, which is seen to conserve energy well at zero temperature. At finite temperature a damping of $\alpha = 0.1$ is included, together with the stochastic fields $\mathbf{b}_{s,i}(\tau)$. Data are saved at a sampling interval of $\Delta\tau_s = 10^3 \Delta\tau$, after relaxing the system over some 100 initial samples. A minimum of $N_s = 4000$ data samples were used for calculating averages, however, the simulation is allowed to run even longer and average over more samples until the percent error in the averaged system magnetic moment is less than 0.1%, if needed. Thus, the final dimensionless time variable (at each temperature) is at least $\tau_{\text{end}} = 40\,000$. The system has a 16×16 grid of primitive cells (two dipoles per cell) with open boundaries, at lattice parameter $a = 320$ nm. Because it is a square grid, the dipolar interactions can be accelerated using fast Fourier transform (FFT) techniques, see section 3.4.

Results for the energy and specific heat are shown in figure 11.10(a), for scaled temperatures $0.001 \leqslant \mathcal{T} \leqslant 0.1$, a range that is primarily determined by the smallness of the coupling constants $k_1 \approx 0.0897$ and $d \approx 8.35 \times 10^{-4}$. For comparison, the corresponding results for different components of magnetic susceptibility are seen in figure 11.10(b). Note that although $k_3 \approx 0.200$ is larger, it mainly serves to maintain the dipoles pointing within the xy-plane. The primary feature in $C(\mathcal{T})$ is a broad peak near $\mathcal{T} \sim 0.02$, which is a similar order of magnitude as k_1. A similar peak appears at a slightly lower temperature in the in-plane susceptibility components, which measure in-plane magnetic fluctuations; there is no similar feature in χ_{zz}. These peaks are associated with the fluctuations that become more likely when the thermal energy is becoming greater than k_1, allowing dipoles to undergo reversals. However, the location of 300 K room temperature on the graph is all the way to the left on the abscissa, marked by arrows. This means that for all practical purposes, the interesting features would take place greatly above room temperature. At such high temperatures many of the assumptions of the spin ice model will be violated, so the features found really would not be accessible, without first destroying the spin ice. Even so, it is good to discuss the further results, with the motivation being the possibility to change the scale of parameters in future novel spin ice materials.

The monopole density for this model is shown in figure 11.11. One can see that having initiated the system at higher temperature, as the temperature is lowered, both the discrete and continuous monopole densities do not tend towards zero as $T \to 0$. Instead, their limiting value is quite large, $\rho(0) \sim \rho^*(0) \sim 0.44$, showing that there is considerable frozen-in disorder, which is due to frustration. As a whole, the

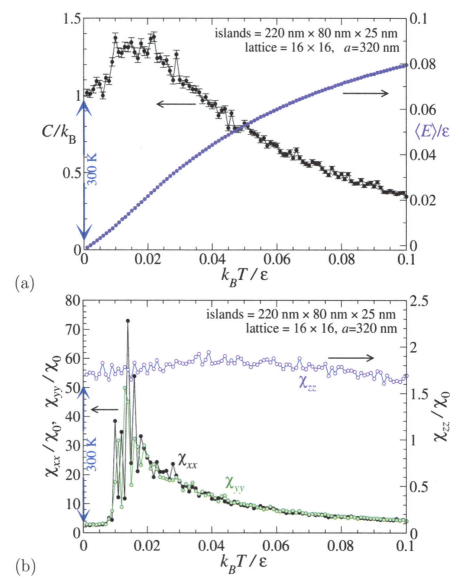

Figure 11.10. Equilibrium properties for a 16 × 16 array of elliptical Py particles, sizes 220 nm × 80 nm × 25 nm, using parameters as given in (11.41) and (11.42, as functions of temperature scaled by the energy unit $\varepsilon = 0.321$ fJ. C and $\langle E \rangle$ are the averaged specific heat and energy per island, respectively. The χ_{ab} are the components of the magnetic susceptibility, in units of χ_0 given in expression (11.45). The blue arrows mark the location of room temperature, essentially not accessible.

dipoles have not been able to move towards the ground state in a dynamic process. This is in great contrast to the results mentioned [5] in the Ising limit for this model at low temperature. For higher temperatures, the monopole density does increase rather sharply around $\mathcal{T} \sim 0.02$, with the discrete definition tending already rather

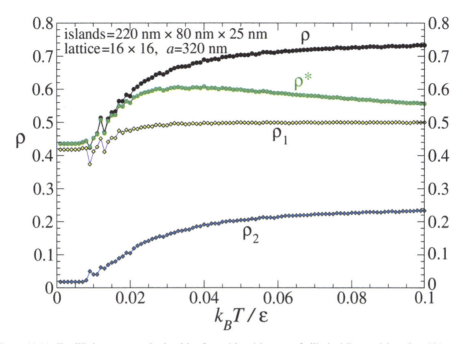

Figure 11.11. Equilibrium monopole densities for a 16 × 16 array of elliptical Py particles, sizes 220 nm × 80 nm × 25 nm, using parameters as given in (11.41) and (11.42), as functions of temperature scaled by the energy unit $\varepsilon = 0.321$ fJ. Note that while the discrete definition ρ moves upward towards its high temperature limiting value, $\rho \to 3/4$, with increasing temperatures, the continuous definition actually falls downward towards its limiting value $\rho^* \to 7/15$. Both definitions remain non-zero at $T \to 0$, showing the result of frustration and lack of ability to reach a ground state. At high temperature, the discrete density of single (double) charges is close to the limiting value $\rho_1 \to \frac{1}{2}$ ($\rho_2 \to \frac{1}{4}$).

close to the limit $\rho \to 0.75$ even at $\mathcal{T} = 0.1$. The densities of single and double charges each tend to their expected asymptotic high-T values, $\rho_1 \to \frac{1}{2}$ and $\rho_2 \to \frac{1}{4}$. At low temperature, there is even a small density of double charges, $\rho_2 > 0$. The continuous definition $\rho^*(T)$ has an unusual behavior, first reaching a value $\rho^* \approx 0.6$ around $\mathcal{T} \sim 0.03$, and then falling gradually for higher temperatures. Over most of the temperature range studied, ρ^* stays above the expected high temperature limit $\rho^* \to 7/15$. A much higher temperature is needed to see the value expected in the fully disordered limit.

Finally for this system we also show the order parameter Z versus temperature in figure 11.12. In this case, Z is close to zero at high T, as expected, but it only takes on a very small value at $T \to 0$. This confirms that the system does not fall into any configuration even vaguely close to a ground state. Also, figure 11.12 displays an rms average value of the total system magnetic moment, e.g.

$$m_{\text{rms}} = \frac{1}{N}\sqrt{\left(\sum_{i=1}^{N} \hat{\mu}_i\right)^2}. \tag{11.46}$$

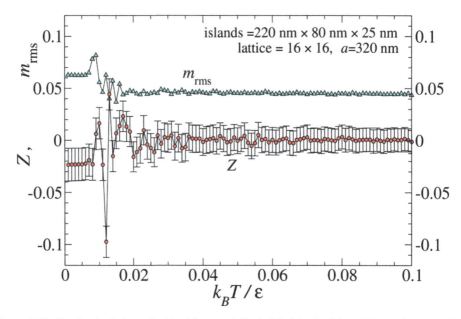

Figure 11.12. For the simulations of a 16 × 16 array of elliptical Py islands of sizes 220 nm × 80 nm × 25 nm, the order parameter Z defined in (11.22) and the rms average of dipole moment per site (see definition (11.46)) versus temperature. The system was cooled starting from the highest temperature shown. There is only a limited ordering at low T, and the system does not tend to a ground state.

This is expected to be zero in a disordered state and also in a ground state, as there is no net magnetic moment in the system. That is found more or less to be the case in the simulations.

11.4.3 Simulations of square ice with artificial model parameters

The first set of simulations of the previous section does not display a thermalization; the system was greatly frustrated and was not able to move towards a ground state at low temperature. Clearly that was due to the over-dominance of the easy-axis anisotropy k_1, especially relative to the temperature, and also relative to the dipolar interactions. One could suppose smaller islands, which would reduce K_1, but at the same time this would also reduce $\varepsilon = \mu_0 \mu M_s$, as both these factors should be roughly proportional to the island volume, leaving k_1 nearly unchanged. This would reduce ε and increase the scaled temperature $\mathcal{T}$, however, the dipolar interactions will become much weaker as μ is reduced with the island volume. For typical real material values, it is difficult to avoid the dominance of K_1. Perhaps another approach would be to consider islands of other shapes, and/or different aspect ratios.

In this second set of simulations, we show the differences that take place for a fictitious model, where the parameters are given assumed values:

$$k_1 = 0.1, \quad d = 0.1, \quad k_3 = 0.1. \tag{11.47}$$

This choice of parameters is called *model C* in [6]. The equal values of k_1 and k_3 tend to maintain the dipoles close to pointing within the xy-plane[2], while the equal values of k_1 and d allow them both to produce an active dynamics within the xy-plane. These are the energy couplings relative to the energy scale $\varepsilon = \mu_0 \mu M_s$, which is not specified. While it is possibly unrealistic to obtain such parameters for typical materials, from the theoretical viewpoint it is interesting to discuss the ability of this model to evolve in the phase space. It is expected to have a smoother dynamics where the frustration effects are moderated. There could be possibilities for the development of future materials that might be described by these more moderate spin ice parameters. The simulations can be carried out as just described for the model of Wang *et al* islands [1], but with a minor modification. Here the dipolar coupling is much stronger, which requires the choice of a smaller Heun time step, $\Delta \tau = 0.001$, to give correct dynamics at finite temperature, and energy-conserving dynamics at zero temperature. Averages were made with data at sampling time interval $\Delta \tau_s = 10^3 \Delta \tau$, using a minimum of $N_s = 4000$ data samples, or until the percent error in total system magnetic moment became less than 0.1%.

A 16 × 16 system size is used, which has 256 primitive cells of two dipoles per cell. A typical high temperature state from the course of the simulation at $\mathcal{T} = 0.50$ is shown in figure 11.13. Arrows drawn as →(–▷) indicate positive (negative) μ_z-components. $|q| = 1$ ($|q| = 2$) monopoles are indicated with smaller (larger) circles at the vertices. Green (red) filled circles indicate positive (negative) signs of the charges. For this higher temperature, there is a large number of charges at the vertices, and the system is generally disordered. A configuration at $\mathcal{T} = 0.22$, close to the transition, is shown in figure 11.14. Now although there is considerable charge present, it is mostly in the form of single charges. The order parameter is $Z = 0.605$ and the system remains otherwise rather disordered. Finally when the temperature is reduced to $\mathcal{T} = 0.02$, as seen in figure 11.15, no monopole charges of any kind are left. The system can be seen to be well ordered, and mostly bears a strong correlation with a ground state. The dipoles tend to have alternating directions from site to site, with only minor fluctuations from a perfect order. The state shown has $Z = 0.982$, close to that expected in a ground state.

The temperature-dependence of average energy, specific heat and magnetic susceptibilities per island are displayed in figure 11.16. There is a strong peak in C near a critical temperature $\mathcal{T}_c \approx 0.22$, and the corresponding peak in the in-plane susceptibility components $\chi_{xx} \approx \chi_{yy}$ at a slightly higher temperature. The fluctuations of the island dipoles out of the xy-plane are small, as testified by the much smaller values of χ_{zz}, and the lack of any significant feature in $\chi_{zz}(T)$.

In figure 11.17, the rms magnetic moment per site and the order parameter Z are displayed. The order parameter again is close to zero at high temperature, however, as $T \to 0$, one sees that Z tends to a finite value close to 1. This indicates strongly that the system is moving towards one of the ground states for very low temperature.

---

[2] The energy cost to tilt out of the xy-plane towards the z-direction by angle θ is $\Delta E = K_1 (1 - \cos^2 \theta) + K_3 \sin^2 \theta = (K_1 + K_3) \sin^2 \theta$.

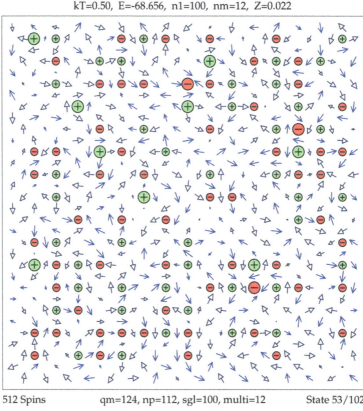

Figure 11.13. A 16 × 16 square spin ice configuration for model C with $d = k_1 = k_3 = 0.1$, at scaled temperature $\mathcal{T} = 0.50$, well above the transition temperature $\mathcal{T}_c \approx 0.22$. Arrows are projections of the xy dipole components on the plane. Those drawn as →(-▷) indicate positive (negative) μ_z-components. $|q| = 1$ ($|q| = 2$) monopoles are indicated with smaller (larger) circles at the vertices. Green (red) filled circles indicate positive (negative) signs of the charges. The system is mostly disordered with a considerable number of single and double monopole charges.

It was a random choice; the system could have just as well tended towards the opposite ground state with the value $Z = -1$. Remember that the system was initiated at higher temperature and relaxed towards lower temperature. The rms magnetic moment also has a weak peak in the same region that χ_{xx} and χ_{yy} have a peak; $m_{\rm rms}$ has been amplified by 10 × in the plot due to its small values. At higher temperature there is a non-zero value. As $T \to 0$, $m_{\rm rms}$ tends towards zero, which would be expected for a ground state where all islands alternate in direction along any row or column of the system.

The local order distributions $p_a(z_a)$ and $p(z)$ for this model are shown in figure 11.18, for a sequence of different temperatures above and below the transition temperature $\mathcal{T}_c \approx 0.22$. For high temperature, the distribution of $p_a(z_a)$ is close to uniform, however, with obvious thermal fluctuations, and also a weak parabolic form. The parabolic form can be seen to be due to the competition of the in-plane

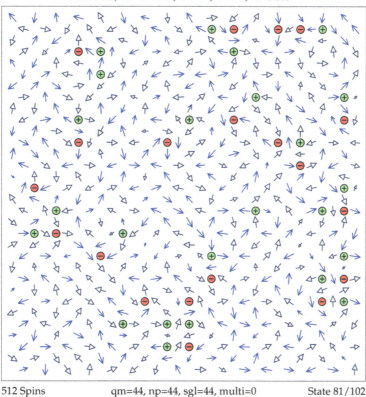

Figure 11.14. A 16 × 16 square spin ice configuration for model C (equation (11.47)) at $\mathcal{T} = 0.22$, close to the transition temperature $\mathcal{T}_c$. See the figure 11.13 caption for the notation. Most of the charges present at this temperature are single charges. The system is somewhat ordered, with order parameter $Z = 0.605$, but there are still strong fluctuations present.

anisotropy K_1 with the temperature. This makes the high temperature probability follow from a Boltzmann factor,

$$p_a(z_a) \propto \exp(K_1 z_a^2 / k_B T), \quad T \gg T_c. \tag{11.48}$$

One would have to go to quite high temperatures for the distribution to appear close to uniform. As the temperature is lowered, the distribution of z_a becomes quite skewed with extra probability at the $z_a = +1$ end, and reduced probability at $z_a = -1$. The changes take place most strongly close to the transition temperature. For very low temperature, $p_a(z_a)$ becomes very strongly peaked at $z_a = +1$. The distribution of z_b also becomes concentrated near the point $z_b = -1$ (not shown here). This is clear evidence that the system moves into the $s = +1$ ground state. This is also reflected in the derived $p(z)$ probability distribution. At high temperature, it takes a triangular shape, peaked at $z = 0$, which is to be expected when both z_a and z_b are uniformly distributed. At low temperature the probability becomes

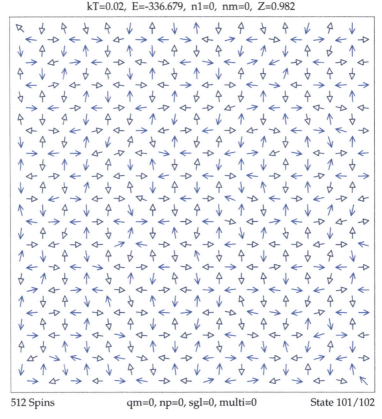

Figure 11.15. A 16 × 16 square spin ice configuration for model C (equation (11.47)) at $\mathcal{T} = 0.02$, well below the transition temperature $\mathcal{T}_c \approx 0.22$. See figure 11.13 caption for the notation. There are no monopole charges of any kind present, and the dipoles exhibit alternating directions from site to site. The order parameter is $Z = 0.982$, which shows the proximity of this state to a ground state, although there are still some weak angular fluctuations.

concentrated at $z = +1$, which confirms that the system relaxes into the $s = +1$ ground state. Therefore, although square lattice spin ice exhibits frustration, that does not prevent the system from moving adequately through phase space, and the system is able to reach a ground state in dynamic processes. In this sense the frustration could be considered a weak form of frustration. The fact that there are only two ground states may help to alleviate the frustration, when dynamics is considered.

The discrete monopole densities versus temperature for single charges, ρ_1, and double charges, ρ_2, are shown in figure 11.19, along with the total discrete charge density ρ and the continuous definition ρ^*. The continuous ρ^* is the only one to remain above zero for very low temperature, because it can measure fluctuations of the dipoles away from the island axes, even in the absence of discrete monopole charges. Single discrete charges are present only for $\mathcal{T}$ greater than about 0.1, and double charges are present for $\mathcal{T} > 0.2$. In contrast to the model for Wang islands,

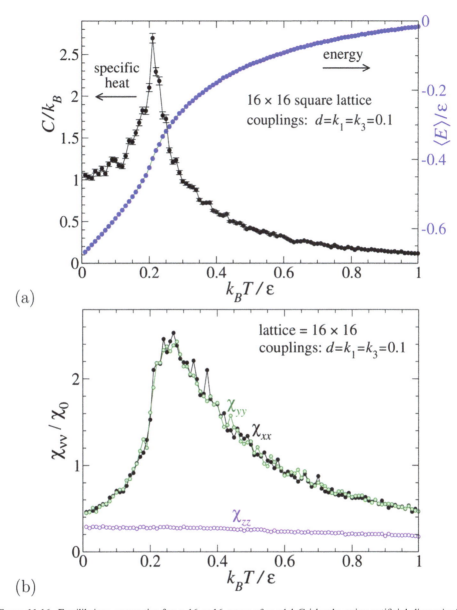

Figure 11.16. Equilibrium properties for a 16 × 16 array of model C islands, using artificial dimensionless model parameters $d = k_1 = k_3 = 0.1$, versus the scaled temperature. In (a) the specific heat and average energy per island are shown. $C(T)$ exhibits a strong peak at a critical temperature $T_c \approx 0.22$, due to a transition towards the ground state for low temperature. In (b) components of the magnetic susceptibility are plotted; there is a peak only for the in-plane components.

this confirms that this model has a thermalized dynamics, and is able to move towards a ground state at low temperature. At higher temperatures, these density values shown are still rather far from the high temperature limits, but further simulations do confirm that eventually the system tends towards the values discussed in section 11.2, $\rho \to \frac{3}{4}$, $\rho_1 \to \frac{1}{2}$ and $\rho_2 \to \frac{1}{4}$.

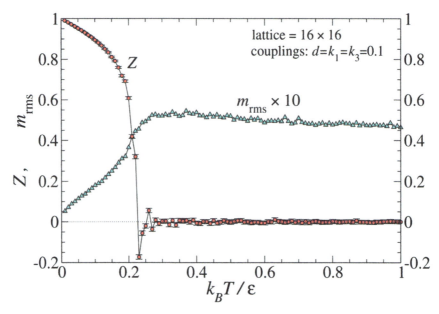

Figure 11.17. Equilibrium properties for a 16 × 16 array of model C islands, using artificial dimensionless model parameters $d = k_1 = k_3 = 0.1$, versus the scaled temperature. The order parameter Z is defined in (11.22) and the rms average of dipole moment per site is defined in (11.46). As Z tends to 1 for low temperature, it shows a collapse of the system to a state near one of the ground states. The rms magnetic moment per site, which has been magnified by 10×, also shows a feature, and goes close to zero as $T \to 0$.

11.4.4 Magnetic hysteresis in model C with artificial parameters

The calculations of the previous section can easily be extended to include an externally applied magnetic field. By looking at the response while the magnetic field strength $\mathbf{h}_{\text{ext}}$ is varied, for the system at a fixed temperature, hysteresis curves can be obtained. Figure 11.20 shows some results, having scanned the magnetic field starting from a large positive value along the $\hat{\mathbf{x}}$-axis, bringing it to a large value in the opposite direction, and finally returning to the original maximum strength and direction. The system is started with the island dipoles randomly oriented. Averages are taken over $N_s = 10^3$ samples at each applied field, after which the field is changed slightly. For the hysteresis curves calculated here, a total of 160 steps in the magnetic field brought it from the maximum value in one direction to its maximum strength in the opposite direction. The direction of the field is indicated by the angle θ_h from the $\hat{\mathbf{x}}$-axis to the field direction. However, the results depend very little on θ_h, so only results for $\mathbf{H}_{\text{ext}}$ along $\hat{\mathbf{x}}$ are shown.

For the higher temperature $\mathcal{T} = 0.1$, the maximum field strength used is not enough to totally saturate the magnetization, and the $m(h_{\text{ext}})$ response is close to linear. Some nonlinearity starts to appear for $\mathcal{T} \sim 0.5$ and below, but there is nothing unusual in the curves. For $\mathcal{T} = 0.1$, a central plateau appears in $m(h_{\text{ext}})$, and the overall response is highly nonlinear.

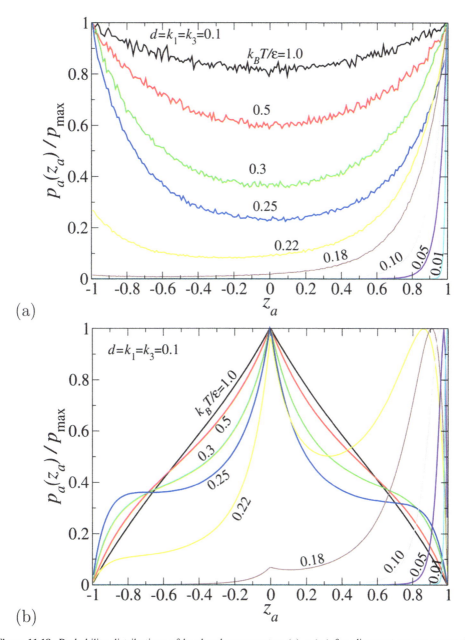

Figure 11.18. Probability distributions of local order parameters (a) $p_a(z_a)$ for alignment on one magnetic sublattice and (b) $p(z)$ that measures the local ordering such as that in one of the ground states, see definition of z in (11.24). Curves are labeled by scaled temperature. The functions are scaled by the maximum values so details of the different curves are visible. At low temperature the distributions become concentrated near $z_a = +1$ and $z = +1$, showing the evolution of the system towards the $s = +1$ ground state. An associated plot of $p_b(z_b)$ is similar to $p_a(z_a)$ but peaks near $z_b = -1$.

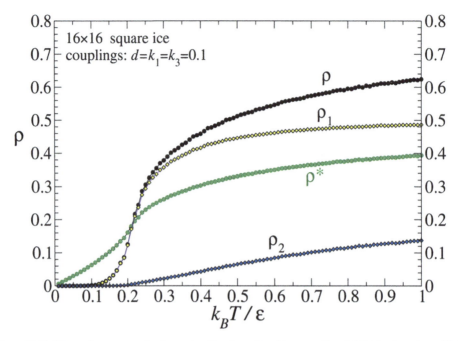

Figure 11.19. The various monopole charge densities for a 16×16 array of model C islands, using artificial dimensionless model parameters $d = k_1 = k_3 = 0.1$, versus the scaled temperature. See (11.28) and (11.29) for definitions. Note how the density from the continuous definition, ρ^*, which measures angular deviations from island long axes, remains non-zero at low temperature while very few discrete single (ρ_1) or double monopoles (ρ_2) are present. For much higher temperatures all the densities do tend towards their asymptotic high-T values, as listed in figure 11.11.

The reason for the plateau can be brought to light by looking at corresponding curves for the responses of order parameter Z in figure 11.21 and total discrete monopole density ρ in figure 11.22, with respect to changing h_{ext}. It is seen that the plateau region of $m(h_{\text{ext}})$ corresponds to $Z \approx \pm 1$ and a very small value of ρ, demonstrating that the system is close to its ground state. The variation of the behavior of $\rho(h_{\text{ext}})$ with decreasing temperature is quite striking. At higher temperatures, the monopole density remains rather large, regardless of the field strength. While in the plateau region, there is close to a linear relation in $m(h_{\text{ext}})$, as the dipoles are being tilted slightly away from the ground state directions by the field. Once the field strength surpasses $h_{\text{ext}} \approx 0.2$ the system is taken away from the ground state configuration, and returns to a state with a small value of Z and large non-zero values of monopole density. The field plays a strong role in taking the system out of its ground states. By using the artificial parameters in this model, one finds a greater reversibility of the system (lesser frustration), as it is able to remain in a reasonable equilibrium-like state. Little hysteresis is present in any of the quantities measured.

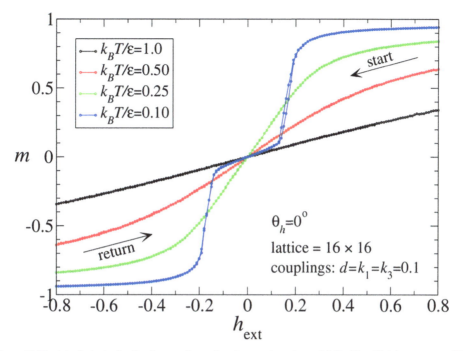

Figure 11.20. A typical set of $m(h_{ext})$ curves for various temperatures in model C with artificial parameters. The field is applied parallel to $\hat{x}$, initially at strength $h_{ext} = +0.8$, then reduced to $h_{ext} = -0.8$ in 160 steps, followed by a return to the original value. Only the component of **m** along the field is displayed. There is only minimum hysteresis seen at the lowest temperature $\mathcal{T} = 0.1$, which is where the system has fallen close to a ground state.

One can compare this *model C* with artificial parameters to the Ising limit and to the model of Wang *et al* islands. Obviously the Ising limit does not have any real dynamics, and the MC simulations do not take into account the energy barriers caused by K_1 working in combination with K_3. As a result, the peak in specific heat for the Ising limit is present at a relatively high temperature, $k_B T \approx 7.2 D$. For the Wang *et al* islands the location of the specific heat peak near $k_B T/\varepsilon \approx 0.01 - 0.02$ is determined instead by the value of $k_1 = 0.0897$. This is a more realistic model, which correctly includes the energy barriers that cause frustration. However, the weak value of d does not help to facilitate an efficient dynamics. Finally, in model C, again $k_1 = 0.1$ is probably most important in determining the peak in C near $k_B T/\varepsilon \approx 0.22$, but in this case with $d = 0.1$ being the same as k_1, the dynamics is more fluid. Energy barriers have become weak and smoothed out and the dynamics is easily able to surpass them. This also suggests that model C is able to maintain the system more closely as if in an equilibrium configuration, that is, not in strongly metastable states. The Wang *et al* islands are much more likely to be held in high energy metastable states that cannot easily fall towards the ground states without special protocols, such as using varying applied fields to move the system towards a ground state.

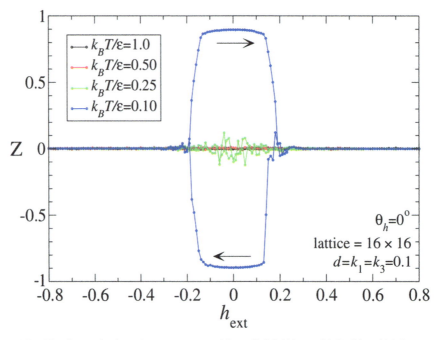

Figure 11.21. The changes in the order parameter Z with applied field in model C with artificial parameters, associated with the $m(h_{ext})$ curves of figure 11.20. For the highest temperature $\mathcal{T} = 1.0$, note that Z remains close to zero for all h_{ext}. The plateau in figure 11.20 for $\mathcal{T} = 0.1$ corresponds here to large values of Z, which by chance have opposite signs for increasing and decreasing h_{ext}.

11.5 Triangular lattice artificial spin ice

Other 2D lattices can be used to build an artificial spin ice. One interesting example is a triangular lattice of vertices where monopoles can be measured, figure 11.23. At each vertex there are six dipoles that are set along the nearest neighbor radii from that vertex to the neighboring vertices. The lattice can be composed from primitive rhombohedral cells with three dipoles per cell (magnetic sublattices). A model for spin ice on this lattice has been considered by Mól et al [8], analyzed through MC simulations of the Ising spin ice model.

This model bears a strong resemblance to the model for square lattice artificial spin ice, because each vertex has an even number of adjacent dipoles. One would expect that the lowest energy of a vertex will be obtained when the dipoles follow a three-in/three-out rule. The configuration in figure 11.23 satisfies this rule; there is another equivalent configuration obtained by reversing all dipoles, which suggests a two-fold degenerate ground state. A rotation through 60° around any vertex also reverses all spins, which supports the statement that there are two ground states.

11.5.1 Counting and energetics of vertex configurations

In the Ising limit there are $2^6 = 64$ possible vertex configurations, but due to symmetry they can be grouped into eight different types. The counting is

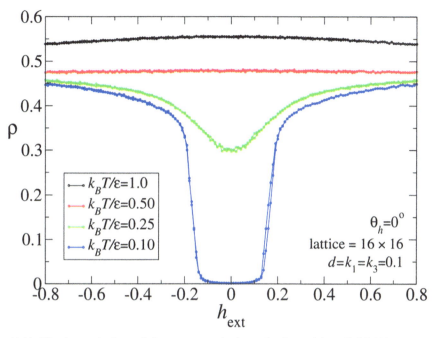

Figure 11.22. The changes in the total discrete monopole charge density ρ with applied field in model C with artificial parameters, associated with the $m(h_{ext})$ curves of figure 11.20. The plateau in figure 11.20 for $\mathcal{T} = 0.1$ corresponds here to small values of ρ, as expected for the system being close to a ground state configuration. At higher temperature the system stays far from a ground state, with large monopole density.

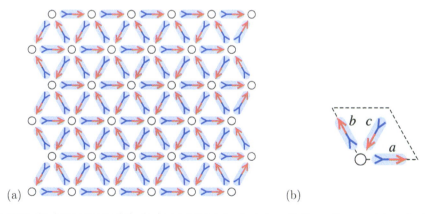

Figure 11.23. (a) Arrangement of dipoles in a ground state of spin ice with the vertices on a triangular lattice. The six dipoles surrounding each vertex satisfy a three-in/three out rule. Note also that all 'up' ('down') triangles have a counterclockwise (clockwise) circulation of the dipoles around their perimeters. (b) A rhombohedral primitive cell from which the lattice can be built, showing the three magnetic sublattices a, b, c, with dipoles at 0° (a), 120° (b) and 240° (c) to the $\hat{x}$-axis.

straightforward. For instance, the number of ways to obtain three inward dipoles and three outward dipoles is $(6 \times 5 \times 4)/(3 \times 2) = 20$. Out of those 20 configurations, there are three different topologies: one with alternating in/out dipoles as one follows the path around the vertex (two ways), one with a neighboring inward pair and a neighboring outward pair (12 ways), and one with three neighboring inward dipoles and three neighboring outward dipoles (six ways). Besides those three topologies, there are another five topologies with higher energies and unequal numbers of inward versus outward dipoles. Then, these latter five topologies are the ones that contribute monopole charge density to the system.

A vertex with its six dipoles is shown in figure 11.24, indicating a notation for the dipolar exchange-like interactions within that vertex. In the Ising limit for triangular spin ice, the energetics is determined by Hamiltonian (11.15) as for square ice. The lattice constant a is the distance between neighboring vertices. The center to center distance between nearest neighbor dipoles is $r_{12} = a/2$, between second nearest neighbors is $r_{13} = (\sqrt{3}/2)a$ and between third nearest neighbors is $r_{14} = a$. The dipoles are drawn in the directions that would correspond to each site's $\hat{u}_i$ vector. Equivalently, the state of that vertex as shown has the Ising variables $\sigma_i = +1$, $i = 1 - 6$. With a somewhat more complicated geometry, one has the following vectors needed for calculation of the exchange constants:

$$\hat{u}_1 = \hat{x}, \quad \hat{u}_2 = \frac{1}{2}\hat{x} + \frac{\sqrt{3}}{2}\hat{y}, \quad \hat{u}_3 = -\frac{1}{2}\hat{x} + \frac{\sqrt{3}}{2}\hat{y}, \quad \hat{u}_4 = -\hat{x}.$$

$$\mathbf{r}_{12} = \frac{a}{2}\left(-\frac{1}{2}\hat{x} + \frac{\sqrt{3}}{2}\hat{y}\right), \quad \mathbf{r}_{13} = \frac{\sqrt{3}\,a}{2}\left(-\frac{\sqrt{3}}{2}\hat{x} + \frac{1}{2}\hat{y}\right), \quad \mathbf{r}_{14} = -a\hat{x}. \quad (11.49)$$

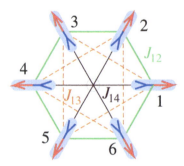

Figure 11.24. Sketch of notation for sites and Ising-limit interactions in a vertex in triangular spin ice. The six full green lines show the nearest neighbor interactions (J_{12}, distance $r_{12} = \frac{a}{2}$), the six orange dashed lines are second nearest neighbor interactions (J_{13}, distance $r_{13} = \frac{\sqrt{3}}{2}a$), and the three thin brown lines are third nearest neighbor interactions (J_{14}, distance $r_{14} = a$). See (11.50) for the values of J_{ij}. The dipole directions shown correspond to all $\sigma_i = +1$ on this vertex.

Then the three exchange constants between nearest, second nearest and third nearest neighbors are found from (11.15) to be

$$J_{12} = 10D, \qquad J_{13} = \frac{14}{3\sqrt{3}}D \approx 2.69D, \qquad J_{14} = 2D. \qquad (11.50)$$

Now that these are determined, they can be used to evaluate the isolated energies of the eight types of vertex topologies. An accounting of their energies is expected to be useful to consider the statistics with their relative probabilities of generation in thermal equilibrium. Following [8] we can name the eight types as type I, II, III, ..., VIII, according to increasing isolated energy. All of the vertices in the ground state in figure 11.23 are in their lowest configuration; they are all type I.

Consider the calculation of the vertex energies, for the topologies summarized in figure 11.25. For a vertex of type I, the Ising variables alternate around the vertex: $\sigma_1 = \sigma_3 = \sigma_5 = +1$, $\sigma_2 = \sigma_4 = \sigma_6 = -1$, for example. This is two-fold degenerate. Then summing over the 15 pair interactions in the vertex, its energy in notation $E(\sigma_1, \sigma_2, \sigma_3, \sigma_4, \sigma_5, \sigma_6)$ is

$$\begin{aligned} E_I &= E(1, -1, 1, -1, 1, -1) = -6J_{12} + 6J_{13} - 3J_{14} \\ E_I &= -6(10D) + 6(2.69D) - 3(2D) \approx -49.8D. \end{aligned} \qquad (11.51)$$

Reversing σ_2 and σ_3, for example, there are 12 equivalent configurations that also satisfy the three-in/three-out rule, with a higher energy

$$\begin{aligned} E_{III} &= E(1, 1, -1, -1, 1, -1) = (2-4)J_{12} + (2-4)J_{13} + (2-1)J_{14} \\ &= -2(10D) - 2(2.69D) + 2D \approx -23.4D. \end{aligned} \qquad (11.52)$$

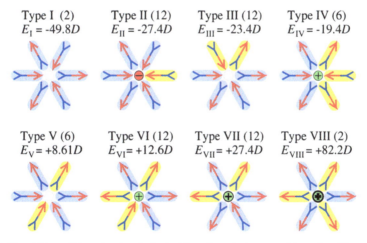

Figure 11.25. The eight possible vertex topologies in triangular spin ice, named in order of increasing energy. The energies are calculated from first, second and third nearest neighbor dipolar interactions, see the text. The multiplicities are given in parenthesis. Islands in blue are in the same direction as in type I; islands in yellow show the reversals relative to type I. Sizes of monopolar charges are $q = \pm 1$ for types II, IV, V, $q = \pm 2$ for type VII and $q = \pm 3$ for type VIII.

Starting from type I, but instead reversing σ_2 and σ_5, there are another six equivalent three-in/three-out configurations, with considerably higher energy,

$$E_V = E(1, 1, 1, -1, -1, -1) = (4 - 2)J_{12} + (2 - 4)J_{13} - 3J_{14}$$
$$= 2(10D) - 2(2.69D) - 6D \approx 8.61D. \quad (11.53)$$

Next, there are 12 ways to obtain four-in/two-out or two-in/four-out that have ± 1 unit of monopole charge, obtained from type I by reversing only one dipole, say, σ_1. The energies are

$$E_{II} = E(-1, -1, 1, -1, 1, -1) = (2 - 4)J_{12} + (2 - 4)J_{13} + (1 - 2)J_{14}$$
$$= -2(10D) - 2(2.69D) - (2D) \approx -27.4D. \quad (11.54)$$

It is important that this is the second lowest energy of a vertex. A single dipole flip out of the ground state already creates a pair of monopoles at relatively little energy cost. One can also obtain ± 1 unit of monopole charge by reversing three neighboring dipoles in type I, say σ_6, σ_1, σ_2, six total ways, with a slightly higher energy

$$E_{IV} = E(-1, 1, 1, -1, 1, 1) = (2 - 4)J_{12} + (2 - 4)J_{13} + 3J_{14}$$
$$= -2(10D) - 2(2.69D) + 3(2D) \approx -19.4D. \quad (11.55)$$

Another way to obtain ± 1 unit of monopole charge is to have all four aligned dipoles neighboring each other, 12 ways, with energy

$$E_{VI} = E(1, 1, 1, 1, -1, -1) = (4 - 2)J_{12} + (2 - 4)J_{13} + (1 - 2)J_{14}$$
$$= 2(10D) - 2(2.69D) - (2D) \approx 12.6D, \quad (11.56)$$

but this is a relatively high energy topology. One could instead start with type I and reverse two dipoles to generate ± 2 units of monopole charge, say, σ_2 and σ_4. There are 12 ways to do this, and the energy is

$$E_{VII} = E(1, 1, 1, 1, 1, -1) = (4 - 2)J_{12} + (4 - 2)J_{13} + (2 - 1)J_{14}$$
$$= 2(10D) + 2(2.69D) + (2D) \approx 27.4D. \quad (11.57)$$

Finally all dipoles could point inward or outward, so there are two ways to obtain ± 3 units of monopole charge, with the highest energy

$$E_{VIII} = E(1, 1, 1, 1, 1, 1) = 6J_{12} + 6J_{13} + 3J_{14}$$
$$= 6(10D) + 6(2.69D) + 3(2D) \approx 82.2D. \quad (11.58)$$

The total multiplicities add up to the 64 possibilities. It is clear that the energetics works in such a way that concentrations such as poles in the vertex greatly increases the energy. Triply charged monopoles will be strongly restricted by the energetics, compared to the type I–IV vertices, which all have negative energies. The results are summarized in figure 11.25.

11.5.2 Comparing frustration in square and triangular spin ice

Triangular spin ice has more variety of vertex states than square ice. It is interesting to compare the probabilities of the different possible vertices in the two systems.

For square ice, there are six dipoles pair interactions in one vertex. Frustration comes about because they cannot all be minimized simultaneously, even in the states obeying the ice rule. The lowest energy, type I vertex, minimizes the four nearest neighbor pairs ($-4J_{12}$) at the expense of the second nearest neighbor pairs, which are not minimized ($+2J_{13}$), see (11.17a). In type II vertices, only two of the nearest neighbors are satisfied while both second nearest neighbor pairs are now minimized, which happens to give a higher energy. Both types have four satisfied dipolar interactions. Mól *et al* [8] suggest that the degree of frustration might be measured by the fraction of satisfied bonds, 4/6 for the vertex states that satisfy the ice rule in square ice.

For triangular ice, there are 15 dipole pairs in a vertex. In those states obeying the three-in/three-out ice rule, which are types I, III and V, there are nine dipoles pairs in their lowest energy directions. Then the fraction of satisfied bonds is 9/15, slightly lower than in square ice. This suggests a stronger degree of frustration in triangular ice, because a greater part of the bonds will never be satisfied.

For square ice out of the 16 vertex configurations, six satisfy the ice rule (fraction 3/8). In triangular ice with 64 vertex states, 20 satisfy the ice rule (fraction 5/16). Of course, the fraction appearing in thermal equilibrium depends on their energies and the temperature via Boltzmann factors, in addition to the topological constraints. But that aside, this suggests that statistically speaking, the greater possibilities in phase space lead also to more frustration effects in triangular ice.

One other important difference is the energy ordering of the vertices with monopole charge. For square ice, charge-free type II vertices are the lowest energy excitation above the ground state, although as in figure 11.3 they cannot be formed without also some type III charged vertices. But this is in contrast to triangular ice, where the lowest energy excited vertex state, type II already contains a unit monopole charge. It is not clear what this implies for the generation of monopole charge in triangular ice. However, triangular ice has the three different types of monopole charges, which might affect the charge densities in equilibrium.

Exercise 11.7. Consider spin ice on a triangular lattice initially in its ground state as in figure 11.23. Calculate the energy change in only the affected vertices if (a) a single dipole is reversed, or (b) three dipoles forming a small triangular loop are reversed. (c) Find the leading corrections due to the next nearest dipolar interactions in other nearby vertices.

11.5.3 Thermal equilibrium in triangular spin ice

The model has been studied within the Ising limit by MC calculations [8] on a range of system sizes. It is found that the specific heat exhibits a sharp peak near $T \approx 15D$, and the height of the peak increases with system size. That suggests a type of phase transition, which can be expected to be associated with the formation of monopole charge density ρ, such as that seen in square ice. The average monopole charge density also rises strongly in the same temperature region. As can be expected, most of the contribution to ρ is due to single charges; there are considerably weaker contributions from the double and triple charges.

Another interesting result found from the simulations was a measurement of the mean separation distance between positive and negative singly charged monopoles as a function of temperature. The opposite signed charges might be expected to be somewhat bound together, similar to systems with bound vortices. It is found that the separation exhibits a broad peak at the same temperature as the sharp peak in specific heat. For still higher temperatures the separation reduces, which can partly be attributed to the generation of a greater density of charges.

This has only been a brief introduction to triangular spin ice. We have seen that triangular spin ice bears some resemblance to square spin ice, especially because both systems have an even number of islands surrounding each vertex.

11.6 Kagomé lattice artificial spin ice

Next we consider artificial spin ice on a Kagomé lattice, where the vertices have an odd number of neighboring islands. This makes the discussion of monopole charges quite different, because now the vertices always have unequal numbers of inward versus outward dipoles. Therefore the vertices are always charged, even in a ground state. With three dipoles per vertex never capable of minimizing the nearest neighbor dipole interactions, the frustration might be expected to be even stronger than in square and triangular ice. Indeed, within each vertex, there are only nearest neighbor dipole interactions, and none of them will be minimized. A model of Kagomé spin ice with ferromagnetic exchange interactions between neighboring spins was first introduced by Willis *et al* [9]; in our discussion here we do not include exchange interactions, but only consider the dipolar and local anisotropy interactions.

Kagomé spin ice is formed with the vertices on a honeycomb grid and the dipoles spanning the spaces between the vertices [10], see figure 11.26 which shows one of the six possible ground states. The colored equilateral triangular cells indicate three sets of primitive cells denoted *ABC*. Within each colored cell the vertex has one dipole inward and two outward, giving a $q = +\frac{1}{2}$ monopole charge; at the vertices connecting the primitive cells there are two inward and one outward, making $q = -\frac{1}{2}$ monopoles. This is in the same charge units as applied to square and triangular ice. A north (or south) pole contributes $+\frac{1}{2}$ (or $-\frac{1}{2}$) unit of charge. If all three dipoles point outward (or inward) at a vertex, then the monopole charge is $+\frac{3}{2}$ (or $-\frac{3}{2}$). This is in particular contrast to an ice model where vertices are surrounded by an even number of dipoles. Generally we may still refer to $\pm\frac{1}{2}$ charges as *single charges*, and $\pm\frac{3}{2}$ charges as multiple charges or as *triple charges*.

With an odd number of dipoles at a vertex, there is no ice rule. Based on energetics, which will be described further below, one can expect that the closest thing to an ice rule will be that the system attempts to make minimal charges ($\pm\frac{1}{2}$) at every vertex. Thus, a ground state is a configuration where there must be minimum charges at all

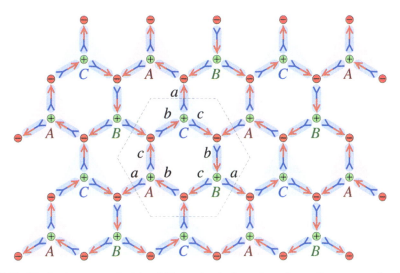

Figure 11.26. Dipoles in a ground state of Kagomé spin ice, where the vertices occupy the sites of a honeycomb lattice, and the dipoles connect between neighboring vertices. Every vertex has a single unit of monopole charge. There are three sets of triangular primitive cells, A, B, C. Each has magnetic basis sites a, b, c. Rotation of the A (or B or C) cells through 120° around the $\hat{\mathbf{z}}$-axis brings their magnetic sublattices to align with those on the B (or C or A) cells. Note that only the b-sublattice dipole is inward within all the ABC cells. The dashed hexagon shows a larger nine-dipole primitive cell from which to generate the lattice.

vertices. In addition, one might expect that the lowest energy will result when each positively charged monopole is surrounded by three negatively charged monopoles, and vice versa. The geometry and symmetry of the lattice force topological constraints as to how to arrange the dipoles to obtain only single monopoles satisfying these requirements at every vertex. A state such as that in figure 11.26 results.

For a ground state configuration, figure 11.26, a 120° rotation of the A (or B or C) primitive cells around the $\hat{\mathbf{z}}$-axis brings those cells into the state of the B (or C or A) cells. This is similar to the dipole reversal in a square ice ground state that interchanges the states of its AB cells. In Kagomé ice, the entire cell is rotated. One can see that each primitive cell has a three-dipole basis, or magnetic sublattices, abc. In the state shown in figure 11.26, the b-sublattice dipoles are inward in all cells, regardless of whether on the A, B or C cells. This *misaligned* or *minority* sublattice could just as well have been the a- or the c-sublattice. Also b could have been outward with a and c inward, which would reverse the signs of all the monopoles. By the choice of any of three sublattices abc as misaligned, and an overall reversal of all dipoles, it is seen that the ground state is six-fold degenerate.

One can see that if only the b-sites in figure 11.26 have their dipoles reversed, then every primitive cell will have three outward dipoles, forming $+\frac{3}{2}$ charges inside the cells and, additionally, forming $-\frac{3}{2}$ charges at the junctions between primitive cells. This appears to be a state of maximum charges and it is probable that it is a state of maximum (dipolar) energy. This would be a two-fold degenerate state (by reversing all dipoles).

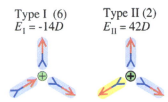

Figure 11.27. The two vertex types in Kagomé spin ice, together with their isolated energies and multiplicities (in parenthesis). The possible charge values on type I are $\pm\frac{1}{2}$; on type II the possibilities are $\pm\frac{3}{2}$.

11.6.1 Kagomé vertex configurations and energetics

Although the ground state is six-fold degenerate, the states of individual vertices are very limited. With only three dipoles per vertex, in the Ising limit there are only $2^3 = 8$ possible vertex configurations. These are split into only two topologies: type I, with single monopole charges (six ways) and type II, with triple monopole charges (two ways), see figure 11.27. The energies of these are determined solely by the nearest neighbor effective exchange constant, which can be found using (11.15) in the Ising limit. Generally, we consider that the Ising model is set up so that $\sigma_i = +1$ for dipoles pointing outward from the vertices in the colored ABC cells, and inward towards the vertices at the corners of those cells. For example, the ground state of figure 11.26 has $\sigma_i = +1$ on a- and c-sublattices and $\sigma_i = -1$ on the b-sublattice.

The lattice constant a is the distance between neighboring vertices. The separation between a pair of dipoles in the same triangular cell is seen to be $r_{12} = a/2$, with a 120° angle between their directions. Then application of (11.15) gives the effective nearest neighbor exchange constant resulting from dipolar interactions,

$$J_{12} = 14D. \tag{11.59}$$

One can check that this is the same as J_{13} for triangular spin ice with the replacement of $r_{13} = \frac{\sqrt{3}}{2}a$ there by $r_{12} = \frac{a}{2}$ for the Kagomé system. Due to the closeness of the dipoles in the Kagomé lattice, the dipole interactions are rather strong. Then, the two vertex topologies have the following energies:

$$E_I = -J_{12} = -14D \tag{11.60a}$$

$$E_{II} = 3J_{12} = 42D. \tag{11.60b}$$

Starting from any state, a single dipole reversal will move an amount of monopole charge $\Delta q = \pm 1$ from one vertex to a neighboring one. On a type I vertex with one inward dipole and two outward, a reversal of the inward dipole (the *minority* one) changes the vertex to type II. The energy cost within the one vertex is already very large: $\Delta E = 4J_{12} = +56D$. On the other hand, if one of the *majority* dipoles was flipped, the type I vertex remains type I, but still some monopole charge was exchanged with a neighboring vertex. Thus one can envision that the single monopole charges even starting in a ground state can move around from one vertex

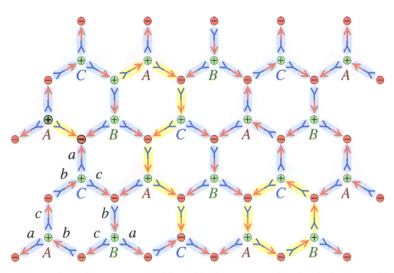

Figure 11.28. Starting from the ground state of figure 11.26, some examples of excitations caused by reversing dipoles (highlighted in yellow). The single reversal of a minority dipole at the left is of very high energy and produces a pair of triple charges as charge $\Delta q = -1$ is transferred in the final direction of the reversed dipole. The string of reversed dipoles in the middle moves charge $\Delta q = -1$ from the top end to the bottom end. The energy increases with the length of the string. The hexagonal loop of reversed dipoles causes no changes in the charges; it is of very low energy.

to another with relatively little cost in energy (at least when measured in individual vertices). This neglects the longer range dipolar interactions, which should contribute something. As long as dipole flips are changing only the majority dipoles at the vertices, no triple charges will be formed, and the local energy changes will be relatively small.

Examples of excitations out of a ground state are shown in figure 11.28. Any dipoles that have been reversed have their islands highlighted in yellow. At the left of figure 11.28, a single minority dipole was flipped in an A cell, creating triple charges at its connected vertices. Note that charge $\Delta q = -1$ has flowed in the final direction of the dipole arrow. This excitation is very high energy, on the order of at least $2 \times 56D$. In the center of figure 11.28, a long string of islands have had their dipoles flipped. All vertices remain type I. At the ends of the string, a majority dipole was flipped, which only changes the sign of the charge in that particular vertex. This results again in a transfer of charge $\Delta q = -1$ in the final direction of the arrows along the string. The C cell in the bottom row now has all negative single charges; at the upper terminus of the string there is an excess of positive single charges. The energy associated with this long string is expected to depend on its length. Finally, on the right in figure 11.28, the dipoles in a hexagonal loop have been flipped. At each cell, a majority/minority pair was reversed. This causes no change in the charges. This should be an excitation of very low energy. The possibility of these and other kinds of low energy excitations will probably enhance the frustration of the system.

Exercise 11.8. Consider some longer range interactions within the dashed hexagonal unit cell in figure 11.26, that take place between different ABC cells. There is a second nearest neighbor effective exchange J_{bb} from Ab to Bb or Cb, and there is a third nearest neighbor effective exchange J_{bc} directly across the hexagon, such as Ab to Cc. Evaluate these coupling constants based on the dipolar interaction (11.15), in units of D.

Exercise 11.9. Starting from the ground state of Kagomé spin ice, figure 11.26, find the energy change (in units of D) in the affected vertices when (a) a single dipole is reversed, or (b) a hexagonal loop of dipoles is reversed. (c) Find the next order corrections due to the second and third nearest neighbor exchange terms acting between different ABC cells, as in the previous exercise.

11.6.2 Order parameters based on ground state configurations

Next we consider how to define some kind of order parameters that measure the closeness of an arbitrary state to a ground state. Of course, with six possible ground states, the degree of frustration in the system is enhanced, because there is such a multiplicity from which the system can select. In square ice, we saw that it is simple to define an order parameter Z_a and Z_b on each magnetic sublattice, and use the combination of the two order parameters to make one overall parameter Z. With six energetically equivalent states for Kagomé ice, this is also possible, but there are some different ways to come up with a measure of the system's closeness to the different ground states.

Complex order parameter

A complex order parameter has been introduced by Chern and Tchernyshyov [10]. With three types of primitive cells and three sublattices in each cell, the Kagomé system effectively has a nine-dipole magnetic unit cell, combining neighboring A, B and C cells into one greater hexagonal primitive cell, see figure 11.26. With the appropriate definition of the Ising variables σ_i (positive when outward in the colored cells and inward at the cell junctions), the complex order parameter is defined by[3]

$$M = \frac{1}{N} \sum_i \sigma_i \exp(-i\mathbf{Q} \cdot \mathbf{r}), \quad \mathbf{Q} \equiv \frac{4\pi}{3a}\hat{\mathbf{x}}. \quad (11.61)$$

N is the number of dipoles used in the sum. This gives a particular phase to each Ising variable. It is designed so that in each of the six perfect ground states, M itself will have the form

---

[3] The sign in the exponential has been changed to -1 here for convenience. It is $+1$ in [10].

$$M = |M|e^{i\phi}, \quad \phi = \frac{n\pi}{3}, \quad n = 0, 1, 2, \ldots 5 \tag{11.62}$$

In simulations, this can be calculated, and plots of the distribution of M in the complex plane then give a good sense of the order present in the system as a whole. For low temperature, the distribution becomes discretized around the possible ground state values. As the temperature is raised, this becomes smeared out. The definition of M averages over all cells and all the abc-sublattices.

Exercise 11.10. Consider the ground states in the Ising limit with the sublattices specified by $\sigma_a, \sigma_b, \sigma_c = \pm 1$, where positive values are outward dipoles in the colored cells. Summing over the hexagonal primitive cell of nine dipoles indicated in figure 11.26, show that the order parameter in (11.61) evaluates to the expression,

$$M = \frac{1}{3}\left(\sigma_a + \sigma_b e^{i\frac{2\pi}{3}} + \sigma_c e^{-i\frac{2\pi}{3}}\right). \tag{11.63}$$

Hint: Use the center of the hexagonal cell as the origin. Note that this result transforms $\sigma_a, \sigma_b, \sigma_c$ into numbers in the complex plane, combining the three sublattices with rotations of 0, $2\pi/3$ and $-2\pi/3$, respectively, see figure 11.29. Then, determine the amplitude and phase of M for the ground state configuration in figure 11.26.

Vector and scalar order parameters

Another approach to defining a set of order parameters, one for each sublattice a, b, c, has been developed in [11]. It is based on looking at a transformation of the B and C cells that brings them to the same state as the A cells, when the system is in a ground state. If the B (or C) cells and their dipoles in figure 11.26 are rotated through $-120°$ (or $+120°$) around the $\hat{z}$-axis, they become oriented just like the A cells. Then, the A, B and C cells are cyclically equivalent under the $120°$ rotation operation.

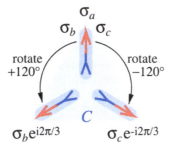

Figure 11.29. Illustration of the calculation of complex order parameter M via (11.63), within one C cell at the top of the nine-dipole magnetic sublattice, using Ising variables $\sigma_a, \sigma_b, \sigma_c$. All three dipoles can be considered to start in the vertical direction. Then b and c are rotated through $+120°$ and $-120°$, respectively, to their final directions. The complex combination of the abc-sublattice vectors then produces M.

We consider the A-cells as reference cells, and denote the described operations as R_i, this brings the B and C cells into coincidence with the A cells. It is the following process on a unit dipole:

$$\hat{\boldsymbol{\mu}}'_i = R_i \hat{\boldsymbol{\mu}}_i = \begin{cases} R(0°) \cdot \hat{\boldsymbol{\mu}}_i, & i \in A \\ R(-120°) \cdot \hat{\boldsymbol{\mu}}_i, & i \in B \\ R(+120°) \cdot \hat{\boldsymbol{\mu}}_i, & i \in C \end{cases}. \quad (11.64)$$

The operator $R(\phi)$ is the rotation operation through angle ϕ. Then these transformed dipoles $\hat{\boldsymbol{\mu}}'_i$ can now be averaged separately on the three magnetic sublattices a, b and c, forming three vector order parameters,

$$\mathbf{w}_a = \frac{1}{N_c} \sum_k \hat{\boldsymbol{\mu}}'_{k,a}, \quad \mathbf{w}_b = \frac{1}{N_c} \sum_k \hat{\boldsymbol{\mu}}'_{k,b}, \quad \mathbf{w}_c = \frac{1}{N_c} \sum_k \hat{\boldsymbol{\mu}}'_{k,c}. \quad (11.65)$$

The sums are over all the N_c colored cells. These are the sublattice average magnetizations. For the model with Heisenberg-like dipoles, these averages are free to point in any direction. In a ground state, however, each of the three vector order parameters takes the particular direction of a, b or c dipoles in the A cells, which were the reference cells. For the A primitive cells, the direction vectors *outward* from the central vertex are[4]

$$\hat{\mathbf{v}}_a \equiv -\frac{\sqrt{3}}{2}\hat{\mathbf{x}} - \frac{1}{2}\hat{\mathbf{y}}, \quad \hat{\mathbf{v}}_b \equiv +\frac{\sqrt{3}}{2}\hat{\mathbf{x}} - \frac{1}{2}\hat{\mathbf{y}}, \quad \hat{\mathbf{v}}_c \equiv \hat{\mathbf{y}}. \quad (11.66)$$

These are not independent, but satisfy

$$\hat{\mathbf{v}}_a + \hat{\mathbf{v}}_b + \hat{\mathbf{v}}_c = 0. \quad (11.67)$$

For the ground state of figure 11.26, the order parameters are

$$(\mathbf{w}_a, \mathbf{w}_b, \mathbf{w}_c) = (+\hat{\mathbf{v}}_a, -\hat{\mathbf{v}}_b, +\hat{\mathbf{v}}_c). \quad (11.68)$$

The three signs here $(+1, -1, +1)$ can be used to specify which ground state. Indeed, these three signs can be written in a general case as *scalar order parameters* for each magnetic sublattice:

$$Z_a \equiv \mathbf{w}_a \cdot \hat{\mathbf{v}}_a, \quad Z_b \equiv \mathbf{w}_b \cdot \hat{\mathbf{v}}_b, \quad Z_c \equiv \mathbf{w}_c \cdot \hat{\mathbf{v}}_c. \quad (11.69)$$

In a ground state, each of these can be ± 1. The state of figure 11.26 has the triplet $(Z_a, Z_b, Z_c) = (+1, -1, +1)$. In a ground state, one must have two majority dipoles and one minority dipole at each vertex, which gives the restriction,

$$Z_a + Z_b + Z_c = \pm 1. \quad (11.70)$$

Only six of the eight choices of Z_a, Z_b, Z_c can satisfy this, and these choices correspond to the six possible ground states.

---

[4] The directions of $\hat{\mathbf{v}}_a, \hat{\mathbf{v}}_b, \hat{\mathbf{v}}_c$ on the B- and C-sublattices are obtained by cyclic permutation of the vectors in (11.66), see figure 11.26.

A *net vector order parameter* is the sum of those on the sublattices,

$$\mathbf{w} = \frac{1}{2}(\mathbf{w}_a + \mathbf{w}_b + \mathbf{w}_c). \tag{11.71}$$

In a ground state, one has each sublattice in the form $\mathbf{w}_a = Z_a \hat{\mathbf{v}}_a$, *etc*, so that the resulting net vector parameter is

$$\mathbf{w} = \frac{1}{2}(Z_a \hat{\mathbf{v}}_a + Z_b \hat{\mathbf{v}}_b + Z_c \hat{\mathbf{v}}_c). \tag{11.72}$$

Out of the triplet (Z_a, Z_b, Z_c), one is minority (for example, $Z_b = -1$) and two are majority ($Z_a = Z_c = +1$). In the state of figure 11.26, with b the minority sublattice,

$$\mathbf{w} = \frac{1}{2}(\hat{\mathbf{v}}_a - \hat{\mathbf{v}}_b + \hat{\mathbf{v}}_c) = \hat{\mathbf{v}}_a + \hat{\mathbf{v}}_c = Z_b \hat{\mathbf{v}}_b. \tag{11.73}$$

The net result is a unit vector in the majority's net direction, which is *the same as the minority's direction*. That is the direction of the b-sublattice (inward in the colored cells) for this example. In the six ground states, the possible values are then $\mathbf{w} = \pm\hat{\mathbf{v}}_a, \pm\hat{\mathbf{v}}_b, \pm\hat{\mathbf{v}}_c$. The six directions are similar to the six ground state directions for the complex order parameter M. The direction of $\mathbf{w}$ in a ground state, however, is exactly the same as the direction of the minority dipole in the A cells. Thus it has an obvious geometrical interpretation.

Scalar ground state order indicators

The three parameters Z_a, Z_b, Z_c are over-complete for determining state of the sublattices in a ground state. Even so, they could be used to specify an effective state vector of the system,

$$\Psi = \frac{1}{\sqrt{3}}(Z_a, Z_b, Z_c) = \frac{1}{\sqrt{3}}(Z_a \Psi_a + Z_b \Psi_b + Z_c \Psi_c). \tag{11.74}$$

This is written in an abstract notation, with the three basis vectors defined from single dipole states on each sublattice, as sketched in figure 11.30 for A cells,

$$\Psi_a \equiv (1, 0, 0), \quad \Psi_b \equiv (0, 1, 0), \quad \Psi_c \equiv (0, 0, 1). \tag{11.75}$$

Based on this, there are state vectors for the ground states, in a notation focused on the minority sublattice,

$$\Psi_{gs}^{1+} \equiv \frac{1}{\sqrt{3}}(1, -1, -1) = \frac{1}{\sqrt{3}}(\Psi_a - \Psi_b - \Psi_c), \tag{11.76a}$$

$$\Psi_{gs}^{2+} \equiv \frac{1}{\sqrt{3}}(-1, 1, -1) = \frac{1}{\sqrt{3}}(-\Psi_a + \Psi_b - \Psi_c), \tag{11.76b}$$

$$\Psi_{gs}^{3+} \equiv \frac{1}{\sqrt{3}}(-1, -1, 1) = \frac{1}{\sqrt{3}}(-\Psi_a - \Psi_b + \Psi_c). \tag{11.76c}$$

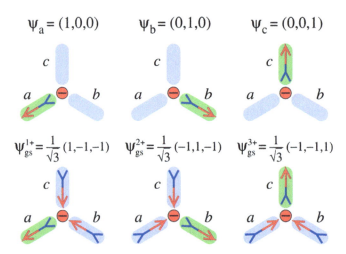

Figure 11.30. For the A cells, notation of the single dipole states Ψ_a, Ψ_b, Ψ_c, and the three principal ground state vectors Ψ_{gs}^{1+}, Ψ_{gs}^{2+}, Ψ_{gs}^{3+}. Islands in green highlight the outward dipoles. The other three ground states, Ψ_{gs}^{1-}, Ψ_{gs}^{2-}, Ψ_{gs}^{3-}, are obtained from these by reversing all dipoles.

These are drawn in figure 11.30 for the A cells of the lattice. The other three ground states, Ψ_{gs}^{1-}, Ψ_{gs}^{2-}, Ψ_{gs}^{1-}, are obtained by reversing the dipoles. The states do not form an orthonormal set. There are overlaps among these states such as $\langle \Phi_{gs}^{1+} | \Phi_{gs}^{1+} \rangle = 1$ and $\langle \Phi_{gs}^{1+} | \Phi_{gs}^{2+} \rangle = \langle \Phi_{gs}^{1+} | \Phi_{gs}^{3+} \rangle = -1/3$. The state in figure 11.26 is Φ_{gs}^{2-}, because the minority b-sublattice is inward in all cells. Then any arbitrary state (11.74) can be expanded in terms of these ground states. To achieve this it is helpful to have the identities of the following exercise.

Exercise 11.11. Show that the single dipole states on the sublattices can be written in terms of the ground states as

$$\Psi_a = -\frac{\sqrt{3}}{2}\left(\Psi_{gs}^{2+} + \Psi_{gs}^{3+}\right) \tag{11.77a}$$

$$\Psi_b = -\frac{\sqrt{3}}{2}\left(\Psi_{gs}^{1+} + \Psi_{gs}^{3+}\right) \tag{11.77b}$$

$$\Psi_c = -\frac{\sqrt{3}}{2}\left(\Psi_{gs}^{1+} + \Psi_{gs}^{2+}\right). \tag{11.77c}$$

Also check that these single-dipole states are mutually orthogonal, using the overlaps among the ground states.

Then the arbitrary state in (11.74) can be expressed alternatively in terms of the ground states, as found in another short exercise.

> **Exercise 11.12.** Show that an arbitrary state can be written in terms of the ground states as
>
> $$\Psi = Z_1 \Psi_{gs}^{1+} + Z_2 \Psi_{gs}^{2+} + Z_3 \Psi_{gs}^{3+} \qquad (11.78)$$
>
> where the coefficients are given by
>
> $$Z_1 = -\frac{Z_b + Z_c}{2}, \quad Z_2 = -\frac{Z_a + Z_c}{2}, \quad Z_3 = -\frac{Z_a + Z_b}{2}. \qquad (11.79)$$
>
> This applies to an arbitrary state, even if the magnitudes of Z_a, Z_b, Z_c are not unity.

In the result of the previous exercise, the parameters Z_1, Z_2, Z_3 give the amplitudes for the three primary ground states. Negative values correspond to the reversed ground states. When the system is in a ground state, one can verify that only one of the parameters is non-zero. Calculation of these in the course of a simulation may give an indication of the degree of condensation of the system towards the ground states.

Nearest neighbor pair correlations
Other quantities of interest are the nearest neighbor pair correlations. These are averages of the form $\langle \hat{\mu}_i \cdot \hat{\mu}_j \rangle$, where i, j are any neighboring sites. The pairs in these averages are always between two of the magnetic sublattices, a, b, c.

Suppose the system is in a ground state. The a-sublattice dipoles can be written as $\hat{\mu}_i = \hat{\mu}_a = Z_a \hat{v}_a$; a similar expression applies to the b- and c-sites. Noting that the unit vectors of any neighboring dipoles always have a 120° angle between them, with for instance, $\hat{v}_a \cdot \hat{v}_b = -\frac{1}{2}$, the correlations are

$$C_{ab} \equiv \langle \hat{\mu}_a \cdot \hat{\mu}_b \rangle_{nn} = -\frac{1}{2} Z_a Z_b. \qquad (11.80)$$

This will have possible values $\pm\frac{1}{2}$. One then realizes that a combination of two such correlations will have a range from 0 to ± 1, for example,

$$C_1 \equiv C_{ab} + C_{ac} = \langle \hat{\mu}_a \cdot \hat{\mu}_b \rangle_{nn} + \langle \hat{\mu}_a \cdot \hat{\mu}_c \rangle_{nn} = -\frac{1}{2} Z_a (Z_b + Z_c). \qquad (11.81)$$

In the states $\Psi_{gs}^{2\pm}$ and $\Psi_{gs}^{3\pm}$ this sum is zero. In state Ψ_{gs}^{1+} with $(Z_a, Z_b, Z_c) = (1, -1, -1)$ one has $C_1 = 1$. In state Ψ_{gs}^{1-} with $(Z_a, Z_b, Z_c) = (-1, 1, 1)$ one also has $C_1 = 1$. The parameter therefore cannot distinguish two states different by an overall reversal, but it can distinguish the ground states into the three classes.

For a general state not necessarily a ground state, we also define the other correlations, all using only nearest neighbor pairs

$$C_1 = C_{ab} + C_{ac} = \langle \hat{\boldsymbol{\mu}}_a \cdot (\hat{\boldsymbol{\mu}}_b + \hat{\boldsymbol{\mu}}_c) \rangle_{nn} \qquad (11.82a)$$

$$C_2 = C_{ab} + C_{bc} = \langle \hat{\boldsymbol{\mu}}_b \cdot (\hat{\boldsymbol{\mu}}_a + \hat{\boldsymbol{\mu}}_c) \rangle_{nn} \qquad (11.82b)$$

$$C_3 = C_{ac} + C_{bc} = \langle \hat{\boldsymbol{\mu}}_c \cdot (\hat{\boldsymbol{\mu}}_a + \hat{\boldsymbol{\mu}}_b) \rangle_{nn}. \qquad (11.82c)$$

One has $C_2 = 1$ when collapsed into one of the states $\Psi_{gs}^{2\pm}$ and $C_3 = 1$ when in one of the states $\Psi_{gs}^{3\pm}$. As the system is very frustrated, this may be very difficult to achieve if the system is relaxed towards low temperature, starting from high temperature.

Probability distributions of ordering

The variables Z_a, Z_b, Z_c give averages over the whole system of some local variables in the individual cells, which are the projections of the dipoles on the local axes $\hat{\mathbf{v}}_a$, $\hat{\mathbf{v}}_b$, $\hat{\mathbf{v}}_c$, see (11.65). We can define those local variables in some triangular cell by

$$z_a = \hat{\boldsymbol{\mu}}'_{k,a} \cdot \hat{\mathbf{v}}_a, \quad z_b = \hat{\boldsymbol{\mu}}'_{k,b} \cdot \hat{\mathbf{v}}_b, \quad z_c = \hat{\boldsymbol{\mu}}'_{k,c} \cdot \hat{\mathbf{v}}_c. \qquad (11.83)$$

Each of these has some probability distribution over the range from -1 to $+1$. At low temperatures the distributions $p_a(z_a)$, $p_b(z_b)$ and $p_c(z_c)$ are expected to be strongly peaked near the limits of this range. In a ground state, the distributions should become delta function spikes at the values $z_a = Z_a$, $z_b = Z_b$, $z_c = Z_c$ for the particular ground state. For higher temperature, one should expect a greatly spread out distribution.

One can also make derived probability distributions that point more directly to the particular ground states. This can be achieved by extending the definitions (11.79) of the Z_1, Z_2, Z_3 order parameters to local definitions. Thus, define the following local variables,

$$z_1 = -\frac{z_b + z_c}{2}, \quad z_2 = -\frac{z_a + z_c}{2}, \quad z_3 = -\frac{z_a + z_b}{2}. \qquad (11.84)$$

If the distributions of z_a, z_b, z_c have been calculated directly in a simulation, by making histograms of the values, then the probabilities for these derived quantities can also be found, indirectly. This is accomplished by using a delta function to pick out the constrained quantity while integrating over all possibilities, as in (11.25) for square ice. Consider the calculation of the probability distribution of z_1, which only depends on the distributions of z_b and z_c:

$$p_1(z_1) = \int_{-1}^{1} dz_b\, p_b(z_b) \int_{-1}^{1} dz_c\, p_c(z_c)\, \delta\left[z_1 + \frac{1}{2}(z_b + z_c) \right]. \qquad (11.85)$$

The delta function picks out only the points in the $z_b z_c$-plane that contribute to a chosen value of z_1. That is a locus of points along a line with a slope of -1, see figure 11.31. In actual calculation, the values of z_1 are partitioned into some number of bins B with a narrow width over the allowed range $-1 \leqslant z_1 \leqslant +1$.

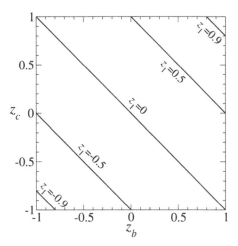

Figure 11.31. Illustration of the $z_b z_c$-plane for the calculation of the derived probability distribution $p_1(z_1)$ using (11.85) from raw probability distributions for z_b and z_c. The solid lines result from (11.84) for the indicated values of z_1.

Values of z_1 within a bin fall into a narrow strip in the $z_b z_c$-plane. Counting the relative frequencies to fall into different strips leads to the calculated probabilities for each bin.

The same process applies to calculation of $p_2(z_2)$ and $p_3(z_3)$. The three distributions taken together can give a graphic view of the condensation of the system towards a ground state (or not).

11.6.3 Monopole charge densities in Kagomé spin ice

Each outward (inward) pole of a dipole at a vertex contributes $+\frac{1}{2}(-\frac{1}{2})$ unit of monopole charge to that vertex, regardless of the type of lattice. These are the half-monopole contributions already presented in the discrete charge definition (11.26). As a result, in Kagomé ice each vertex is always charged, with possible values $q_k = \pm\frac{1}{2}, \pm\frac{3}{2}$ (single and triple charges). This is in strong contrast to spin ice on square and triangular lattices. The continuum charge definition (11.27) gives q_k^* in the same range, from $-\frac{3}{2}$ to $+\frac{3}{2}$. One expects that its limiting values could carry very little statistical weight.

In a simulation if the system is found to have n_1 single charges per vertex and n_3 triple charges per vertex, then these partial densities each give the related charge densities, including the charge magnitudes,

$$\rho_1 = \frac{1}{2}\langle n_1 \rangle, \quad \rho_3 = \frac{3}{2}\langle n_3 \rangle. \tag{11.86}$$

The combination gives the total monopole charge density,

$$\rho = \rho_1 + \rho_3 = \left\langle \frac{1}{2}n_1 + \frac{3}{2}n_3 \right\rangle. \tag{11.87}$$

In a ground state, we have $\rho_1 = \frac{1}{2}$ and $\rho_3 = 0$, as every vertex is occupied by a single (half-unit) charge. This is the maximum value of ρ_1. The continuous definitions will have the same values. As the temperature is raised, the density from single charges must decrease, while triple charges are being generated at some vertices. We noted earlier that when a dipole reverses, an amount of discrete charge $\Delta q = -1$ flows between its vertices in the final direction of the flipped dipole. The discrete charges are still quantized in integer units of charge. Thus, although the vertices always have half-integer amounts of charge, the excitations move integer charge values around in the lattice.

High temperature limit

At very high temperature, the system with Heisenberg-like dipoles will eventually become very disordered, with dipoles pointing in arbitrary directions on the unit sphere. Then any dipole has equal probabilities of pointing inward or outward at a vertex. For the discrete definition of charge, out of the eight states at a vertex, there are six with $|q| = \frac{1}{2}$ and two with $|q| = \frac{3}{2}$. Then we obtain the high temperature limiting values for number and charge densities,

$$\langle n_1 \rangle = \frac{3}{4}, \quad \Longrightarrow \quad \rho_1 = \frac{1}{2}\langle n_1 \rangle = \frac{3}{8}, \tag{11.88a}$$

$$\langle n_3 \rangle = \frac{1}{4}, \quad \Longrightarrow \quad \rho_3 = \frac{3}{2}\langle n_3 \rangle = \frac{3}{8}, \tag{11.88b}$$

$$\rho = \rho_1 + \rho_3 = \frac{3}{4}. \tag{11.88c}$$

Even though the number density of multiple charges is lower, the charge densities from single and multiple charges will become equal. In addition, the limiting value of total charge density, $\rho \to \frac{3}{4}$, is the same for Kagomé ice as for square ice. It is an interesting coincidence; the reader can check that for triangular ice the limit is $\rho \to \frac{15}{16}$, with the largest part of the charge density due to single charges.

Exercise 11.13. Using the definition of continuous charge (11.27), show that the high temperature limiting charge density in Kagomé ice is

$$\rho^* = \left\langle |q_k^*| \right\rangle = \left\langle \left| \frac{1}{2} \sum_{i_k=1}^{3} x_{i_k} \right| \right\rangle = \frac{13}{32} = 0.40625 \tag{11.89}$$

11.7 Langevin dynamics for Kagomé ice in thermal equilibrium

The Langevin-LLG equations (11.40) for Heisenberg model spin ice on 16×16 and 32×32 Kagomé lattices (approximate sizes) with open boundaries have been

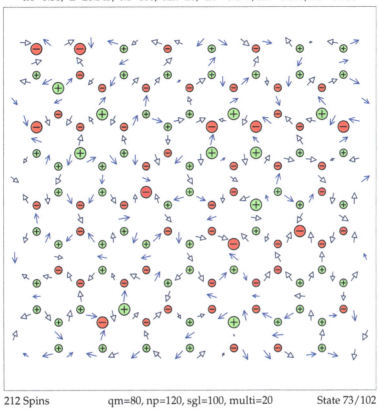

Figure 11.32. A 16 × 16 Kagomé spin ice configuration for a model with $d = k_1 = 01$, $k_3 = 0.5$, at scaled temperature $\mathcal{T} = 0.30$, well above the transition temperature $\mathcal{T}_c \approx 0.08$. Arrows are projections of the xy dipole components on the plane. Those drawn as $\rightarrow$($\dashrightarrow$) indicate positive (negative) μ_z-components. $|q| = \frac{1}{2}$ ($|q| = \frac{3}{2}$) charges are indicated with smaller (larger) circles at the vertices. Green (red) filled circles indicate positive (negative) signs of the charges. The system is mostly disordered with a considerable number of triple monopole charges, and single monopoles in random locations.

used to investigate the system in thermal equilibrium [11]. The 16 × 16 system at scaled temperature $\mathcal{T} = k_B T/\varepsilon = 0.30$ is shown in figure 11.32. Due to the structure of Kagomé the system has an 8 × 8 grid of hexagonal cells of vertices, with 212 dipoles. Vertices with single charges $|q| = \frac{1}{2}$ are shown as smaller circles; those with triple charges $|q| = \frac{3}{2}$ are shown with larger circles. Green (red) indicates a positive (negative) sign. At this relatively high temperature, there is considerable disorder associated with the presence of a relatively large number of triple charges. The dipoles are not tending to align along the axes between neighboring vertices.

The calculational procedure for obtaining equilibrium states is the same as described for square lattice spin ice. The calculations were performed using parameters similar to those in model C for square ice but with stronger easy-plane anisotropy k_3,

$$d = 0.1, \quad k_1 = 0.1, \quad k_3 = 0.5. \tag{11.90}$$

The stronger k_3 will be expected to maintain the dipoles more strongly within the xy-plane, but otherwise it should not have a strong effect on the in-plane dynamics. As for model C, with relatively strong dipolar interactions, the Heun time step used is $\Delta \tau = 0.001$, to ensure correct dynamics at finite temperature, and energy-conserving dynamics at zero temperature. Averages were made by sampling at time interval $\Delta \tau_s = 10^3 \Delta \tau$, using at least $N_s = 4000$ data samples, or until the percent error in total system magnetic moment became less than 0.1%.

For the dipolar interactions, the more complex lattice structure makes acceleration via an FFT method cumbersome. Instead, for small lattices the dipolar terms can simply be summed over the lattice. The coding is simpler but then the execution is slower. With open boundaries and smaller systems the is no artificial cutoff placed on the sums. The system is initiated at a higher temperature, in a random state. After relaxing a temperature, the final state is used as the initial state at the next lower temperature.

The temperature-dependence of average energy and specific heat per dipole are displayed in figure 11.33. There is a strong but wide peak in $C(T)$, but at a temperature around $\mathcal{T}_c \approx 0.08$, about one third of the peak location in square ice. To be sure on this point, the results for the *same model parameters* (11.90) but on square ice are also shown in figure 11.33. In square ice the transition is around $\mathcal{T}_c \approx 0.22$. The peak is assumed to be associated with a transition to greater order at low temperature for both systems.

A state of the Kagomé system at $\mathcal{T} = 0.08$, which is where the specific heat peaks, is shown in figure 11.34. There is more order than at $\mathcal{T} = 0.30$, as can be seen by the lack of triple charges at the vertices. All vertices for this particular state have single charges. There is still considerable disorder, however, in the directions of the dipoles. Mostly, they do not have an organized alignment into or out of their neighboring vertices. There are considerable thermal fluctuations in the system.

When the system is cooled further, to $\mathcal{T} = 0.01$, states as shown in figure 11.35 result. Again, all vertices have single charges, which is expected for a ground state. However, both the dipole directions and the monopole arrangement here remain very disordered. Not only are the dipoles not pointing very close to the island axes (towards/away from vertices), but the arrangement of the monopole charges does not alternate as expected in a ground state, such as the one in figure 11.26. At this low temperature, well below $\mathcal{T}_c$, the system apparently has not relaxed into anything close to a ground state configuration. The effects of frustration are very strong, leaving a large residual anisotropy energy. This is confirmed by other averages, discussed below.

The lower transition temperature $\mathcal{T}_c \approx 0.08$ in Kagomé ice might be explained by different reasons. For one, the lower number of nearest neighbors in Kagomé could effectively weaken the long-range dipolar effects. We already saw that the nearest neighbor effective exchange couplings are actually stronger in Kagomé than in

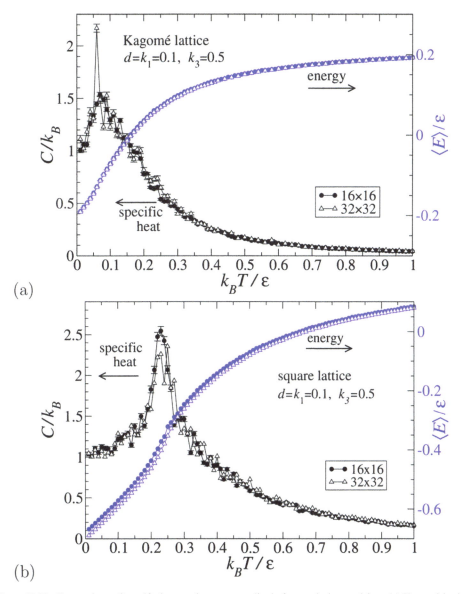

Figure 11.33. Comparison of specific heat and energy per dipole for a spin ice model on (a) Kagomé lattice and the same model on (b) square lattice. The peak location in square ice ($\mathcal{T} \approx 0.22$) for this increased value of k_3 compared to model C (see figure 11.16) is about the same. For Kagomé ice the peak is at a much lower temperature, $\mathcal{T}_c \approx 0.08$, possibly due to very low energy excitations in Kagomé ice that are not present in square ice.

square ice, so without further analysis, the long-range effects need to be studied more carefully to see if that really affects $\mathcal{T}_c$. Another reason to consider is that frustration effects in Kagomé are expected to be stronger, and while this might prevent the system from moving towards the ground states, it is not clear how it should affect $\mathcal{T}_c$.

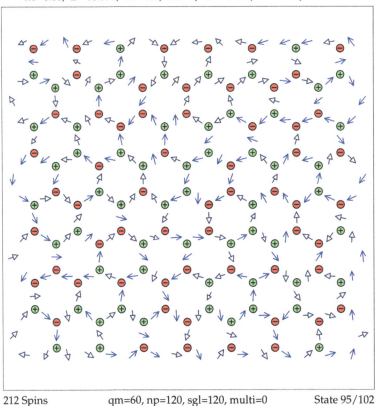

Figure 11.34. A 16 × 16 Kagomé spin ice configuration for model parameters given in (11.90), at $\mathcal{T} = 0.08$, close to the transition temperature $\mathcal{T}_c$. See figure 11.32 caption for the notation. All of the charges present at this temperature are single charges. There are still strong fluctuations present, and the ground state order parameters Z_1, Z_2, Z_3 are all rather small, which indicates a great distance in phase space from any ground state.

One effect could be that the large number of ground states causes greater fluctuations at a lower temperature, thereby moving the peak in $C(T)$ to a lower temperature. A third possible reason, which probably makes even more clear sense, is the presence of very low energy excitations in Kagomé ice, which are not present in square ice. We saw in figure 11.28 that there are excitations away from a ground state that cost little energy, such as the hexagonal closed string, and even the longer strings that simply move charge around. Those keep the vertices as type I in Kagomé ice, thus costing only some longer-range dipolar energies. There are similar closed loop excitations in square ice, but they change vertices from type I to type II, which produces local dipolar energies that are considerably larger.

Regardless of the reason for its lower $\mathcal{T}_c$, in Kagomé ice we also see evidence that the low temperature states do not tend very close to ground states. This is in contrast to square ice. In particular, consider the nearest neighbor correlation

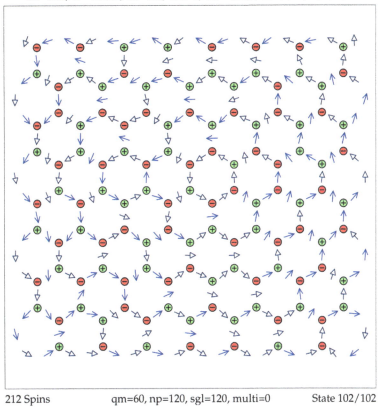

Figure 11.35. A 16×16 Kagomé spin ice configuration for model parameters given in (11.90), at $\mathcal{T} = 0.01$, well below the transition temperature $\mathcal{T}_c \approx 0.08$. See the figure 11.32 caption for the notation. All of the charges present at this temperature are single charges, however, they are not in an organized arrangement which would be expected for a ground state. Z_1, Z_2, Z_3 are all small, which indicates a great distance in phase space from any ground state. Also, the dipoles are not generally pointing inward or outward relative to the vertices, which is an extra anisotropy energy and frustration is apparent.

functions C_1, C_2 and C_3. The function $C_1(T)$ obtained from the simulations is shown in figure 11.36; the results for $C_2(T)$ and $C_3(T)$ are essentially the same. There is a smooth behavior, with a peak as $\mathcal{T} \to 0$. The limiting value, however, is $C_1 \to 1.38$ as $\mathcal{T} \to 0$. The correlation C_1 is a measure of the projection of the system onto the $\Psi_{gs}^{1\pm}$ ground states. If the system moves into one of those states, one would have $C_1 = 1$, $C_2 = C_3 = 0$. In a ground state, the pairs would be oriented at relative angles $\cos^{-1}(-\frac{1}{2}) = 120°$, which gives $C_1 = 1$. Instead, the system moves into a low temperature state where the average nearest neighbor relative angles must be close to $\cos^{-1}(-\frac{1.38}{2}) \approx 134°$. This is more greatly anti-aligned. That must cost extra anisotropy energy, as the dipoles are not all able to align with the long axes of the islands. Note that also $C_{ab}(T)$, $C_{bc}(T)$ and $C_{ac}(T)$ have the same form as $C_1(T)$,

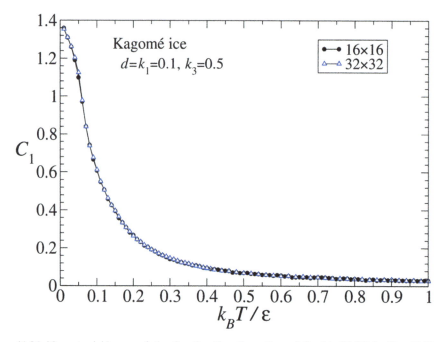

Figure 11.36. Nearest neighbor correlation function $C_1 = C_{ab} + C_{bc}$ as defined in (11.82) for 32 × 32 Kagomé lattice spin ice model, versus scaled temperature. $C_1 \to 1$ together with $C_2 = C_3 = 0$ would indicate the condensation of the system into one of the $\Psi_{gs}^{1\pm}$ ground states; that is not the case here. The behaviors of $C_2(T)$ and $C_3(T)$ are nearly the same as $C_1(T)$; the partial terms are $C_{ab}(T) \approx C_{bc}(T) \approx \frac{1}{2}C_1(T)$.

but with half the amplitude. Therefore this measure demonstrates that the system does not move into a ground state for very low temperature.

The difficulty of reaching a clear ground state in Kagomé spin ice has been seen in experiments by Zhang *et al* [12], where it was found that a state of ordered monopole charges can appear, described as magnetic charge crystallization. This state is similar to ones such as that in figure 11.35, where although the monopoles are somewhat ordered, there is an underlying spin disorder. Obviously the presence of charge disorder is topologically coupled to spin disorder, as we have seen when considering excitations out of a ground state. Once the charges are mostly fixed, they remain in place, not allowing the spins to make a final relaxation into a ground state, even in a local length scale.

The results for the various charge densities are shown in figure 11.37. As expected, the density from single charges tends towards $\rho_1 \approx 0.5$ at low temperature, and diminishes with increasing $\mathcal{T}$, eventually approaching the high temperature asymptotic value $\rho_1 \to \frac{3}{8}$ of (11.88) (this is confirmed by higher temperature data, see [11]). The charge density ρ_3 from triple charges is very close to zero for $\mathcal{T} < 0.08$, above which it rises smoothly and eventually approaches the high temperature limit, $\rho_3 \to \frac{3}{8}$. The continuous charge definition gives the density ρ^* quite large in the low temperature regime, and exhibits a sudden shift in slope near $\mathcal{T} \approx 0.07$. As for square ice, ρ^* can easily make deviations away from zero for low $\mathcal{T}$ as it is readily

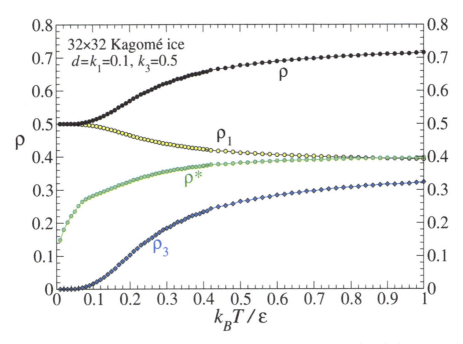

Figure 11.37. The various monopole charge densities for a 32 × 32 Kagomé lattice spin ice versus scaled temperature, using artificial model parameters. The total discrete charge density ρ is the sum of single (ρ_1) and triple (ρ_3) monopole contributions. The continuous charge density based on definition (11.27) is ρ^*.

affected by any small spin deviations of the island dipoles away from their long axes. This takes place at low $\mathcal{T}$ where the discrete monopole charges have not yet been created.

The order parameters Z_1, Z_2 and Z_3 that measure the average projection onto the different ground states are plotted versus temperature in figure 11.38, for a 32 × 32 Kagomé system. The definitions were given in (11.79). For high temperatures all three are zero; there is no particular ordering. For low temperatures, there are only small deviations away from zero, hardly significant relative to the errors. If the system were to move into the Ψ_{gs}^{1+} ground state, as an example, these parameter would become $Z_1 = 1$, $Z_2 = Z_3 = 0$. This gives further confirmation that the system as a whole does not order into a ground state configuration as the temperature is lowered.

To make this last point clearer, one can look at the probability distributions of the local variables such as z_a, z_b, z_c, for projections onto the three magnetic sublattice directions, or the variables z_1, z_2, z_3, for projections onto the different ground states. These give a clear overview of the local dipolar ordering. Figure 11.39 shows $p_a(z_a)$ and $p_1(z_1)$ for this model on 32 × 32 Kagomé ice at various temperatures. The distributions for $p_b(z_b)$ and $p_c(z_c)$ are very similar to that for $p_a(z_a)$; these were calculated from the raw simulation data. The distributions for $p_2(z_2)$ and $p_3(z_3)$ are also very similar to $p_1(z_1)$, which was derived from $p_b(z_b)$ and $p_c(z_c)$ using (11.85). At high temperature (well above the transition temperature $\mathcal{T}_c \approx 0.08$), z_a is distributed

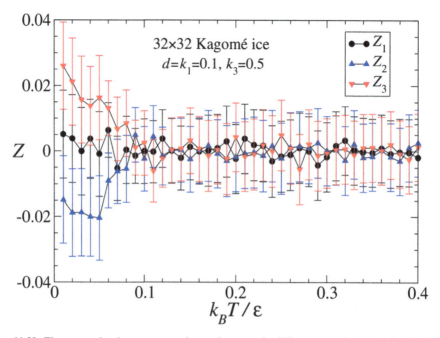

Figure 11.38. The averaged order parameters that project onto the different ground states, defined in (11.79), for 32×32 Kagomé spin ice, versus scaled temperature. The parameters give an indication of condensation into a ground state. For example, the values in the Ψ_{gs}^{1+} ground state would be $Z_1 = 1$, $Z_2 = Z_3 = 0$. At high temperatures with $Z_1 \approx Z_2 \approx Z_3 \approx 0$ there is no ordering. At low temperature the ordering is very slight.

in a parabolic form, as would be dictated by a Boltzmann factor where the in-plane uniaxial anisotropy competes with the temperature,

$$p_a(z_a) \propto \exp(K_1 z_a^2 / k_B T), \quad T \gg T_c. \tag{11.91}$$

As the temperature is lowered, there is *not* a collapse of the distribution $p_a(z_a)$ to one side or the other, as takes place in square ice, compare figure 11.18. Instead, the extreme values $z_a \approx \pm 1$ become equally probable, even as $\mathcal{T}$ becomes very low, making a nearly symmetric distribution. This is a signal of the greater frustration present in Kagomé ice. The system is not condensing into one of the ground states.

The corresponding distribution of z_1 in figure 11.39 also remains fairly symmetrical even at low temperature. For high temperature, the distribution of z_1 is close to triangular. A perfect triangle would result from (11.85) if z_b and z_c were uniformly distributed from -1 to $+1$. The deviation from straight lines is caused by the K_1 anisotropy competing with temperature, which enhances the probability slightly at the extremes $z_1 \approx \pm 1$. The strong peak at $z_1 = 0$ shows a large deviation of the system away from any ground state. For example, if the system were to collapse into the Ψ_{gs}^{1+} (Ψ_{gs}^{1-}) ground state, one would obtain a very skewed distribution, concentrated around $z_1 = 1(z_1 = -1)$, together with $z_2 = 0$ and $z_3 = 0$. In fact, the main thing that appears at low $\mathcal{T}$ is z_1 is concentrated mostly near $z_1 \approx 0$, with weaker

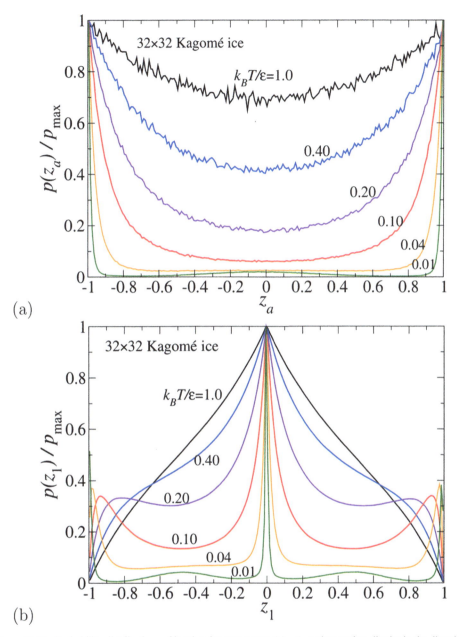

Figure 11.39. Probability distributions of local order parameters (a) $p(z_a)$, that project dipoles in the direction for that sublattice, and (b) $p_1(z_1)$, that projects onto the ground states $\Psi_{gs}^{1\pm}$. The different curves are scaled relative to their maximum values to better show the details; they are labeled by the scaled temperature. $p(z_a)$ was evaluated from raw simulation data; $p(z_b)$ and $p(z_c)$ are similar. $p_1(z_1)$ was derived from combining results for $p(z_b)$ and $p(z_c)$, see (11.85). Together with similar results for $p_2(z_2)$ and $p_3(z_3)$, these indicate that even for very low temperature the system does not condense globally into one of the ground states.

peaks also at $z_1 \approx \pm 1$. With similar distributions resulting for z_2 and z_3, it can be seen that there is amplitude present in all of the six ground states. Clearly the system does not collapse into one particular ground state, as in square ice.

Frustration comparison—Kagomé versus square ice
The above results show that in Kagomé spin ice, geometric frustration greatly prohibits ergodic evolution of the system under LLG dynamics. Relaxing the system from higher towards lower temperature, frustration prevents a uniform exploration of the entire phase space, preventing the system as whole from arriving at an individual ground state. The frustration is apparently stronger than in square lattice spin ice. The Kagomé system has some very low energy excitations from the ground state, which is also six-fold degenerate in a perfect system. This multiplicity of low energy states is likely to be important for explaining why frustration is so effective at preventing the system from finding one of the ground states. Zhang *et al* [12] give a view of how to produce thermal ordering in real square and Kagomé systems, for example, by tuning the lattice constant to control the interactions.

These results highlight the differences one can expect for a realistic dynamical evolution compared to making averages by MC simulations.

About using Heisenberg-like dipoles for the islands
Use of Heisenberg-like dipoles and Langevin–LLG dynamics for the study of spin ice is motivated by the fact that it describes a real dynamics. This is in contrast to the Ising limit. We note that even if there is very strong shape anisotropy in the individual islands, their magnetic moments still must make temporal fluctuations described by an LLG equation. To leading order, the total magnetic moment of the island dipole is constant in magnitude and only changes directions. This was found to be a reasonable approximation in [7], where the effective anisotropy constants K_1 and K_3 were estimated for thin islands. Then the Langevin–LLG dynamics should be a good way to find the time evolution of a system of islands. This approach does ignore some effects, for example, it is assumed that each island is exposed to a uniform field from each other island. Obviously this is not correct for the closest islands, and this defect is probably the greatest error in this approach. However, these near-field effects probably act to facilitate the collective reversals of islands that are misaligned. They might lead to a more thermalized dynamics, that could reduce the frustration effects.

This chapter has only touched on an introduction to some problems in artificial spin ices. Some other topics of interest such as frozen-in entropy [13] are of great current interest and examples can be found in the literature.

Bibliography

[1] Wang R F *et al* 2006 Artificial spin ice in a geometrically frustrated lattice of nanoscale ferromagnetic islands *Nature* **439** 303

[2] Morgan J P, Stein A, Langridge S and Marrows C 2011 Thermal ground-state ordering and elementary excitations in artificial magnetic square ice *Nat. Phys.* **7** 75

[3] Castelnovo C, Moessner R and Sondhi S L 2008 Magnetic monopoles in spin ice *Nature* **451** 42

[4] Mól L A S, Silva R L, Silva R C, Pereira A R, Moura-Melo W A and Costa B V 2009 Magnetic monopole and string excitations in a two-dimensional spin ice *J. Appl. Phys.* **106** 063913

[5] Silva R C, Nascimento F S, Mól L A S, Moura-Melo W A and Pereira A R 2012 Thermodynamics of elementary excitations in artificial magnetic square ice *New J. Phys.* **14** 015008

[6] Wysin G M, Moura-Melo W A, Mól L A S and Pereira A R 2013 Dynamics and hysteresis in square lattice artificial spin ice *New J. Phys.* **15** 045029

[7] Wysin G M, Moura-Melo W A, Mól L A S and Pereira A R 2012 Magnetic anisotropy of elongated thin ferromagnetic nano-islands for artificial spin ice arrays *J. Phys.: Condens. Matter* **24** 296001

[8] Mól L A S, Pereira A R and Moura-Melo W A 2012 Extending spin ice concepts to another geometry: The artificial triangular spin ice *Phys. Rev.* B **85** 184410

[9] Wills A S, Ballou R and Lacroix C 2002 Model of localized highly frustrated ferromagnetism: The Kagomé spin ice *Phys. Rev.* B **66** 144407

[10] Chern G-W and Tchernyshyov O 2012 Magnetic charge and ordering in Kagomé spin ice *Phil. Trans. R. Soc.* **370** 5718

[11] Wysin G M, Pereira A R, Moura-Melo W A and de Araujo C I L 2015 Order and thermalized dynamics in Heisenberg-like square and Kagomé spin ices *J. Phys.: Condens. Matter* **27** 076004

[12] Zhang S, Gilbert I, Nisoli C, Chern G-W, Erickson M J, O'Brien L, Leighton C, Lammert P E, Crespi V H and Schiffer P 2013 Crystallites of magnetic charges in artificial spin ice *Nature* **500** 553

[13] Lammert Paul E, Ke Xianglin, Li Jie, Nisoli Cristiano, Garand David M, Crespi Vincent H and Schiffer Peter 2010 Direct entropy determination and application to artificial spin ice *Nat. Phys.* **6** 786

Lightning Source UK Ltd.
Milton Keynes UK
UKOW07n1809120116

266257UK00003B/29/P